DIGITAL PROCESSING OF RANDOM SIGNALS

Prentice Hall Information and System Sciences Series

Thomas Kailath, Editor

DIGITAL PROCESSING
OF RANDOM SIGNALS
THEORY AND METHODS

Boaz Porat

Department of Electrical Engineering
Technion, Israel Institute of Technology
Haifa, Israel

PRENTICE-HALL, INC., Englewood Cliffs, New Jersey 07632

Library of Congress Cataloging-in-Publication Data

```
Porat, Boaz.
     Digital processing of random signals : theory and methods / Boaz Porat.
        p. cm. -- (Prentice-Hall information and system sciences series)
     Includes bibliographical references and index.
     ISBN 0-13-063751-3
     1. Signal processing--Digital techniques.  2. Signal processing--
     statistical methods.  3. Stochastic processes.  I. Title.
     II. Series.
     TK5102.9.P67    1993
     621.382'2'015195--dc20                        93-38074
                                                       CIP
```

Editorial/production supervision: *Dit Mosco*
Cover design: *Lundgren Graphics, Ltd.*
Manufacturing buyer: *Alexis Heydt*
Acquisitions editor: *Karen Gettman*

 ©1994 by P T R Prentice Hall
Prentice-Hall, Inc.
A Paramount Communications Company
Englewood Cliffs, New Jersey 07632

The publisher offers discounts on this book when ordered in bulk quantities. For more information, contact:

Corporate Sales Department
PTR Prentice Hall
113 Sylvan Avenue
Englewood Cliffs, NJ 07632

Phone: 201-592-2863
Fax: 201-592-2249

Mathematica is a registered trademark of Wolfarm Research, Inc.

Printed in the United States of America
10 9 8 7 6 5 4 3 2 1

ISBN 0-13-063751-3

Prentice-Hall International (UK) Limited, *London*
Prentice-Hall of Australia Pty. Limited, *Sydney*
Prentice-Hall Canada Inc., *Toronto*
Prentice-Hall Hispanoamericana, S.A., *Mexico*
Prentice-Hall of India Private Limited, *New Delhi*
Prentice-Hall of Japan, Inc., *Tokyo*
Simon & Schuster Asia Pte. Ltd., *Singapore*
Editora Prentice-Hall do Brasil, Ltda., *Rio de Janeiro*

Contents

Preface

A large part of electrical engineering involves some form of signal processing. Examples include (but are not limited to) communication, radar, sonar, audio (voice and music), image processing, control systems, and biomedical engineering. This is not to say that electrical engineering has a monopoly on signal processing. Most physical sciences, as well as many social sciences, have their own needs for *time-series analysis*, which is closely related to engineering's signal processing.

As the name implies, *random signal processing* involves signals whose behavior is random in whole or in part. Signals generated by nature are often random. Even when they are not so, our limited understanding of the mechanisms by which they were generated often behooves us to regard them as random. Noise in communication systems, radar clutter, external disturbances to control systems, and reverberation and interference in public address systems, these are some examples of random signals the electrical engineer has to contend with. Random signal processing includes signal classification, modeling, estimation of signal parameters, prediction of future signal behavior, filtering, enhancement, and more.

This book presents the subject of random signal processing at an advanced graduate level suitable for engineering students as well as for practicing engineers. Many books on this subject have been written from the statisticians' point of view, and their mathematical level makes them unsuitable for typical engineering curricula. There are a few books on an elementary graduate level, which make minimal use of advanced mathematics. This book takes the middle ground. The approach is rigorous, but the mathematical prerequisites are not beyond what a graduate engineering student should know. It was written from the point of view of an electrical engineer, but most of the material is sufficiently general to be suitable to other engineering fields. The main topics of the book are as follows.

Chapter 2 introduces stationary processes and discusses their structure and main properties. Both time domain (Wold decomposition, ergodicity, correlation functions, parametric models) and frequency domain (spectral distribution and density, parametric spectra) characterizations are included, with specialization to Gaussian processes.

Chapter 3 serves as an introduction to statistical estimation theory. It does not specifically address random processes, but builds the general framework for the rest of the book. Properties of estimators, consistency concepts, estimation bounds, maximum likelihood, the method of moments, least-squares estimation, maximum entropy estimation, and model order selection criteria are the main topics of this chapter.

Chapter 4 deals with classical spectral estimation, based on periodograms. It also presents averaged, smoothed, and windowed periodograms and analyzes their statistical properties.

Chapter 5 introduces parametric estimation for stationary processes and provides the necessary theoretical background. The Cramér-Rao bound is derived in both exact and asymptotic forms. General properties of maximum likelihood estimates are given. Estimation based on the method of moments is discussed, including asymptotic lower bounds.

Chapters 6 and 7 deal with estimation of autoregressive (AR) and autoregressive moving average (ARMA) processes, respectively. The book offers an expanded treatment of these subjects due to their importance to the field. A broad coverage of estimation methods and their performance analysis is given, including both linear and nonlinear methods. Simulation examples are given in which the actual performance is compared with theoretical predictions. Examples of applications to speech compression and biomedical signal analysis are presented.

Chapter 8 continues the treatment of AR and ARMA processes, but centers on adaptive estimation techniques. The main topics are the recursive least-squares algorithm and its extensions (pseudolinear regression and stochastic gradient), recursive maximum likelihood, and lattice algorithms.

Chapter 9 discusses the estimation of multiple sinusoidal signals in noise. The main topics are maximum likelihood and its approximations, linear prediction methods and their variants (such as the Tufts-Kumaresan method), and covariance matrix-based methods, such as MUSIC and ESPRIT.

Chapter 10 introduces estimation methods for non-Gaussian signals based on high-order statistics. This is a rapidly developing field, and this book is one of the first to cover it in considerable detail. The chapter introduces the concepts of cumulants and polyspectra and presents their statistical properties. Then attention is given to parametric estimation methods for moving average and ARMA processes. Both linear and nonlinear methods are treated, and their performance is analyzed.

Chapters 11 and 12 deal with time-frequency analysis of nonstationary signals, which is also a rapidly developing field. Chapter 11 is devoted to linear transforms, and introduces the short-time Fourier transform, the Gabor representation, and the wavelet transform. Chapter 12 is devoted to quadratic and other nonlinear transforms. It discusses the Wigner-Ville distribution, the ambiguity function, the Cohen class of distributions and its numerous special cases, and a high-order extension of the ambiguity function.

The book includes four appendixes. Appendix A summarizes miscellaneous definitions and facts. Appendix B gives a general background on Hilbert spaces. Appendix C contains some background on asymptotic statistical theory. Appendix

D gives a brief discussion of Kronecker products and Liapunov matrix equations.

References are given at the end of each chapter, listed alphabetically by authors' names. References indicated by [Ax], where x is a number, point to the corresponding item in Appendix A.

The book includes numerous homework problems of various levels of difficulty. The problems are an integral part of the text, and a good part of them should be attempted by the reader who wishes to master the material. Ample hints are provided for the more difficult problems, and solutions are given whenever the result is used in a later chapter.

A unique feature of the book is the inclusion of an extensive software package, written in Mathematica. Mathematica is, at the time this book is published, the most advanced language for scientific computations. It has made it possible to include a rich set of algorithms, some of which are rather sophisticated, in a manageable size of code and documentation. It is my hope that the reader not yet familiar with Mathematica will use this opportunity to learn it. Nevertheless, the book *does not depend* on the software pedagogically, so the instructor has the freedom to exclude this part from the course syllabus. It should be emphasized that these are not "toy programs." The software is suitable for use in serious applications, and engineers using this book are encouraged to use the software not just for study. Naturally, the publisher is not responsible for the software, but I would welcome any comments, bug reports, and suggestions. To provide extra motivation for this, I will send every reader who reports a bug or makes a useful suggestion an updated version of the software.

Prerequisites for the book include probability, random processes, linear algebra, linear systems, Fourier transforms, and some real analysis. Most of all, some mathematical maturity is essential. The reader should be able to follow "epsilon-delta" arguments and understand limit and convergence concepts (both deterministic and probabilistic).

The material in the book can be covered in two semesters (or quarters). It is best to cover the material in its order of appearance. I have taught a one-semester graduate course at the department of electrical engineering of Technion, based on Chapters 2 through 7, for several years. The material in Chapters 8 through 12 is typically covered at Technion in "special topics" courses.

I thank the following persons for their generous help. Petre Stoica and Mats Viberg thoroughly read the entire manuscript. In their wisdom and experience, they greatly helped to make it a better book. Ofer Porat read the first half of the book, helped polish mathematical derivations and proofs, and suggested many clarifications. Haim Barak, Shmuel Farkash, Bert Hochwald, and Ofer Zeitouni read various parts of the book and offered numerous corrections and suggestions. Bill Thomas did a superb copyediting job. The remaining errors are solely my responsibility. Tom Kailath and Martin Morf gave me my first education in the field and are responsible for my continuing enthusiasm for it. And, finally, special thanks to Ben Friedlander, my friend and colleague, for his guidance and support over the years.

In summary, I hope that the reader of this book, whether a student or engineer, will benefit from both the theoretical and the applied parts of the book. Sound theoretical basis is essential for the demanding nature of today's electrical engineering. Familiarity with working software packages will help bridge the gap between theory and applications.

Boaz Porat

Haifa, Israel
October 1993

To Aliza, Ofer, and Noga

and

In Memory of David, Tova, and Ruth Freud

CHAPTER 1

Introduction

A *discrete-time random signal* is simply a (potentially infinite) sequence of random numbers. By *random* we mean that the numbers obey a certain probabilistic law. This law may be completely known, known in part, or completely unknown. Discrete-time random signals are often obtained by sampling some physical signal, generated artificially or by nature. Signals in radar, sonar, audio, communication, control systems, and biomedical engineering are a few examples. By comparison, nonrandom signals are ones whose shape is completely known and can usually be reproduced at will. In some cases, the distinction between the two types of signals is blurred. For example, consider a binary-coded message transmitted as a sequence of bits over a digital communication channel. Such a sequence can appear as random to an uninformed observer, while in fact it is completely known to the author of the message. Signals of this kind are sometimes called *pseudorandom*.

A deep philosophical question, which has been under debate for centuries, is whether randomness is built into nature or whether our "randomness premise" is just a reflection of our ignorance of nature's fundamental mechanisms. In any case, there is no denying that random models have proved very successful in many disciplines involving either natural or artificial signals.

Complete knowledge of the probabilistic law governing the signal is rare. Typically, we only have partial knowledge, or none at all. On the other hand, we often have the ability to *measure* the signal, at least during certain periods. Measurement of the signal can be under controlled circumstances (e.g., in lab experiments) or uncontrolled ones, depending on the specific application. Once we have collected some measurements, we can use them to improve our knowledge of the probabilistic model that the signal is assumed to obey. There are many reasons for wanting to do this, some of which are as follows.

- To get a better understanding of the physical mechanism generating the signal. For example, ocean waves are a random phenomenon that, however, is governed by an underlying physical model (involving gravitational and hydro-dynamic laws). By measuring wave height as a function of time at various sea states, we can improve our understanding of ocean wave phenomena.

- To infer about some of the signal parameters. For example, a radar echo from a moving target contains information about the target motion. The arrival time of the echo gives information about its range, while its Doppler frequency shift gives information about its speed. The extraction of these two parameters from the echo can be done in an optimal manner if the probabilistic signal model is known.

- To track changes in the signal's source and help identify their cause. For example, certain diseases affect the electrical signals generated by the human brain (these signals are called electroencephalograms, or EEG for short). EEG signal analysis may help identify a pathological condition, track its development, and give indications about the effect of medications.

- To predict the signal's future behavior. For example, a good probabilistic model of stock market behavior may help one to predict its future trends and take advantage of them.

- To improve the quality of the signal. For example, suppose we wish to improve the quality of a public address voice system by reducing noise and reverberation. A good model for the noise and reverberation may help in attenuating these effects in an "optimal" manner (i.e., without too much harm to the desired sound signal).

- To achieve data compression for storage or transmission. For example, speech signals are typically sampled at a rate of 8000 Hz, using 8 bits per sample. Thus, 64,000 bits are needed to transmit or store each second of speech. Compression methods, built on probabilistic models of the human voice, can reduce this number by an order of magnitude without too much degradation of the voice quality. Much greater compression can be typically achieved with images (which are, essentially, two-dimensional signals).

- To synthesize artificial signals similar to the natural ones, for example, artificial speech.

Statistics is the science dealing with inference about probabilistic models from measurements. Therefore, any serious treatment of random signals must rely on statistical theory. The branch of statistics dealing with random signals is called *time series analysis*. The references at the end of this chapter include some of the better known books on time series analysis: Anderson [1971], Box and Jenkins [1970], Brillinger [1981], Brockwell and Davis [1991], Grenander and Rosenblatt [1957], Priestley [1981], and Rosenblatt [1985]. Some of these books, however, are not well suited to the needs of the electrical engineer for two reasons: their mathematical level is typically rather advanced, and they concentrate on topics of interest mostly to statisticians. The reference list also includes some recent electrical engineering books on this subject: Marple [1987], Therrien [1992], Kay [1988], and Scharf [1991]. Of those, the former two are more elementary, while the latter are more advanced.

This book was written for the graduate electrical engineering student, as well as for the professional engineer. The exposition is reasonably rigorous, relying on medium-level probability theory and mathematical analysis. Proofs (or partial proofs in some cases) are provided for most of the theorems. Topics of interest to electrical engineering are emphasized. In general, the book is of a more advanced level than the four electrical engineering books mentioned above, but is less advanced than most of the statistics books. The following paragraphs contain brief descriptions of the topics covered in the book.

THE STRUCTURE OF STATIONARY PROCESSES

Loosely speaking, a stationary process is one whose probabilistic law does not change over time. A striking property of stationary processes is that their probabilistic law can often be inferred from a single, but infinite, sequence of measurements. The condition under which this property holds is called ergodicity. Stationarity and ergodicity together enable us to get consistent estimates of the signal model, that is, estimates whose quality improves (in a properly defined sense) as the size of the measured sequence increases.

Many random processes encountered in real life are stationary to a good approximation. Even when they are not, they can often be regarded as approximately stationary over a certain time interval, which, with luck, may be sufficient for analysis purposes. For example, the noise in a radar receiver is typically stationary over very long periods (e.g., from the time the unit is turned on and reaches a stable temperature till it is turned off). On the other hand, a human speech signal can be regarded as approximately stationary over a period of a few tens of milliseconds only. Nevertheless, those few tens of milliseconds are usually enough to perform a reasonable estimation of the signal model for the specific period.

Chapter 2 lays out the foundations for treatment of stationary signals. It starts with some basic definitions, which are assumed to be known to the reader, and are given for completeness (Secs. 2.1 and 2.2). Then the notions of stationarity and ergodicity are explained (Sec. 2.3), followed by the introduction of the covariances of stationary processes and their properties. For a stationary Gaussian process, the covariances (plus the mean value, if it is not zero), completely define the probabilistic law. After some examples (in Sec. 2.5), we turn our attention to the Hilbert space description of stationary processes and the Wold decomposition. Wold's theorem is a fundamental statement on the time domain structure of a (wide sense) stationary process. Sections 2.8 and 2.9 are devoted to frequency domain representations of stationary processes: spectral distributions and spectral densities. Frequency domain methods have been central to signal analysis since the path-breaking work of Fourier. Classical random signal analysis has put a strong emphasis on frequency domain methods; on the other hand, time domain is emphasized in modern treatments. Section 2.10 continues the discussion of linear prediction started in Sec. 2.6. After that, ergodicity results are given for Gaussian processes. The final section in this chapter introduces parametric models for stationary processes, in particular models involving linear time-invariant rational systems, driven by white noise. This is a very important section, since the estimation of the parameters of such models occupies a large part of the book.

PARAMETER ESTIMATION THEORY

Chapter 3 provides background on the statistical parameter estimation necessary for the statistical treatment of random signals. The coverage is rather broad: the material has been chosen on the basis of its interest to random signal analysis in general, not necessarily to parts covered in the book. On the other hand, only point estimation is included. Interval estimation, as well as hypothesis testing, has been omitted. The former has been sacrificed mainly due to space considerations. The latter is important for signal detection and classification, topics that are outside the scope of the book.

The chapter starts with a definition of the basic parameter estimation problem in a probabilistic framework. It then introduces the notions of statistic, estimates, and information (in the sense of Fisher). Properties of estimates are described in Sec. 3.2, where we give various measures by which an estimate can be judged to be "good." Two fundamental properties of an estimate are its bias and variance. The former measures the average deviation of the estimate from the true value, while the latter measures its dispersion from the average value. It so happens that the variance of bias-free estimates is bounded from below, depending on the probabilistic structure of the problem. Section 3.3 gives several lower bounds on the variances of estimates. Of these, the most important is the Cramér-Rao bound, but other bounds are given for completeness and for their general interest.

Section 3.4 extends the definition of estimates and introduces sequential estimation. In dealing with sequential estimation, the properties of interest are consistency, asymptotic efficiency, and asymptotic normality. A consistent estimate is one that approaches the true value of the estimated parameter as the number of measurements tends to infinity (in a properly defined sense). An asymptotically efficient estimate is one whose variance approaches the Cramér-Rao lower bound (the reader is warned that this is a very loose definition). An asymptotically normal estimate is one whose probability distribution, after proper normalization, approaches a Gaussian distribution. Estimates having all three properties are usually considered "best," at least from a classical viewpoint.

Sections 3.5 and 3.6 discuss the two most important methods for parameter estimation: maximum likelihood and the method of moments. Maximum likelihood uses the measurements directly for obtaining the estimates. The method of moments relies on preliminary data reduction: it computes certain statistics first and uses those statistics to obtain the estimates. The difference in the properties of the estimates resulting from these different approaches is investigated in the two sections. Both methods lead, under proper assumptions, to consistent asymptotically normal estimates. However, while maximum likelihood has the potential for asymptotic efficiency, this is not necessarily the case for the method of moments.

The rest of Chapter 3 is devoted to topics of secondary importance to the book (but not to the field in general): linear least-squares estimation, maximum entropy estimation, and the problem of model order selection.

NONPARAMETRIC SPECTRUM ESTIMATION

Traditional random signal analysis has been based almost exclusively on frequency domain methods. For this reason, the name *spectral estimation* has become almost synonymous to random signal analysis. Chapter 4 is devoted to the traditional methods. Foremost of those is the periodogram, which is simply the square magnitude of the discrete Fourier transform of the measured signal. This topic is introduced in Sec. 4.3 after a preliminary discussion on estimation of the covariances of stationary processes in Sec. 4.2. The statistical properties of the periodogram are analyzed, in particular its lack of consistency. The remaining sections are devoted to several modifications of the periodogram, which aim at restoring its consistency. Periodogram averaging methods are described in Sec. 4.4, periodogram smoothing in Sec. 4.5, and windowed periodograms in Sec. 4.6. In the two latter cases, consistency of the estimates is established and their asymptotic variances are computed. Examples are given to illustrate the performance of the various spectral estimates.

PARAMETER ESTIMATION FOR GAUSSIAN PROCESSES

Chapter 5 is the first of four chapters devoted to parametric estimation methods for stationary Gaussian processes. The chapter deals with the basic theory common to all types of parametric models. It first derives general formulas for the Cramér-Rao lower bound for Gaussian process in both exact and asymptotic forms. The main result is Whittle's asymptotic formula, which expresses the asymptotic Cramér-Rao bound in terms of the spectral density of the process. This formula is then specialized to processes with rational transfer function models.

Maximum likelihood estimation for Gaussian processes is discussed next, and its consistency, efficiency, and asymptotic normality are established subject to certain conditions. Finally, the chapter explores application of the method of moments to this problem, with specialization to estimation based on the sample covariances. An asymptotic lower bound on the variance of estimates based on the sample covariances is derived and compared with the Cramér-Rao bound.

AUTOREGRESSIVE PARAMETER ESTIMATION

Autoregressive processes are a special case of processes generated by linear rational filters in which the filter is all-pole. They are particularly simple for analysis and estimation, and their theory is rich and elegant. Chapter 6, which is devoted to autoregressive processes, starts with the celebrated Yule-Walker method and an analysis of its properties. It continues, in Sec. 6.2, with the Levinson-Durbin algorithm, which provides a computationally efficient solution of the Yule-Walker equations. This section also introduces the concept of partial correlation coefficients, which are very important to autoregressive parameter estimation.

Section 6.3 describes some algorithms related to Levinson-Durbin, in particular the Schur algorithm. Section 6.4 introduces lattice filters, a topic that is further elaborated in Chapter 8. Section 6.5 is of secondary importance, making the connection between autoregressive models and maximum entropy parameter estimation. Sections 6.6, 6.7, and 6.8 discuss various alternatives to the Yule-Walker methods: least-squares estimation, maximum likelihood, and direct estimation of the partial correlation coefficients. These methods, while asymptotically equivalent to the Yule-Walker estimate, offer certain advantages in estimation from short data records.

AUTOREGRESSIVE MOVING AVERAGE PARAMETER ESTIMATION

Autoregressive moving average processes correspond to linear time-invariant filters with both poles and zeros. In the special case of zeros only, they are called moving average processes. Chapter 7 treats the parameter estimation problem for both kinds of processes. Relative to the case of autoregressive processes, estimation of linear filter models containing zeros is difficult. The estimation methods described in the chapter can be classified into two groups: linear and nonlinear. Typically, linear methods are simpler to implement, but their statistical accuracy is relatively poor. Nonlinear methods are computationally intensive and difficult to implement, but they have the potential to achieve better accuracy (relative to the Cramér-Rao bound).

Sections 7.2 and 7.3 are devoted to the two types of estimation methods for moving average processes. Algorithms based on the method of moments are described. Their performance is analyzed and illustrated by numerical examples. Sections 7.4 and 7.5 similarly treat autoregressive moving average processes. Sections 7.6 and 7.7 describe the maximum likelihood approach and give both approximate and exact maximum likelihood algorithms. Finally, an application of autoregressive moving average modeling to biomedical signal analysis is discussed in Sec. 7.8.

ADAPTIVE ESTIMATION

Some applications of random signal processing are in real time. Data are collected continuously at a certain rate, and it is desired to update the estimates as more data are collected. Sometimes it is desired to gradually forget the effect of old data as well, especially when the signal parameters are not constant, but slowly time varying. In such cases, it is reasonable to compute the new estimate as some function of the old estimate and the new data. If properly implemented, such a scheme can considerably reduce the total computational load and yield smoothly varying estimates. Adaptive estimation is the name given to implementations of this kind. Chapter 8 is devoted to this subject.

The chapter starts with the recursive least-squares algorithm, in Sec. 8.2. This algorithm is suitable for linear models and computes a new estimate at each data point. Its "memory" of old data can be made to grow or to fade, depending on a certain internal parameter. In the context of random signals, this algorithm is suitable for autoregressive models, but not for the more general ones. Sections 8.3 and 8.4 extend the recursive least-squares algorithm to handle autoregressive moving average models. The former presents the extended least-squares algorithm, and the latter the recursive maximum likelihood algorithm. Of the two, the first is simpler, but the second has better convergence properties and higher statistical accuracy. Section 8.5 is devoted to the stochastic gradient approach to adaptive estimation. Stochastic gradient algorithms require the least number of computations per data point, but their convergence may be very slow relative to the previous algorithms.

About half the chapter is devoted to a special class of adaptive algorithms, lattice algorithms. Lattice algorithms are computationally efficient versions of the algorithms presented in Secs. 8.2 and 8.3. Typically, they have better numerical behavior as well. Section 8.6 provides the necessary mathematical background to lattice algorithms. Section 8.7 presents algorithms for autoregressive models and Sec. 8.8 for autoregressive moving average models. Section 8.9 gives a brief survey of some extensions to lattice algorithms.

ESTIMATION OF DETERMINISTIC PROCESSES

Some random signals are called deterministic, because they can be predicted with zero error from sufficiently many measurements. Sinusoidal signals and sums of such provide a typical example. When noise is added to a deterministic signal, the perfect prediction property is lost, and estimation becomes more difficult. Chapter 9 deals with the estimation of parameters of sinusoidal signals when the measurements contain additive white noise. This problem can be handled with reasonable success by the methods presented in Chapter 4. Nevertheless, recent research has led to impressive improvements over the methods of Chapter 4, in cases where the number of measurements is small and the signal-to-noise ratio is not high.

Section 9.2 derives the Cramér-Rao bound for the sinusoids-in-white-noise model in both exact and asymptotic forms. Section 9.3 presents the maximum likelihood estimate of the signal parameters. Maximum likelihood is difficult to implement for this problem, and is rarely applied. The remaining sections discuss some alternatives to maximum likelihood. Sections 9.4 and 9.5 present two linear methods: Prony's method and its modification due to Tufts and Kumaresan. Sections 9.6 and 9.7 deal with application of the method of moments to sinusoidal parameter estimation.

HIGH-ORDER STATISTICAL ANALYSIS

The mean and the covariances of a stationary Gaussian process are sufficient to characterize its probabilistic law. This is not the case for non-Gaussian processes.

The moments of orders higher than 2 provide additional information about the process. Utilization of the high-order moments in the estimation procedure can lead to improved accuracy. Additionally, high-order moments can be used to attenuate (to some extent) the effect of Gaussian noise added to the desired non-Gaussian signal. Great progress has been made in recent years in this subject. Estimation methods based on high-order moments have been developed, accompanied by methods of analysis of their performance. Chapter 10 is devoted to this subject.

The first topic in the chapter is the concept of cumulants and their properties. Cumulants are functions of the high-order moments and are often more convenient to deal with than the moments themselves. In particular, cumulants of Gaussian processes of orders higher than 2 are identically zero. Section 10.2 introduces cumulants in general, and Sec. 10.3 specializes the discussion to stationary linear processes. Section 10.4 further specializes the discussion to autoregressive moving average processes and shows how to compute the cumulants of such processes.

Section 10.5 begins the subject of estimation. It presents the standard estimates of cumulants, and derives their asymptotic variances. Section 10.6 treats the estimation of moving average processes from cumulants, using linear methods. Section 10.7 similarly treats autoregressive moving average processes. Section 10.8 treats both kinds of processes by nonlinear methods. Section 10.9 offers a digression: it shows how to reconstruct the input signal to a linear process from the output process and the estimated model and illustrates the use of this procedure in a communication problem. Finally, Sec. 10.10 discusses the estimation of sinusoidal signal parameters using high-order cumulants.

TIME-FREQUENCY SIGNAL ANALYSIS

The methods presented in Chapters 2 through 10 are limited to processes that are either stationary or approximately so over sufficiently long intervals. Fast time variations of the signal's characteristics prevent efficient time averaging and render most algorithms based on the method of moments useless. Likelihood-based methods can be used in principle, but the lack of general and robust models for nonstationary processes severely limits their usefulness.

In recent years there has been an intensive effort to develop analysis methods for nonstationary signals. This has come to be known as time-frequency analysis, since the main goal has been the characterization of the instantaneous frequency contents of the signal versus time. The two classes of time-frequency methods are linear and nonlinear. Chapters 11 and 12 are devoted to these two classes, respectively. Admittedly, none of these methods has proved particularly useful for *random* signals so far. This subject is still under active research, and its inclusion in the book was partly in order to encourage the readers to eventually make their own contributions to such research.

The main linear methods for time-frequency analysis are the short-time Fourier transform, its close relative, the Gabor representation, and the wavelet transform. Chapter 11 develops the basic theory for these three methods. Section 11.2

presents the short-time Fourier transform for continuous-time signals. Sections 11.3 and 11.4 present the Gabor representation, first on an elementary level and then on an advanced one. The advanced treatment of the Gabor representation requires some functional analysis background, in particular the concept of a frame. The wavelet transform is treated in Secs. 11.5 and 11.6. Special attention is given to orthonormal wavelets based on the multiresolution analysis of Mallat. Section 11.7 discusses the implementation of the Gabor and wavelet transforms for discrete-time signals.

Of the nonlinear time-frequency representations, the most common are quadratic. Chapter 12 starts with a discussion of the fundamental representation of this type, the Wigner-Ville distribution. The basic distribution is given in Sec. 12.2 and some of its extensions (including discrete-time versions) in Sec. 12.3. The ambiguity function, which is the inverse two-dimensional Fourier transform of the Wigner-Ville distribution, is described in Sec. 12.4. Section 12.5 introduces a general class of quadratic time-frequency distributions, the Cohen class.

Sections 12.6 and 12.7 introduce a nonquadratic time-frequency representation, the high-order ambiguity function. This function is useful for the treatment of a special class of nonstationary signals, complex signals with constant amplitude and continuous phase. Algorithms based on the high-order ambiguity function are described, and an example illustrating its use for radar signal processing is given.

THE SOFTWARE

The book includes an extensive software package, which implements most of the algorithms described in the book. The software is written in Mathematica [Wolfram, 1991] and is available on a diskette enclosed in the book. At the end of each chapter there is a section describing the procedures relevant to that chapter. For each procedure, we give the calling sequence (name and input parameters), the returned values, and a brief description of its function. The types of the input and output variables are not specified formally, but can be inferred from the procedures themselves.

REFERENCES

Anderson, T. W., *The Statistical Analysis of Time Series*, Wiley, New York, 1971.

Box, G. E. P., and Jenkins, G. M., *Time Series Analysis: Forecasting and Control*, Holden-Day, San Francisco, 1970.

Brillinger, D. R., *Time Series, Data Analysis and Theory*, Holden-Day, San Francisco, 1981.

Brockwell, P. J., and Davis, R. A., *Time Series: Theory and Methods,* 2nd ed., Springer-Verlag, New York, 1991.

Grenander, U., and Rosenblatt, M., *Statistical Analysis of Stationary Time Series,* Wiley, New York, 1957.

Kay, S. M., *Modern Spectral Estimation, Theory and Applications,* Prentice Hall, Englewood Cliffs, NJ, 1988.

Marple, S. L., *Digital Spectral Analysis with Applications,* Prentice Hall, Englewood Cliffs, NJ, 1987.

Priestley, M. B., *Spectral Analysis and Time Series,* Academic Press, New York, 1981.

Rosenblatt, M., *Stationary Sequences and Random Fields,* Birkhäuser, Boston, 1985.

Scharf, L. L., *Statistical Signal Processing,* Addison-Wesley, Reading, MA, 1991.

Therrien, C. W., *Discrete Random Signals and Statistical Signal Processing,* Prentice Hall, Englewood Cliffs, NJ, 1992.

Wolfram, S., *Mathematica, A System for Doing Mathematics by Computer,* 2nd ed., Addison-Wesley, Redwood City, CA, 1991.

CHAPTER 2

The Structure of Stationary Processes

2.1. DISCRETE-TIME RANDOM PROCESSES

This chapter is devoted to the basic theory of discrete-time stationary processes. This material is classical—it has been around for over 50 years. Most of the chapter is concerned with the two fundamental descriptions of stationary processes: the time domain and the frequency domain characterizations. These two descriptions complement each other and are equally important. The material in this chapter forms the basis on which the entire book is built.

We begin our discussion of stationary processes with a brief reminder of some basic facts from probability theory with which the reader is assumed to be familiar. A *probability space* is a triplet $(\Omega, \mathcal{A}, P)$, where:

(1) Ω is a nonempty set, whose members are called *outcomes*.
(2) $\mathcal{A}$ is a σ-algebra of subsets of Ω, whose members are called *events*. A σ-algebra is defined by the following axioms:
  a. The set Ω belongs to $\mathcal{A}$.
  b. For every set A in $\mathcal{A}$, the complement set $\Omega - A$ also belongs to $\mathcal{A}$.
  c. For every countable family of sets in $\mathcal{A}$, say $\{A_n, 1 \leq n < \infty\}$, the union $\bigcup_{n=1}^{\infty} A_n$ belongs to $\mathcal{A}$.
(3) P is a *probability measure* on $\mathcal{A}$, that is, a function that assigns to each $A \in \mathcal{A}$ a real number $P(A)$ in the range $[0, 1]$, and has the following properties:
  a. $P(\Omega) = 1$.
  b. If $\{A_n, 1 \leq n < \infty\}$ is a countable family of disjoint subsets of Ω, then

$$P\left(\bigcup_{n=1}^{\infty} A_n\right) = \sum_{n=1}^{\infty} P(A_n).$$

11

Remarks:

1. Finite unions are regarded as a special case of countable unions.
2. It follows from the axioms that the empty set $\emptyset$ belongs to $\mathcal{A}$ and that count-able (or finite) intersections of members of $\mathcal{A}$ belong to $\mathcal{A}$.
3. It follows from the axioms that $P(\emptyset) = 0$ and $P(\Omega - A) = 1 - P(A)$.

A *real scalar random variable* X is a function from Ω to the real line $\boldsymbol{R}$ such that, for all $x \in \boldsymbol{R}$, the set $\{\omega : \omega \in \Omega, X(\omega) \leq x\}$ belongs to $\mathcal{A}$ (i.e., this set is an event and therefore it has a probability). A *complex scalar random variable* is a function from Ω to the complex plane $\boldsymbol{C}$ such that its real and imaginary parts are real scalar random variables. A *real (complex) vector random variable* is a finite ordered collection of real (complex) random variables, say $[X_1, X_2, \ldots, X_n]^T$ [A1].

Let $[x_1, x_2, \ldots, x_n]^T$ be a vector of real numbers and $[X_1, X_2, \ldots, X_n]^T$ a real vector random variable. The set $\bigcap_{i=1}^{n} \{\omega : X_i(\omega) \leq x_i\}$ is an event (being a finite intersection of events), so it has a probability. Let this probability be denoted by $F_{X_1, \ldots, X_n}(x_1, \ldots, x_n)$. If we let each x_i take all the values in $\boldsymbol{R}$, we get a function from $\boldsymbol{R}^n$ to $[0, 1]$. This function is called the *cumulative distribution function* (c.d.f.) of the vector random variable $[X_1, X_2, \ldots, X_n]^T$.

The *probability density function* (p.d.f.) of a vector random variable is defined as

$$f_{X_1, \ldots, X_n}(x_1, \ldots, x_n) = \frac{\partial^n F_{X_1, \ldots, X_n}(x_1, \ldots, x_n)}{\partial x_1 \partial x_2 \ldots \partial x_n} \qquad (2.1.1)$$

provided the right side exists everywhere on the support of the c.d.f. When there is no ambiguity, it is customary to omit the subscripts from the c.d.f. and the p.d.f. and denote them simply by $F(x_1, \ldots, x_n)$ and $f(x_1, \ldots, x_n)$, respectively. On the other hand, we can use the *random variables themselves* as arguments and get new random variables $F(X_1, \ldots, X_n)$ and $f(X_1, \ldots, X_n)$, respectively. The distinction between the two kinds of argument substitution is important.

Let $\boldsymbol{Z}$ denote the set of integers. A family of random variables indexed on $\boldsymbol{Z}$, say $\{x_t, t \in \boldsymbol{Z}\}$, is called a *discrete-time random process*. Note that we use lower-case letters for random processes, while preserving uppercase letters for random variables. We regard the index t as representing time. In many applications, the discrete-time process is obtained by sampling a continuous-time process. In this case, the index denotes the tth sample point, with $t = 0$ corresponding to time zero.

A discrete-time random process can be either real or complex and either scalar or vector, depending on the nature of the random variables $\{x_t\}$. This book is mainly concerned with real scalar random processes, so we usually omit the "real scalar" qualifier. When we come to discuss the other types (complex scalar, real or complex vector), we will qualify them explicitly.

Every finite subset of $\boldsymbol{Z}$, say $\{t_1, t_2, \ldots, t_n\}$, yields a corresponding vector random variable $[x_{t_1}, x_{t_2}, \ldots, x_{t_n}]^T$, so it has a c.d.f. The random process $\{x_t\}$ is fully characterized by the distribution functions corresponding to all $n \geq 1$ and all $\{t_1, t_2, \ldots, t_n\}$. Note that infinite subsets of $\boldsymbol{Z}$ do not possess distribution functions.

2.2. GAUSSIAN RANDOM PROCESSES

We first recall the definition of the multivariate Gaussian distribution. Let $\boldsymbol{\mu} = [\mu_1, \mu_2, \ldots, \mu_n]^T$ be a real vector, and let Γ be a real symmetric positive definite matrix. Define [A2]

$$f(\boldsymbol{x}) = (2\pi)^{-n/2}|\Gamma|^{-1/2}\exp\{-\frac{1}{2}(\boldsymbol{x} - \boldsymbol{\mu})^T\Gamma^{-1}(\boldsymbol{x} - \boldsymbol{\mu})\}, \quad \forall \boldsymbol{x} \in \boldsymbol{R}^n. \qquad (2.2.1)$$

The function $f(\boldsymbol{x})$ is called the *multivariate Gaussian* (or *normal*) probability density function. A real vector random variable is said to be Gaussian if its p.d.f. is multivariate Gaussian.

It can be shown that (where E denotes expectation)

(a) $E(X_k) = \mu_k, 1 \le k \le n$; that is, $\boldsymbol{\mu}$ is the *mean* of the vector random variable.

(b) $E(X_k - \mu_k)(X_\ell - \mu_\ell) = \Gamma_{k\ell}, 1 \le k, \ell \le n$; that is, Γ is the *covariance matrix* of the vector random variable.

We remark that the definition of a vector Gaussian random variable can be extended to the case where the covariance matrix Γ is singular (i.e., only positive semidefinite), but we shall not discuss this extension here.

A discrete-time random process $\{x_t\}$ is said to be Gaussian if every finite subset of the $\{x_t\}$ forms a vector Gaussian random variable. Gaussian processes are of great interest from both theoretical and practical points of view. Many random processes encountered in real-life applications are Gaussian, or nearly so. This phenomenon is not coincidental, but follows from the central limit theorem. Naturally, Gaussian processes will play an important role in this book as well. However, not all our theory requires the Gaussian assumption, so we shall take care to use it only when necessary. Also, parts of the theory of random processes specifically require the process in question to be *non-Gaussian*, as we shall see in Chapter 10.

2.3. STATIONARITY AND ERGODICITY

Loosely speaking, a stationary random process is one whose probabilistic properties do not change in time. More formally, the process $\{x_t\}$ will be called *stationary in the strict sense* if, for all $n \ge 1$, for all integers $t_1, t_2, \ldots, t_n, \tau$, and for all $\boldsymbol{x} \in \boldsymbol{R}^n$,

$$F_{x_{t_1}, \ldots, x_{t_n}}(\boldsymbol{x}) = F_{x_{t_1+\tau}, \ldots, x_{t_n+\tau}}(\boldsymbol{x}). \qquad (2.3.1)$$

Weaker forms of stationarity are useful in many circumstances. We consider real scalar processes, but the following definitions can be extended to other types of processes. Define the kth-order moments of the process $\{x_t\}$ by

$$\mu(t_1, t_2, \ldots, t_k) = E(x_{t_1}x_{t_2}\ldots x_{t_k}). \qquad (2.3.2)$$

The process $\{x_t\}$ will be called *stationary in the moments up to order n* if, for all

$1 \leq k \leq n$ and for all integers $t_1, t_2, \ldots, t_k, \tau$, the kth-order moments are finite, and

$$\mu(t_1, t_2, \ldots, t_k) = \mu(t_1 + \tau, t_2 + \tau, \ldots, t_k + \tau). \qquad (2.3.3)$$

Stationarity in the strict sense implies stationarity of all finite moments. A process that is stationary in the moments up to second order is also called *stationary in the wide sense*. A Gaussian process that is stationary in the wide sense is also stationary in the strict sense. This property does not extend to non-Gaussian processes.

Ergodicity is concerned with the problem of estimating the statistical properties of a random process from time averages of a single realization (measurement) of the process. A proper treatment of ergodicity requires deep knowledge of probability theory. Here we limit ourselves to one form of ergodicity, which uses only mean-square convergence.

Let $\{x_t\}$ be a process that is stationary in the moments up to order n. The process will be called *mean-square (MS) ergodic in the moments up to order n* if, for all $1 \leq k \leq n$ and for all integers $t_1, t_2, \ldots, t_k$,

$$\lim_{N \to \infty} E \left| \frac{1}{2N+1} \sum_{t=-N}^{N} x_{t+t_1} x_{t+t_2} \cdots x_{t+t_k} - \mu(t_1, t_2, \ldots, t_k) \right|^2 = 0. \qquad (2.3.4)$$

Other forms of ergodicity involve different forms of convergence of the time averages (e.g., convergence with probability 1). A process that is MS ergodic in the moments of first and second order is also called *ergodic in the wide sense*.

2.4. THE COVARIANCES OF STATIONARY PROCESSES

Let $\{x_t\}$ be a real scalar wide sense stationary process. The mean of the process, $E(x_t)$, is independent of the time parameter t, and the second-order moment $E(x_{t_1} x_{t_2})$ depends only on the time difference $t_1 - t_2$. Denote

$$\mu = E(x_t) \qquad (2.4.1)$$

$$r_k = E(x_{t+k} - \mu)(x_t - \mu) = E(x_{t+k} x_t) - \mu^2. \qquad (2.4.2)$$

r_k is called the *covariance* of lag k, or simply the kth covariance. The doubly infinite sequence $\{r_k, k \in \mathbf{Z}\}$ is called the *covariance sequence* of the process. The covariance sequence of real scalar processes is symmetric; that is, $r_k = r_{-k}$. The covariance matrices of a vector process $\{x_t\}$ are defined by

$$r_k = E(x_{t+k} - \mu)(x_t - \mu)^T = E(x_{t+k} x_t^T) - \mu \mu^T. \qquad (2.4.3)$$

These covariances satisfy the symmetry property $r_k = r_{-k}^T$.

Let X denote a vector of n consecutive elements of the wide sense stationary random process $\{x_t\}$. The covariance matrix of X has the structure

$$R_n = E(X - \mu)(X - \mu)^T = \begin{bmatrix} r_0 & r_{-1} & \cdots & r_{-n+1} \\ r_1 & r_0 & \cdots & r_{-n+2} \\ \vdots & \vdots & \ddots & \vdots \\ r_{n-1} & r_{n-2} & \cdots & r_0 \end{bmatrix} \qquad (2.4.4)$$

The matrix R_n is seen to be constant along its diagonals. Such a matrix is called *Toeplitz*. When $\{x_t\}$ is a vector process, the matrix is constant along its *block diagonals*. Such a matrix is called *block Toeplitz*.

The positive semidefiniteness of the matrices R_n implies that, for every set of complex numbers $\{\alpha_k, 0 \le k \le n-1\}$, we have

$$\sum_{k=0}^{n-1} \sum_{\ell=0}^{n-1} \alpha_k^* r_{k-\ell} \alpha_\ell \ge 0, \qquad (2.4.5)$$

where $(\cdot)^*$ denotes complex conjugation. The sequence $\{r_k\}$ is said to be *positive semidefinite*.

Stationary Gaussian processes are fully characterized by their mean μ and their covariance sequence $\{r_k\}$. For non-Gaussian processes, moments of higher order play an important role. This issue will be taken up in detail in Chapter 10. For now we just introduce the notion of fourth-order cumulants, which will be needed in Chapter 4. Let $\{X_1, X_2, X_3, X_4\}$ be four zero mean real scalar random variables, not necessarily distinct. The *fourth-order cumulant* of these random variables is

$$\begin{aligned} \kappa_4(X_1, X_2, X_3, X_4) =& E(X_1 X_2 X_3 X_4) - E(X_1 X_2)E(X_3 X_4) \\ & - E(X_1 X_3)E(X_2 X_4) - E(X_1 X_4)E(X_2 X_3). \end{aligned} \qquad (2.4.6)$$

The fourth-order cumulant of jointly Gaussian random variables is identically zero; otherwise, it is generally nonzero. The general definition of cumulants and their properties is deferred to Chapter 10.

2.5. SOME EXAMPLES

In this section we illustrate the concepts introduced so far by several examples.

Example 2.1. White noise.
    Let $\{w_t\}$ be a sequence of zero mean, identically distributed random variables. We consider three cases:

(1) The distribution is non-Gaussian, and the $\{w_t\}$ are statistically independent.
(2) The distribution is non-Gaussian, and the $\{w_t\}$ are uncorrelated [i.e., $E(w_{t_1} w_{t_2}) = 0$ for $t_1 \ne t_2$], but not necessarily independent.

(3) The distribution is Gaussian, and the $\{w_t\}$ are uncorrelated and therefore independent.

The engineering literature commonly refers to all three cases as *white noise*, although they are different. The precise terminology is *independent identically distributed* (i.i.d.) process for case (1) and *uncorrelated* process for case (2). The process in case (3) is commonly called *white Gaussian noise*. The covariance sequence of white noise is $\{\sigma_w^2 \delta(k)\}$, where σ_w^2 is the variance of w_t and $\delta(k)$ is the Kronecker delta [A3].

An i.i.d. process is ergodic in all its moments (provided they are finite). Similarly, an uncorrelated process is ergodic in the wide sense. Both properties can be proved by applying the strong law of large numbers to the left side of (2.3.4), but we will not do it here.   ∎

Example 2.2. DC process.
Let α be a Gaussian random variable with zero mean and unit variance, and let

$$x_t = \alpha, \quad \forall t. \tag{2.5.1}$$

This process is easily checked to be Gaussian, stationary, and zero mean. Its covariance sequence is given by

$$r_k = 1, \quad \forall k. \tag{2.5.2}$$

Every realization of this process looks like a dc signal, but the dc level is random and therefore different for each realization.

The dc process is not ergodic in any of its moments. For example, when $k = 1$, the left side of (2.3.4) is equal to 1. A similar argument applies to moments of higher orders.   ∎

Example 2.3. Harmonic (sinusoidal) process.
Let

$$x_t = A\cos(\omega_0 t - \phi), \tag{2.5.3}$$

where ω_0 is a fixed real number and ϕ random and uniformly distributed in $[0, 2\pi]$. We distinguish between two cases:

(1) A is a fixed positive real number.
(2) A is a Rayleigh distributed random variable, statistically independent of ϕ; that is,

$$f_A(a) = \frac{a}{\sigma^2}\exp\left(-a^2/2\sigma^2\right), \quad a \geq 0. \tag{2.5.4}$$

In both cases the process is wide sense stationary. The mean is easily checked to be zero. The kth covariance is given by

$$r_k = E\{A^2 \cos(\omega_0 t + \omega_0 k - \phi)\cos(\omega_0 t - \phi)\}$$

$$= \frac{1}{2} E\{A^2 \cos(2\omega_0 t + \omega_0 k - 2\phi) + A^2 \cos\omega_0 k\}$$

$$= \frac{1}{2} E(A^2) \cos\omega_0 k. \tag{2.5.5}$$

In case (1), $E(A^2) = A^2$, while in case (2), $E(A^2) = 2\sigma^2$. In both cases, the covariance sequence is harmonic, having the same frequency as the given process.

The process in case (1) is non-Gaussian. On the other hand, the process in case (2) can be shown to be Gaussian (Prob. 2.1). In this case, the process also can be written as

$$x_t = A\cos\phi\cos\omega_0 t + A\sin\phi\sin\omega_0 t = A_I\cos\omega_0 t + A_Q\sin\omega_0 t, \tag{2.5.6}$$

where A_I and A_Q, the *in-phase* and *quadrature* components, are independent Gaussian random variables with zero means and variances σ^2.

The process in case (1) is ergodic in all its moments. An intuitive argument for this property can be given as follows. A single realization enables the exact calculation of the frequency ω_0 and the amplitude A. From these two parameters, we can compute the moments of all orders. On the other hand, the process in case (2) is not ergodic in its second order moments, because the value of σ^2 cannot be inferred from a single realization.   ∎

2.6. HILBERT SPACES OF STATIONARY PROCESSES

Hilbert spaces form a convenient framework for the analysis of wide sense stationary random processes. Appendix B contains some facts about Hilbert spaces, which will be needed here. We will restrict our discussion to Hilbert spaces over the real field.

Assume we are given a probability space $(\Omega, \mathcal{A}, P)$ and consider *all* real scalar random variables on this space satisfying

$$E(X) = 0, \quad E(X^2) < \infty. \tag{2.6.1}$$

Let $L_2(\Omega, \mathcal{A}, P)$ denote the set of all such random variables [A4]. Finite sums of random variables from this set obviously belong to the set, and so do multiples by real scalars. Furthermore, the group axioms and the scalar multiplication axioms are easily checked to hold (cf. App. B). Therefore, $L_2(\Omega, \mathcal{A}, P)$ is a linear space over the real field.

For any two random variables in $L_2(\Omega, \mathcal{A}, P)$, let their inner product be defined as their covariance; that is,

$$\langle X, Y \rangle = E(XY). \tag{2.6.2}$$

To comply with axiom (11) of the inner product, we define two random variables X_1 and X_2 to be identical if they differ at most on a set of probability zero. Then, it is easy to check that the inner product axioms are satisfied (cf. App. B). Therefore,

$L_2(\Omega, \mathcal{A}, P)$ is an inner product space. The norm of an element X in this space is its standard deviation $\sqrt{\text{var}(X)}$.

The only issue that remains to be settled is whether $L_2(\Omega, \mathcal{A}, P)$ is complete, that is, whether every Cauchy sequence of random variables in $L_2(\Omega, \mathcal{A}, P)$ converges to a random variable in this space. The answer is affirmative and is, in fact, a deep result in measure theory [Royden, 1968, Ch. 11]. We thus have [A4]

Theorem 2.1. $L_2(\Omega, \mathcal{A}, P)$ is a Hilbert space. ∎

Suppose we are given a real scalar wide sense stationary random process $\{y_t\}$ in $(\Omega, \mathcal{A}, P)$ with zero mean and finite variance. The zero mean assumption is not restrictive, since we can easily modify any process with finite mean by subtracting the mean and deal with $\{y_t - \mu\}$. A basic property of Hilbert spaces is that every subset in the space spans a subspace that is, by definition, the smallest Hilbert space containing all the elements of the spanning set. Let us introduce the following subspaces of $L_2(\Omega, \mathcal{A}, P)$.

$$\mathcal{H}_\infty^{(y)} = \overline{Sp}\{y_t, -\infty < t < \infty\} \tag{2.6.3a}$$

$$\mathcal{H}_t^{(y)} = \overline{Sp}\{y_s, -\infty < s \le t\} \tag{2.6.3b}$$

$$\mathcal{H}_{t,p}^{(y)} = \overline{Sp}\{y_s, t-p+1 \le s \le t\} \tag{2.6.3c}$$

$$\mathcal{H}_{-\infty}^{(y)} = \bigcap_{t=-\infty}^{\infty} \mathcal{H}_t^{(y)}. \tag{2.6.3d}$$

Loosely speaking, $\mathcal{H}_\infty^{(y)}$ is the Hilbert space spanned by the entire process $\{y_t\}$, $\mathcal{H}_t^{(y)}$ is the space spanned by the "present and past" of the process, and $\mathcal{H}_{-\infty}^{(y)}$ is the space spanned by the "remote past." The following holds for all $-\infty < s < t < \infty$ and for all $p \ge 0$:

$$\mathcal{H}_{-\infty}^{(y)} \subseteq \mathcal{H}_s^{(y)} \subseteq \mathcal{H}_t^{(y)} \subseteq \mathcal{H}_\infty^{(y)} \tag{2.6.4a}$$

$$\mathcal{H}_{t,p}^{(y)} \subseteq \mathcal{H}_{t,p+1}^{(y)} \subseteq \mathcal{H}_t^{(y)} \tag{2.6.4b}$$

$$\overline{\bigcup_{p=0}^{\infty} \mathcal{H}_{t,p}^{(y)}} = \mathcal{H}_t^{(y)}. \tag{2.6.4c}$$

Consider a specific element y_t of the given process. By definition, this element always belongs to $\mathcal{H}_t^{(y)}$, but it does not belong to $\mathcal{H}_{t-1}^{(y)}$ in general. By the Projection Theorem B.1, there exists a unique element $\hat{y}_t$ in $\mathcal{H}_{t-1}^{(y)}$ such that the error $y_t - \hat{y}_t$ has minimum norm (i.e., minimum variance). Similarly, there exists a unique element $\hat{y}_{t,p}$ in $\mathcal{H}_{t-1,p}^{(y)}$ such that $y_t - \hat{y}_{t,p}$ has minimum norm. The random variables $\hat{y}_t$ and $\hat{y}_{t,p}$ are called *the best linear predictors* of y_t with respect to $\mathcal{H}_{t-1}^{(y)}$ and $\mathcal{H}_{t-1,p}^{(y)}$, respectively.

Let us denote

$$u_t = y_t - \hat{y}_t, \quad u_{t,p} = y_t - \hat{y}_{t,p}. \tag{2.6.5}$$

The collection $\{u_t, -\infty < t < \infty\}$ is a random process on $L_2(\Omega, \mathcal{A}, P)$, which is called the *innovation process* of $\{y_t\}$. The random variable $u_{t,p}$ is called the pth-order *partial innovation* of y_t. The following relations among the above entities are of great importance.

Theorem 2.2.

$$\underset{p\to\infty}{\text{l.i.m.}}\,\hat{y}_{t,p} = \hat{y}_t, \quad \underset{p\to\infty}{\text{l.i.m.}}\,u_{t,p} = u_t. \tag{2.6.6}$$

Proof. The first limit follows from Theorem B.4. The second follows immediately from the first. ∎

Let us assume, from now on, that u_t is not identically zero; therefore, the same holds for $u_{t,p}$. The prediction $\hat{y}_{t,p}$ is a linear combination of $\{y_s, t-p \leq s \leq t-1\}$. Denote this linear combination as

$$\hat{y}_{t,p} = -\sum_{k=1}^{p} \alpha_{p,k}(t) y_{t-k}. \tag{2.6.7}$$

Then we have:

Theorem 2.3. The coefficients of the linear combination in (2.6.7) are independent of t and are given by

$$\begin{bmatrix} \alpha_{p,1} \\ \alpha_{p,2} \\ \vdots \\ \alpha_{p,p} \end{bmatrix} = -R_p^{-1} \begin{bmatrix} r_1 \\ r_2 \\ \vdots \\ r_p \end{bmatrix}, \tag{2.6.8}$$

where R_p is the matrix given in (2.4.4).

Proof. Write the partial innovation as

$$u_{t,p} = y_t - \hat{y}_{t,p} = y_t + \sum_{k=1}^{p} \alpha_{p,k}(t) y_{t-k}.$$

Multiply both sides by $y_{t-\ell}$ (where $1 \leq \ell \leq p$) and take expected values. By the orthogonality property of the best linear predictor (Theorem B.1), the left side is identically zero. Hence

$$r_\ell + \sum_{k=1}^{p} \alpha_{p,k}(t) r_{\ell-k} = 0, \quad 1 \leq \ell \leq p. \tag{2.6.9}$$

The matrix R_p is nonsingular (Prob. 2.2). Therefore, the solution of this set of equations is precisely the one given by (2.6.8). In particular, it is seen that the coefficients $\{\alpha_{p,k}, 1 \leq k \leq p\}$ are independent of t. ∎

Remark. The $\alpha_{p,k}$ are defined even when u_t is zero. In this case, R_p may be singular and then these coefficients will not be unique. However, by (2.6.9) they still will be independent of t.

Example 2.4. Let $\{y_t\}$ be white noise. Then the right side of (2.6.8) is identically zero for all p, so $\hat{y}_{t,p}$ is identically zero. Therefore, white noise cannot be predicted from its own past. The process itself is its own innovation. ∎

Example 2.5. Let $\{y_t\}$ be the harmonic process in (2.5.3), assume that ω_0 is not an integer multiple of π, and let $p = 2$. Then

$$\begin{bmatrix} \alpha_{2,1} \\ \alpha_{2,2} \end{bmatrix} = -\begin{bmatrix} 1 & \cos\omega_0 \\ \cos\omega_0 & 1 \end{bmatrix}^{-1} \begin{bmatrix} \cos\omega_0 \\ \cos 2\omega_0 \end{bmatrix} = \begin{bmatrix} -2\cos\omega_0 \\ 1 \end{bmatrix}.$$

Therefore,

$$\hat{y}_{t,2} = 2\cos\omega_0 y_{t-1} - y_{t-2}.$$

Problem 2.3 continues to explore this example. ∎

The following theorem summarizes the basic properties of the innovation process $\{u_t\}$.

Theorem 2.4. $\{u_t\}$ is a wide sense stationary, uncorrelated process, with zero mean and finite variance.

Proof. It is clear from (2.6.4) and (2.6.5) that $u_t \in \mathcal{H}_t^{(y)}$; hence it has zero mean and finite variance. To show that $\{u_t\}$ is wide sense stationary, observe first that

$$\lim_{p\to\infty} \langle u_{t,p}, u_{s,p} \rangle = \langle u_t, u_s \rangle.$$

This follows from Theorem 2.2 and Lemma B.2. Now, by Theorem 2.3 we can write (with $\alpha_{p,0} = 1$)

$$u_{t,p} = \sum_{k=0}^{p} \alpha_{p,k} y_{t-k}, \quad u_{s,p} = \sum_{\ell=0}^{p} \alpha_{p,\ell} y_{s-\ell}.$$

Hence,

$$E(u_{t,p} u_{s,p}) = E\left\{ \left(\sum_{k=0}^{p} \alpha_{p,k} y_{t-k} \right) \left(\sum_{\ell=0}^{p} \alpha_{p,\ell} y_{s-\ell} \right) \right\} = \sum_{k=0}^{p} \sum_{\ell=0}^{p} \alpha_{p,k} \alpha_{p,\ell} r_{t-s-k+l}.$$

The right side depends only on the difference $t - s$ for all p. Therefore, this remains true as $p \to \infty$, so $E(u_t u_s)$ is a function of $t - s$. This proves that $\{u_t\}$ is wide sense stationary.

To show that $\{u_t\}$ is an uncorrelated process, let $s < t$. By the orthogonality property, $u_t \perp \mathcal{H}_{t-1}^{(y)}$; hence $u_t \perp \mathcal{H}_s^{(y)}$. But $u_s \in \mathcal{H}_s^{(y)}$, so $u_t \perp u_s$; that is, $E(u_t u_s) = 0$. By symmetry, this holds for $s > t$ as well. ∎

Let us denote the variance of u_t by σ_u^2. If $\sigma_u^2 = 0$, then $u_t = 0$ almost surely. In this case, the process $\{y_t\}$ is said to be *deterministic*. Otherwise, the process is said to be *regular*. Deterministic processes can be perfectly predicted from their past, while for regular processes there is nonzero prediction error. This prediction error (the innovation process) is white noise "in the wide sense;" that is, it is uncorrelated, but not necessarily i.i.d.

All the concepts introduced in this section can be generalized to (real or complex) vector processes. We will not need these generalizations, so we skip the details and refer the interested reader to [Caines, 1988].

2.7. THE WOLD DECOMPOSITION

In this section we continue to explore the relationship between a given wide sense stationary process $\{y_t\}$ and its innovation process $\{u_t\}$. First, let us define

$$\beta_s = \frac{E(y_t u_{t-s})}{\sigma_u^2}, \quad 0 \le s < \infty. \tag{2.7.1}$$

The right side does not depend on t because of the joint stationarity of $\{y_t\}$ and $\{u_t\}$. This can be proved in the same manner as the stationarity of $\{u_t\}$ in Theorem 2.4. We have:

Lemma 2.1. The sequence $\{\beta_s\}$ satisfies

$$\beta_0 = 1, \quad \sum_{s=0}^{\infty} \beta_s^2 < \infty. \tag{2.7.2}$$

Proof. We have seen that u_t is uncorrelated with $\hat{y}_t$; hence

$$\beta_0 = \frac{1}{\sigma_u^2} E\{u_t(u_t + \hat{y}_t)\} = \frac{\sigma_u^2 + 0}{\sigma_u^2} = 1.$$

Also, for all $m \ge 0$ we have

$$0 \le E\left(y_t - \sum_{s=0}^{m} \beta_s u_{t-s}\right)^2 = E(y_t^2) - 2\sum_{s=0}^{m} \beta_s E(y_t u_{t-s}) + \sum_{s=0}^{m}\sum_{r=0}^{m} \beta_s \beta_r E(u_{t-s} u_{t-r})$$

$$= E(y_t^2) - 2\sum_{s=0}^{m} \sigma_u^2 \beta_s^2 + \sum_{s=0}^{m} \sigma_u^2 \beta_s^2 = r_0 - \sigma_u^2 \sum_{s=0}^{m} \beta_s^2.$$

Therefore,

$$\sum_{s=0}^{m} \beta_s^2 \le \frac{r_0}{\sigma_u^2}, \quad \forall m.$$

So, finally,

$$\sum_{s=0}^{\infty} \beta_s^2 < \infty. \qquad \blacksquare$$

Next, let $\{x_t\}$ be the random process

$$x_t = \sum_{s=0}^{\infty} \beta_s u_{t-s}. \qquad (2.7.3)$$

It is straightforward to show that, by Lemma 2.1, the sum on the right side of (2.7.3) converges in the mean square, so $\{x_t\}$ is well defined (Prob. 2.4). Similarly, it is straightforward to show that x_t is wide sense stationary. The next theorem, known as the *Wold decomposition theorem* [Wold, 1938, 1954], is perhaps the most important result on wide sense stationary processes.

Theorem 2.5. Any regular wide sense stationary process $\{y_t\}$ can be expressed as

$$y_t = x_t + v_t, \qquad (2.7.4)$$

where x_t is given by (2.7.3), and
 (a) The process v_t is uncorrelated with u_t; that is, $E(u_t v_s) = 0$, $\forall s, t$.
 (b) $v_t \in \mathcal{H}_{-\infty}^{(y)}$, $\forall t$.
 (c) The process $\{x_t\}$ is regular.
 (d) The process $\{v_t\}$ is deterministic.

Proof. The existence of the decomposition (2.7.4) is trivial; just *define* v_t as

$$v_t = y_t - x_t = y_t - \sum_{s=0}^{\infty} \beta_s u_{t-s}.$$

The interest is, of course, in the four properties of this decomposition, proved as follows.
 (a) x_s, by its definition, belongs to $\mathcal{H}_s^{(y)}$; therefore, the same holds for v_s. On the other hand, $u_t \perp \mathcal{H}_s^{(y)}$ for all $t > s$, so $E(u_t v_s) = 0$ for all $t > s$. Also, for $t \le s$ we have

$$E(u_t v_s) = E\{u_t(y_s - x_s)\} = \sigma_u^2 \beta_{s-t} - \sum_{k=0}^{\infty} \beta_k E(u_t u_{s-k})$$

$$= \sigma_u^2 \beta_{s-t} - \sigma_u^2 \beta_{s-t} = 0.$$

 (b) Since $v_t \in \mathcal{H}_t^{(y)}$ and $v_t \perp u_t$, necessarily $v_t \in \mathcal{H}_{t-1}^{(y)}$. Carrying this argument inductively, we see that $v_t \in \mathcal{H}_{t-s}^{(y)}$ for all $s \ge 0$. Hence,

$$v_t \in \bigcap_{s=0}^{\infty} \mathcal{H}_{t-s}^{(y)} = \mathcal{H}_{-\infty}^{(y)}.$$

(c) Let us first show that $u_t \in \mathcal{H}_t^{(x)}$. Since u_t is orthogonal to $\mathcal{H}_{t-1}^{(y)}$, it follows that all $\{u_t\}$ are orthogonal to $\mathcal{H}_{-\infty}^{(y)}$. Since x_t is a linear combination of the $\{u_t\}$, it follows that all $\{x_t\}$ are orthogonal to $\mathcal{H}_{-\infty}^{(y)}$. Let us denote $\mathcal{L}_t = \mathcal{H}_t^{(x)} \oplus \mathcal{H}_{-\infty}^{(y)}$ (see App. B for the definition and properties of the direct sum operation $\oplus$). Then by (2.7.4) and (b) it follows that $y_t \in \mathcal{L}_t$, hence $u_t \in \mathcal{L}_t$. But, since u_t is orthogonal to $\mathcal{H}_{-\infty}^{(y)}$, necessarily $u_t \in \mathcal{H}_t^{(x)}$. Now, since $\beta_0 = 1$, we have

$$x_t = u_t + \sum_{s=1}^{\infty} \beta_s u_{t-s},$$

where $u_t \perp \mathcal{H}_{t-1}^{(x)}$ and $\sum_{s=1}^{\infty} \beta_s u_{t-s} \in \mathcal{H}_{t-1}^{(x)}$. Therefore, the second term is just $\hat{x}_t$, and u_t is the innovation of x_t. Hence x_t is regular.

(d) We have, by (2.7.3), (2.7.4), (a) and (b),

$$v_t \in \mathcal{H}_{t-1}^{(y)} \subseteq \mathcal{H}_{t-1}^{(u)} \oplus \mathcal{H}_{t-1}^{(v)}.$$

Hence

$$\hat{v}_t = v_t | \mathcal{H}_{t-1}^{(v)} = v_t | (\mathcal{H}_{t-1}^{(u)} \oplus \mathcal{H}_{t-1}^{(v)}) = v_t.$$

Therefore v_t is deterministic. ∎

A process $\{y_t\}$ for which $\mathcal{H}_{-\infty}^{(y)} = \{0\}$ is said to be *purely indeterministic*. For a process of this kind, the deterministic part of the Wold decomposition is identically zero. The part x_t in the Wold decomposition is purely indeterministic (Prob. 2.5). Wold's theorem therefore expresses any wide sense stationary process as a sum of two orthogonal processes, the first of which is purely indeterministic, while the second is deterministic. The purely indeterministic component is expressed as a convolution of the innovation process $\{u_t\}$ with a linear time-invariant system having the impulse response $\{\beta_s\}$ (see Figure 2.1). This impulse response has a finite energy (i.e., it is square summable), but is not necessarily stable in the BIBO sense (i.e., it is not necessarily absolutely summable).

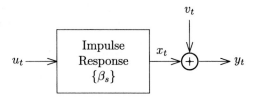

Figure 2.1. The Wold decomposition of a wide sense stationary process.

The simplest example of a purely indeterministic process is that of white noise (either Gaussian or not), as was shown in Example 2.4. Such a process is its own innovation and has $\beta_s = \delta(s)$. The simplest example of a deterministic process is the dc process of Example 2.2. This process clearly satisfies $y_t = y_{t-1} = \hat{y}_t$; hence

$u_t = 0$. The harmonic process of Example 2.3 is also deterministic, since it satisfies (Prob. 2.3)

$$y_t = 2\cos\omega_0 y_{t-1} - y_{t-2} = \hat{y}_t;$$

hence $u_t = 0$.

As an introduction to our next topic, consider the power series

$$f(z) = \sum_{s=0}^{\infty} a_s z^s,$$

where z is a complex variable and $\{a_s\}$ is a square summable sequence. Since $\{a_s\}$ is necessarily bounded, $f(z)$ is analytic on the open unit disk $|z| < 1$. Furthermore, the square summability of $\{a_s\}$ implies that $f(e^{j\omega})$ exists almost everywhere on $-\pi \le \omega \le \pi$ and ([A5], [A6])

$$\frac{1}{2\pi}\int_{-\pi}^{\pi} |f(e^{j\omega})|^2 d\omega < \infty.$$

We have:

Lemma 2.2.

$$\log|f(0)| \le \frac{1}{2\pi}\int_{-\pi}^{\pi} \log|f(e^{j\omega})|d\omega < \infty. \tag{2.7.5}$$

If the left inequality of (2.7.5) holds with equality, then $f(z)$ is nonzero inside the unit circle. ∎

This lemma is a consequence of Theorem 17.17 in [Rudin, 1986].[†] We remark that, if (2.7.5) holds with equality, $f(z)$ is called an *outer* function; see the discussion in [Rudin, 1986, Ch. 17].

Now let $\beta(z)$ be the z transform of $\{\beta_s\}$ [where β_s is defined in (2.7.1)]; that is [A7],

$$\beta(z) = \sum_{s=0}^{\infty} \beta_s z^{-s}. \tag{2.7.6}$$

Since $\{\beta_s\}$ is square summable, the function $\beta(z)$ is analytic outside the unit circle and

$$\frac{1}{2\pi}\int_{-\pi}^{\pi} |\beta(e^{j\omega})|^2 < \infty.$$

This function has an interesting and important property, which is expressed by the following theorem.

[†] The following technical point should be mentioned in this context. In Rudin [1986], $f(e^{j\omega})$ is defined as the nontangential limit of $f(z)$, while here it is defined as the L_2 limit $\lim_{n\to\infty}\sum_{s=0}^{n} a_s e^{j\omega s}$. The two can be shown to be equal almost everywhere on $[-\pi, \pi]$ by Theorems 17.11 and 17.12 in [Rudin, 1986].

Theorem 2.6. The function $\beta(z)$ is nonzero outside the unit circle. Consequently, the function $\log \beta(z)$ is analytic outside the unit circle.

Proof. Let $\{y_t\}$ be a purely indeterministic process with parameters σ_u^2 and $\{\beta_s\}$. Let $\alpha_p(z)$ be defined by

$$\alpha_p(z) = 1 + \sum_{k=1}^{p} \alpha_{p,k} z^{-k},$$

where the $\{\alpha_{p,k}\}$ were defined in Theorem 2.3. Recall that the partial innovation sequence $\{u_{t,p}\}$ is obtained by passing $\{y_t\}$ through the time-invariant filter $\alpha_p(z)$, and $\{y_t\}$, in turn, is obtained by passing $\{u_t\}$ through the time-invariant filter $\beta(z)$. Hence, using the whiteness of $\{u_t\}$ and standard properties of linear time-invariant systems, we have

$$E(u_{t,p}^2) = \frac{\sigma_u^2}{2\pi} \int_{-\pi}^{\pi} |\alpha_p(e^{j\omega})|^2 |\beta(e^{j\omega})|^2 d\omega.$$

Now, by Theorem 2.2, $\lim_{p\to\infty} E(u_{t,p}^2) = \sigma_u^2$. Also, both functions $\beta(z^{-1})$ and $\alpha_p(z^{-1})$ satisfy the conditions of Lemma 2.2, and $\log |\beta(z^{-1})||_{z=0} = \log |\alpha_p(z^{-1})||_{z=0} = 0$. Hence,

$$\log \sigma_u^2 = \lim_{p\to\infty} \log\{E(u_{t,p}^2)\} = \lim_{p\to\infty} \log\left\{ \frac{\sigma_u^2}{2\pi} \int_{-\pi}^{\pi} |\alpha_p(e^{-j\omega})|^2 |\beta(e^{-j\omega})|^2 d\omega \right\}$$

$$\geq \log \sigma_u^2 + \lim_{p\to\infty} \left\{ \frac{1}{2\pi} \int_{-\pi}^{\pi} \log \left[|\alpha_p(e^{-j\omega})|^2 |\beta(e^{-j\omega})|^2 \right] d\omega \right\} \qquad (*)$$

$$= \log \sigma_u^2 + \frac{1}{2\pi} \int_{-\pi}^{\pi} 2 \log |\beta(e^{-j\omega})| d\omega + \lim_{p\to\infty} \left\{ \frac{1}{2\pi} \int_{-\pi}^{\pi} 2 \log |\alpha_p(e^{-j\omega})| d\omega \right\}.$$

$(*)$ follows from Jensen's inequality [A8] (actually its dual for concave functions) and the concavity of the logarithmic function. By Lemma 2.2, $\int_{-\pi}^{\pi} 2 \log |\beta(e^{-j\omega})| d\omega \geq 0$ and $\int_{-\pi}^{\pi} 2 \log |\alpha_p(e^{-j\omega})| d\omega \geq 0$ for all p, hence also in the limit as $p \to \infty$. Therefore, the above inequality is necessarily an equality. In particular,

$$\frac{1}{2\pi} \int_{-\pi}^{\pi} \log |\beta(e^{-j\omega})| d\omega = 0, \quad \lim_{p\to\infty} \left\{ \frac{1}{2\pi} \int_{-\pi}^{\pi} \log |\alpha_p(e^{-j\omega})| d\omega \right\} = 0.$$

Therefore, by Lemma 2.2, $\beta(z^{-1})$ is nonzero inside the unit circle, so $\beta(z)$ is nonzero outside it. ∎

Suppose $\{y_t\}$ is a purely indeterministic process, so it can be represented as

$$y_t = \sum_{s=0}^{\infty} \beta_s u_{t-s}. \qquad (2.7.7)$$

We may ask whether this representation is *invertible*, that is, whether there exists a square summable sequence $\{\alpha_s\}$ such that

$$u_t = \sum_{s=0}^{\infty} \alpha_s y_{t-s}. \tag{2.7.8}$$

The answer to this question depends on the behavior of $\beta(z)$ on the unit circle. More precisely:

Theorem 2.7. Assume that $|\beta(e^{j\omega})|$ is bounded on $[-\pi, \pi]$ and that

$$\frac{1}{2\pi} \int_{-\pi}^{\pi} \frac{1}{|\beta(e^{j\omega})|^2} d\omega < \infty. \tag{2.7.9}$$

Then the relation (2.7.8) holds with $\sum_{s=0}^{\infty} \alpha_s^2 < \infty$, and the sequence $\{\alpha_s\}$ is the inverse Fourier transform of $1/\beta(e^{j\omega})$ [A9]; that is,

$$\alpha_s = \frac{1}{2\pi} \int_{-\pi}^{\pi} \frac{1}{\beta(e^{j\omega})} e^{j\omega s} d\omega. \tag{2.7.10}$$

Proof. Define $\alpha(z) = 1/\beta(z)$. By Theorem 2.6, $\alpha(z)$ is analytic outside the unit circle. By (2.7.9), $\alpha(e^{j\omega})$ exists as an L_2 limit on the unit circle, and its Fourier coefficients $\{\alpha_s\}$ are square summable. To show that these coefficients are identically zero for $s < 0$, recall the inverse z transform formula

$$\alpha_s = \frac{1}{2\pi j} \oint \frac{z^{s-1}}{\beta(z)} dz = \frac{1}{2\pi j} \oint \frac{z^{-(s+1)}}{\beta(z^{-1})} dz,$$

where the integration is over any closed contour enclosing the origin and lying in the region of convergence of $1/\beta(z^{-1})$. $\beta(z^{-1})$ is analytic and nonzero inside the unit circle; therefore, the integrand is analytic inside the unit circle for all $s < 0$; hence, by the Cauchy integral formula, $\alpha_s = 0$ for $s < 0$. By Prob. 2.17, the square summability of $\{\alpha_s\}$ and the boundedness of $|\beta(e^{j\omega})|$ guarantee the existence of the right side of (2.7.8). Now the equality in (2.7.8) follows from the fact that $\alpha(z)\beta(z) = 1$. ∎

Corollary. The conditions

$$\sup_{\omega \in [-\pi,\pi]} |\beta(e^{j\omega})| < \infty, \qquad \inf_{\omega \in [-\pi,\pi]} |\beta(e^{j\omega})| > 0$$

are sufficient for (2.7.8). ∎

The form of the right side of (2.7.7) is called *infinite moving average*. An infinite moving average is a convolution of white noise with a square summable sequence. Thus, every purely indeterministic process has an infinite moving average representation in terms of its innovation process. This is called the *innovation representation* of the process, and the sequence $\{\beta_s\}$ is said to be the *impulse response* of the innovation representation.

The expression (2.7.8) can be written as

$$y_t = -\sum_{s=1}^{\infty} \alpha_s y_{t-s} + u_t. \qquad (2.7.11)$$

This form is called an *infinite autoregression*. Thus, a purely indeterministic process has an infinite autoregressive representation whenever the condition of Theorem 2.7 holds.

We conclude this section with a remark on terminology. Let $\{\gamma_s, s \in \mathbf{Z}\}$ be a doubly infinite, absolutely summable sequence and let $\{w_t\}$ be an i.i.d. process. The random process

$$y_t = \sum_{s=-\infty}^{\infty} \gamma_s w_{t-s} \qquad (2.7.12)$$

is said to be a *linear process*. The summability of $\{|\gamma_s|\}$ is sufficient to prove the existence and strict sense stationarity of the right side of (2.7.12). This, however, is beyond the scope of this book. A purely indeterministic process is not necessarily linear. However, a purely indeterministic Gaussian process is linear if the sequence $\{\beta_s\}$ is absolutely summable. In this case, the representation (2.7.12) can be made *causal*; that is, the summation can be made to extend from $s = 0$ to ∞.

2.8. SPECTRAL DISTRIBUTIONS AND SPECTRAL DENSITIES

So far we have concerned ourselves with time domain descriptions of stationary random processes. Now we wish to explore their frequency domain descriptions. Signal analysis in the frequency domain is also known as *spectral analysis*. Signal analysis by means of Fourier transforms is widely used in many scientific disciplines. It is therefore important to see how the Fourier transform can be applied to random signals.

We may be tempted to introduce spectral techniques to random process theory by considering the Fourier transform

$$Y(w) = \sum_{t=-\infty}^{\infty} y_t e^{-j\omega t}. \qquad (2.8.1)$$

Unfortunately, the right side of (2.8.1) fails to converge in the usual sense when $\{y_t\}$ is a stationary process, because such a process is not square summable (with probability 1).

It turns out that this difficulty can be overcome, and there exists a *spectral representation* for stationary processes. This topic is beyond the scope of this book. The interested reader is referred, for example, to Caines [1988, Sec. 1.4]. We settle for a discussion of the spectra of the *covariances* of stationary processes, the *power spectra*.

Let us start with a special case. Suppose the covariances $\{r_k\}$ are absolutely summable; that is, $\sum_{k=-\infty}^{\infty} |r_k| < \infty$. Then the Fourier transform

$$S(\omega) = \sum_{k=-\infty}^{\infty} r_k e^{-j\omega k} \tag{2.8.2}$$

exists and is uniformly continuous on $\omega \in [-\pi, \pi]$. The function $S(\omega)$ is called the *power spectral density* of the process. Other common names for this function are the *power spectrum, energy spectrum*, or just the *spectrum*. The covariances can be expressed in terms of the power spectrum by the inverse Fourier transform

$$r_k = \frac{1}{2\pi} \int_{-\pi}^{\pi} S(\omega) e^{j\omega k} d\omega. \tag{2.8.3}$$

The basic properties of the power spectral density are given in the following theorem.

Theorem 2.8. The power spectral density of a real stationary process is real, symmetric in ω, and nonnegative.

Proof. The first two properties are easily verified as follows.

$$S^*(\omega) = S(-\omega) = \sum_{k=-\infty}^{\infty} r_k e^{j\omega k} = \sum_{\ell=-\infty}^{\infty} r_\ell e^{-j\omega\ell} = S(\omega).$$

To prove the third property, recall that the matrix R_n defined in (2.4.4) is positive semidefinite for all n. Define

$$\boldsymbol{v} = [1, e^{-j\omega}, \ldots, e^{-j\omega(n-1)}]^H, \quad S_n(\omega) = \frac{1}{n} \boldsymbol{v}^H R_n \boldsymbol{v}.$$

Then, it is easy to verify that

$$0 \le \frac{1}{n} \boldsymbol{v}^H R_n \boldsymbol{v} = S_n(\omega) = \sum_{k=-n+1}^{n-1} \left(1 - \frac{|k|}{n}\right) r_k e^{-j\omega k}.$$

$S_n(\omega)$ is the *Cesàro sum* of $\{r_k e^{-j\omega k}\}$. By the Cesàro summability theorem [A10], it converges to $S(\omega)$ as $n \to \infty$. Hence $S(\omega)$ is nonnegative, as stated. ∎

The power spectral density may exist even when the sequence of covariances is not absolutely summable. Consider a purely indeterministic process $\{y_t\}$, and define

$$S(\omega) = \sigma_u^2 \beta(e^{j\omega})\beta(e^{-j\omega}) = \sigma_u^2 |\beta(e^{j\omega})|^2. \tag{2.8.4}$$

We then get, by Parseval's identity,

$$\frac{1}{2\pi}\int_{-\pi}^{\pi}S(\omega)e^{j\omega k}d\omega = \frac{\sigma_u^2}{2\pi}\int_{-\pi}^{\pi}\beta(e^{j\omega})[\beta(e^{j\omega})e^{-j\omega k}]^*d\omega = \sigma_u^2\sum_{s=-\infty}^{\infty}\beta_s\beta_{s-k}.$$

Also,

$$r_k = E(y_{t+k}y_t) = E\left(\sum_{s=-\infty}^{\infty}\beta_s u_{t+k-s}\right)\left(\sum_{r=-\infty}^{\infty}\beta_r u_{t-r}\right)$$

$$= \sigma_u^2\sum_{s=-\infty}^{\infty}\sum_{r=-\infty}^{\infty}\beta_s\beta_r\delta(k-s+r) = \sigma_u^2\sum_{s=-\infty}^{\infty}\beta_s\beta_{s-k}.$$

Hence

$$\frac{1}{2\pi}\int_{-\pi}^{\pi}S(\omega)e^{j\omega k}d\omega = r_k.$$

Therefore, $S(\omega)$ is a spectral density, in the sense that its inverse Fourier transform exists and is equal to $\{r_k\}$. However, $S(\omega)$ is not necessarily continuous, since the sequence $\{r_k\}$ is not necessarily absolutely summable.

Next we consider the general case, where the covariances of the process are such that the right side of (2.8.2) does not converge in any sense. For a simple example, recall the dc process with $r_k = 1, \forall k$. In this case it is tempting to define the spectral density by the Dirac delta function [A11], that is, to let $S(\omega) = 2\pi\delta_D(\omega)$. With this definition, the inversion formula (2.8.3) will hold for this example. Unfortunately, the use of Dirac delta functions is not sufficient to handle the most general covariance sequences, so we shall not pursue this approach.

It turns out that the proper generalization of (2.8.3) is obtained when the power spectral density is replaced by the *power spectral distribution* (or just *spectral distribution* for short). A (real) spectral distribution function $\Psi(\omega)$ is a real-valued function on $[-\pi, \pi]$ that is bounded, monotone nondecreasing, and continuous from the right and with $\Psi(-\pi) = 0$. The celebrated theorem of Herglotz expresses the covariance sequence of any wide sense stationary process in terms of a spectral distribution function.

Theorem 2.9. To every covariance sequence $\{r_k\}$ there corresponds a spectral distribution function $\Psi(\omega)$ such that

$$r_k = \frac{1}{2\pi}\int_{-\pi}^{\pi}e^{j\omega k}d\Psi(\omega), \tag{2.8.5}$$

where the right side is a Stieltjes integral.

Proof. Consider the functions $S_n(\omega)$ used in the proof of Theorem 2.8. These functions are continuous on $[-\pi, \pi]$ for every n, are nonnegative, and are related to

$\{r_k\}$ through

$$\left(1 - \frac{|k|}{n}\right) r_k = \frac{1}{2\pi} \int_{-\pi}^{\pi} e^{j\omega k} S_n(\omega) d\omega, \quad -n+1 \leq k \leq n-1.$$

Define

$$\Psi_n(\omega) = \int_{-\pi}^{\omega} S_n(\lambda) d\lambda.$$

Then

$$\left(1 - \frac{|k|}{n}\right) r_k = \frac{1}{2\pi} \int_{-\pi}^{\pi} e^{j\omega k} d\Psi_n(\omega), \quad -n+1 \leq k \leq n-1.$$

The functions $\Psi_n(\omega)$ are monotone nondecreasing and satisfy $\Psi_n(-\pi) = 0, \Psi_n(\pi) = 2\pi r_0$. Therefore, they are bounded and have bounded variation, so they satisfy the conditions of Helly's convergence theorem [Kolmogorov and Fomin, 1975, Sec. 36]. By that theorem, there exists a subsequence $\{n_1, n_2, \ldots\}$ such that

$$\lim_{i \to \infty} n_i = \infty, \quad \lim_{i \to \infty} \Psi_{n_i}(\omega) = \Psi(\omega)$$

and

$$r_k = \frac{1}{2\pi} \int_{-\pi}^{\pi} e^{j\omega k} d\Psi(\omega).$$

The function $\Psi(\omega)$, being a limit of spectral distribution functions, is also a spectral distribution function. ∎

Example 2.6. Recall the examples in Sec. 2.5. The white noise has the spectral density $S(\omega) = \sigma_w^2, \forall \omega$. The dc process has the spectral distribution $\Psi(\omega) = 2\pi \mathbf{u}(\omega)$, where $\mathbf{u}(x)$ denotes a unit step function at $x = 0$. The Gaussian harmonic process has the spectral distribution (Prob. 2.7)

$$\Psi(\omega) = \pi\sigma^2 [\mathbf{u}(\omega + \omega_0) + \mathbf{u}(\omega - \omega_0)]. \quad ∎$$

The next result is a theorem of Kolmogorov, which expresses the innovation variance of a wide sense stationary regular process as a function of its spectral distribution. Recall that a monotone nondecreasing function on a closed interval is differentiable, except on a set of measure zero [Kolmogorov and Fomin, 1975, Sec. 31]. Let $\Psi'(\omega)$ denote the derivative of $\Psi(\omega)$. We arbitrarily assign the value 1 to $\Psi'(\omega)$ at the points where the derivative of $\Psi(\omega)$ does not exist. Kolmogorov's theorem is then:

Theorem 2.10. The innovation variance of a process with spectral distribution $\Psi(\omega)$ is given by

$$\sigma_u^2 = \exp\left\{ \frac{1}{2\pi} \int_{-\pi}^{\pi} \log \Psi'(\omega) d\omega \right\}. \tag{2.8.6}$$

Proof. We proceed as in the proof of Theorem 2.6. This time, however, the process is not necessarily purely indeterministic, so $E(u_{t,p}^2)$ is given by

$$E(u_{t,p}^2) = \frac{1}{2\pi} \int_{-\pi}^{\pi} |\alpha_p(e^{j\omega})|^2 d\Psi(\omega) \geq \frac{1}{2\pi} \int_{-\pi}^{\pi} |\alpha_p(e^{j\omega})|^2 \Psi'(\omega) d\omega.$$

Carrying the same chain of inequalities as in the proof of Theorem 2.6, we get

$$\log \sigma_u^2 \geq \frac{1}{2\pi} \int_{-\pi}^{\pi} \log \Psi'(\omega) d\omega. \tag{2.8.7}$$

To prove the reverse inequality, let $\{x_t\}$, $\{v_t\}$ be the two components of the Wold decomposition of the given process and $\Psi_x(\omega)$, $\Psi_v(\omega)$ their respective spectral distributions. Observe that, by the orthogonality of $\{x_t\}$ and $\{v_t\}$, $\Psi(\omega) = \Psi_x(\omega) + \Psi_v(\omega)$, so $\Psi'(\omega) \geq \Psi_x'(\omega) = \sigma_u^2 |\beta(e^{j\omega})|^2$. Also, by Theorem 2.6,

$$\frac{1}{2\pi} \int_{-\pi}^{\pi} \log |\beta(e^{j\omega})|^2 d\omega = 0.$$

Therefore,

$$\frac{1}{2\pi} \int_{-\pi}^{\pi} \log \Psi'(\omega) d\omega \geq \log \sigma_u^2 + \frac{1}{2\pi} \int_{-\pi}^{\pi} \log |\beta(e^{j\omega})|^2 d\omega = \log \sigma_u^2. \tag{2.8.8}$$

Finally, (2.8.7) and (2.8.8) together imply (2.8.6). ∎

Corollary. The process is regular if

$$\frac{1}{2\pi} \int_{-\pi}^{\pi} \log \Psi'(\omega) d\omega > -\infty \tag{2.8.9a}$$

and is deterministic if

$$\frac{1}{2\pi} \int_{-\pi}^{\pi} \log \Psi'(\omega) d\omega = -\infty. \tag{2.8.9b}$$

∎

Next we wish to explore the general structure of the spectral distribution function. To this end, we summon a famous result known as the Lebesgue decomposition theorem [Kolmogorov and Fomin, 1975, Sec. 33].

Theorem 2.11. Every monotone nondecreasing bounded function $\Psi(\omega)$ on the closed interval $[-\pi, \pi]$ is a sum of three components:

$$\Psi(\omega) = \Psi_{ac}(\omega) + \Psi_{ju}(\omega) + \Psi_{si}(\omega), \tag{2.8.10}$$

where
(1) $\Psi_{ac}(\omega)$ is an *absolutely continuous* function, that is, a function having the following property: for every $\varepsilon > 0$ there exists $\delta > 0$ such that

$$\sum_{i=1}^{n} |\Psi_{ac}(\overline{\omega}_i) - \Psi_{ac}(\underline{\omega}_i)| < \varepsilon$$

for every finite collection $\{(\underline{\omega}_i, \overline{\omega}_i)\}$ of nonoverlapping intervals satisfying

$$\sum_{i=1}^{n} |\overline{\omega}_i - \underline{\omega}_i| < \delta.$$

(2) $\Psi_{ju}(\omega)$ is a *jump* function, that is, a function of the form

$$\Psi_{ju}(\omega) = \sum_{\omega_i < \omega} [\Psi(\omega_i^+) - \Psi(\omega_i^-)],$$

where $\{\omega_i\}$ is the set of discontinuity points of $\Psi(\omega)$, and $[\Psi(\omega_i^+) - \Psi(\omega_i^-)]$ are the jumps of the function at the discontinuity points. The function $\Psi(\omega)$ can have a countable number of discontinuities at most, and the jumps are finite and summable.

(3) $\Psi_{si}(\omega)$ is a *singular* function, that is, a function that is continuous, and its derivative is zero almost everywhere (and may be nonexistent on a set of measure zero). ∎

Now we have two different decompositions for a given spectral distribution, one based on the decomposition of the process to its deterministic and purely indeterministic parts and the other based on Lebesgue's theorem:

$$\Psi(\omega) = \Psi_{ac}(\omega) + \Psi_{ju}(\omega) + \Psi_{si}(\omega) = \Psi_x(\omega) + \Psi_v(\omega). \qquad (2.8.11)$$

Since $\{x_t\}$ possesses a spectral density, it has an absolutely continuous spectral distribution. Therefore, clearly, $\Psi_{ac}(\omega) \geq \Psi_x(\omega)$. The two are in fact equal, as asserted by the following theorem.

Theorem 2.12. For any regular stationary process,

$$\Psi_{ac}(\omega) = \Psi_x(\omega), \quad \Psi_{ju}(\omega) + \Psi_{si}(\omega) = \Psi_v(\omega). \qquad (2.8.12)$$

Proof. Note that $\Psi_x'(\omega)$ is nonzero almost everywhere by (2.8.9a). By Kolmogorov's theorem,

$$\log \sigma_u^2 = \frac{1}{2\pi} \int_{-\pi}^{\pi} \log \Psi'(\omega) d\omega = \frac{1}{2\pi} \int_{-\pi}^{\pi} \log[\Psi_x'(\omega) + \Psi_v'(\omega)] d\omega$$

$$= \frac{1}{2\pi} \int_{-\pi}^{\pi} \log \Psi_x'(\omega) d\omega + \frac{1}{2\pi} \int_{-\pi}^{\pi} \log \left[1 + \frac{\Psi_v'(\omega)}{\Psi_x'(\omega)}\right] d\omega$$

$$= \log \sigma_u^2 + \frac{1}{2\pi} \int_{-\pi}^{\pi} \log \left[1 + \frac{\Psi_v'(\omega)}{\Psi_x'(\omega)}\right] d\omega.$$

Hence $\log \left[1 + \frac{\Psi_v'(\omega)}{\Psi_x'(\omega)}\right] = 0$ almost everywhere, implying that $\Psi_v'(\omega) = 0$ almost everywhere. This shows that $\Psi_v(\omega)$ contains only jump and singular components, from which we deduce (2.8.12). ∎

A word of caution is in order here. Theorem 2.12 is valid only for *regular* processes, that is, processes that have nonvanishing innovation. A deterministic process can have an absolutely continuous spectral distribution provided (2.8.9b) holds. The following example illustrates this possibility.

Example 2.7. Consider a Gaussian process with the spectral density function

$$S(\omega) = \begin{cases} 1, & \text{if } |\omega| \leq \omega_0 \\ 0, & \text{otherwise} \end{cases}$$

for some $0 < \omega_0 < \pi$. This process satisfies (2.8.9b); therefore, it is deterministic. Its spectral distribution is absolutely continuous, and its covariances are given by $r_k = \sin(\omega_0 k)/(\pi k)$. ∎

In general, the spectral distribution of a deterministic process can include all three components of the Lebesgue decomposition. It is only for regular processes that the purely indeterministic component "absorbs" the absolutely continuous part of the spectral distribution.

2.9. SPECTRAL FACTORIZATION

In this section we will be concerned with the factorization of a given spectral density $S(\omega)$ of a purely indeterministic process as

$$S(\omega) = U(\omega)U(-\omega) \quad \text{a.e.,}$$

where $U(\omega)$ is a complex-valued function on $[-\pi, \pi]$. Any such $U(\omega)$ is called a *spectral factor* of $S(\omega)$. A given spectral density function may have many spectral factors. Of special interest are spectral factors satisfying the following five properties.

(1) $S(\omega) = U(\omega)U(-\omega)$ almost everywhere.
(2) $(1/2\pi) \int_{-\pi}^{\pi} |U(\omega)|^2 d\omega < \infty$.
(3) The inverse Fourier transform of $U(\omega)$ vanishes for negative indexes; that is, $U(\omega) = \sum_{s=0}^{\infty} \gamma_s e^{-j\omega s}$.
(4) The coefficient γ_0 is real and positive.
(5) The function $\overline{U}(z) = \sum_{s=0}^{\infty} \gamma_s z^{-s}$ is nonzero outside the unit circle.

Property (1) is just the defining property of a spectral factor. Property (2) is needed to guarantee that the inverse Fourier transform mentioned in property (3) exists. Property (3) implies that $U(\omega)$, regarded as the frequency response of a time-invariant filter, corresponds to a *causal* filter. Property (4) is needed to eliminate the trivial sign ambiguity of $U(\omega)$. Property (5) is the crucial one; it is sometimes referred to as the *minimum phase* property. Correspondingly, a function $U(\omega)$ satisfying all five properties is sometimes called a *minimum phase spectral factor*.

The existence of a minimum phase spectral factor was established in Sec. 2.7; the function

$$S^+(\omega) = \sigma_u \beta(e^{j\omega}) \tag{2.9.1}$$

satisfies all five properties (Prob. 2.10), hence it is a minimum phase spectral factor.

A sixth property is sometimes of interest:

(6) The function $U(\omega)$ satisfies $(1/2\pi) \int_{-\pi}^{\pi} |U(\omega)|^{-2} d\omega < \infty$.

A function $U(\omega)$ satisfying all six properties is called a *strong spectral factor* of $S(\omega)$ [this name is used for $\overline{U}(z)$ as well]. The minimum-phase spectral factor $S^+(\omega)$ does not always satisfy property (6). As was explained in Sec. 2.7 (Theorem 2.7 and the related discussion), property (6) implies the representation (2.7.8) of the innovation process of $\{y_t\}$. If this property holds, the uniqueness of the strong spectral factor $S^+(\omega)$ can be established subject to further conditions. We will not discuss this subject here; see [Caines, 1988, Sec. 4.1].

2.10. PROPERTIES OF THE BEST LINEAR PREDICTOR

In Sec. 2.6 we introduced the finite-order best linear predictors $\{\alpha_{p,k}\}$ and used them to derive some of the results related to the Wold decomposition. These predictors have many interesting properties, some of which will be used in later chapters. In this section we explore the algebraic and analytic properties of the finite-order predictors. We assume that the process is regular throughout. Whenever the spectral density $S(\omega)$ is mentioned, the process will be understood to be purely indeterministic.

The coefficients of the best linear predictor $\{\alpha_{p,k}\}$ were defined in Eq. (2.6.7). We add to it the definition $\alpha_{p,0} = 1$ for all p. Denote the variance of the partial innovation $u_{t,p}$ by d_p. Then we get, as in the proofs of Theorems 2.3 and 2.4,

$$d_p = E(u_{t,p}^2) = E(u_{t,p} y_t) = \sum_{k=0}^{p} \alpha_{p,k} r_{-k}. \tag{2.10.1}$$

Combine this with (2.6.8) to get

$$
\begin{bmatrix}
r_0 & r_{-1} & \cdots & r_{-p} \\
r_1 & r_0 & \cdots & r_{-p+1} \\
\vdots & \vdots & \ddots & \vdots \\
r_p & r_{p-1} & \cdots & r_0
\end{bmatrix}
\begin{bmatrix}
\alpha_{p,0} \\
\alpha_{p,1} \\
\vdots \\
\alpha_{p,p}
\end{bmatrix}
=
\begin{bmatrix}
d_p \\
0 \\
\vdots \\
0
\end{bmatrix}. \tag{2.10.2}
$$

This is known as the *Yule-Walker equation*. It should be read as an equation for the unknowns $\{\alpha_{p,k}, 1 \le k \le p\}$ and d_p, the covariances r_k being given. Recall that the matrix on the left side of (2.10.2) is symmetric Toeplitz, to be denoted by R_{p+1} [cf. (2.4.4)].

The predictor coefficients provide a decomposition of R_{p+1} as a product of lower-diagonal-upper factors. Define

$$A_{p+1} = \begin{bmatrix} \alpha_{0,0} & 0 & \cdots & 0 \\ \alpha_{1,1} & \alpha_{1,0} & \cdots & 0 \\ \vdots & \vdots & \ddots & 0 \\ \alpha_{p,p} & \alpha_{p,p-1} & \cdots & \alpha_{p,0} \end{bmatrix}, \quad D_{p+1} = \begin{bmatrix} d_0 & 0 & \cdots & 0 \\ 0 & d_1 & \cdots & 0 \\ \vdots & \vdots & \ddots & 0 \\ 0 & 0 & \cdots & d_p \end{bmatrix}. \quad (2.10.3)$$

Then:

Theorem 2.13. R_{p+1} can be expressed as

$$R_{p+1} = A_{p+1}^{-1} D_{p+1} (A_{p+1}^T)^{-1}. \quad (2.10.4)$$

Proof. From the Yule-Walker equation it is easily seen that the product $A_{p+1} R_{p+1}$ has the form

$$A_{p+1} R_{p+1} = \begin{bmatrix} d_0 & x & \cdots & x \\ 0 & d_1 & \cdots & x \\ \vdots & \vdots & \ddots & x \\ 0 & 0 & \cdots & d_p \end{bmatrix} = D_{p+1} X,$$

where the x's denote unspecified terms and X is an upper triangular matrix with 1's on the main diagonal and unspecified terms above it. Therefore,

$$R_{p+1} = A_{p+1}^{-1} D_{p+1} X.$$

Now, due to the symmetry of R_{p+1}, X must be the transpose of A_{p+1}^{-1}, which yields (2.10.4). ∎

Corollary 1. If the process $\{y_t\}$ is regular, the variance of the partial innovation d_k is positive for all k; hence the matrix R_{p+1} is positive definite for all p.

Corollary 2. If the process $\{y_t\}$ is regular, the inverse of R_{p+1} admits the upper-diagonal-lower decomposition

$$R_{p+1}^{-1} = A_{p+1}^T D_{p+1}^{-1} A_{p+1}. \quad (2.10.5)$$

∎

Some properties of the finite-order predictors can be conveniently expressed using the polynomials constructed from the predictor coefficients. For each order p we define

$$\alpha_p(z) = \sum_{k=0}^{p} \alpha_{p,k} z^{-k}, \quad \overline{\alpha}_p(z) = z^p \alpha_p(z) = \sum_{k=0}^{p} \alpha_{p,p-k} z^k. \quad (2.10.6)$$

The monic polynomials $\overline{\alpha}_p(z)$ are called the *predictor polynomials*.

Theorem 2.14. The polynomials $\overline{\alpha}_p(z)$ satisfy the orthogonality relationship

$$\frac{1}{2\pi} \int_{-\pi}^{\pi} \overline{\alpha}_n(e^{j\omega}) S(\omega) \overline{\alpha}_m(e^{-j\omega}) d\omega = d_n \delta(n-m), \quad \forall n, m. \tag{2.10.7}$$

Proof. The decomposition (2.10.4) implies, for all n, m,

$$d_n \delta(n-m) = \sum_{k=0}^{n} \sum_{\ell=0}^{m} \alpha_{n,n-k} \, r_{k-\ell} \, \alpha_{m,m-\ell}$$

$$= \sum_{k=0}^{n} \sum_{\ell=0}^{m} \alpha_{n,n-k} \, \alpha_{m,m-\ell} \, \frac{1}{2\pi} \int_{-\pi}^{\pi} S(\omega) e^{j\omega(k-\ell)} d\omega$$

$$= \frac{1}{2\pi} \int_{-\pi}^{\pi} S(\omega) \left[\sum_{k=0}^{n} \alpha_{n,n-k} e^{j\omega k} \right] \left[\sum_{\ell=0}^{m} \alpha_{m,m-\ell} e^{-j\omega\ell} \right] d\omega$$

$$= \frac{1}{2\pi} \int_{-\pi}^{\pi} \overline{\alpha}_n(e^{j\omega}) S(\omega) \overline{\alpha}_m(e^{-j\omega}) d\omega. \qquad \blacksquare$$

Theorem 2.14 has the following interesting interpretation. Let $S^+(\omega)$ be the minimum phase spectral factor of $S(\omega)$ introduced in Sec. 2.9. Then (2.10.7) can be written as

$$\frac{1}{2\pi} \int_{-\pi}^{\pi} [d_n^{-1/2} \overline{\alpha}_n(e^{j\omega}) S^+(\omega)][d_n^{-1/2} \overline{\alpha}_m(e^{j\omega}) S^+(\omega)]^* d\omega = \delta(n-m).$$

Therefore, the functions $d_n^{-1/2} \overline{\alpha}_n(e^{j\omega}) S^+(\omega)$ are orthonormal on the interval $[-\pi, \pi]$.

Theorem 2.15. The predictor polynomial $\overline{\alpha}_p(z)$ of a regular process is nonzero outside and on the unit circle; that is, its zeros are inside the unit circle.

Proof. We will first prove that there are no zeros on the unit circle and then that there are no zeros outside it.

(1) Suppose $\overline{\alpha}_p(z)$ has a zero at $z = -1$ so that $\overline{\alpha}_p(z) = (z+1)\xi(z)$ for some monic polynomial $\xi(z) = \sum_{k=0}^{p-1} \xi_k z^{p-1-k}$. Using standard linear system theory, we can express the partial innovation by the cascaded operations

$$v_t = \sum_{k=0}^{p-1} \xi_k y_{t-k}, \quad u_{t,p} = v_t + v_{t-1},$$

where the process $\{v_t\}$ is wide sense stationary. Let $\{\rho_k\}$ be the covariance sequence of $\{v_t\}$. Then these covariances satisfy the Yule-Walker equation

$$\begin{bmatrix} \rho_0 & \rho_1 \\ \rho_1 & \rho_0 \end{bmatrix} \begin{bmatrix} 1 \\ 1 \end{bmatrix} = \begin{bmatrix} d_p \\ 0 \end{bmatrix},$$

which is possible only if $d_p = 0$, in contradiction to the regularity assumption. This proves that $z = -1$ cannot be a zero of $\overline{\alpha}_p(z)$. In exactly the same manner it is shown that $z = 1$ cannot be a zero of $\overline{\alpha}_p(z)$. The last case that we need to consider is a pair of zeros at $z = e^{\pm j\omega_0}$. Then $\overline{\alpha}_p(z) = (z^2 - 2\cos\omega_0 z + 1)\xi(z)$ for some monic polynomial $\xi(z) = \sum_{k=0}^{p-2} \xi_k z^{p-2-k}$. Correspondingly, the partial innovation can be expressed as

$$v_t = \sum_{k=0}^{p-2} \xi_k y_{t-k}, \quad u_{t,p} = v_t - 2\cos\omega_0 v_{t-1} + v_{t-2}.$$

Hence

$$\begin{bmatrix} \rho_0 & \rho_1 & \rho_2 \\ \rho_1 & \rho_0 & \rho_1 \\ \rho_2 & \rho_1 & \rho_0 \end{bmatrix} \begin{bmatrix} 1 \\ -2\cos\omega_0 \\ 1 \end{bmatrix} = \begin{bmatrix} d_p \\ 0 \\ 0 \end{bmatrix},$$

which is again possible only if $d_p = 0$.

(2) Suppose $\overline{\alpha}_p(z)$ has m zeros outside the unit circle, say $\{\mu_i, 0 \leq i \leq m\}$, and $p - m$ zeros inside it, say $\{\nu_i, 0 \leq i \leq p - m\}$. Then

$$\overline{\alpha}_p(z) = \left[\prod_{i=1}^{m}(z - \mu_i)\right]\left[\prod_{i=1}^{p-m}(z - \nu_i)\right].$$

Let

$$\xi(z) = \left[\prod_{i=1}^{m}(z - (\mu_i^*)^{-1})\right]\left[\prod_{i=1}^{p-m}(z - \nu_i)\right].$$

It is easy to check that

$$|e^{j\omega} - (\mu_i^*)^{-1}|^2 = |\mu_i|^{-2}|e^{j\omega} - \mu_i|^2,$$

and $|\mu_i|^{-2} < 1$. Therefore,

$$|\xi(e^{j\omega})|^2 = \left(\prod_{i=1}^{m}|\mu_i|^{-2}\right)|\overline{\alpha}_p(e^{j\omega})|^2 < |\overline{\alpha}_p(e^{j\omega})|^2,$$

so

$$\frac{1}{2\pi}\int_{-\pi}^{\pi}|\xi(e^{j\omega})|^2 S(\omega)d\omega < \frac{1}{2\pi}\int_{-\pi}^{\pi}|\overline{\alpha}_p(e^{j\omega})|^2 S(\omega)d\omega = d_p.$$

This inequality means that the monic polynomial $\xi(z)$, when acting as a filter on $\{y_t\}$, would yield a prediction error whose variance is smaller than d_p. This contradicts the definition of $\overline{\alpha}_p(z)$ as the best linear predictor of order p. The contradiction proves that $\overline{\alpha}_p(z)$ has no zeros outside the unit circle. ∎

Finally, we explore the limiting behavior of $\alpha_p(e^{j\omega})$ as p tends to infinity.

Theorem 2.16. Assume that $\inf_\omega S(\omega) = B > 0$. Then the sequence $\{\alpha_p(e^{j\omega})\}$ converges to $\alpha(e^{j\omega})$ in $L_2[-\pi, \pi]$, where $\alpha(e^{j\omega})$ was defined in (2.7.8).

Proof. First note that

$$\frac{1}{2\pi} \int_{-\pi}^{\pi} |\alpha(e^{j\omega}) - \alpha_p(e^{j\omega})|^2 S(\omega) d\omega = d_p - \sigma_u^2. \qquad (2.10.8)$$

The proof of this identity is left as an exercise to the reader (Prob. 2.11). From Theorem 2.2 it therefore follows that

$$\lim_{p\to\infty} \frac{1}{2\pi} \int_{-\pi}^{\pi} |\alpha(e^{j\omega}) - \alpha_p(e^{j\omega})|^2 S(\omega) d\omega = 0.$$

But, by the theorem's assumption,

$$\frac{1}{2\pi} \int_{-\pi}^{\pi} |\alpha(e^{j\omega}) - \alpha_p(e^{j\omega})|^2 d\omega \leq \frac{1}{2\pi B} \int_{-\pi}^{\pi} |\alpha(e^{j\omega}) - \alpha_p(e^{j\omega})|^2 S(\omega) d\omega.$$

Therefore,

$$\lim_{p\to\infty} \frac{1}{2\pi} \int_{-\pi}^{\pi} |\alpha(e^{j\omega}) - \alpha_p(e^{j\omega})|^2 d\omega = 0. \qquad \blacksquare$$

2.11. WIDE SENSE ERGODICITY OF GAUSSIAN PROCESSES

In this section we give necessary and sufficient conditions for wide sense ergodicity of stationary Gaussian processes, both in terms of the covariance sequence and of the spectral distribution of the process. More general results about ergodicity of stationary processes can be found in Doob [1953, Ch. X] or Karlin and Taylor [1975, Ch. 9].

Our first result concerns the ergodicity of the mean.

Theorem 2.17. A wide sense stationary process $\{y_t\}$ is MS ergodic in the mean if and only if its covariance sequence satisfies

$$\lim_{N\to\infty} \frac{1}{2N+1} \sum_{n=-N}^{N} r_n = 0. \qquad (2.11.1)$$

Proof. We can assume, without loss of generality, that the process has zero mean. Define

$$W_N = \frac{1}{2N+1} \sum_{n=-N}^{N} y_n.$$

Assume first that the process is MS ergodic in the mean so that $\lim_{N\to\infty} E(W_N)^2 = 0$. We get, using the Cauchy-Schwarz inequality,

$$\left|\frac{1}{2N+1}\sum_{n=-N}^{N} r_n\right|^2 = \left|\frac{1}{2N+1}\sum_{n=-N}^{N} E(y_n y_0)\right|^2 = |E(y_0 W_N)|^2$$

$$\leq E(y_0)^2 E(W_N)^2 = r_0 E(W_N)^2.$$

Therefore, MS ergodicity of the mean implies (2.11.1). To show the converse, compute $E(W_N)^2$ as follows.

$$E(W_N)^2 = \frac{1}{(2N+1)^2}\sum_{n=-N}^{N}\sum_{m=-N}^{N} E(y_n y_m) = \frac{1}{(2N+1)^2}\sum_{n=-N}^{N}\sum_{m=-N}^{N} r_{n-m}$$

$$= \frac{2}{(2N+1)^2}\sum_{m=0}^{2N}\sum_{k=0}^{m} r_k - \frac{r_0}{2N+1}.$$

Since we are interested in the limit as $N \to \infty$, the second term is immaterial, and we consider only the first term.

Choose $\varepsilon > 0$, and let M be such that

$$\left|\sum_{k=0}^{m} r_k\right| < m\varepsilon, \quad \forall m > M.$$

The existence of M follows from (2.11.1). Now, for $N > M$ we have

$$\left|\frac{1}{(2N+1)^2}\sum_{m=0}^{2N}\sum_{k=0}^{m} r_k\right| \leq \frac{1}{(2N+1)^2}\left|\sum_{m=0}^{M}\sum_{k=0}^{m} r_k\right| + \frac{1}{(2N+1)^2}\left|\sum_{m=M+1}^{2N}\sum_{k=0}^{m} r_k\right|$$

$$\leq \frac{1}{(2N+1)^2}\left|\sum_{m=0}^{M}\sum_{k=0}^{m} r_k\right| + \frac{1}{(2N+1)^2}\sum_{m=M+1}^{2N} m\varepsilon \leq \frac{1}{(2N+1)^2}\left|\sum_{m=0}^{M}\sum_{k=0}^{m} r_k\right| + \varepsilon.$$

Now let $N \to \infty$. Since M is fixed, the right side is smaller than 2ε for large enough N. Hence the left side approaches zero as $N \to \infty$, proving that $\lim_{N\to\infty} E(W_N)^2 = 0$. ∎

Note that we have not used the Gaussian assumption to prove this theorem; therefore, it is true for non-Gaussian processes as well. However, the next theorem makes an explicit use of the Gaussian assumption.

Theorem 2.18. A wide sense stationary Gaussian process is MS ergodic in the second order moments if and only if

$$\lim_{N\to\infty} \frac{1}{2N+1}\sum_{n=-N}^{N} r_n^2 = 0. \tag{2.11.2}$$

Proof. As in the previous theorem, the process $\{y_t\}$ can be assumed to have zero mean without loss of generality. Define, for a given m,

$$z_t^{(m)} = y_{t+m}y_t - r_m.$$

Then $\{z_t^{(m)}\}$ is zero mean wide sense stationary, and ergodicity of the covariances of $\{y_t\}$ is equivalent to the ergodicity of the mean of $\{z_t^{(m)}\}$ for all m. Now, Theorem 2.17 applies to $\{z_t^{(m)}\}$, so this process is ergodic if and only if

$$\lim_{N\to\infty} \frac{1}{2N+1} \sum_{n=-N}^{N} \rho_n^{(m)} = 0,$$

where $\{\rho_n^{(m)}\}$ are the covariances of $\{z_t^{(m)}\}$. But, since $\{y_t\}$ is Gaussian,

$$\rho_n^{(m)} = E(z_{t+n}^{(m)} z_t^{(m)}) = E\left\{(y_{t+n+m}y_{t+n} - r_m)(y_{t+m}y_t - r_m)\right\}$$
$$= E(y_{t+n+m}y_{t+n}y_{t+m}y_t) - r_m^2 = r_n^2 + r_{n+m}r_{n-m}.$$

Hence ergodicity of the covariances of $\{y_t\}$ is equivalent to

$$\lim_{N\to\infty} \frac{1}{2N+1} \sum_{n=-N}^{N} (r_n^2 + r_{n+m}r_{n-m}) = 0, \quad \forall m. \tag{2.11.3}$$

It remains to show the equivalence of (2.11.2) to (2.11.3). The latter clearly implies the former, by substituting $m = 0$. To show that (2.11.2) implies (2.11.3), recall the inequality $2|ab| \leq (a^2 + b^2)$, yielding

$$\left| \frac{1}{2N+1} \sum_{n=-N}^{N} r_{n+m}r_{n-m} \right| \leq \left| \frac{1}{2(2N+1)} \sum_{n=-N}^{N} (r_{n+m}^2 + r_{n-m}^2) \right|$$

$$= \frac{1}{2(2N+1)} \left| \sum_{n=-N+m}^{N+m} r_n^2 + \sum_{n=-N-m}^{N-m} r_n^2 \right| \leq \frac{1}{2N+1} \left| \sum_{n=-N}^{N} r_n^2 \right| + \frac{mr_0^2}{2N+1}.$$

As $N \to \infty$, the right side goes to zero by (2.11.2). ■

The next theorem expresses the ergodicity conditions in terms of the spectral distribution of the process.

Theorem 2.19. Let $\{y_t\}$ be a stationary Gaussian process with spectral distribution $\Psi(\omega)$. Then:

(a) The process is MS ergodic in the mean if and only if $\Psi(\omega)$ is continuous at $\omega = 0$.

(b) The process is MS ergodic in the covariances if and only if $\Psi(\omega)$ is continuous for all $\omega \in [-\pi, \pi]$.

Proof.

(a) We have

$$\frac{1}{2N+1} \sum_{n=-N}^{N} r_n = \frac{1}{2N+1} \sum_{n=-N}^{N} \frac{1}{2\pi} \int_{-\pi}^{\pi} e^{j\omega n} d\Psi(\omega)$$

$$= \frac{1}{2\pi} \int_{-\pi}^{\pi} \left[\frac{1}{2N+1} \sum_{n=-N}^{N} e^{j\omega n} \right] d\Psi(\omega) = \frac{1}{2\pi} \int_{-\pi}^{\pi} \frac{\sin \frac{(2N+1)\omega}{2}}{(2N+1)\sin \frac{\omega}{2}} d\Psi(\omega).$$

Now, the integrand is bounded by 1 in absolute value and, as $N \to \infty$, approaches 1 at $\omega = 0$ and zero elsewhere. Hence, by the bounded convergence theorem [Royden, 1968, p. 81], the limit as $N \to \infty$ of the right side is $(1/2\pi)[\Psi(0^+) - \Psi(0^-)]$. This is zero if and only if $\Psi(\omega)$ is continuous at $\omega = 0$.

(b) We have

$$\frac{1}{2N+1} \sum_{n=-N}^{N} r_n^2 = \frac{1}{2N+1} \sum_{n=-N}^{N} \frac{1}{(2\pi)^2} \int_{-\pi}^{\pi} \int_{-\pi}^{\pi} e^{j\omega n} e^{-j\lambda n} d\Psi(\omega) d\Psi(\lambda)$$

$$= \frac{1}{(2\pi)^2} \int_{-\pi}^{\pi} \int_{-\pi}^{\pi} \left[\frac{1}{2N+1} \sum_{n=-N}^{N} e^{j(\omega-\lambda)n} \right] d\Psi(\omega) d\Psi(\lambda)$$

$$= \frac{1}{(2\pi)^2} \int_{-\pi}^{\pi} \int_{-\pi}^{\pi} \frac{\sin \frac{(2N+1)(\omega-\lambda)}{2}}{(2N+1)\sin \frac{(\omega-\lambda)}{2}} d\Psi(\omega) d\Psi(\lambda).$$

As before, the integrand approaches 1 for $\omega = \lambda$ and zero elsewhere. Hence

$$\lim_{N\to\infty} \left\{ \frac{1}{2N+1} \sum_{n=-N}^{N} r_n^2 \right\} = \frac{1}{(2\pi)^2} \sum_i [\Psi(\omega_i^+) - \Psi(\omega_i^-)]^2,$$

where the sum extends over all the jumps of $\Psi(\omega)$. In particular, the limit is zero if and only if the spectral distribution is continuous at all the points of $[-\pi, \pi]$. ∎

It follows from Theorem 2.19 that, for a Gaussian stationary process to be ergodic, its spectral distribution is not allowed to have jumps. A purely indeterministic Gaussian process is thus ergodic. A deterministic process is not ergodic if its spectral distribution contains jumps; otherwise, it is ergodic. For example, the process described at the end of Sec. 2.8 is ergodic.

2.12. PARAMETRIC MODELS FOR STATIONARY PROCESSES

The descriptions of wide sense stationary processes by Wold decomposition and the spectral distribution are most general. However, in preparation for the practical aspects of random signal analysis, we shall now compromise generality and discuss some special cases. Our interest in this section will be in processes that can be modeled by a *finite number of parameters*. First, we consider purely indeterministic processes and then deterministic ones.

MOVING AVERAGE MODELS

As we saw in Sec. 2.7, a purely indeterministic process can be modeled as the output of a linear time-invariant filter with square summable impulse response, whose input is an uncorrelated process—the innovation of the output process. This motivates the introduction of a finite *moving average* (MA) model defined by

$$y_t = \sum_{k=0}^{q} b_k w_{t-k}, \tag{2.12.1}$$

where
- $\{w_t\}$ is an uncorrelated process with zero mean and variance σ_w^2.
- $\{b_k, 0 \le k \le q\}$ is a finite impulse response, with $b_0 = 1, b_q \ne 0$.

Such a process is denoted by MA(q), and q is called the *order* of the process. When $\{w_t\}$ is Gaussian, so is $\{y_t\}$.

The spectral density of a MA process is given by

$$S(\omega) = \sigma_w^2 |b(e^{j\omega})|^2. \tag{2.12.2}$$

The covariances of the process are given by

$$r_k = \sigma_w^2 \sum_{s=0}^{q-|k|} b_s b_{s+|k|}. \tag{2.12.3}$$

A moving average process is *finitely correlated*, that is, $r_k = 0$ for $|k| > q$.

The impulse response $\{b_k\}$, being finite, is square summable, and $b(z)$ is analytic outside the unit circle (in fact, on the entire z plane except $z = 0$). This does not imply, however, that $\{w_t\}$ is the innovation process of $\{y_t\}$. As we saw in Sec. 2.7, the filter generating a purely indeterministic process from its innovation is nonzero outside the unit circle. We have, in fact:

Theorem 2.20.
  (a) Suppose $b(z)$ of the MA(q) process (2.12.1) has no zeros outside the unit circle. Then $\{w_t\}$ is the innovation process of $\{y_t\}$.
  (b) Suppose $b(z)$ has m zeros outside the unit circle, say $\{\mu_i, 0 \le i \le m\}$, and $q - m$ zeros on or inside the unit circle, say $\{\nu_i, 0 \le i \le q - m\}$. Let

$$\tilde{b}(z) = \left[\prod_{i=1}^{m} (1 - \mu_i^{-1} z^{-1}) \right] \left[\prod_{i=1}^{q-m} (1 - \nu_i z^{-1}) \right]. \tag{2.12.4}$$

Then the coefficients of $\tilde{b}(z)$ are the impulse response of the innovation representation of $\{y_t\}$.

Proof. Let

$$\overline{S}(z) = \sigma_w^2 b(z)b(z^{-1}) = \sum_{k=-q}^{q} r_k z^{-k}.$$

Then $\overline{S}(z)$ has $2q$ zeros: the q zeros of $b(z)$ and their reciprocals. Let α be a zero of $b(z)$ of multiplicity r. If $|\alpha| \neq 1$, then $\overline{S}(z)$ will have zeros at α and α^{-1}, each of multiplicity r. If $|\alpha| = 1$, then $\overline{S}(z)$ will have a zero of multiplicity $2r$ at α (Prob. 2.19). Therefore, $\overline{S}(z)$ can be factored in 2^q ways (including permutations of multiple zeros) as $\overline{S}(z) = C\tilde{b}(z)\tilde{b}(z^{-1})$, where $\tilde{b}(z)$ is a monic polynomial of degree q. Out of these, exactly one factorization (disregarding permutations of multiple zeros) has all the zeros of $\tilde{b}(z)$ on or inside the unit circle, and the corresponding $\tilde{b}(z)$ is given by (2.12.4). In particular, if $b(z)$ has no zeros outside the unit circle, then $\tilde{b}(z) = b(z)$. Finally, Theorem 2.6 and the uniqueness of $\tilde{b}(z)$ imply that $\tilde{b}(z)$ is indeed the innovation representation of the process.   ∎

Example 2.8. Let

$$y_t = w_t + 1.5w_{t-1} - w_{t-2}, \quad \sigma_w = 1.$$

The corresponding polynomial is

$$b(z) = 1 + 1.5z^{-1} - z^{-2} = (1 - 0.5z^{-1})(1 + 2z^{-1}).$$

Therefore,

$$\tilde{b}(z) = (1 - 0.5z^{-1})(1 + 0.5z^{-1}) = 1 - 0.25z^{-2}$$

and

$$y_t = u_t - 0.25u_{t-2}, \quad \sigma_u = 2.   ∎$$

As we see, the innovation representation of a MA(q) process is also MA(q). As long as we are interested only in the output process $\{y_t\}$, it is convenient to replace the original model by its spectrally equivalent innovation model. However, when we are interested in the input process $\{w_t\}$, such a replacement is not permitted.

AUTOREGRESSIVE MODELS

Next we approach the modeling issue from a prediction point of view. According to Theorem 2.7, a large class of purely indeterministic processes can be represented by an infinite autoregression as in (2.7.11). This motivates the introduction of *finite autoregressive* (AR) processes defined by

$$y_t = -\sum_{k=1}^{p} a_k y_{t-k} + u_t, \qquad (2.12.5)$$

where
- $\{u_t\}$ is an uncorrelated process with zero mean and variance σ_u^2.
- $\{a_k, 0 \le k \le p\}$ is a finite sequence with $a_0 = 1, a_p \ne 0$.

Such a process is denoted by $\text{AR}(p)$, and p is called the *order* of the process. When $\{u_t\}$ is Gaussian, so is $\{y_t\}$.

Let us denote the z transform of the sequence $\{a_k\}$ by $a(z)$. Since $\{y_t\}$ is generated from $\{u_t\}$ by passing it through the filter $1/a(z)$, this filter must be stable for $\{y_t\}$ to be wide sense stationary. Hence we *impose* on the model (2.12.5) the condition that $a(z)$ have no zeros outside or on the unit circle.

The spectral density of $\{y_t\}$ is given by

$$S(\omega) = \frac{\sigma_u^2}{|a(e^{j\omega})|^2}. \tag{2.12.6}$$

The notation $\{u_t\}$ for the input process is not coincidental. In fact, we have:

Theorem 2.21. The input process of an autoregressive process is always its innovation process.

Proof. It is easy to check, by the requirements imposed on $a(z)$, that $\sigma_u/a(e^{j\omega})$ is a minimum-phase spectral factor of $S(\omega)$ (in fact, a strong one in this case). Since $z^p a(z)$ is a polynomial, we can establish the uniqueness of the minimum-phase spectral factor exactly as in the proof of Theorem 2.20. Therefore, $a(z) = 1/\beta(z)$, and it follows that $\{u_t\}$ is indeed the innovation process of $\{y_t\}$.

Corollary. The parameters $\{a_k\}$ are the coefficients of the pth-order best linear predictor of the process.   ∎

The covariances of the process can be computed in the following manner. Multiply (2.12.5) by $y_{t-\ell}$, where $\ell \ge 0$, and take expected values. By the properties of the innovation process, we get

$$r_\ell = -\sum_{k=1}^p a_k r_{\ell-k} + \sigma_u^2 \delta(\ell).$$

Using the symmetry of r_k, we can collect these equations for $0 \le \ell \le p$ to get

$$\left\{ \begin{bmatrix} 1 & \cdots & a_{p-1} & a_p \\ a_1 & \cdots & a_p & 0 \\ \vdots & & \vdots & \vdots \\ a_p & \cdots & 0 & 0 \end{bmatrix} + \begin{bmatrix} 1 & 0 & \cdots & 0 \\ a_1 & 1 & \cdots & 0 \\ \vdots & \vdots & \ddots & \vdots \\ a_p & a_{p-1} & \cdots & 1 \end{bmatrix} \right\} \begin{bmatrix} 0.5 r_0 \\ r_1 \\ \vdots \\ r_p \end{bmatrix} = \begin{bmatrix} \sigma_u^2 \\ 0 \\ \vdots \\ 0 \end{bmatrix}. \tag{2.12.7}$$

The solution of this set of equations yields $\{r_k, 0 \le k \le p\}$. The coefficient matrix in the left side of (2.12.7) is guaranteed to be nonsingular when the zeros of $a(z)$ are strictly inside the unit circle (the proof is not difficult, but will not be given here). The covariances of higher lags can be obtained from the recursion

$$r_\ell = -\sum_{k=1}^{p} a_k r_{\ell-k}, \quad \ell > p. \tag{2.12.8}$$

In Chapter 6 we will learn a more efficient way of computing the $\{r_k\}$ from the AR coefficients.

Example 2.9. The models

$$y_t = -a_1 y_{t-1} + u_t, \quad y_t = -a_1 y_{t-1} - a_2 y_{t-2} + u_t$$

describe autoregressive processes of first and second order respectively. For the former case a_1 has to be in the range $|a_1| < 1$. For the latter case, $\{a_1, a_2\}$ have to be in the open triangle defined by

$$|a_2| < 1, 1 + a_2 + a_1 > 0, 1 + a_2 - a_1 > 0. \qquad \blacksquare$$

AR processes are very important in practical signal analysis, and we will have much more to say about them later. However, neither AR nor MA models are entirely satisfactory from a practical point of view. Real-life processes are often characterized by spectral densities that have both high peaks and deep "valleys." AR models can well approximate high peaks, by locating their poles near the unit circle, at the proper frequencies. Similarly, MA models can well approximate deep valleys, by locating their zeros near the unit circle, at the proper frequencies. However, an AR model is not good at approximating deep valleys, while an MA model is not good at approximating high peaks. By "not good" we mean that very high order may be necessary to achieve a good approximation, so the model is "wasteful" in its number of parameters.

AUTOREGRESSIVE MOVING AVERAGE MODELS

By combining the AR and the MA models, we may hope to be able to model both peaks and valleys in the power spectrum. This motivates the introduction of the following model. An *autoregressive moving average* (ARMA) process is a wide sense stationary process defined by

$$y_t = -\sum_{k=1}^{p} a_k y_{t-k} + \sum_{k=0}^{q} b_k w_{t-k}, \tag{2.12.9}$$

where
- $\{w_t\}$ is an uncorrelated process with zero mean and variance σ_w^2.
- $\{a_k, 0 \le k \le p\}$ is a finite sequence with $a_0 = 1, a_p \ne 0$.
- $\{b_k, 0 \le k \le q\}$ is a finite sequence with $b_0 = 1, b_q \ne 0$.
- The polynomials $z^p a(z)$ and $q^z b(z)$ are coprime.

Such a process is denoted by ARMA(p, q). The integers p and q are, respectively, the *AR order* and the *MA order*. The sequences $\{a_k, 0 \le k \le p\}$ and $\{b_k, 0 \le k \le q\}$ are, respectively, the *AR parameters* and the *MA parameters*. When $\{u_t\}$ is Gaussian, so is $\{y_t\}$.

Similarly to the case of the AR model, we require that $a(z)$ have no zeros on or outside the unit circle so that $\{y_t\}$ be indeed wide sense stationary.

The spectral density of an ARMA process is given by

$$S(\omega) = \frac{\sigma_w^2 |b(e^{j\omega})|^2}{|a(e^{j\omega})|^2}. \tag{2.12.10}$$

This expression explains why we required $z^p a(z)$ and $z^q b(z)$ to be coprime, that is, to avoid trivial cancellations in (2.12.10). There may still be cancellations of zeros of $z^p a(z)$ with zeros of $b(z^{-1})$, and vice versa. Such cancellations will affect the innovation model of the process, as we will show below.

The following theorem relates the ARMA model to the innovation representation of the process.

Theorem 2.22. The process $\{w_t\}$ is the innovation of $\{y_t\}$ if and only if $b(z)$ has no zeros outside the unit circle. If this condition is not satisfied and we define $\tilde{b}(z)$ as in (2.12.4), then $\{y_t\}$ satisfies

$$y_t = -\sum_{k=1}^{p} a_k y_{t-k} + \sum_{k=0}^{q} \tilde{b}_k u_{t-k}, \tag{2.12.11}$$

where $\{u_t\}$ is the innovation of $\{y_t\}$.

Proof. The proof of this theorem combines the arguments given in the proofs of Theorems 2.20 and 2.21. ∎

We note that the replacement of $b(z)$ by $\tilde{b}(z)$ may cause further cancellations—those of zeros of $z^p a(z)$ with zeros of $b(z^{-1})$. In this case, the AR and MA orders of the innovation model will be smaller than p and q, respectively.

Example 2.10. Let

$$y_t = -0.4y_{t-1} + 0.05y_{t-2} + w_t + 2.2w_{t-1} + 0.4w_{t-2}, \quad \sigma_w = 1.$$

This is an ARMA$(2, 2)$ process, with

$$a(z) = 1 + 0.4z^{-1} - 0.05z^{-2} = (1 + 0.5z^{-1})(1 - 0.1z^{-1})$$
$$b(z) = 1 + 2.2z^{-1} + 0.4z^{-2} = (1 + 2z^{-1})(1 + 0.2z^{-1}).$$

When passing from $b(z)$ to $\tilde{b}(z)$, the common factor $(1 + 0.5z^{-1})$ will be canceled. Therefore, the innovation representation of the process is the ARMA$(1, 1)$ model

$$y_t = 0.1y_{t-1} + u_t + 0.2u_{t-1}, \quad \sigma_u = 2. \quad ∎$$

As in the case of an MA process, it is preferable to use the innovation model (2.12.11) when only the output process is of interest. When the input process $\{w_t\}$ is of interest, conversion to the innovation model is not permitted.

The covariances of the process can be computed as follows. Let $\{c_k\}$ denote the covariances of an AR process having the polynomial $a(z)$ and the innovation variance σ_w^2. These covariances can be computed as in (2.12.7) and (2.12.8). Then, by (2.12.10), the covariances of the ARMA process are given by

$$r_m = \sum_{k=0}^{q}\sum_{\ell=0}^{q} b_k b_\ell c_{m-k+\ell}. \qquad (2.12.12)$$

From a practical point of view, ARMA models combine the advantages of the AR and MA models: they can well approximate both peaks and valleys in the spectral density. ARMA models will be discussed in detail in Chapter 7.

STATE-SPACE MODELS

The ARMA model can be converted to an equivalent state-space model. Let n be the maximum of p and $q+1$, and define $a_{p+1} = \cdots = a_n = 0$ if $n > p$, or $b_{q+1} = \cdots = b_{n-1} = 0$ if $n > q+1$. Let A, B, C be the matrices

$$A = \begin{bmatrix} -a_1 & -a_2 & \cdots & -a_{n-1} & -a_n \\ 1 & 0 & \cdots & 0 & 0 \\ 0 & 1 & \cdots & 0 & 0 \\ \vdots & \vdots & \ddots & \vdots & \vdots \\ 0 & 0 & \cdots & 1 & 0 \end{bmatrix}, B = \begin{bmatrix} 1 \\ 0 \\ 0 \\ \vdots \\ 0 \end{bmatrix}, C = \begin{bmatrix} b_0 & b_1 & \cdots & b_{n-1} \end{bmatrix}.$$

$$(2.12.13)$$

Let the n-dimensional vector random process x_t be defined by the vector difference equation

$$x_t = A x_{t-1} + B u_t, \qquad (2.12.14a)$$

and let y_t be related to x_t by

$$y_t = C x_t. \qquad (2.12.14b)$$

Then the input-output relationship described by (2.12.13) and (2.12.14) is identical to that described by (2.12.9). The proof of this construction is given in books on linear systems (e.g., [Kailath, 1981]). The matrix A is called the *top-row companion matrix*, and (2.12.13) is called the *controller canonical form*. Conversely, any n-dimensional state equation of the form (2.12.14) has an equivalent ARMA$(n, n-1)$ representation, the transfer function of which is [A12]

$$\frac{b(z)}{a(z)} = C(\mathbf{1}_n - z^{-1}A)^{-1}B. \qquad (2.12.15)$$

For details see Kailath [1981]. Problems 2.23 and 2.24 discuss some important properties of the state-space model (2.12.13), (2.12.14). These properties will be used in Chapter 7.

MODELS FOR DETERMINISTIC PROCESSES

Modeling of deterministic processes is a difficult problem. None of the above models (AR, MA, or ARMA) applies to such processes, since their innovation is identically zero. In fact, the absolutely continuous and the singular components of deterministic processes defy any attempt for general modeling. The situation is somewhat easier with the jump component. Since the number of jumps of any spectral distribution is at most countable and their sum is finite, we can approximate it by a finite number of jumps. Thus, we arrive at the model

$$y_t = \sum_{m=1}^{M} A_m \cos(\omega_m t - \phi_m). \tag{2.12.16}$$

Each component in this sum represents two (symmetrical) jumps in the spectral distribution function. The phases $\{\phi_m\}$ are assumed independent and uniformly distributed on $[0, 2\pi]$. The amplitudes $\{A_m\}$ can be either fixed or random. In the latter case, they may be taken to be independent and Rayleigh distributed, and then $\{y_t\}$ will be Gaussian (this is a straightforward extension of Prob. 2.1). The covariances of this process are given by

$$r_k = \frac{1}{2} \sum_{m=1}^{M} E(A_m^2) \cos \omega_m k. \tag{2.12.17}$$

Deterministic processes of this type will be discussed in Chapter 9.

2.13. MATHEMATICA PACKAGES

The packages `Spectrum.m`, `ArmaData.m`, and `Rarma.m` implement some of the numerical procedures encountered in this chapter. The package `Misc.m` includes miscellaneous auxiliary functions for all the packages described in this book. Here is a brief description of the functions in these packages.

`Spectrum[a_, b_, sigu2_, omega_]` computes the spectral density of an ARMA(p, q) process with AR coefficients in the list `a`, MA coefficients in the list `b`, and innovation variance `sigu2`. An AR process can be handled as a special case by inputting `b = {1}`. Similarly, an MA process can be handled as a special case by inputting `a = {1}`. The parameter `omega` can be either an integer or a list of frequency values (in radians). In the former case, the function returns a list of spectral density values at the given number of frequency points, equally spaced in the range $[0, \pi]$. In the latter case the function returns the spectral density values at the given frequencies.

`ArmaData[a_, b_, sigu2_, Nsam_]` generates `Nsam` data points of a Gaussian ARMA process with parameters `a`, `b` and `sigu2`. The input is a random Gaussian i.i.d. sequence. The initial conditions are automatically computed to make the data wide sense stationary. This is accomplished by the internal routine InCon (not directly accessible to the user).

Rarma[a_, b_, sigu2_, m_] implements Eq. (2.12.12), computing the covariances of an ARMA process with parameters a, b and sigu2, of orders 0 through m. The covariance sequences of AR and MA processes can be computed as special cases. The function uses the internal function Rar to compute the covariances of the AR process with parameters a and sigu2, as needed in (2.12.12). Rar uses an algorithm to be studied in Chapter 6. For didactical reasons, we have included the internal function Rar1, which does the same based on (2.12.7) and (2.12.8). However, Rar1 is less efficient than Rar and there is no reason to prefer it.

Zeros[n_] returns a vector of n zeros.

ZeroPad[seq_, n_] pads a given sequence with zeros to yield a sequence of length n.

Toeplitz[col_, row_] builds a Toeplitz matrix from a given first column col and a first row row.

Hankel[col_, row_] builds a Hankel matrix (a matrix constant along its antidiagonals) from a given first column col and a last row row.

Convolution[a_, b_] convolves the sequences a and b.

Horner[p_, x_] evaluates the polynomial with coefficients p at a point (or a list of points) x, using Horner's method.

Proots[p_] finds the roots of the polynomial with coefficients p. Unlike built-in Mathematica routines, it returns the list of roots, not a list of rules for the roots.

StableQ[p_] returns True if the polynomial with coefficients p is stable and False if it is not.

Stabilize[p_] stabilizes the polynomial with coefficients p by reflecting all unstable roots into the unit disk. This function can be used to obtain the innovation models of MA and ARMA processes, as explained in Sec. 2.12.

LsqSqr[A_, b_] Solves the least-squares problem $Ax = b$; that is, it computes $x = (A^T A)^{-1} A^T b$, see Secs. 3.7 and 7.5. The computation is by Q-R decomposition of A.

PosIntQ[n_] is a predicate returning True if n is a positive integer and False otherwise. NnIntQ[n_] is a predicate returning True if n is a nonnegative integer and False otherwise. NorSQ[x_] is a predicate returning True if x is a number or a symbol and False otherwise.

The functions CovSeq (in ArmaData.m) and Garma (in Rarma.m) will be explained in later chapters.

REFERENCES

Caines, P. E., *Linear Stochastic Systems*, Wiley, New York, 1988.

Doob, J. L., *Stochastic Processes*, Wiley, New York, 1953.

Kailath, T., *Linear Systems*, Prentice Hall, Englewood Cliffs, NJ, 1981.

Karlin, S., and Taylor, H. M., *A First Course in Stochastic Processes,* 2nd ed., Academic Press, New York, 1975.

Kolmogorov, A. N., and Fomin, S. V., *Introductory Real Analysis,* Dover Publications, New York, 1975.

Royden, H. L., *Real Analysis,* Macmillan, New York, 1968.

Rudin, W., *Real and Complex Analysis,* 3rd ed., McGraw-Hill, New York, 1986.

Söderström, T., and Stoica, P., *System Identification,* Prentice Hall, Englewood Cliffs, NJ, 1989.

Wold, H., *The Analysis of Stationary Time Series,* 2nd ed., Almquist and Wicksell, Uppsala, Sweden, 1954 (originally published in 1938).

PROBLEMS

2.1. Use (2.5.4) to show that A_I and A_Q in (2.5.6) are independent Gaussian random variables and hence conclude that x_t is Gaussian.

2.2. Prove that the assumption $u_t \neq 0$ implies that the matrix R_p in (2.6.8) is nonsingular for all p. Hint: Assume the contrary and let p be the minimum integer for which R_p is singular. Show that this implies that y_t is a linear combination of $\{y_{t-k}, 1 \leq k \leq p-1\}$. Therefore, $u_{t,p-1} = 0$; hence $u_t = 0$, in contradiction to the assumption.

2.3. Show that the innovation of the harmonic process in Example 2.5 is identically zero, so the process can be perfectly predicted from its two past values.

2.4. Prove that the right side of (2.7.3) converges in the mean square and defines a wide sense stationary process.

2.5. Prove that the process $\{x_t\}$ in Wold decomposition is purely indeterministic.

2.6. Prove that, in the notation of Wold decomposition, $\mathcal{H}_t^{(u)} = \mathcal{H}_t^{(x)}$.

2.7. Show, by performing the integral (2.8.5), that the harmonic process (2.5.3) has the spectral distribution given in Example 2.6.

2.8. Let v_t have the power spectral density given in Example 2.7, let x_t be white noise independent of v_t, and $y_t = x_t + v_t$. What is the Wold decomposition of y_t?

2.9. Can a sum of two deterministic processes be purely indeterministic? If not, prove it. If yes, give an example.

2.10. Explain why $\sigma_u \beta(e^{j\omega})$ satisfies conditions (1) to (5) in Sec. 2.9 (summarize what was shown in this regard in previous sections).

2.11. Prove formula (2.10.8).

2.12. The next five problems explore some properties of Toeplitz matrices, which will be needed in later chapters.

Assume that the covariance sequence $\{r_k\}$ is absolutely summable and let $S(\omega)$ be its spectral density. Let R_n be the Toeplitz matrix defined in (2.4.4). For each n, let $\bar{\lambda}_n$ and $\underline{\lambda}_n$ be the largest and smallest eigenvalues of R_n, respectively. Also, let

$$\bar{S} = \max_{\omega \in [-\pi,\pi]} S(\omega), \quad \underline{S} = \min_{\omega \in [-\pi,\pi]} S(\omega).$$

(a) Prove that $\{\bar{\lambda}_n\}$ is monotone nondecreasing and $\{\underline{\lambda}_n\}$ is monotone nonincreasing. Hint: Recall that

$$\bar{\lambda}_n = \max_{v^H v = 1} v^H R_n v, \quad \underline{\lambda}_n = \min_{v^H v = 1} v^H R_n v.$$

(b) For any vector $v \in \boldsymbol{C}^n$, define $v(\omega) = \sum_{k=0}^{n-1} v_k e^{j\omega k}$. Show that

$$v^H R_n v = \frac{1}{2\pi} \int_{-\pi}^{\pi} |v(\omega)|^2 S(\omega) d\omega.$$

(c) Use (b) to show that

$$\bar{\lambda}_n \leq \bar{S}, \quad \underline{\lambda}_n \geq \underline{S}.$$

(d) Show that

$$\lim_{n\to\infty} \bar{\lambda}_n = \bar{S}, \quad \lim_{n\to\infty} \underline{\lambda}_n = \underline{S}.$$

Hint: Let $\bar{\omega}$ and $\underline{\omega}$ be the points of global maximum and global minimum of $S(\omega)$, respectively. Define

$$\bar{v}_n = n^{-1/2}[1, e^{-j\bar{\omega}}, \ldots, e^{-j\bar{\omega}(n-1)}]^T,$$
$$\underline{v}_n = n^{-1/2}[1, e^{-j\underline{\omega}}, \ldots, e^{-j\underline{\omega}(n-1)}]^T,$$

and show that

$$\lim_{n\to\infty} \bar{v}_n^H R_n \bar{v}_n = \bar{S}, \quad \lim_{n\to\infty} \underline{v}_n^H R_n \underline{v}_n = \underline{S}.$$

(e) Show that all the above results continue to hold if $\{r_n\}$ is an absolutely summable sequence that is not positive semidefinite. In this case, $S(\omega)$ is not necessarily nonnegative, and neither are $\bar{S}$ and $\underline{S}$.

2.13. Let R be a positive definite matrix and $R^{1/2}$ its lower triangular square root [A13]. Let A be a symmetric matrix, not necessarily positive definite, and define $B = R^{T/2} A R^{1/2}$. For any symmetric matrix X, denote the maximum and minimum eigenvalues of X by $\bar{\lambda}(X)$ and $\underline{\lambda}(X)$, respectively.

(a) Prove that

$$\bar{\lambda}(B) \leq \begin{cases} \bar{\lambda}(A)\bar{\lambda}(R), & \text{if } \bar{\lambda}(A) \geq 0 \\ \bar{\lambda}(A)\underline{\lambda}(R), & \text{if } \bar{\lambda}(A) < 0 \end{cases},$$

$$\underline{\lambda}(B) \geq \begin{cases} \underline{\lambda}(A)\underline{\lambda}(R), & \text{if } \underline{\lambda}(A) \geq 0 \\ \underline{\lambda}(A)\bar{\lambda}(R), & \text{if } \underline{\lambda}(A) < 0 \end{cases}.$$

(b) Use (a) to show that

$$\max_i |\lambda_i(B)| \le \bar{\lambda}(R) \max_i |\lambda_i(A)|.$$

(c) State and prove the corresponding results for $C = (R^{-1/2})^T A R^{-1/2}$.

(d) Let $\{r_n, -\infty < n < \infty\}$ be a symmetric positive definite absolutely summable sequence and $\{a_n, -\infty < n < \infty\}$ a symmetric absolutely summable sequence that is not necessarily positive definite. Let $B_n = R_n^{T/2} A_n R_n^{1/2}$ and $C_n = R_n^{-T/2} A_n R_n^{-1/2}$. Using (a), (b), and Prob. 2.12, prove that $\{|\lambda_i(B_n)|, 1 \le n < \infty, 1 \le i \le n\}$ are uniformly bounded, and give an expression for the upper bound. If the spectral density of $\{r_n\}$ is strictly positive, prove that $\{|\lambda_i(C_n)|, 1 \le n < \infty, 1 \le i \le n\}$ are uniformly bounded, and give an expression for the upper bound.

2.14. Let $\{a_k\}$ and $\{b_k\}$ be two symmetric sequences satisfying $a_k = b_k = 0$ for $|k| > K$ and arbitrary otherwise. For $n > K$, let A_n and B_n be the symmetric banded Toeplitz matrices $A_{n,ij} = a_{i-j}$, $B_{n,ij} = b_{i-j}$. Prove that

$$\lim_{n\to\infty} \frac{1}{n} \mathrm{tr}\{A_n B_n A_n B_n\} = \sum_{i=-2K}^{2K} \left[\sum_j a_j b_{i+j}\right]^2,$$

where the inner sum extends from $\max\{-K, -K - i\}$ to $\min\{K, K - i\}$. Prove that the right side is positive, except when one or both sequences are identically zero. Hint: Proceed through the following steps.

(a) Let $Z_{n,k}$ be the $n \times n$ shift matrix $[Z_{n,k}]_{i,j} = \delta(i - j - k)$. Show that

$$\mathrm{tr}\{A_n B_n A_n B_n\} =$$
$$\sum_{i=-K}^{K} \sum_{j=-K}^{K} \sum_{k=-K}^{K} \sum_{\ell=-K}^{K} a_i b_j a_k b_\ell \, \mathrm{tr}\{Z_{n,i} Z_{n,j} Z_{n,k} Z_{n,\ell}\}.$$

(b) Show that

$$\lim_{n\to\infty} \frac{1}{n} \mathrm{tr}\{Z_{n,i} Z_{n,j} Z_{n,k} Z_{n,\ell}\} = \delta(i + j + k + \ell).$$

(c) Using the symmetry of the two sequences, show that

$$\sum_{i=-K}^{K} \sum_{j=-K}^{K} \sum_{k=-K}^{K} \sum_{\ell=-K}^{K} a_i b_j a_k b_\ell \delta(i + j + k + \ell) = \sum_{i=-2K}^{2K} \left[\sum_j a_j b_{i+j}\right]^2.$$

(d) Show that the right side is zero if and only if one (or both) of the sequences is identically zero.

2.15. Extend the previous problem to the case where $\{a_k\}$ and $\{b_k\}$ are infinite and square summable. Prove that

$$\lim_{n\to\infty} \frac{1}{n} \mathrm{tr}\{A_n B_n A_n B_n\} = \sum_{i=-\infty}^{\infty} \left[\sum_{j=-\infty}^{\infty} a_j b_{i+j}\right]^2.$$

Hint: Let $A_n^{(K)}$ and $B_n^{(K)}$ be constructed from the K-truncated sequences. First show that there exists K such that

$$\frac{1}{n}\left|\text{tr}\{A_n B_n A_n B_n\} - \text{tr}\{A_n^{(K)} B_n^{(K)} A_n^{(K)} B_n^{(K)}\}\right| < \frac{1}{3}\epsilon, \quad \forall n > K,$$

and

$$\left|\sum_{i=-\infty}^{\infty}\left[\sum_{j=-\infty}^{\infty} a_j b_{i+j}\right]^2 - \sum_{i=-2K}^{2K}\left[\sum_j a_j b_{i+j}\right]^2\right| < \frac{1}{3}\epsilon.$$

Then use the previous problem to show that there exists $N > K$ such that

$$\left|\frac{1}{n}\text{tr}\{A_n B_n A_n B_n\} - \sum_{i=-2K}^{2K}\left[\sum_j a_j b_{i+j}\right]^2\right| < \frac{1}{3}\epsilon, \quad \forall n > N.$$

2.16. Let $\{a_k\}$ and $\{b_k\}$ be symmetric sequences not identically zero. Assume that $\{a_k\}$ is absolutely summable and positive definite. Prove that there exist $C > 0$ and n_0 such that

$$\frac{1}{n}\text{tr}\{A_n^{-1} B_n A_n^{-1} B_n\} \geq C, \quad \forall n > n_0.$$

Hint: Proceed through the following steps.

(a) Let $\lambda_{n,k}$ be the eigenvalues of A_n and $U_{n,k}$ its eigenvectors. Show that

$$\text{tr}\{A_n^{-1} B_n A_n^{-1} B_n\} = \sum_{k=1}^{n}\sum_{\ell=1}^{n}\lambda_{n,k}^{-1}\lambda_{n,\ell}^{-1}(U_{n,k}^T B_n U_{n,\ell})^2.$$

(b) Use Prob. 2.2 to find a lower bound C_1 on $\lambda_{n,k}^{-1}\lambda_{n,\ell}^{-1}$. Then show that

$$\text{tr}\{A_n^{-1} B_n A_n^{-1} B_n\} \geq C_1 \text{tr}\{B_n^2\}.$$

(c) Find n_0 and a positive C_2 such that $n^{-1}\text{tr}\{B_n^2\} \geq C_2$ for all $n > n_0$. Then $C = C_1 C_2$ is the desired lower bound.

2.17. Let $\{y_t\}$ be a purely indeterministic process such that $|\beta(e^{j\omega})|$ is bounded. Prove that the norm of the right side of (2.7.8) is finite for any square summable sequence $\{\alpha_s\}$. Hint: Consider the truncated sum $x_{t,r} = \sum_{s=0}^{r} \alpha_s y_{t-s}$. Use Parseval's identity to show that

$$E(x_{t,r}^2) = \frac{\sigma_u^2}{2\pi}\int_{-\pi}^{\pi}|\alpha_r(e^{j\omega})|^2|\beta(e^{j\omega})|^2 d\omega,$$

where $\alpha_r(e^{j\omega}) = \sum_{s=0}^{r}\alpha_s e^{-j\omega s}$. Show that this implies

$$E(x_{t,r}^2) \leq \sigma_u^2\left(\sum_{s=0}^{r}\alpha_s^2\right)\sup_{\omega\in[-\pi,\pi]}|\beta(e^{j\omega})|^2.$$

Finally, take $r \to \infty$ to establish the stated result.

2.18. The corollary to Theorem 2.10 is sometimes expressed in the following form: A process is regular if and only if

$$\frac{1}{2\pi} \int_{-\pi}^{\pi} |\log \Psi'(\omega)| d\omega < \infty.$$

This is known as the *Paley-Wiener condition*. Show that the Paley-Wiener condition is equivalent to (2.8.9a). Hint: Express $\log \Psi'(\omega)$ as $\Gamma^+(\omega) + \Gamma^-(\omega)$ where $\Gamma^+(\omega)$ is equal to $\log \Psi'(\omega)$ when the latter is nonnegative and zero otherwise, and $\Gamma^-(\omega)$ is defined in a dual manner. Then show that $\int_{-\pi}^{\pi} \Gamma^+(\omega) d\omega < \infty$, and use it to deduce the equivalence of the Paley-Wiener condition to (2.8.9a).

2.19. In the proof of Theorem 2.20, explain why the multiplicity of the zero of $\overline{S}(z)$ at $z = \alpha$ is $2r$ if $|\alpha| = 1$. Hint: Consider the cases of real α and complex α separately, and observe that $\alpha^{-1} = \alpha^*$.

2.20. Let $\{x_t\}$ be an AR(p) process and let $\{u_t\}$ be its innovation process, σ_u^2 its innovation variance, and $a(z)$ its characteristic polynomial. Let $\{v_t\}$ be a white noise process, independent of $\{x_t\}$, and let σ_v^2 be its variance. Let $y_t = x_t + v_t$. The process $\{y_t\}$ is called *AR in white noise*.
   (a) Is $\{u_t\}$ the innovation process of $\{y_t\}$?
   (b) Express the spectral density of $\{y_t\}$ in terms of σ_u^2, $a(z)$, and σ_v^2.
   (c) Suggest a procedure for obtaining the innovation model of $\{y_t\}$. Hint: Use the material in Secs. 2.9 and 2.12.

2.21. Consider the MA(1) process $y_t = u_t - u_{t-1}$.
   (a) What is the covariance sequence of this process?
   (b) Use the Yule-Walker equation to compute the coefficients of the best linear predictors of orders 2 and 3.
   (c) Extend part (b) to the best linear predictor of order n. Try to guess the solution; then use the Yule-Walker equation to prove your guess.
   (d) What happens when n goes to infinity? Interpret the result in the context of Sec. 2.7.

2.22. Let $\{y_t\}$ be the ARMA process

$$y_t = -0.5y_{t-1} + w_t + 2w_{t-1}.$$

What is the innovation process of $\{y_t\}$?

2.23. Using (2.12.13) and (2.12.14), prove that the first component of the vector x_t is an autoregressive process whose innovation and characteristic polynomial are $\{u_t\}$ and $a(z)$, respectively. What are the other components of this vector?

2.24. Let $\overline{R}_t = E(x_t x_t^T)$.
   (a) Using (2.12.14a), prove that

$$\overline{R}_t = A\overline{R}_{t-1}A^T + \sigma_u^2 BB^T.$$

(b) From the stationarity of $\{x_t\}$ we have $\overline{R}_t = \overline{R}_{t-1} = \overline{R}$; hence $\overline{R}$ satisfies the Liapunov equation

$$\overline{R} - A\overline{R}A^T = \sigma_u^2 BB^T.$$

Use Theorems D.7 and D.8 to conclude about the structure of $\overline{R}$. Reconcile your conclusion with the result of the previous problem.

2.25. Express the impulse response coefficients $\{\beta_s\}$ in terms of the matrices A, B, and C in (2.12.15).

2.26. Suggest a state-space model for the deterministic process defined in (2.12.16). What is the fundamental difference between this model and the one for an ARMA process? Hint: Find a state-space model for a single sinusoid first.

CHAPTER 3

Parameter Estimation Theory

3.1. PRINCIPLES OF PARAMETER ESTIMATION

Statistical decision theory is concerned with two major types of problems: hypothesis testing and estimation. Estimation theory, in turn, has developed in two main directions: the parametric and nonparametric approaches. The parametric approach assumes that the data obey a probabilistic model of a known structure, but the model contains some unknown *parameters*. These parameters are usually real valued, and their number is finite. The data are used to estimate the unknown parameters, thus completing our knowledge of the model (up to the estimation error). The main goal of the parametric approach is *optimality*, that is, the achievement of the best possible accuracy in the estimated parameters. By comparison, the nonparametric approach attempts to build a model for the data based on as few prior assumptions as possible. The main goal of this approach is *robustness*, that is, the building of models whose goodness is not highly sensitive to the true properties of the data and that can accommodate large classes of problems equally well. The largest part of this book is concerned with the parametric approach, so we devote this chapter to the study of the basic principles of parameter estimation theory.

Assume we are given a family of distribution functions of the N-dimensional vector Y, indexed by a *vector of parameters* θ. We denote this family by $\{F_\theta(Y), \theta \in \Theta\}$. Throughout this book, Θ is assumed to be a subset of R^M, so θ is a real vector of dimension $M \leq N$. We will also assume throughout this book that the densities $\{f_\theta(Y), \theta \in \Theta\}$ exist.[†]

The notation a.s. (θ) will stand for almost surely (or with probability 1) with respect to the specific probability distribution $F_\theta(Y)$. The notation a.e. will stand

---

[†] In a more general setup, the probability measures associated with $\{F_\theta(Y)\}$ are assumed to be dominated by (i.e., absolutely continuous with respect to) a common σ-finite measure $\mu(Y)$. In the special case treated here, $\mu(Y)$ is the Lebesgue measure on R^N.

for almost everywhere on $\boldsymbol{R}^N$ (with respect to the Lebesgue measure). By the above assumption, any property that holds a.e. also holds a.s. (θ) for all θ, but the converse is not necessarily true.

Parameter estimation theory is concerned with the following problem. Suppose the random variable Y is known to have a cumulative distribution function that is a member of $\{F_\theta(Y)\}$, but we do not know which one. We make an experiment during which we measure a *realization* (also *sample* or *observation*) of Y. From this realization, we wish to guess the value of θ governing the distribution of Y. Such a guess is called an *estimate* of θ.

In general, a *statistic* $T(Y)$ is a measurable vector function of the realization Y, which does not depend on θ. For example, Y^2 is a statistic (assuming that Y is scalar), but $Y + \theta$ is not. Since $T(Y)$ is measurable, it is a random variable, so it has a distribution function. The distribution function of the statistic is indexed by θ and will be denoted by $G_\theta(T)$. The dimension of $T(Y)$, which we denote by L, is generally different from N. In most cases of interest, $L \leq N$.

A family of distributions $\{F_\theta(Y), \theta \in \Theta\}$ is said to be *complete* if, given a function $h(Y)$ such that

$$E_\theta h(Y) = 0, \quad \forall \theta \in \Theta,$$

then necessarily $h(Y) = 0$ a.e. The notation E_θ means expectation with respect to the distribution function $F_\theta(Y)$.

Example 3.1. Let $\{F_\theta(Y), -\infty < \theta < \infty\}$ be the family of univariate Gaussian distributions with mean θ and variance 1. We will show that this family is complete. We have

$$E_\theta h(Y) = (2\pi)^{-1/2} \int_{-\infty}^{\infty} h(y) \exp\{-0.5(y-\theta)^2\} dy$$

$$= (2\pi)^{-1/2} \exp(-0.5\theta^2) \int_{-\infty}^{\infty} h(y) \exp(-0.5y^2) \exp(\theta y) dy.$$

The integral in the right side is the two-sided Laplace transform of the function $h(y) \exp(-0.5y^2)$ evaluated at $-\theta$. If the Laplace transform is identically zero for all θ, then $h(y) \exp(-0.5y^2) = 0$ a.e.; hence, $h(y) = 0$ a.e. Therefore, the family $\{F_\theta(Y), -\infty < \theta < \infty\}$ is complete.

Let us continue this example and consider the family of bivariate Gaussian distributions with means $[\theta, \theta]^T$ and covariance matrix $\mathbf{1}_2$. Let $h(Y) = Y_1 - Y_2$. It is easy to show that $E_\theta(Y_1 - Y_2) = 0$ for all θ although $h(Y) \neq 0$; hence this family is not complete. ∎

An *estimate* $\hat{\theta}(Y)$ is a statistic of dimension M, such that $\hat{\theta}(Y) \in \Theta$, and is close to θ in some sense. Of course, "close" is not a well-defined term. Parameter estimation theory is concerned, on the one hand, with quantitative definitions of closeness (or goodness) of estimates and, on the other hand, with estimation methods and their statistical analysis.

A statistic $T(Y)$ is said to be *sufficient* for the parameter θ if the conditional distribution of Y, given $T(Y)$, does not depend on θ. The motivation for this definition is as follows. The distribution of Y can be expressed as $F_\theta(Y) = \int F(Y|T)dG_\theta(T)$. If $T(Y)$ is sufficient, then $F(Y|T)$ does not depend on θ. Therefore, the parameter θ affects the distribution of Y only through $G_\theta(T)$. Stated differently, knowledge of $G_\theta(T)$ is sufficient for determining $F_\theta(Y)$.

A sufficient statistic $T(Y)$ is called *complete sufficient* if the induced family $\{G_\theta(T), \theta \in \Theta\}$ is complete. A statistic $T(Y)$ is said to be *minimal sufficient* for the parameter θ if (1) it is sufficient, and (2) it is a function of any other sufficient statistic for θ. In other words, if $T'(Y)$ is any sufficient statistic for θ, then there exists a function $h(\cdot)$ such that $T(Y) = h(T'(Y))$.

The observation Y itself is a trivial sufficient statistic for θ, of dimension N. The vector Y_s, obtained from Y by sorting its elements in an increasing order, is a statistic of dimension N, called the *order statistic*. It may happen that a given family of distributions does not provide any sufficient statistic of dimension less than N. This means that any data reduction inevitably causes a loss of information about the parameter.

Example 3.2. Let $Y = [Y_1, Y_2, \ldots, Y_N]^T$ be a vector of N independent Gaussian random variables, each with mean θ and variance 1. Let $T(Y)$ be the sample mean

$$T(Y) = \frac{1}{N} \sum_{n=1}^{N} Y_n.$$

Then

$$f_\theta(Y) = (2\pi)^{-N/2} \exp\left\{-\frac{1}{2} \sum_{n=1}^{N} (Y_n - \theta)^2\right\}$$

$$= (2\pi)^{-N/2} \exp\left\{-\frac{1}{2}\left[\sum_{n=1}^{N} Y_n^2 - 2N\theta T + N\theta^2\right]\right\}.$$

The sample mean $T(Y)$ is known to be Gaussian with mean θ and variance $1/N$. Hence its p.d.f. is

$$g_\theta(T) = (2\pi)^{-1/2} N^{1/2} \exp\{-(N/2)(T - \theta)^2\}$$

$$= (2\pi)^{-1/2} N^{1/2} \exp\{-(1/2)[NT^2 - 2N\theta T + N\theta^2]\}.$$

So

$$f_\theta(Y|T) = \frac{f_\theta(T|Y)f_\theta(Y)}{g_\theta(T)}$$

$$= (2\pi)^{-(N-1)/2} N^{-1/2} \delta_D\left(T - \frac{1}{N}\sum_{n=1}^{N} Y_n\right) \exp\left\{\frac{1}{2}\left[NT^2 - \sum_{n=1}^{N} Y_n^2\right]\right\}.$$

The right side does not depend on θ, so $T(Y)$ is a sufficient statistic for θ. The use of the Dirac delta function is necessary here, since the conditional density is nonzero only on the hyperplane $N^{-1} \sum_{n=1}^{N} Y_n = T$.   ∎

The next theorem, known as Neyman-Fisher factorization theorem, gives a simple criterion for sufficiency.

Theorem 3.1. The statistic $T(Y)$ is sufficient for θ if and only if there exists a factorization of $f_\theta(Y)$ of the form

$$f_\theta(Y) = g_\theta(T(Y)) \, h(Y) \quad \text{a.e.} \tag{3.1.1}$$

with $g_\theta(T(Y)) \geq 0$ and $h(Y) \geq 0$ a.e.   ∎

Note that the factorization (3.1.1) means that $f_\theta(Y)$ depends on θ only through the statistic $T(Y)$, while $h(Y)$, which depends on the complete realization, is independent of θ. The factorization theorem is elementary in the case of discrete distributions (i.e., when the "densities" involved are probability mass functions). However, the case of absolutely continuous distributions, which is of interest to us here, is quite difficult, and its proof is beyond the scope of this book. A rigorous proof is given, for example, in Lehmann [1986, Sec. 2.6].

Example 3.3. Let $Y = [Y_1, Y_2, \ldots, Y_N]^T$ be a vector of N independent Gaussian random variables, each with mean θ_1 and variance θ_2. Let $\theta = [\theta_1, \theta_2]^T$, and define

$$T_1(Y) = \frac{1}{N} \sum_{n=1}^{N} Y_n, \quad T_2(Y) = \frac{1}{N} \sum_{n=1}^{N} (Y_n - T_1(Y))^2.$$

$T_1(Y)$ is the sample mean, while $T_2(Y)$ is the *sample variance*. We will show that these two are, together, a sufficient statistic for θ. We have

$$f_\theta(Y) = (2\pi\theta_2)^{-N/2} \exp\left\{ -\frac{1}{2\theta_2} \sum_{n=1}^{N} (Y_n - T_1(Y) + T_1(Y) - \theta_1)^2 \right\}$$

$$= (2\pi\theta_2)^{-N/2} \exp\left\{ -\frac{1}{2\theta_2} \left[\sum_{n=1}^{N} (Y_n - T_1(Y))^2 + N(T_1(Y) - \theta_1)^2 \right] \right\}$$

$$= (2\pi\theta_2)^{-N/2} \exp\left\{ -\frac{N}{2\theta_2} [T_2(Y) + (T_1(Y) - \theta_1)^2] \right\}.$$

This has the form (3.1.1), with $h(Y) = 1$. Hence, by the factorization theorem, $[T_1(Y), T_2(Y)]^T$ is a sufficient statistic for θ.   ∎

Next we introduce the concept of *information* in the sense of Fisher. Fisher's idea was to express the information carried by the density $f_\theta(Y)$ about the parameter θ in quantitative terms. An adequate definition of information should possess the following properties:

- The larger the sensitivity of $f_\theta(Y)$ to changes in θ, the larger should be the information.
- The information should be *additive* in the following sense: if Y_1 and Y_2 are two independent observations from the same distribution, the information corresponding to the combined observation $[Y_1, Y_2]^T$ should be the sum of the informations of the two observations.
- The information should be insensitive to the sign of the change in θ and preferably positive.
- The information should be a *deterministic* quantity; that is, it should not depend on the specific (random) observation.

Consider first the case of scalar θ. In this case, Fisher's information is defined as

$$J(\theta) = E_\theta \left(\frac{\partial \log f_\theta(Y)}{\partial \theta} \right)^2 = -E_\theta \left(\frac{\partial^2 \log f_\theta(Y)}{\partial \theta^2} \right). \tag{3.1.2}$$

For $J(\theta)$ to exist, the partial derivative of $\log f_\theta(Y)$ with respect to θ has to exist and to have a finite second moment. For the second form in (3.1.2), the second partial derivative has to exist and to have a finite first moment. The equality of the two forms holds if (Prob. 3.3, [A15])

$$0 = \frac{\partial^2}{\partial \theta^2} \int_{-\infty}^{\infty} f_\theta(Y) dY = \int_{-\infty}^{\infty} \frac{\partial^2 f_\theta(Y)}{\partial \theta^2} dY. \tag{3.1.3}$$

Fisher's information satisfies all four properties listed above. The partial derivative measures the sensitivity to changes in the value of the parameter. The logarithmic function implies additivity for independent observations. The square guarantees insensitivity to sign, and the expectation guarantees that $J(\theta)$ be a function of θ only, not of Y.

Fisher's information has a natural extension to vector parameters. When θ is an M-dimensional vector, $J(\theta)$ is an $M \times M$ matrix whose (k, ℓ)th entry is defined by

$$J_{k\ell}(\theta) = E_\theta \left(\frac{\partial \log f_\theta(Y)}{\partial \theta_k} \cdot \frac{\partial \log f_\theta(Y)}{\partial \theta_\ell} \right) = -E_\theta \left(\frac{\partial^2 \log f_\theta(Y)}{\partial \theta_k \partial \theta_\ell} \right). \tag{3.1.4}$$

$J(\theta)$ is called the *Fisher information matrix*. Clearly, this matrix (when it exists) is positive semidefinite. See the remarks following (3.1.2) for the existence and equality of the two forms.

The main importance of Fisher's information is its relation to the variances of estimates of θ. This subject will be taken up in Sec 3.3.

Example 3.4. Consider the Gaussian random vector in Example 3.3. We have

$$\log f_\theta(Y) = -(N/2)\log(2\pi) - (N/2)\log\theta_2 - (1/2\theta_2)\sum_{n=1}^{N}(Y_n - \theta_1)^2$$

$$\frac{\partial \log f_\theta(Y)}{\partial \theta_1} = (1/\theta_2)\sum_{n=1}^{N}(Y_n - \theta_1)$$

$$\frac{\partial \log f_\theta(Y)}{\partial \theta_2} = -(N/2\theta_2) + (1/2\theta_2^2)\sum_{n=1}^{N}(Y_n - \theta_1)^2.$$

Carrying out the various expectations yields

$$J(\theta) = \begin{bmatrix} N/\theta_2 & 0 \\ 0 & N/2\theta_2^2 \end{bmatrix}.$$

3.2. PROPERTIES OF ESTIMATES

When we estimate the parameter θ by some $\hat{\theta}(Y)$, usually there will be a nonzero estimation error $\theta - \hat{\theta}(Y)$. The quality of the estimate can be judged by how small this error is. Our first task is to introduce some measure of the size of the error. Such a measure is commonly called a *loss function* and is denoted by $L(\theta - \hat{\theta}(Y))$. The loss function should be positive and in some sense proportional to the size of the error. In this book, we will only use the *quadratic loss*, which is the square of the Euclidean norm of the error

$$L(\theta - \hat{\theta}(Y)) = \|\theta - \hat{\theta}(Y)\|^2. \tag{3.2.1}$$

The loss, being a function of the random variable Y, is itself random. The average loss with respect to the distribution of Y is called the *risk* and is given by

$$R(\theta, \hat{\theta}) = E_\theta(L(\theta - \hat{\theta}(Y))). \tag{3.2.2}$$

This notation should be interpreted as follows: the risk is a function of the parameter θ and of the *estimation law* $\hat{\theta}(\cdot)$, but not of the random observation Y.

Ideally, we would like to have an estimate whose risk is minimized for all $\theta \in \Theta$. This, however, is an unachievable goal, as is clear from the following example. Consider the estimate $\hat{\theta}(Y) = \theta_0$, where θ_0 is some fixed point of Θ. This estimate simply ignores the observation; hence it is certainly not a very good one. However, its risk (using the quadratic loss function) is zero at the point $\theta = \theta_0$, so no estimate can be better than it at this particular point.

The above example shows that we cannot achieve a uniformly minimum risk estimate in general. We therefore discuss other desirable properties of estimates, each of which is considered "good" in some sense.

BIAS

The *bias* of an estimate $\hat{\theta}(Y)$ is its average deviation from the true value of the parameter; that is,

$$b(\theta) = E_\theta(\hat{\theta}(Y)) - \theta. \tag{3.2.3}$$

An estimate whose bias is identically zero for all $\theta \in \Theta$ is said to be *unbiased*. Unbiasedness is a desirable, though not a very strong property. An estimate can be unbiased and still have relatively large errors. Conversely, good estimates are not necessarily unbiased.

In general, an unbiased estimate is not guaranteed to exist. In cases where it does exist, it is generally nonunique. There is an important case, though, when existence implies uniqueness, as expressed by the following theorem.

Theorem 3.2. Assume that the family of distributions $\{F_\theta(Y), \theta \in \Theta\}$ is complete and that there exists an unbiased estimate of θ. Then this estimate is essentially unique.

Proof. Assume $\hat{\theta}_1(Y)$ and $\hat{\theta}_2(Y)$ are two unbiased estimates. Then

$$E_\theta(\hat{\theta}_1(Y) - \hat{\theta}_2(Y)) = \theta - \theta = 0.$$

Hence, by the completeness of the distributions, $\hat{\theta}_1(Y) - \hat{\theta}_2(Y) = 0$ a.e.   ∎

ADMISSIBILITY

An estimate $\hat{\theta}(Y)$ is said to be *inadmissible* if there exists another estimate, say $\hat{\theta}'(Y)$, whose risk is uniformly lower than that of $\hat{\theta}(Y)$; that is,

$$R(\theta, \hat{\theta}') \leq R(\theta, \hat{\theta}), \quad \forall \theta \in \Theta, \tag{3.2.4a}$$

and

$$R(\theta_0, \hat{\theta}') < R(\theta_0, \hat{\theta}) \tag{3.2.4b}$$

for at least one $\theta_0 \in \Theta$. If there is no such $\hat{\theta}'(Y)$, the estimate $\hat{\theta}(Y)$ is called *admissible*. Admissibility is a rather weak property; it is important in theoretical statistics, but in this book we shall not consider it further.

THE MINIMAX PROPERTY

An estimate $\hat{\theta}(Y)$ is said to be *minimax* if it satisfies

$$\sup_{\theta \in \Theta} R(\theta, \hat{\theta}) = \inf_{\hat{\theta}'}[\sup_{\theta \in \Theta} R(\theta, \hat{\theta}')]. \tag{3.2.5}$$

A minimax estimate thus minimizes the worst possible risk over Θ. Minimax estimates play an important role in parameter estimation theory; see, for example,

Lehmann [1983]. We shall not use them in this book, since they are very difficult to obtain in the types of problems of interest to us.

UNIFORMLY MINIMUM VARIANCE UNBIASED ESTIMATES

Since a uniformly minimum risk estimate does not exist, it is sensible to restrict the class of estimates for a given problem and seek a uniformly minimum risk estimate in this class. The class of unbiased estimates is particularly attractive for this purpose, since the requirement for unbiasedness would eliminate estimates that are good locally (i.e., for particular values of θ), but with unsatisfactory global behavior.

The estimate $\hat{\theta}(Y)$ is said to be *uniformly minimum variance unbiased* (UMVU), if it is unbiased, and for any other unbiased estimate $\hat{\theta}'(Y)$

$$E_\theta\{(\hat{\theta}(Y) - \theta)(\hat{\theta}(Y) - \theta)^T\} \le E_\theta\{(\hat{\theta}'(Y) - \theta)(\hat{\theta}'(Y) - \theta)^T\}, \quad \forall \theta \in \Theta. \quad (3.2.6)$$

Note that when the parameter is a vector (3.2.6) is an inequality between matrices [A14].

UMVU estimates are highly desirable. Unfortunately, for many estimation problems such estimates do not exist or are very difficult to compute. Also, UMVU estimates are not necessarily admissible; see Lehmann [1983, p. 112] for a counterexample.

A special case of great theoretical importance is when the family of distributions admits a complete sufficient statistic for θ. According to the following theorem, due to Rao, Blackwell, Lehmann, and Scheffe, a UMVU estimate exists whenever there is at least one unbiased estimate.

Theorem 3.3. Let $\{F_\theta(Y), \theta \in \Theta\}$ be a family of distributions for which there exist a complete sufficient statistic $T(Y)$ and an unbiased estimate $\hat{\theta}'(Y)$. Then $\hat{\theta}(Y) = E(\hat{\theta}'(Y)|T)$ is a UMVU estimate of θ. This estimate is essentially unique.

Proof.
  (a) Since $T(Y)$ is sufficient, the conditional distribution of Y on T does not depend on θ; hence $\hat{\theta}(Y)$ is indeed a function of the measurement only, not of θ.
  (b) $\hat{\theta}(Y)$ is unbiased, since

$$E_\theta(\hat{\theta}(Y)) = E_\theta E(\hat{\theta}'(Y)|T) = E_\theta(\hat{\theta}'(Y)) = \theta.$$

  (c) We will show that the variance of $\hat{\theta}(Y)$ is less or equal to that of $\hat{\theta}'(Y)$. Observe that, by the definition of $\hat{\theta}(Y)$,

$$E((\hat{\theta}'(Y) - \hat{\theta}(Y))|T) = 0.$$

Therefore,

$$
\begin{aligned}
&E\{(\hat{\theta}'(Y) - \theta)(\hat{\theta}'(Y) - \theta)^T | T\} \\
&= E\{(\hat{\theta}'(Y) - \hat{\theta}(Y) + \hat{\theta}(Y) - \theta)(\hat{\theta}'(Y) - \hat{\theta}(Y) + \hat{\theta}(Y) - \theta)^T | T\} \\
&= E\{(\hat{\theta}'(Y) - \hat{\theta}(Y))(\hat{\theta}'(Y) - \hat{\theta}(Y))^T | T\} + E\{(\hat{\theta}(Y) - \theta)(\hat{\theta}(Y) - \theta)^T | T\} \\
&\quad + E\{(\hat{\theta}'(Y) - \hat{\theta}(Y))(\hat{\theta}(Y) - \theta)^T | T\} + E\{(\hat{\theta}(Y) - \theta)(\hat{\theta}'(Y) - \hat{\theta}(Y))^T | T\} \\
&= E\{(\hat{\theta}'(Y) - \hat{\theta}(Y))(\hat{\theta}'(Y) - \hat{\theta}(Y))^T | T\} + E\{(\hat{\theta}(Y) - \theta)(\hat{\theta}(Y) - \theta)^T | T\} \\
&\geq E\{(\hat{\theta}(Y) - \theta)(\hat{\theta}(Y) - \theta)^T | T\}.
\end{aligned}
$$

Now remove the conditioning on T by taking the expectation E_θ to get

$$
E_\theta\{(\hat{\theta}'(Y) - \theta)(\hat{\theta}'(Y) - \theta)^T\} \geq E_\theta\{(\hat{\theta}(Y) - \theta)(\hat{\theta}(Y) - \theta)^T\}
$$

with equality if and only if $\hat{\theta}'(Y) = \hat{\theta}(Y)$ a.s. (θ).

(d) We have seen that $\hat{\theta}(Y)$ is unbiased. But, since the family of distributions of $T(Y)$ is complete, $\hat{\theta}(Y)$ is essentially unique by Theorem 3.2. Thus, it is independent of the choice of $\hat{\theta}'(Y)$ and its variance is smaller or equal to the variances of all $\hat{\theta}'(Y)$ leading to it. ∎

ESTIMATES OF FUNCTIONS OF THE PARAMETER

So far we have concerned ourselves with estimates of the parameter θ governing the distribution of the measurement. Sometimes we are interested in estimating a *function* of θ, say $g(\theta)$. Some of the concepts discussed above apply to this case as well. An estimate $\hat{g}(Y)$ is said to be unbiased if

$$
E_\theta(\hat{g}(Y) - g(\theta)) = 0, \quad \forall \theta \in \Theta.
$$

If an unbiased estimate $\hat{g}(Y)$ exists, $g(\theta)$ is said to be *estimable*. The concepts of admissibility, minimax, and UMVU, are similarly defined. Theorems 3.2 and 3.3 hold for such estimates and are proved in a similar manner. Later we will encounter some uses for estimates of functions of the parameter θ.

The following is an example of a nonestimable function.

Example 3.5. Let Y be Gaussian with mean θ and variance 1, and suppose we wish to estimate $g(\theta) = \exp(-0.5\theta^2)$. Assume that there exists an unbiased estimate $\hat{g}(Y)$. Then

$$
E_\theta \hat{g}(Y) = \frac{1}{\sqrt{2\pi}} \int_{-\infty}^{\infty} \hat{g}(y) \exp(-0.5y^2 + \theta y - 0.5\theta^2) dy = \exp(-0.5\theta^2).
$$

Hence

$$
\int_{-\infty}^{\infty} [\hat{g}(y) \exp(-0.5y^2)] \exp(\theta y) dy = \sqrt{2\pi}.
$$

But then $\hat{g}(y)\exp(-0.5y^2)$ must be the inverse bilateral Laplace transform of $\sqrt{2\pi}$ evaluated at $-\theta$. Since the inverse bilateral Laplace transform of a constant does not exist, the above $g(\theta)$ is not estimable. We might argue that $\hat{g}(Y) = (1/\sqrt{2\pi})\delta_D(Y)$ is unbiased, but this is not a valid statistic (it is not a measurable function of Y). ∎

3.3. COVARIANCE INEQUALITIES AND EFFICIENCY

In this section we present several general inequalities pertaining to variances of estimates. We begin our discussion with a general matrix inequality, which will serve as a basis for the subsequent cases. We then present the celebrated Cramér-Rao inequality and introduce the notion of efficiency. Then we present some generalizations and alternatives to the Cramér-Rao inequality. Throughout this section the family of distributions is assumed to be absolutely continuous (implying the existence of densities), and the parameter set Θ is assumed to be an open subset of $\boldsymbol{R}^M$ (possibly the entire space).

THE BASIC MATRIX INEQUALITY

Let ξ_1 and ξ_2 be two zero-mean random vectors and denote $\Gamma_{ij} = E\xi_i\xi_j^T, 1 \leq i,j \leq 2$. Assume that Γ_{22} is nonsingular. Then, since a covariance matrix is always positive semidefinite, we have

$$E(\xi_1 - \Gamma_{12}\Gamma_{22}^{-1}\xi_2)(\xi_1 - \Gamma_{12}\Gamma_{22}^{-1}\xi_2)^T = \Gamma_{11} - 2\Gamma_{12}\Gamma_{22}^{-1}\Gamma_{12}^T + \Gamma_{12}\Gamma_{22}^{-1}\Gamma_{22}\Gamma_{22}^{-1}\Gamma_{12}^T$$
$$= \Gamma_{11} - \Gamma_{12}\Gamma_{22}^{-1}\Gamma_{12}^T \geq 0. \qquad (3.3.1)$$

This is the basic inequality. Equality holds if and only if

$$\xi_1 = \Gamma_{12}\Gamma_{22}^{-1}\xi_2 \quad \text{a.s.} \qquad (3.3.2)$$

THE CRAMÉR-RAO INEQUALITY

Suppose we wish to estimate a K-dimensional function of θ, say $g(\theta)$, where $K \leq M$. Assume that $g(\theta)$ is estimable. Subject to some regularity conditions, the Cramér-Rao inequality gives a lower bound on the variance of *any* unbiased estimate of $g(\theta)$, say $\hat{g}(Y)$. The regularity conditions are as follows. Some of them involve the change of order of integration with respect to Y and differentiation with respect to θ. See [A15] for a sufficient condition for such a change.

(1) The set of Y on which $f_\theta(Y)$ is strictly positive is independent of θ.
(2) The partial derivatives $\{\partial f_\theta(Y)/\partial\theta_m, 1 \leq m \leq M\}$ exist for all $\theta \in \Theta$ and satisfy

$$0 = \frac{\partial}{\partial\theta_m}\int f_\theta(Y)dY = \int \frac{\partial f_\theta(Y)}{\partial\theta_m}dY.$$

(3) The Fisher information matrix $J(\theta)$ exists and is positive definite for all $\theta \in \Theta$.

(4) The function $g(\theta)$ is differentiable for all $\theta \in \Theta$. Its Jacobian (a matrix of dimensions $K \times M$) will be denoted by $G(\theta)$.

(5) The estimate $\hat{g}(Y)$ satisfies

$$\frac{\partial g(\theta)}{\partial \theta_m} = \frac{\partial}{\partial \theta_m} \int \hat{g}(Y) f_\theta(Y) dY = \int \hat{g}(Y) \frac{\partial f_\theta(Y)}{\partial \theta_m} dY.$$

The Cramér-Rao theorem is then as follows ([Cramér, 1946; Rao, 1945]).

Theorem 3.4. The variance of $\hat{g}(Y)$ satisfies

$$V(\theta) = E_\theta(\hat{g}(Y) - g(\theta))(\hat{g}(Y) - g(\theta))^T \geq G(\theta) J^{-1}(\theta) G^T(\theta). \qquad (3.3.3)$$

A necessary and sufficient condition for equality is

$$\hat{g}(Y) - g(\theta) = G(\theta) J^{-1}(\theta) \frac{\partial \log f_\theta(Y)}{\partial \theta} \quad \text{a.s. } (\theta). \qquad (3.3.4)$$

Proof. Let $\xi_1 = \hat{g}(Y) - g(\theta)$, $\xi_2 = \partial \log f_\theta(Y)/\partial \theta$, and $\xi = [\xi_1^T, \xi_2^T]^T$. By the regularity condition (2),

$$E_\theta \xi_2 = E_\theta \frac{1}{f_\theta(Y)} \frac{\partial f_\theta(Y)}{\partial \theta} = \int \frac{\partial f_\theta(Y)}{\partial \theta} dY = 0,$$

and obviously $E_\theta \xi_1 = 0$. Also, by regularity conditions (2), (4), and (5),

$$E_\theta(\xi_1 \xi_2^T) = E_\theta \left(\hat{g}(Y) - g(\theta)\right) \left(\frac{1}{f_\theta(Y)} \frac{\partial f_\theta(Y)}{\partial \theta}\right)^T$$

$$= \int \hat{g}(Y) \left(\frac{\partial f_\theta(Y)}{\partial \theta}\right)^T dY - 0 = G(\theta).$$

By the definition of the Fisher information matrix, $E_\theta(\xi_2 \xi_2^T) = J(\theta)$. Invoking the basic matrix inequality, we get

$$V(\theta) - G(\theta) J^{-1}(\theta) G^T(\theta) \geq 0.$$

This proves (3.3.3), while (3.3.4) follows from (3.3.2).   ∎

The right side of (3.3.3) is called *the Cramér-Rao lower bound* (CRB).

SPECIAL CASES OF THE CRAMÉR-RAO INEQUALITY

Consider first the case where $g(\theta) = \theta$, so $G(\theta) = \mathbf{1}$. The CRB for an unbiased estimate $\hat{\theta}(Y)$ is

$$E_\theta(\hat{\theta}(Y) - \theta)(\hat{\theta}(Y) - \theta)^T \geq J^{-1}(\theta) \qquad (3.3.5)$$

with equality if and only if

$$\hat{\theta}(Y) - \theta = J^{-1}(\theta)\frac{\partial \log f_\theta(Y)}{\partial \theta} \quad \text{a.s. } (\theta). \tag{3.3.6}$$

If equality holds in (3.3.5), the estimate $\hat{\theta}(Y)$ is said to be *efficient*. Thus, an efficient estimate is an unbiased estimate that achieves the CRB with equality.

If (3.3.6) holds not only a.s. (θ), but for all Y and all θ, then we can integrate (3.3.6) to obtain

$$f_\theta(Y) = h(Y)\exp\{c_1^T(\theta)T(Y) - c_0(\theta)\}, \tag{3.3.7}$$

where $T(Y) = \hat{\theta}(Y)$, $\partial c_1(\theta)/\partial \theta = J(\theta)$, and $\partial c_0(\theta)/\partial \theta = J(\theta)\theta$. This implies that the densities $\{f_\theta(Y), \theta \in \Theta\}$ form an *exponential family*. By Theorem 3.1, $\hat{\theta}(Y)$ is then a sufficient statistic for the given family of distributions.

Next consider the case where $\hat{\theta}(Y)$ is a *biased* estimate of θ, with bias $b(\theta)$. Then $\hat{\theta}(Y)$ is an unbiased estimate of $g(\theta) = \theta + b(\theta)$. Denote $B(\theta) = \partial b(\theta)/\partial \theta$ (assuming that the Jacobian exists). Then we get from Theorem 3.4:

Theorem 3.5. The variance of $\hat{\theta}(Y)$ satisfies [A12]

$$E_\theta(\hat{\theta}(Y) - \theta - b(\theta))(\hat{\theta}(Y) - \theta - b(\theta))^T \geq [\mathbf{1}_M + B(\theta)]J^{-1}(\theta)[\mathbf{1}_M + B(\theta)]^T. \tag{3.3.8}$$

∎

This is known as the CRB for biased estimates. It is much less useful than the CRB (3.3.5) because the bias function $b(\theta)$ is only rarely computable. Furthermore, while (3.3.5) is independent of the actual estimate, (3.3.8) depends on the actual estimate through the bias function, so it is not a universal bound.

THE CHAPMAN-ROBBINS INEQUALITY

The Chapman-Robbins inequality is based on an idea similar to the one underlying the Cramér-Rao inequality. However, instead of using the derivatives of the log density, it uses finite differences. This way, the need for the regularity conditions (2) to (5) is avoided. On the other hand, the resulting bound is difficult to compute, as we will presently show.

As before, let $\hat{g}(Y)$ be an unbiased estimate of $g(\theta)$. Let ξ_1 be as in the Cramér-Rao inequality. Let $\{\theta_1, \ldots, \theta_M\}$ be a set of points in Θ, all different from θ (where θ is the point at which we wish to compute the bound). Let ξ_2 be the M-dimensional vector whose mth component is $f_{\theta_m}(Y)/f_\theta(Y) - 1$. Then

$$E_\theta \xi_{2,m} = \int f_{\theta_m}(Y)dY - \int f_\theta(Y)dY = 0.$$

Let $J_\Delta(\theta)$ be the covariance matrix of ξ_2. Note that

$$(J_\Delta)_{mn}(\theta) = \int \frac{f_{\theta_n}(Y)f_{\theta_m}(Y)}{f_\theta(Y)}dY - 1.$$

Let $G_\Delta(\theta) = E_\theta(\xi_1 \xi_2^T)$. The mth column of this matrix is given by

$$E_\theta(\hat{g}(Y) - g(\theta)) \left(\frac{f_{\theta_m}(Y)}{f_\theta(Y)} - 1 \right)$$

$$= \int \hat{g}(Y) f_{\theta_m}(Y) dY - \int \hat{g}(Y) f_\theta(Y) dY = g(\theta_m) - g(\theta).$$

Theorem 3.6. If $J_\Delta(\theta)$ is positive definite, the variance of $\hat{g}(Y)$ satisfies

$$V(\theta) = E_\theta(\hat{g}(Y) - g(\theta))(\hat{g}(Y) - g(\theta))^T \geq G_\Delta(\theta) J_\Delta^{-1}(\theta) G_\Delta^T(\theta). \qquad (3.3.9)$$

Equation (3.3.9) is the Chapman-Robbins bound [Chapman and Robbins, 1951]. It follows directly from the basic matrix inequality, similarly to the Cramér-Rao inequality. Since (3.3.9) holds for any choice of $\{\theta_1, \ldots, \theta_M\}$, we get:

Corollary. If $K = 1$ [i.e., if $g(\theta)$ is scalar], then

$$V(\theta) \geq \sup_{\theta_1, \ldots, \theta_M \in \Theta} G_\Delta(\theta) J_\Delta^{-1}(\theta) G_\Delta^T(\theta). \qquad (3.3.10)$$

■

For any specific choice of $\{\theta_1, \ldots, \theta_M\}$, the right side of (3.3.9) may be smaller, greater or incomparable to the CRB, so improvement over the CRB may require much experimentation. In the case of scalar $g(\theta)$, the supremum in (3.3.10) is greater or equal to the CRB (Prob. 3.13). The supremum is extremely difficult to compute in general, since it requires maximization over an M^2-dimensional space (there are M vectors, each of dimension M). Only the case of scalar θ ($M = 1$) is perhaps computationally feasible.

THE BARANKIN INEQUALITY

The Barankin inequality can be viewed as an extension of the Chapman-Robbins inequality. Suppose that, instead of choosing M points in Θ, we choose $M+1$ probability distribution functions on Θ, say $\{P_m(\psi), 1 \leq m \leq M+1, \psi \in \Theta\}$. Let ξ_1 be as before, and assume that $f_\psi(Y)$, as a function of ψ, is integrable with respect to each of the distributions $P_m(\psi)$. Define ξ_2 as

$$\xi_{2,m} = \frac{1}{f_\theta(Y)} \int f_\psi(Y)[dP_m(\psi) - dP_{M+1}(\psi)].$$

Then

$$E_\theta \xi_{2,m} = \int [dP_m(\psi) - dP_{M+1}(\psi)] \int f_\psi(Y) dY = \int [dP_m(\psi) - dP_{M+1}(\psi)] = 0.$$

Let $G_B(\theta) = E_\theta(\xi_1 \xi_2^T)$. The mth column of this matrix is given by

$$E_\theta(\hat{g}(Y) - g(\theta)) \left(\frac{1}{f_\theta(Y)} \int f_\psi(Y) [dP_m(\psi) - dP_{M+1}(\psi)] \right)$$

$$= \int [dP_m(\psi) - dP_{M+1}(\psi)] \int \hat{g}(Y) f_\psi(Y) dY = \int g(\psi) [dP_m(\psi) - dP_{M+1}(\psi)].$$

Let $J_B(\theta)$ be the covariance matrix of ξ_2. Then we get, similarly to the CRB and the Chapman-Robbins bound:

Theorem 3.7. If $J_B(\theta)$ is positive definite, the variance of $\hat{g}(Y)$ satisfies

$$V(\theta) \geq G_B(\theta) J_B^{-1}(\theta) G_B^T(\theta). \tag{3.3.11}$$

∎

　　　The right side of (3.3.11) is the Barankin bound [Barankin, 1949]. In the scalar case we can replace the right side of (3.3.11) by the supremum over all $\{P_m(\psi), 1 \leq m \leq M+1\}$. Note that the Chapman-Robbins bound is a special case of the Barankin bound, where $P_m(\psi)$ is a unit-step function at $\psi = \theta_m$ for $1 \leq m \leq M$, and $P_{M+1}(\psi)$ is a unit-step function at $\psi = \theta$. Therefore, the supremum of the Barankin bound is greater or equal to the supremum of the Chapman-Robbins bound. The Barankin bound is even more difficult to compute than the Chapman-Robbins bound, since it requires maximization over a function space.

THE BHATTACHARYYA SYSTEM OF INEQUALITIES

　　　As we have seen, the Fisher information measures the sensitivity of the density function to small deviations of θ from its true value. We can therefore view the CRB as a "local bound." On the other hand, both the Chapman-Robbins information $J_\Delta(\theta)$ and the Barankin information $J_B(\theta)$ measure the sensitivity of $f_\theta(Y)$ to large deviations and can be viewed as "global bounds."

　　　The Bhattacharyya system of informations is similar to the Fisher information in that it measures the local sensitivity of the density function. It extends the Fisher information by adding higher-order derivatives and consequently strengthens the Cramér-Rao bound. To present the Bhattacharyya system, we will need the following notation. For any integer k, let $\partial^k f_\theta(Y)/\partial\theta^k$ be the column vector consisting of all kth-order *different* partial derivatives of $f_\theta(Y)$ with respect to the components of θ. We assume that these derivatives exist and are invariant to the order of differentiation. Note that the length of this vector is $\binom{M+k-1}{k}$; see Feller [1950, p. 39]. Let $\xi_2^{(k)} = (1/f_\theta(Y))(\partial^k f_\theta(Y)/\partial\theta^k)$. Pick $K \geq 1$ and let $\xi_2 = [(\xi_2^{(1)})^T, \ldots, (\xi_2^{(K)})^T]^T$. The following regularity conditions are required for the Bhattacharyya bounds [A15].

(1') The vector ξ_2 exists for all $\theta \in \Theta$ and

$$0 = \frac{\partial^k}{\partial \theta^k} \int f_\theta(Y) dY = \int \frac{\partial^k f_\theta(Y)}{\partial \theta^k} dY, \quad 1 \le k \le K.$$

(2') The Bhattacharyya matrix $J_K(\theta) = E_\theta \xi_2 \xi_2^T$ exists and is positive definite for all $\theta \in \Theta$.

(3') The function $g(\theta)$ is K-times differentiable for all $\theta \in \Theta$. We denote the matrix of kth-order partial derivatives by $G^{(k)}(\theta)$ [i.e., the ℓth column of this matrix is $\partial^k g_\ell(\theta)/\partial \theta^k$].

(4') The estimate $\hat{g}(Y)$ satisfies

$$\frac{\partial^k g_\ell(\theta)}{\partial \theta^k} = \frac{\partial^k}{\partial \theta^k} \int \hat{g}_\ell(Y) f_\theta(Y) dY = \int \hat{g}_\ell(Y) \frac{\partial^k f_\theta(Y)}{\partial \theta^k} dY.$$

Under these conditions it is easy to show, similarly to the proof of Theorem 3.4, that $E_\theta \xi_2 = 0$ and, if $K > 1$,

$$E_\theta \xi_1 \xi_2^T = [G^{(1)}(\theta), \dots, G^{(K)}(\theta)].$$

Hence we get as a result of the basic inequality:

Theorem 3.8. The variance of $\hat{g}(Y)$ satisfies

$$\begin{aligned} V(\theta) &= E_\theta(\hat{g}(Y) - g(\theta))(\hat{g}(Y) - g(\theta))^T \\ &\ge [G^{(1)}(\theta), \dots, G^{(K)}(\theta)] J_K^{-1}(\theta)[G^{(1)}(\theta), \dots, G^{(K)}(\theta)]^T. \end{aligned} \quad (3.3.12)$$

A necessary and sufficient condition for equality is

$$\hat{g}(Y) - g(\theta) = [G^{(1)}(\theta), \dots, G^{(K)}(\theta)] J_K^{-1}(\theta) \xi_2 \quad \text{a.s. } (\theta). \quad (3.3.13)$$

∎

The system of inequalities (3.3.12) for $K = 1, 2, 3, \dots$ is called the *Bhattacharyya system* [Bhattacharyya, 1946]. For $K = 1$, we get the Cramér-Rao inequality. Obviously, the Bhattacharyya bounds are all tighter (greater or equal) than the CRB and monotonically increase with K. The tightness of the Bhattacharyya bounds relative to the Chapman-Robbins bounds cannot be determined in general. An estimate that satisfies (3.3.13) for some K, but for no $k < K$, is called *Kth-order efficient*. Such an estimate is necessarily UMVU. Kth-order efficient estimates, similarly to (Fisher) efficient estimates, do not necessarily exist.

3.4. SEQUENCES OF ESTIMATES

The estimation problem discussed so far concerns a random variable Y of a fixed dimension. In the context of random processes, we are often interested in a situation where the dimension of Y is allowed to grow. Suppose we have a sequence of random variables $\{y_n\}$, and we let Y_N be the vector of the N consecutive points $\{y_n, 1 \leq n \leq N\}$. Suppose, further, that for every N the random variable Y_N is governed by the distribution $F_\theta(Y_N)$. We thus have a family of distributions $\{F_\theta(Y_N), \theta \in \Theta, 1 \leq N < \infty\}$ indexed by θ and N. It should be emphasized that, when the number of points N grows, the distribution $F_\theta(Y_N)$ changes, but it still depends on the same vector parameter θ of a *fixed* dimension M.

The estimation problem can be now modified as follows. For each N, find an estimate $\hat{\theta}_N(Y_N)$ such that the *sequence of estimates* $\{\hat{\theta}_N(Y_N), 1 \leq N < \infty\}$ will approach θ in some sense. This is called the *sequential estimation* problem. It is common to simplify the terminology and notation by referring to $\{\hat{\theta}_N(Y_N), 1 \leq N < \infty\}$ as an estimate (instead of a sequence of estimates) and omitting the braces. That we are dealing with a sequence will be understood from the subscript N.

A special case of the sequential estimation problem is when the $\{y_n\}$ are i.i.d. This case is treated extensively in the statistical literature, since many problems in applied statistics concern independent experiments on a given population. In this book, however, we shall mostly be interested in cases where the $\{y_n\}$ are not independent.

An estimate $\hat{\theta}_N(Y_N)$ is said to be *asymptotically unbiased* if

$$\lim_{N \to \infty} E_\theta(\hat{\theta}_N(Y_N)) - \theta = 0, \quad \forall \theta \in \Theta. \tag{3.4.1}$$

While bias is often tolerable in fixed sample-size estimation problems, it is not so for sequential estimation. Sequential estimates are usually required to be asymptotically unbiased.

The most important property of sequential estimates is consistency. This property concerns the limiting behavior of $\hat{\theta}_N(Y_N)$ as N tends to infinity. Loosely speaking, a sequence of estimates is consistent if it converges to θ in some sense. The various forms of consistency correspond to different modes of convergence of sequences of random variables.

The estimate $\hat{\theta}_N(Y_N)$ is said to be *weakly consistent* if

$$\lim_{N \to \infty} P_\theta\{\|\hat{\theta}_N(Y_N) - \theta\| < \delta\} = 1, \quad \forall \delta > 0, \quad \forall \theta \in \Theta. \tag{3.4.2}$$

The estimate is said to be *mean-square consistent* (or *consistent in quadratic mean*) if

$$\lim_{N \to \infty} E_\theta\{\|\hat{\theta}_N(Y) - \theta\|^2\} = 0, \quad \forall \theta \in \Theta. \tag{3.4.3}$$

The estimate is said to be *strongly consistent* if

$$P_\theta\{\lim_{N \to \infty} \hat{\theta}_N(Y_N) = \theta\} = 1, \quad \forall \theta \in \Theta. \tag{3.4.4}$$

Strong consistency and mean-square consistency do not imply each other in general. Either strong consistency or mean-square consistency implies weak consistency, but not conversely. An important exception is when the parameter set Θ [hence $\hat{\theta}_N(Y_N)$] is bounded. In this case, weak consistency implies mean-square consistency. This is a straightforward consequence of the fact that convergence in probability implies convergence in quadratic mean whenever the sequence of random variables is dominated by a random variable with a finite second moment; see Chung [1974, p. 67].

The decay rate of the errors of sequential estimates is of great practical importance. Since the error of a consistent estimate goes to zero in the appropriate sense as $N \to \infty$, it is reasonable to normalize $(\hat{\theta}_N(Y_N) - \theta)$ by an appropriate function of N to make its "size" nearly independent of N as $N \to \infty$. This argument can be expressed in exact mathematical terms as follows. Assume that there exists a positive monotone sequence $d(N)$, tending to infinity, such that $d(N)(\hat{\theta}_N(Y_N) - \theta)$ converges in distribution to a random variable ξ. Then the distribution of ξ can serve as a measure of the limiting behavior of $\hat{\theta}_N(Y_N)$. In most cases of interest, the distribution of ξ will be multivariate Gaussian, with zero mean and covariance matrix Ξ. A weakly consistent estimate $\hat{\theta}_N(Y_N)$ for which $d(N)(\hat{\theta}_N(Y_N) - \theta)$ converges in distribution to a multivariate Gaussian random variable is called *consistent asymptotically normal* (CAN). The covariance of the limiting distribution, $\Xi(\theta)$, is called the *normalized asymptotic covariance* of $\hat{\theta}(Y_N)$. In most cases of interest, the proper normalizing sequence is $d(N) = \sqrt{N}$.

A clear distinction should be made between the normalized asymptotic covariance and the quantity

$$V(\theta) = \lim_{N \to \infty} d^2(N) E_\theta (\hat{\theta}_N(Y_N) - \theta)(\hat{\theta}_N(Y_N) - \theta)^T. \qquad (3.4.5)$$

In fact, $V(\theta)$ may not even exist for a given CAN estimate. This happens, for example, when $\hat{\theta}_N(Y_N)$ is weakly consistent, but not mean-square consistent. Even when $\hat{\theta}(Y_N)$ is mean-square consistent, its limiting normalized covariance $V(\theta)$ is not necessarily equal to $\Xi(\theta)$.

The concept of Fisher information can be extended to the sequential case; we simply assign an information $J_N(\theta)$ to each of the $F_\theta(Y_N)$. We then get a sequence of informations (or information matrices) $\{J_N(\theta), 1 \le N < \infty\}$. If there exists a positive monotone sequence $d(N)$, tending to infinity, and a matrix $J_0(\theta)$ such that

$$\lim_{N \to \infty} d^{-2}(N) J_N(\theta) = J_0(\theta), \qquad (3.4.6)$$

we call $J_0(\theta)$ the *normalized asymptotic information matrix*.

A CAN estimate $\hat{\theta}_N(Y_N)$ for which $\Xi(\theta) = J_0^{-1}(\theta)$ [with the same $d(N)$ being used for both] is said to be *best asymptotically normal* (BAN). Such an estimate is also called *asymptotically efficient*.

Remark. An alternative definition is to call an estimate asymptotically efficient if

$$V(\theta) = J_0^{-1}(\theta). \qquad (3.4.7)$$

This definition requires the estimate to be mean-square consistent and is stronger than the first definition. Since the first definition is the one used in most standard texts, we will continue to use it. However, we will still be interested to examine whether (3.4.7) applies in specific situations.

The above definition of asymptotic efficiency is sometimes referred to as *Fisher efficiency*. Alternative definitions of efficiency have been proposed. In particular, *efficiency in the sense of Bahadur* [Bahadur, 1960] is perhaps more important than Fisher efficiency and is emphasized in modern texts, but is outside the scope of this book.

The reader may have noticed that the bias sequence $b_N(\theta)$ did not appear in the definition of asymptotic efficiency. As we have said before, a necessary condition for consistency is that $\lim_{N\to\infty} b_N(\theta) = 0$. However, this does not mean that $\lim_{N\to\infty} \partial b_N(\theta)/\partial\theta = 0$ or that the bias derivative even exists. Indeed, it is possible by carefully controlling the bias behavior to construct estimates that are *superefficient*. Such estimates have normalized asymptotic variances that are smaller than $J_0^{-1}(\theta)$ at some values of θ. For examples of superefficient estimates, see Zacks [1971, p. 208].

3.5. MAXIMUM LIKELIHOOD ESTIMATION

Many parameter estimation problems lend themselves to ad hoc estimation techniques suitable only for the problem at hand. However, there are only few *general* estimation techniques applicable to large classes of problems. Here we shall present the two most common ones, the method of maximum likelihood and the method of moments. The first will be discussed in this section and the second in the next. A third method, that of *Bayes estimation*, will not be treated in this book.

Consider the problem of estimating θ when the probability densities $\{f_\theta(Y), \theta \in \Theta\}$ exist and are bounded. The *maximum likelihood estimate* of θ (ML for short) is the point of global maximum of $f_\theta(Y)$, when viewed as a function of θ, keeping Y fixed and equal to the observed value of this random variable. We will use the notation

$$\hat{\theta}_{\mathrm{ML}}(Y) = \underset{\theta\in\Theta}{\mathrm{argmax}}\, f_\theta(Y). \qquad (3.5.1)$$

The density $f_\theta(Y)$, when viewed as a function of θ, with Y serving as a parameter, is called the *likelihood function*. In strict statistical terminology, any multiple of $f_\theta(Y)$ by an arbitrary function of Y is called a likelihood function; however, for our purposes it is adequate to refer only to $f_\theta(Y)$ by this name. Obviously, different realizations Y yield different functions $f_\theta(Y)$, so the point of global maximum depends on Y. So $\hat{\theta}_{\mathrm{ML}}(Y)$ is a legitimate estimator.

Since the logarithmic function is strictly monotone, the maximum point of $f_\theta(Y)$ coincides with the maximum point of $\log f_\theta(Y)$. This is called the *log likelihood function* and is often used in place of $f_\theta(Y)$.

Example 3.6. Consider the model in Example 3.3. The ML estimate of θ_1 and θ_2 can be found by differentiating $\log f_\theta(Y)$ with respect to θ and equating the derivatives to zero. Straightforward calculation gives

$$\hat{\theta}_{1,\text{ML}} = T_1(Y), \quad \hat{\theta}_{2,\text{ML}} = T_2(Y).$$

It is interesting to note that the ML estimate of θ_2 is biased (see Prob. 3.7). ∎

An important insight to maximum likelihood is gained by examining it in the case of sequential estimation from i.i.d. measurements. Let θ be the true value of the parameter and ψ any other value. Let $f_\theta(y)$ be the probability density of any single measurement, so the joint probability density of N measurements is $\prod_{n=1}^{N} f_\theta(y_n)$. Define $L(\psi, Y_N) = (1/N) \sum_{n=1}^{N} \log f_\psi(y_n)$. Then

$$L(\psi, Y_N) - L(\theta, Y_N) = \frac{1}{N}\sum_{n=1}^{N}\log f_\psi(y_n) - \frac{1}{N}\sum_{n=1}^{N}\log f_\theta(y_n) = \frac{1}{N}\sum_{n=1}^{N}\log\left\{\frac{f_\psi(y_n)}{f_\theta(y_n)}\right\}.$$

$\hat{\theta}_{\text{ML}}(Y_N)$ is the maximum point of $L(\psi, Y_N)$, where the maximization is with respect to ψ. Equivalently, it is the maximum point of $L(\psi, Y_N) - L(\theta, Y_N)$. Now, since the measurements are i.i.d.,

$$E_\theta[L(\psi, Y_N) - L(\theta, Y_N)] = E_\theta \log\left\{\frac{f_\psi(y)}{f_\theta(y)}\right\}.$$

Since the logarithmic function is concave, we have by Jensen's inequality [A8]

$$E_\theta[L(\psi, Y_N) - L(\theta, Y_N)] \le \log\left\{E_\theta \frac{f_\psi(y)}{f_\theta(y)}\right\} = \log 1 = 0.$$

Equality holds for $\psi = \theta$, so θ is a point of global maximum of $E_\theta[L(\psi, Y_N) - L(\theta, Y_N)]$. By the strong law of large numbers, $L(\psi, Y_N) - L(\theta, Y_N)$ converges to its expected value almost surely (θ). Therefore, we expect the maximum point of the former to converge to the maximum point of the latter, which is θ. This argument can indeed be formalized subject to proper regularity conditions, as discussed in Theorem 3.12 below.

We begin our discussion of the ML estimate by showing its *invariance property*. Let $g(\theta)$ be a function from Θ onto a subset Φ of $\boldsymbol{R}^J$, where $J \le M$. For any $\phi \in \Phi$, let $g^{-1}(\phi) \subseteq \Theta$ be the inverse image of ϕ. Define

$$\ell_\phi(Y) = \sup_{\theta \in g^{-1}(\phi)} f_\theta(Y).$$

The function $\ell_\phi(Y)$ is called the *induced likelihood* of $g(\theta)$. Then we have:

Theorem 3.9. If $\hat{\theta}$ is the ML estimate of θ, then $\hat{\phi} = g(\hat{\theta})$ is the point of global maximum of the induced likelihood $\ell_\phi(Y)$.

Proof.

$$f_{\hat{\theta}}(Y) \leq \sup_{\theta \in g^{-1}(\hat{\phi})} f_{\theta}(Y) = \ell_{\hat{\phi}}(Y) \leq \sup_{\phi \in \Phi} \ell_{\phi}(Y) = \sup_{\theta \in \Theta} f_{\theta}(Y) = f_{\hat{\theta}}(Y).$$

Therefore, $\ell_{\hat{\phi}}(Y) = \sup_{\phi \in \Phi} \ell_{\phi}(Y)$, which proves that $\hat{\phi}$ maximizes the induced likelihood.   ∎

Theorem 3.9 justifies calling $g(\hat{\theta})$ the ML estimate of $g(\theta)$. In the simple case where $g(\theta)$ is one to one, $\phi = g(\theta)$ as an invertible transformation of the parameter space. Thus, Theorem 3.9 implies invariance of the ML estimate to parametrizations of the family of densities that are in one to one correspondence.

The next two theorems show the connection of ML estimation to sufficiency and efficiency.

Theorem 3.10. The ML estimate is a function of any sufficient statistic for $\{f_{\theta}(Y), \theta \in \Theta\}$.

Proof. By the Neyman-Fisher theorem,

$$f_{\theta}(Y) = g_{\theta}(T(Y)) \, h(Y)$$

[recall that, in the extreme case, $T(Y)$ may be Y itself]. Clearly, maximization of $f_{\theta}(Y)$ with respect to θ is equivalent to maximization of $g_{\theta}(T(Y))$, since $h(Y)$ is independent of θ. Therefore, the maximum point can depend on Y only through $T(Y)$, so the ML estimate is a function of $T(Y)$.   ∎

Theorem 3.11. Suppose there exists an efficient estimate for θ such that (3.3.6) holds for all Y and θ. Then this estimate is equal to the ML estimate.

Proof. By the theorem's assumption, the equality

$$\frac{\partial \log f_{\theta}(Y)}{\partial \theta} = J(\theta)(\hat{\theta}(Y) - \theta)$$

holds for all Y and θ. In particular, this equality holds at $\theta = \hat{\theta}_{\mathrm{ML}}(Y)$. Since this is a maximum point, the left side is zero, so necessarily $\hat{\theta}_{\mathrm{ML}}(Y) = \hat{\theta}(Y)$.   ∎

We now arrive at the main reasons for the importance of maximum likelihood estimation: its consistency, asymptotic normality, and asymptotic efficiency. In this chapter we will limit our discussion to the i.i.d. case. The asymptotic properties of maximum likelihood have been a subject of extensive research. In general, none of the three properties can be guaranteed. Therefore, most of the literature in this area deals with sufficient conditions for various modes of consistency and efficiency. A comprehensive survey of these results is beyond the scope of this book. Instead, we will give some results of restricted generality, which are, however, sufficient for our purposes.

First, we address the question of consistency. We will need the regularity conditions (1) to (3) of Sec. 3.3 (the ones used for the Cramér-Rao inequality). In addition, we will make the following assumptions.

(a) The functions $f_\theta(y)$ are differentiable with respect to θ up to a third order.

(b) For each θ, there exists an open sphere $\{\psi : \|\psi - \theta\| < \rho\}$ and positive functions $B_{ijk}(y)$ such that, for all ψ in the sphere,

$$\left| \frac{\partial^3 \log f_\psi(y)}{\partial \psi_i \, \partial \psi_j \, \partial \psi_k} \right| \leq B_{ijk}(y), \quad 1 \leq i, j, k \leq M \qquad (3.5.2a)$$

and

$$b_{ijk} = E_\theta B_{ijk}(y) < \infty, \quad 1 \leq i, j, k \leq M. \qquad (3.5.2b)$$

Theorem 3.12. Under the above assumptions, there exists a sequence of local maxima of the likelihood function that converges to θ almost surely.

Proof. Fix a point θ in Θ. Denote $b = \sum_i \sum_j \sum_k b_{ijk}$. Let λ_1 be the smallest eigenvalue of the Fisher information $J_0(\theta)$ (the information of a single measurement). Let $\epsilon_0 = \min\{\rho, 3\lambda_1/(12 + b)\}$. Choose $0 < \epsilon < \epsilon_0$ and let ψ be a point in Θ such that $\|\psi - \theta\| = \epsilon$. Let $L(\psi, Y_N) = (1/N) \sum_{n=1}^N \log f_\psi(y_n)$. Then, by Taylor's theorem in $\boldsymbol{R}^M$,

$$\begin{aligned} &L(\psi, Y_N) - L(\theta, Y_N) \\ &= L^{(1)}(\theta, Y_N; \psi - \theta) + \frac{1}{2} L^{(2)}(\theta, Y_N; \psi - \theta) + \frac{1}{6} L^{(3)}(\psi_*, Y_N; \psi - \theta), \quad (3.5.3) \end{aligned}$$

where ψ_* lies on the line joining ψ and θ [see App. C for the notation $L^{(k)}(\cdot)$ in the context of Taylor series].

By the strong law of large numbers, $\partial L(\theta, Y_N)/\partial \theta$ converges almost surely to its expected value, which is zero. Therefore, there exists a null set Π_1 in the probability space Ω such that, for all $\omega \in \Omega - \Pi_1$, there exists N_1 (depending on ω and ϵ), such that, for all $N > N_1$, $|\partial L(\theta, Y_N(\omega))/\partial \theta| < \epsilon^2/M$; hence $|L^{(1)}(\theta, Y_N; \psi - \theta)| < \epsilon^3$.

Express the second term in (3.5.3) as

$$\frac{1}{2} L^{(2)}(\theta, Y_N; \psi - \theta) - \frac{1}{2}(\psi - \theta)^T J_0(\theta)(\psi - \theta) + \frac{1}{2}(\psi - \theta)^T J_0(\theta)(\psi - \theta).$$

Observe that

$$-\frac{1}{2}(\psi - \theta)^T J_0(\theta)(\psi - \theta) \leq -\frac{1}{2}\lambda_1 \epsilon^2.$$

Also, by the strong law of large numbers, $\partial^2 L(\theta, Y_N)/\partial \theta^2$ converges almost surely to its expected value, which is $-J_0(\theta)$. As before, this implies the existence of a null set Π_2 such that, for all $\omega \in \Omega - \Pi_2$, there exists N_2, such that, for all $N > N_2$, $0.5|L^{(2)}(\theta, Y_N(\omega); \psi - \theta) + (\psi - \theta)^T J(\theta)(\psi - \theta)| < \epsilon^3$.

Finally, by invoking the strong law of large numbers a third time and using assumption (b), we get that there exists a null set Π_3 such that, for all $\omega \in \Omega - \Pi_3$, there exists N_3, such that, for all $N > N_3$, $|L^{(3)}(\psi_*, Y_N(\omega); \psi - \theta)| < b\epsilon^3$. Summing up, for $\omega \in \Omega - (\Pi_1 \cup \Pi_2 \cup \Pi_3)$ and for all $N > \max\{N_1, N_2, N_3\}$,

$$L(\psi, Y_N) - L(\theta, Y_N) < -\frac{1}{2}\lambda_1 \epsilon^2 + (2 + b/6)\epsilon^3 < 0 \quad \text{if} \quad \epsilon < \frac{3\lambda_1}{(12 + b)}. \qquad (3.5.4)$$

Recall that a continuous function on a compact set attains its maximum in the set. The set $\{\psi : \|\psi - \theta\| \leq \epsilon\}$ is compact, and by (3.5.4) the maximum of $L(\psi, Y_N)$ cannot be on the boundary for ϵ, ω and N as above. Therefore, there is a local maximum of $L(\psi, Y_N)$ in the interior of this set. This sequence of local maxima converges to θ almost surely, as we have just shown. ∎

It should be stressed that Theorem 3.12 does not guarantee consistency of the ML estimate. Recall that we have defined the ML estimate as the global maximizer of the likelihood function. In general, the likelihood function may have multiple local maxima, and Theorem 3.12 does not guarantee that the sequences of global and local maxima coincide. Only when the maximum of the likelihood function is *unique* with probability 1, does the theorem guarantee consistency. For additional conditions under which strong consistency holds, see Zacks [1971, pp. 233–234].

Despite the above remark, the sequence of local maxima may be as good as maximum likelihood from a practical point of view. Suppose we have at our disposal another sequence of estimates, say $\{\tilde{\theta}_N(Y_N)\}$, which is known to be consistent (methods of constructing consistent estimates are discussed in the next section). Numerical maximization methods typically start at some initial point and converge to the nearest local maximum. In searching for a local maximum of $L(\psi, Y_N)$, we can take $\psi = \tilde{\theta}_N(Y_N)$ as an initial point. Since $\tilde{\theta}_N(Y_N)$ is consistent, this method guarantees (with probability approaching 1) that the sequence of local maxima thus constructed will indeed be *the* sequence constructed in Theorem 3.12.

Next we show the asymptotic normality and efficiency of the estimate $\hat{\theta}_N(Y_N)$ given by Theorem 3.12.

Theorem 3.13. Under the assumptions of Theorem 3.12, the estimate $\hat{\theta}_N(Y_N)$ is best asymptotically normal.

Proof. Since $L(\psi, Y_N)$ is differentiable, its partial derivatives at the point $\hat{\theta}_N(Y_N)$ are zero. We have, as in Theorem 3.12,

$$0 = \sqrt{N} \left. \frac{\partial L(\psi, Y_N)}{\partial \psi_m} \right|_{\psi = \hat{\theta}_N} = \sqrt{N} \frac{\partial L(\theta, Y_N)}{\partial \theta_m}$$

$$+ \sum_k \sqrt{N}(\hat{\theta}_{N,k} - \theta_k) \left[\frac{\partial^2 L(\theta, Y_N)}{\partial \theta_m \partial \theta_k} + \frac{1}{2} \sum_\ell \left. \frac{\partial^3 L(\psi, Y_N)}{\partial \psi_m \partial \psi_k \partial \psi_\ell} \right|_{\psi = \psi_*} (\hat{\theta}_{N,\ell} - \theta_\ell) \right].$$

$$(3.5.5)$$

As in the proof of Theorem 3.12, the term in the brackets on the right side can be shown to converge in probability to $[-J_0(\theta)]_{k,m}$. Therefore, (3.5.5) is of the form

$$\sqrt{N}(\hat{\theta}_N - \theta) = A_N^{-1}\left[\sqrt{N}\frac{\partial L(\theta, Y_N)}{\partial \theta}\right], \qquad (3.5.6)$$

where the sequence of random matrices A_N converges in probability to $J_0(\theta)$. The elements of the matrix $J_0^{-1}(\theta)$ are continuous functions of the elements of $J_0(\theta)$; hence, by Theorem C.1, the sequence A_N^{-1} converges in probability to $J_0^{-1}(\theta)$. By the central limit theorem, $\sqrt{N}(\partial L(\theta, Y_N)/\partial \theta)$ is asymptotically normal with zero mean and covariance matrix $J_0(\theta)$. Therefore, by Theorem C.4, the left side of (3.5.6) is asymptotically normal with zero mean and covariance $J_0^{-1}(\theta)$. This shows that $\sqrt{N}(\hat{\theta}_N - \theta)$ is BAN.   ∎

3.6. THE METHOD OF MOMENTS

The method of maximum likelihood, despite its theoretical appeal (chiefly its good asymptotic properties), is often difficult to implement. Analytic solutions to the maximization problem are available only in a few simple cases. Numerical solutions are possible in many cases, but may be difficult to program, computationally intensive, or both. The problem is aggravated when the number of measurements is large and there is no sufficient statistic of a fixed dimension (i.e., whose dimension is independent of the number of measurements, as in the Gaussian i.i.d. case). The idea behind the method of moments is to base the estimate on an *insufficient statistic*, one whose dimension is small (relative to the number of measurements) but that still conveys information about the underlying parameter θ. The operation of using a relatively low dimensional statistic is also known as *data reduction*. Often, the statistic consists of a vector of sample moments; hence the name *method of moments*. Consider, for example, a fixed set of sample covariances $\{\hat{r}_k, 0 \le k \le K-1\}$ computed from a set of consecutive measurements of a wide-sense stationary process. The probability distribution of the process governs that of the sample covariances; hence this statistic contains information about the parameters of the distribution. In this example, the method of moments deals with various ways of estimating the process parameters from the sample covariances. In general, the statistic is required to converge in some sense to a nonrandom vector (in the above example it is the corresponding vector of true covariances), which functionally depends on the parameter vector θ in an invertible manner. The method of moments relies on the functional dependence and on the "closeness" of the statistic to the true nonrandom value to obtain an estimate of θ. Example 3.7 below further illustrates this idea.

Due to the data reduction inherent in the method of moments, efficiency (as defined in Sec. 3.3) is almost always forfeited. Often, asymptotic efficiency is lost as well (there are exceptions though, as we will see in Chapter 6). The chief motivation for using the method of moments is ease of implementation and computational efficiency, especially when the number of measurements is large. In fact, the method of moments is very common in parameter estimation for stationary processes. In this section we provide the theoretical background to the method and analyze the statistical properties of the corresponding estimates.

Consider the sequential parameter estimation problem posed in Sec 3.4. Assume that the parameter set $\Theta \subseteq \boldsymbol{R}^M$ is open. Let $\{s_N(Y_N), 1 \leq N < \infty\}$ be a sequence of statistics of a *fixed* dimension K such that $K \geq M$. Suppose that, for a fixed θ, this sequence converges to a constant s in one or more of the three usual senses (weakly, strongly, or in quadratic mean). Since convergence depends on the underlying distribution, s depends on θ, so this construction yields a function $s(\theta)$ whose domain is $\Theta \subseteq \boldsymbol{R}^M$ and whose range is some $\boldsymbol{S}_0 \subseteq \boldsymbol{R}^K$. Assume that $s(\theta)$ is one to one and continuously differentiable, and its Jacobian $S(\theta)$ is full rank for all θ. Let us also denote $\boldsymbol{S} = \left[\bigcup_{N=1}^{\infty} \{s_N(Y_N) : Y_N \in \boldsymbol{R}^N\} \right] \cup \boldsymbol{S}_0$ and assume that $\boldsymbol{S}$ is open in $\boldsymbol{R}^K$.

Let $g(s)$ be a measurable function $g(s)$ on $\boldsymbol{S}$ into $\boldsymbol{R}^M$ such that the restriction of $g(s)$ on $\boldsymbol{S}_0$ satisfies $g(s(\theta)) = \theta$ for all $\theta \in \Theta$. This means that this restriction is the *inverse function* of $s(\theta)$. $g(s)$ is not necessarily unique, but its restriction on $\boldsymbol{S}_0$ is unique. In particular, for $K > M$ there is an infinite number of such functions. We can use $g(s)$ to construct a sequence of estimates $\{\hat{\theta}_N = g(s_N(Y_N)), 1 \leq N < \infty\}$. To guarantee that $\hat{\theta}_N$ be a legitimate estimate, we need to confine its possible values to the set Θ. This is accomplished by the following slight modification. Let $\hat{\theta}_0 = \theta_0$, an arbitrary (but fixed) point in Θ. For $N \geq 1$, define

$$\hat{\theta}_N = \begin{cases} g(s_N(Y_N)), & \text{if } g(s_N(Y_N)) \in \Theta \\ \hat{\theta}_{N-1}, & \text{otherwise.} \end{cases} \qquad (3.6.1)$$

If $g(s)$ is sufficiently well behaved, we can expect $\hat{\theta}_N$ to be a consistent estimate of θ in the same sense as the convergence of $s_N(Y_N)$ to s. We will formalize this argument after looking at an example.

Example 3.7. Let $\{y_n\}$ be a Gaussian zero mean AR(1) process. The unknown parameter is the autoregression coefficient $a_1 = a$ (see Sec. 2.12), which is in the range $|a| < 1$. The innovation variance σ_u^2 is assumed known and equal to 1. The covariances of the process are given by $r_k = (-a)^{|k|}/(1 - a^2)$. Let $\{\hat{r}_k(N)\}$ be the estimated covariances

$$\hat{r}_k(N) = \frac{1}{N} \sum_{n=1}^{N-k} y_{n+k} y_n, \quad k \geq 0.$$

By Theorem 2.18, the $\{\hat{r}_k(N)\}$ are consistent estimates of the covariances. Consider the following two candidate estimates of a:

$$\hat{a}_1 = g_1(\hat{r}_0(N), \hat{r}_1(N)) = -\frac{\hat{r}_1(N)}{\hat{r}_0(N)}, \quad \hat{a}_2 = g_2(\hat{r}_1(N)) = \frac{-2\hat{r}_1(N)}{1 + \sqrt{1 + 4\hat{r}_1^2(N)}}.$$

It is easy to check that $g_1(r_0, r_1) = g_2(r_1) = a$ and that both functions are continuous and thus consistent estimates of a. This example will be continued. ∎

We first consider the cases of strong and weak consistency. The more difficult question of mean-square consistency will be discussed at the end of this section.

Theorem 3.14. If $s_N(Y_N)$ is a strongly consistent estimate of $s(\theta)$ and $g(s)$ is continuous, then $g(s_N(Y_N))$ is a strongly consistent estimate of θ.

Proof. The sequence $s_N(Y_N)$ converges to $s(\theta)$ on a set $A \subseteq \Omega$ such that the probability of $\Omega - A$ (with respect to the probability measure associated with θ) is zero (Ω is the underlying probability space). By the continuity of g, $g(s_N(Y_N))$ converges to $g(s(\theta))$ on this set. Since $g(s(\theta)) = \theta$, this proves the strong consistency of the estimate. ∎

Theorem 3.15. If $s_N(Y_N)$ is a weakly consistent estimate of $s(\theta)$ and $g(s)$ is continuous, then $g(s_N(Y_N))$ is a weakly consistent estimate of θ.

Proof. The proof is identical to that of Theorem C.1. ∎

The next theorem explores the asymptotic distribution of a consistent estimate $g(s_N(Y_N))$.

Theorem 3.16. Let $d(N)$ be a positive sequence that tends to infinity with N, and assume that $d(N)(s_N(Y_N) - s(\theta))$ converges in distribution to a Gaussian random vector with zero mean and positive definite covariance $\Sigma(\theta)$. Assume that $g(s)$ is continuously differentiable and its Jacobian $G(s)$ is nonsingular for all $s(\theta)$. Then $d(N)(g(s_N(Y_N)) - g(s(\theta)))$ is asymptotically normal with zero mean and covariance $G(s(\theta))\Sigma(\theta)G^T(s(\theta))$.

Proof. By the Cramér-Wold theorem [Billingsley, 1986, p. 397], it is sufficient to show that $d(N)v^T(g(s_N(Y_N)) - g(s(\theta)))$ is asymptotically normal with zero mean and covariance $v^T G(s(\theta))\Sigma(\theta)G^T(s(\theta))v$ for every M-vector v. Since $d(N)(s_N(Y_N) - s(\theta))$ converges in distribution, it is $O_p(1)$ by Lemma C.2. Therefore, $(s_N(Y_N) - s(\theta))$ is $O_p(d^{-1}(N))$. Therefore, by Taylor's theorem for functions of random variables (Theorem C.2),

$$g_m(s_N(Y_N)) - g_m(s(\theta)) = G_m(s(\theta))(s_N(Y_N) - s(\theta)) + o_p(d^{-1}(N)),$$

where $g_m(\cdot)$ is the mth component of $g(\cdot)$ and $G_m(\cdot)$ is the mth row of $G(\cdot)$. Hence

$$d(N)v^T(g(s_N(Y_N)) - g(s(\theta))) = v^T G(s(\theta))[d(N)(s_N(Y_N) - s(\theta))] + O_p(d^{-1}(N))$$
$$= v^T G(s(\theta))[d(N)(s_N(Y_N) - s(\theta))] + o_p(1).$$

Hence, by Theorem C.3, the left side has the same asymptotic distribution as

$v^T G(s(\theta))[d(N)(s_N(Y_N) - s(\theta))]$, which is normal with zero mean and variance $v^T G(s(\theta)) \Sigma(\theta) G^T(s(\theta)) v$. ∎

The "memory" in the definition (3.6.1) does not affect the asymptotic properties of $\hat{\theta}_N$ (consistency, normality, and asymptotic covariance), because the probability of $g(s_N(Y_N))$ falling outside the set Θ decreases to zero as $N \to \infty$.

Example 3.7 (continued). The Jacobian of $g_1(r_0, r_1)$ evaluated at the true point is given by $G_1(a) = -(1 - a^2)[a, 1]$. The Jacobian of $g_2(r_1)$ at the true point is $G_2(a) = -(1 - a^2)^2/(1 + a^2)$. The normalized asymptotic covariance of $\{\hat{r}_0(N), \hat{r}_1(N)\}$ can be shown (using tools to be developed in Chapter 4) to be

$$\Sigma(a) = \frac{1}{(1 - a^2)^3} \begin{bmatrix} 2(1 + a^2) & -4a \\ -4a & 1 + 4a^2 - a^4 \end{bmatrix}.$$

Therefore, the normalized asymptotic variance of $\hat{a}_1$ is

$$V_1 = \frac{1}{(1 - a^2)} [a \quad 1] \begin{bmatrix} 2(1 + a^2) & -4a \\ -4a & 1 + 4a^2 - a^4 \end{bmatrix} \begin{bmatrix} a \\ 1 \end{bmatrix} = (1 - a^2).$$

The normalized asymptotic variance of $\hat{a}_2$ is

$$V_2 = \frac{(1 - a^2)(1 + 4a^2 - a^4)}{(1 + a^2)^2}.$$

It is easy to check that $V_2 \geq V_1$, with equality at $a = 0$ only. Therefore, while both estimates are consistent, the first has smaller asymptotic variance for all a. ∎

Our next result provides a lower bound on the normalized asymptotic covariance of the estimate $g(s_N(Y_N))$. This lower bound is a consequence of the following lemma.

Lemma 3.1. If G is a full rank $M \times K$ matrix with $K \geq M$, S is a right inverse of G (i.e., $GS = 1_M$), and Σ is a $K \times K$ symmetric positive definite matrix, then

$$G\Sigma G^T \geq (S^T \Sigma^{-1} S)^{-1}. \tag{3.6.2}$$

Proof. Since $GS = 1_M$, we get

$$\begin{aligned} 0 &\leq [G - (S^T\Sigma^{-1}S)^{-1}S^T\Sigma^{-1}]\Sigma[G - (S^T\Sigma^{-1}S)^{-1}S^T\Sigma^{-1}]^T \\ &= G\Sigma G^T - GS(S^T\Sigma^{-1}S)^{-1} - (S^T\Sigma^{-1}S)^{-1}S^T G^T + (S^T\Sigma^{-1}S)^{-1} \\ &= G\Sigma G^T - (S^T\Sigma^{-1}S)^{-1}; \end{aligned}$$

hence (3.6.2).

Corollary. Under the conditions of Theorem 3.16, the normalized asymptotic covariance of $g(s_N(Y_N))$ is bounded from below by $(S^T(\theta)\Sigma^{-1}(\theta)S(\theta))^{-1}$, where $S(\theta)$ is the Jacobian of $s(\theta)$.

The corollary follows from the lemma upon observing that, since $g(s)$ is an inverse function of $s(\theta)$, $G(s(\theta))S(\theta) = \mathbf{1}_M$ for all θ. ∎

Note that the lower bound on the variance of the estimate depends only on the properties of the statistic $s_N(Y_N)$, its normalized asymptotic covariance and the Jacobian of its limit with respect to θ.

Example 3.7 (concluded). The inverse of the matrix $\Sigma(a)$ is given by

$$\Sigma^{-1}(a) = \frac{1}{2}\begin{bmatrix} 1 + 4a^2 - a^4 & 4a \\ 4a & 2(1+a^2) \end{bmatrix}.$$

The Jacobian of $[r_0, r_1]^T$ is given by

$$S(a) = \frac{1}{(1-a^2)^2}\begin{bmatrix} 2a \\ -(1+a^2) \end{bmatrix}.$$

Carrying out the multiplication yields

$$(S^T(a)\Sigma^{-1}(a)S(a))^{-1} = (1-a^2).$$

As we see (cf. the result for V_1 above), the estimate $\hat{a}_1$ achieves the least possible asymptotic variance in this case. A similar conclusion holds trivially for $\hat{a}_2$, since both $\hat{r}_1$ and $g_2(\hat{r}_1)$ are scalars. ∎

Since the asymptotic covariance of any consistent estimate depending on $s_N(Y_N)$ and satisfying the conditions of Theorem 3.16 is bounded by the matrix $(S^T(\theta)\Sigma^{-1}(\theta)S(\theta))^{-1}$, it is reasonable to ask whether there exists an estimate that achieves this lower bound. The answer is positive, as we will show next. Let us introduce the nonlinear cost function

$$C(\psi,\xi) = \frac{1}{2}(\xi - s(\psi))^T\Sigma^{-1}(\psi)(\xi - s(\psi)), \quad \xi \in \mathbf{S}, \psi \in \Theta. \tag{3.6.3}$$

The property of interest of $C(\psi,\xi)$ is given by the following lemma.

Lemma 3.2. Let $C(x,y)$ be a function from an open set in $\mathbf{R}^M \times \mathbf{R}^K$ to $\mathbf{R}$, satisfying the following conditions:
(1) $C(x,y)$ is twice continuously differentiable.
(2) There exists a point (x_0, y_0) for which

$$\left.\frac{\partial C(x,y)}{\partial x}\right|_{\substack{x=x_0 \\ y=y_0}} = 0,$$

where the left side is the vector of partial derivatives with respect to the components of x.
(3)

$$\left.\frac{\partial^2 C(x,y)}{\partial x^2}\right|_{\substack{x=x_0 \\ y=y_0}} > 0,$$

where the left side is the matrix of second partial derivatives with respect to the components of x.

Then there exist an open neighborhood of y_0, say $\boldsymbol{N} \in \boldsymbol{R}^K$, and a unique function on $\boldsymbol{N}$ into $\boldsymbol{R}^K$, say $x = g(y)$, having the following property: for each $y \in \boldsymbol{N}$, $C(x, y)$, regarded as a function of x, has a local minimum at $x = g(y)$. The Jacobian of $g(y)$ at $y = y_0$ is given by

$$G(y_0) = - \left(\frac{\partial^2 C(x, y)}{\partial x^2} \right)^{-1} \left(\frac{\partial^2 C(x, y)}{\partial x \partial y} \right) \Bigg|_{\substack{x=x_0 \\ y=y_0}} \tag{3.6.4}$$

Proof. Consider the system of equations

$$\frac{\partial C(x, y)}{\partial x} = 0. \tag{3.6.5}$$

By condition (2), equality holds at (x_0, y_0). By condition (3), the Jacobian of the left side of (3.6.5) is nonsingular at (x_0, y_0). Therefore, by the implicit function theorem [Apostol, 1974, p. 374], there exists a unique function $g(y)$, continuously differentiable on some neighborhood of y_0, say $\boldsymbol{N}_1 \subseteq \boldsymbol{R}^K$, such that (3.6.5) is satisfied for all $\{y \in \boldsymbol{N}_1, x = g(y)\}$. The function $g(y)$ will define a local minimizer of $C(x, y)$ if $\partial^2 C(x, y)/\partial x^2$ is positive definite at $x = g(y)$. Since $\partial^2 C(x, y)/\partial x^2$ is positive definite at (x_0, y_0) and since it is continuous with respect to both x and y, it is positive definite in a neighborhood of (x_0, y_0), say $\boldsymbol{N}_2 \subseteq \boldsymbol{R}^{K+M}$. Since $g(y)$ is continuous, there exists an open set $\boldsymbol{N} \subseteq \boldsymbol{N}_1$ such that $\boldsymbol{N} \times g(\boldsymbol{N}) \subseteq \boldsymbol{N}_2$. Therefore, $x = g(y)$ defines a local minimum of $C(x, y)$ for all $y \in \boldsymbol{N}$.

To derive formula (3.6.4), expand (3.6.5) as

$$\frac{\partial C(x, y)}{\partial x} - \frac{\partial C(x, y)}{\partial x} \Bigg|_{\substack{x=x_0 \\ y=y_0}} = \frac{\partial^2 C(x, y)}{\partial x^2} \Bigg|_{\substack{x=x_0 \\ y=y_0}} (x - x_0)$$
$$+ \frac{\partial^2 C(x, y)}{\partial x \partial y} \Bigg|_{\substack{x=x_0 \\ y=y_0}} (y - y_0) + \epsilon,$$

where ϵ is $o(\|x - x_0\| + \|y - y_0\|)$. The second term in the left side is identically zero. On the manifold $x = g(y)$, the first term is also zero. Expressing $(x - x_0)$ in terms of $(y - y_0)$ and ϵ and taking the limit as $y \to y_0$ yields (3.6.4). ∎

The function $C(\psi, \xi)$ defined by (3.6.3) satisfies the conditions of Lemma 3.2, provided both $s(\theta)$ and $\Sigma(\theta)$ are twice continuously differentiable and $\Sigma(\theta)$ is positive definite. Its first derivative is easily shown to be zero at $(\theta, s(\theta))$, and its second derivative at this point is given by (Prob. 3.17)

$$\frac{\partial^2 C(\psi, \xi)}{\partial \psi^2} \Bigg|_{\substack{\psi=\theta \\ \xi=s(\theta)}} = S^T(\theta) \Sigma^{-1}(\theta) S(\theta). \tag{3.6.6}$$

Therefore, $C(\psi, \xi)$ can be used to construct the following estimate. Let $\hat{\theta}_0$ be an arbitrary point in Θ. Let

$$\hat{\theta}_N = \begin{cases} g(s_N(Y_N)), & \text{if } s_N(Y_N) \in \mathbf{N} \\ \hat{\theta}_{N-1}, & \text{otherwise,} \end{cases} \qquad (3.6.7)$$

where $g(\xi)$ is the local minimizer constructed in Lemma 3.2. Since $s_N(Y_N)$ is consistent, the probability of the first case in (3.6.7) approaches 1 as $N \to \infty$, so the "memory" in this definition does not affect the asymptotic properties of $\hat{\theta}_N$. Since $g(\xi)$ is continuous, $\hat{\theta}_N$ is weakly or strongly consistent, depending on the mode of consistency of $s_N(Y_N)$. Since $g(\xi)$ is continuously differentiable, its normalized asymptotic covariance is given by $G(s(\theta))\Sigma(\theta)G^T(s(\theta))$ (under the assumptions of Theorem 3.16). The Jacobian $G(s(\theta))$ is given by (Prob. 3.17)

$$G(s(\theta)) = (S^T(\theta)\Sigma^{-1}(\theta)S(\theta))^{-1}S^T(\theta)\Sigma^{-1}(\theta). \qquad (3.6.8)$$

Therefore,

$$G(s(\theta))\Sigma(\theta)G^T(s(\theta)) = (S^T(\theta)\Sigma^{-1}(\theta)S(\theta))^{-1}.$$

In conclusion, the estimate $\hat{\theta}_N$ achieves the lowest possible asymptotic covariance among all consistent estimates depending on $s_N(Y_N)$ and satisfying the assumptions of Theorem 3.16.

Computation of the local minimum of (3.6.3) can be quite difficult in practice, because of the need to compute the matrix $\Sigma(\psi)$ (possibly many times during the minimization process). Suppose, however, that we have a *consistent estimate* of this matrix, say $\hat{\Sigma}_N(Y_N)$. Such estimates can be obtained in many examples of interest, and, as we will show next, they facilitate a convenient alternative to (3.6.3).

For any symmetric matrix A, let $\text{vec}(A)$ denote a row vector consisting of the entries of A on and above the main diagonal (the independent entries of A). Let $\eta = [\xi^T, \text{vec}(\hat{\Sigma}_N(Y_N))]^T$ and $\eta_0 = [s^T(\theta), \text{vec}(\Sigma(\theta))]^T$ [both vectors have dimension $K(K+3)/2$]. Consider the function

$$\hat{C}(\psi, \eta) = \frac{1}{2}(\xi - s(\psi))^T \hat{\Sigma}_N^{-1}(Y_N)(\xi - s(\psi)). \qquad (3.6.9)$$

This function satisfies the assumptions of Lemma 3.2. Moreover, the second partial derivatives of $\hat{C}(\psi, \eta)$ at the point (θ, η_0) are equal to those of the function in (3.6.3) (Prob. 3.18). Therefore, the sequence of local minima of $\hat{C}(\psi, [s_N(Y_N)^T, \text{vec}(\hat{\Sigma}_N(Y_N))]^T)$ has the same asymptotic properties as (3.6.7): it is consistent and asymptotically normal and its normalized asymptotic covariance achieves the lower bound. The computational advantage of minimizing (3.6.9) over minimizing (3.6.3) is clear: the matrix $\hat{\Sigma}_N(Y_N)$ needs to be computed and inverted only once. Therefore, in cases where the consistent estimate $\hat{\Sigma}_N(Y_N)$ is available, the estimate based on (3.6.9) is preferable to the one based on (3.6.3).

The remainder of this section is devoted to the problem of the mean-square consistency of $g(s_N(Y_N))$. While weak or strong consistency depends only on the local behavior of g [i.e., its continuity at $s(\theta)$], mean-square consistency depends on its global behavior. As a result, mean-square consistency is considerably more difficult to establish than the two other forms of consistency. The only exception is when the parameter set Θ is bounded. As was already mentioned, weak consistency implies mean-square consistency in this case. In particular, the estimates corresponding to the local minima of (3.6.3) and (3.6.9) are mean-square consistent. When Θ is not bounded, further conditions on $s_N(Y_N)$ and $g(s_N(Y_N))$ need to be imposed in order to guarantee mean-square consistency. Some of these conditions require S to be convex, so we assume that $g(s)$ can be smoothly extended to an open convex set including S. In the sequel we will use the notation $g(s)$ for the extended function. Two sufficient conditions for mean-square consistency are given by the next two theorems.

Theorem 3.17. If $s_N(Y_N)$ is a mean-square consistent estimate of $s(\theta)$ and if $g(s)$ satisfies the Lipschitz condition

$$\|g(s_2) - g(s_1)\| < A\|s_2 - s_1\|, \quad \forall s_1, s_2 \in S,$$

for some positive constant A, then $g(s_N(Y_N))$ is a mean-square consistent estimate of θ.

Proof. It follows from the Lipschitz condition that

$$E_\theta\{\|g(s_N(Y_N)) - g(s(\theta))\|^2\} < A^2 E_\theta\{\|s_N(Y_N) - s(\theta)\|^2\}.$$

The right side converges to zero as $N \to \infty$; hence so does the left side. ∎

The Lipschitz condition is a strong one and is not often satisfied. The next theorem relaxes this condition, at the expense of a more stringent condition on the sequence $s_N(Y_N)$.

Theorem 3.18. Assume that $g(s)$ is differentiable and let $G(s) = \partial g(s)/\partial s$ be its Jacobian. Assume there exist $\alpha > 1$ and $\beta = \alpha/(\alpha - 1) > 1$ such that

(1) $$\lim_{N \to \infty} E_\theta |s_{N,k}(Y_N) - s_k(\theta)|^{2\alpha} = 0, \quad 1 \le k \le K$$

(2) $$E_\theta[\overline{G}_{k\ell}(s_N(Y_N))]^{2\beta} < A < \infty, \quad 1 \le \ell \le K, 1 \le k \le M,$$

where $\overline{G}_{k\ell}(s_N(Y_N)) = \sup_{\gamma \in [0,1]} |G_{k\ell}(\gamma s_N(Y_N) + (1 - \gamma)s(\theta))|.$

Then $g(s_N(Y_N))$ is mean-square consistent.

Proof. Since $g(s)$ is differentiable and S is convex, we have, by the mean-value theorem for functions from $\mathbf{R}^K$ to $\mathbf{R}$,

$$g_m(s_N(Y_N)) - g_m(s(\theta)) = \sum_k G_{mk}(s_*)(s_{N,k}(Y_N) - s_k(\theta))$$

for some s_* on the line joining $s_N(Y_N)$ and $s(\theta)$; that is, for some $0 \le \gamma \le 1$,

$$s_* = \gamma s_N(Y_N) + (1 - \gamma)s(\theta).$$

Hence

$$|g_m(s_N(Y_N)) - g_m(s(\theta))|^2$$
$$= \sum_k \sum_\ell (s_{N,k}(Y_N) - s_k(\theta))(s_{N,\ell}(Y_N) - s_\ell(\theta))G_{mk}(s_*)G_{m\ell}(s_*)$$
$$\leq \sum_k \sum_\ell |(s_{N,k}(Y_N) - s_k(\theta))(s_{N,\ell}(Y_N) - s_\ell(\theta))|\overline{G}_{mk}(s_N(Y_N))\overline{G}_{m\ell}(s_N(Y_N)).$$

Taking expectations and using Hölder's inequality [A16] twice on the right side, once with (α, β) and once with $(2,2)$, yields

$$E_\theta \|g(s_N(Y_N)) - g(s(\theta))\|^2$$
$$\leq \sum_k \sum_\ell \sum_m \left[E_\theta |(s_{N,k}(Y_N) - s_k(\theta))|^{2\alpha}\right]^{1/2\alpha} \left[E_\theta |(s_{N,\ell}(Y_N) - s_\ell(\theta))|^{2\alpha}\right]^{1/2\alpha}$$
$$\left[E_\theta (\overline{G}_{mk}(s_N(Y_N)))^{2\beta}\right]^{1/2\beta} \left[E_\theta (\overline{G}_{m\ell}(s_N(Y_N)))^{2\beta}\right]^{1/2\beta}.$$

Finally, taking the limit as $N \to \infty$ and using (1) and (2) yields zero on the right side. ∎

Theorem 3.18 offers a useful trade-off between the moments of the statistic $s_N(Y_N)$ and those of $\overline{G}$. For example, if the moments of $|s_N(Y_N) - s(\theta)|$ of all orders tend to zero in the limit, then it is sufficient that (2) above hold for some $\beta = 1 + \epsilon$. On the other extreme, if $|G_{k\ell}(s)|$ is bounded on $\mathbf{S}$, then mean-square consistency of $s_N(Y_N)$ is sufficient. This is just a special case of Theorem 3.17, since a bounded Jacobian implies the Lipschitz condition.

Theorem 3.18 can be strengthened subject to some additional assumptions, as expressed by the following theorem.

Theorem 3.19. Assume, in addition to the conditions given in Theorem 3.18, that $g(s)$ is twice differentiable and let $H_i(s) = \partial^2 g_i(s)/\partial s^2$ be the Hessian of $g_i(s)$. Let $d(N)$ be as in Theorem 3.16. Assume that there exist $\alpha > 1$ and $\beta = \alpha/(\alpha-1) > 1$ such that

(3) $\qquad \lim_{N\to\infty} E_\theta |d^2(N)(s_{N,k}(Y_N) - s_k(\theta))^4|^\alpha = 0, \quad 1 \leq k \leq K$

(4) $\qquad E_\theta [\overline{H}_{i,k\ell}(s_N(Y_N))]^{2\beta} < A < \infty, \quad 1 \leq k,\ell \leq K, 1 \leq i \leq M,$

$\qquad$ where $\quad \overline{H}_{i,k\ell}(s_N(Y_N)) = \sup_{\gamma\in[0,1]} |H_{i,k\ell}(\gamma s_N(Y_N) + (1-\gamma)s(\theta))|.$

Then

$$\lim_{N\to\infty} d^2(N)E_\theta \|g(s_N(Y_N)) - g(s(\theta)) - G(s(\theta))(s_N(Y_N) - s(\theta))\|^2 = 0. \quad (3.6.10)$$

Proof. Note that assumption (3) is stronger than (1), while (4) is in addition to (2). The proof proceeds similarly to Theorem 3.18. By Taylor's theorem for functions from $\mathbf{R}^K$ to $\mathbf{R}$, we have

$$g_i(s_N(Y_N)) = g_i(s(\theta)) + G_i(s(\theta))(s_N(Y_N) - s(\theta))$$
$$+ 0.5(s_N(Y_N) - s(\theta))^T H_i(s_*)(s_N(Y_N) - s(\theta)),$$

where

$$s_* = \gamma s_N(Y_N) + (1 - \gamma)s(\theta)), \quad 0 \le \gamma \le 1.$$

Therefore,

$$|g_i(s_N(Y_N)) - g_i(s(\theta)) - G_i(s(\theta))(s_N(Y_N) - s(\theta))|^2 \le 0.25 \sum_k \sum_\ell \sum_m \sum_n$$
$$|(s_{N,k}(Y_N) - s_k(\theta))(s_{N,\ell}(Y_N) - s_\ell(\theta))(s_{N,m}(Y_N) - s_m(\theta))(s_{N,n}(Y_N) - s_n(\theta))|$$
$$\overline{H}_{i,k\ell}(s_N(Y_N))\overline{H}_{i,mn}(s_N(Y_N)).$$

Multiplying by $d^2(N)$, taking expectations, and using Hölder's inequality twice as in the proof of Theorem 3.18 yields

$$d^2(N)E_\theta|g_i(s_N) - g_i(s(\theta)) - G_i(s(\theta))(s_N(Y_N) - s(\theta))|^2 \le 0.25 \sum_k \sum_\ell \sum_m \sum_n$$
$$\left[E_\theta|d^2(N)(s_{N,k}(Y_N) - s_k(\theta))^4|^\alpha\right]^{1/4\alpha} \left[E_\theta|d^2(N)(s_{N,\ell}(Y_N) - s_\ell(\theta))^4|^\alpha\right]^{1/4\alpha}$$
$$\left[E_\theta|d^2(N)(s_{N,m}(Y_N) - s_m(\theta))^4|^\alpha\right]^{1/4\alpha} \left[E_\theta|d^2(N)(s_{N,n}(Y_N) - s_n(\theta))^4|^\alpha\right]^{1/4\alpha}$$
$$\left[E_\theta(\overline{H}_{i,k\ell}(s_N(Y_N)))^{2\beta}\right]^{1/2\beta} \left[E_\theta(\overline{H}_{i,mn}(s_N(Y_N)))^{2\beta}\right]^{1/2\beta}.$$

Finally, taking the limit as $N \to \infty$ and using (3) and (4) yields zero on the right side. Since this is true for all $1 \le i \le K$, it implies (3.6.10). ∎

As before, there is a trade-off between the limiting moments of $s_N(Y_N)$ and those of $\overline{H}$. If the moments of $d(N)|s_N(Y_N) - s(\theta)|^2$ of all orders tend to zero in the limit, then it is sufficient that (4) above hold for some $\beta = 1 + \epsilon$. If $|H_{i,k\ell}(s)|$ is bounded on S, then it is sufficient that (3) hold for $\alpha = 1$.

Finally, make the additional assumption
(5) The statistic $s_N(Y_N)$ satisfies

$$\lim_{N \to \infty} d^2(N)E_\theta(s_N(Y_N) - s(\theta))(s_N(Y_N) - s(\theta))^T = \Sigma(\theta),$$

where $\Sigma(\theta)$ is the normalized asymptotic covariance of $s_N(Y_N)$.
Then $G\Sigma(\theta)G^T$ is the actual limiting covariance of $d(N)(g(s_N(Y_N)) - g(s(\theta)))$, as proved below.

Theorem 3.20. Under assumptions (1) to (5), we have

$$\lim_{N \to \infty} d^2(N)E_\theta(g(s_N(Y_N)) - g(s(\theta)))(g(s_N(Y_N)) - g(s(\theta)))^T$$
$$= G(s(\theta))\Sigma(\theta)G^T(s(\theta)). \tag{3.6.11}$$

Proof. Denote temporarily

$$X = d(N)(g(s_N(Y_N)) - g(s(\theta))), \quad Z = d(N)G(s(\theta))(s_N(Y_N) - s(\theta)).$$

Then

$$|X_k X_\ell - Z_k Z_\ell| = |(X_k - Z_k)X_\ell + (X_\ell - Z_\ell)Z_k|$$
$$\leq |X_k - Z_k||X_\ell| + |X_\ell - Z_\ell||Z_k|.$$

Taking expectations and using the Cauchy-Schwarz inequality gives

$$E_\theta |X_k X_\ell - Z_k Z_\ell| \leq \left[E_\theta |X_k - Z_k|^2\right]^{1/2} \left[E_\theta |X_\ell|^2\right]^{1/2}$$
$$+ \left[E_\theta |X_\ell - Z_\ell|^2\right]^{1/2} \left[E_\theta |Z_k|^2\right]^{1/2} \tag{3.6.12}$$

As N tends to infinity, we have:
  (a) The terms $E_\theta |X_k - Z_k|^2$ and $E_\theta |X_\ell - Z_\ell|^2$ converge to zero by (3.6.10).
  (b) The term $E_\theta |Z_k|^2$ converges to a finite constant by assumption (5).
  (c) The term $E_\theta |X_\ell|^2$ converges to the same value as the limit of $E_\theta |Z_\ell|^2$ (i.e., to a finite constant) due to (3.6.10).
In summary, the limit of both sides of (3.6.12) is zero; hence

$$\lim_{N \to \infty} d^2(N)E_\theta(g_k(s_N(Y_N)) - g_k(s(\theta)))(g_\ell(s_N(Y_N)) - g_\ell(s(\theta)))$$
$$= G_k(s(\theta))\Sigma(\theta)G_\ell(s(\theta)).$$

This is the same as (3.6.11).   ∎

3.7. LEAST-SQUARES ESTIMATION

An important class of estimation problems concerns the following model. Let θ and Y be the parameter and measurement vectors of dimensions M and N, respectively. Suppose that there exist L known functions $\{h_\ell(\theta, Y), 1 \leq \ell \leq L\}$, called the *model functions*, such that *ideally*

$$h_\ell(\theta, Y) = 0, \quad 1 \leq \ell \leq L.$$

However, suppose that the measurements are known to be subject to errors, or the model functions $h_\ell(\theta, Y)$ are only approximate, or both. We can lump all errors of this kind in an L-dimensional vector e, the ℓth component of which represents the error in the ℓth model function; that is,

$$h_\ell(\theta, Y) = e_\ell, \quad 1 \leq \ell \leq L. \tag{3.7.1}$$

The *least-squares estimate* of θ is defined as the global minimizer of $\|e\|^2$, that is, the global minimizer of

$$C(\theta) = \sum_{\ell=1}^{L} h_\ell^2(\theta, Y). \tag{3.7.2}$$

A slightly more general approach is to minimize a weighted norm of the error vector. Let W be an $L \times L$ positive definite matrix and define

$$C_W(\theta) = \boldsymbol{e}^T W \boldsymbol{e} = \sum_{\ell=1}^{L} \sum_{m=1}^{L} W_{\ell,m} h_\ell(\theta, Y) h_m(\theta, Y). \qquad (3.7.3)$$

The *weighted least-squares estimate* of θ is defined as the global minimizer of $C_W(\theta)$ and is dependent on the choice of the weight matrix.

Both (3.7.2) and (3.7.3) generally lead to nonlinear minimization problems. In most cases the minimization problem has to be solved numerically. The literature on nonlinear optimization is extensive (Fletcher [1987] is an excellent example), but is outside the scope of this book. We limit ourselves to a special case in which the problem reduces to the solution of linear equations. Consider the following model, which is a special case of (3.7.1)

$$\boldsymbol{h}(Y) - H(Y)\theta = \boldsymbol{e}, \qquad (3.7.4)$$

where $\boldsymbol{h}(Y)$ is an L-dimensional vector and $H(Y)$ an $L \times M$ matrix, both dependent on Y in general. We will assume that $L \geq M$ and that $H(Y)$ is full rank, since otherwise the solution for θ is not unique. The cost function (3.7.3) takes the form

$$C_W(\theta) = (\boldsymbol{h}(Y) - H(Y)\theta)^T W (\boldsymbol{h}(Y) - H(Y)\theta). \qquad (3.7.5)$$

The global minimizer of (3.7.5) can be easily found by differentiating and equating the Jacobian to zero, yielding

$$[H(Y)^T W H(Y)]\theta = H(Y)^T W \boldsymbol{h}(Y). \qquad (3.7.6)$$

This is known as the *normal equation* for the cost function (3.7.5). Its explicit solution is

$$\hat{\theta} = [H(Y)^T W H(Y)]^{-1} H(Y)^T W \boldsymbol{h}(Y). \qquad (3.7.7)$$

This is called the *weighted linear least-squares estimate* of θ. In the special case where W is the identity matrix, we get

$$\hat{\theta} = [H(Y)^T H(Y)]^{-1} H(Y)^T \boldsymbol{h}(Y). \qquad (3.7.8)$$

This is called the *linear least-squares estimate* of θ. Least-squares estimates are seldom computed by using the explicit formulas, because these are known to be numerically ill conditioned. Many good numerical techniques are available for computing least-square estimates, but these are outside the scope of this book (see, e.g., Golub and Van Loan [1983]).

It should be emphasized that linear least-squares estimates are not based on any probabilistic model. However, in some special cases they are equivalent to maximum likelihood estimates under the Gaussian assumption (see Prob. 3.21).

3.8. MAXIMUM ENTROPY ESTIMATION

Suppose you are asked to guess the head and tail probabilities of a given coin. Given your experience in coin tossing, and in the absence of any other information about the specific coin, you will probably say that the probabilities are $1/2$ for each outcome. Similarly, for a die-throwing experiment, most people will guess equal probabilities of $1/6$ for each outcome, if nothing else is known about the specific die. A fair coin or a fair die is "the most random," in the sense that the outcome is the most difficult to predict before the experiment is done. Now assume, for the sake of illustration, that you are told that the given die has a probability 0.5 of showing 1. Again, most people would guess probabilities of 0.1 for the outcomes 2 through 6. Such a die is the "fairest" among all die satisfying the given constraint, although it is not a fair die.

The *entropy* of a random variable (or of a distribution) is a quantitative measure of the randomness of the corresponding experiment. The more random the experiment, the higher the entropy. The mathematical definition of the entropy of a discrete random variable with probability mass function $\{f_i, 1 \le i \le n\}$ is

$$H = -E \log f = - \sum_{i=1}^{n} f_i \log f_i. \tag{3.8.1}$$

In the coin-tossing experiment, the entropy is

$$H = -p \log p - (1-p) \log(1-p),$$

where p is the probability of head. It is easy to verify that the entropy is maximized for $p = 0.5$, that is, for a fair coin. Similarly, the entropy of a die-throwing experiment is maximized when $f_i = 1/6, i = 1, \ldots, 6$, and the entropy of a die constrained to $f_1 = 0.5$ is maximized when $f_i = 0.1, i = 2, \ldots, 6$. Thus, the definition of entropy given in (3.8.1) conforms to our notion of randomness, and, consequently, to our notion of fairness. The maximum entropy principle can be stated as:

- Given only partial information about the properties of a distribution (or no information at all), choose the distribution whose entropy is the highest among all distributions satisfying the information (or constraints) you are given.

The concept of entropy is extended to continuous random variables by defining the *differential entropy*

$$H = -E \log f = - \int f(y) \log f(y) dy, \tag{3.8.2}$$

where $f(y)$ is the probability density function of the random variable Y. If the distribution depends on the parameter vector θ, say $f_\theta(y)$, the differential entropy becomes a function of θ, say $H(\theta)$.

The concept of entropy can be useful for estimation purposes if we agree to accept the maximum entropy principle. The following is a typical situation.

Suppose the parameter θ is high dimensional and can be partitioned as $[\theta_1, \theta_2]$, where θ_1 is relatively low dimensional. Suppose we are given measurements that enable us to estimate θ_1, but the estimation of θ_2 is difficult or impossible (e.g., because the dimension θ_2 is too high to be practical, or the measurements contain no or little information about it, etc.). We can then estimate θ_1 by, say, $\hat{\theta}_1$ (which is a function of the measurements) and choose θ_2 to maximize $H(\hat{\theta}_1, \theta_2)$.

Maximum entropy estimation will not be used in this book, except for a brief mention in Chapter 6.

3.9. MODEL ORDER SELECTION

The estimation problems discussed so far involve a parameter space of a *fixed* dimension M. This means that all competing models $f_\theta(Y)$ have the same number of free parameters, and the maximum likelihood is performed with respect to a fixed number of parameters. It often happens in practice that we are not sure of the actual number of parameters in the model. For example, in the linear model (3.7.4), we may be interested in fitting vectors θ of different dimensions (while properly adjusting the dimension of $H(Y)$) and exploring the effect on the residual vector e. The problem of deciding among competing models of different dimensions is called the *model order selection* problem. Many approaches have been suggested to this problem. Here we describe two approaches, which have gained popularity in random process estimation and which will be used in later chapters of this book. The first is due to Akaike and is known as the *Akaike information criterion* (AIC) [Akaike, 1974]. The second is due to Rissanen and is known as the *minimum description length* (MDL) [Rissanen, 1978].

Suppose that, for each $M_{\min} \leq M \leq M_{\max}$, we have a parameter space $\Theta(M)$ of dimension M. For each such space, we have a family of densities $f_{\theta(M)}(Y)$. The spaces are assumed to be "nested," in the following sense: to each $\theta(M)$ there corresponds a point $\theta(M+1) \in \Theta(M+1)$ such that $f_{\theta(M)}(Y)$ and $f_{\theta(M+1)}(Y)$ are identical. Our aim is, given an observation Y, to estimate the dimension M and the value of $\theta(M)$ governing the distribution of Y.

A naive attempt to solve the problem would be to let $\hat{\theta}(M)$ be the maximizer of $f_{\theta(M)}(Y)$ for each M and then choose M such that $f_{\hat{\theta}(M)}(Y)$ is maximized over $M_{\min} \leq M \leq M_{\max}$. Unfortunately, this attempt will fail. The reason is that, due to the nesting of the parameter spaces, $f_{\hat{\theta}(M)}(Y)$ will be a monotonically increasing (or at least nondecreasing) function of M, so it will reach its maximum at $M_{\max}$. The "solution" of taking the largest possible parameter space is trivial and uninteresting, since it is tantamount to arbitrarily deciding on the value of M.

Akaike's information criterion is based on the notion of *Kullback-Leibler information* (or *discrepancy measure*) [Kullback, 1959]. This is defined as

$$\mathrm{KL}(\theta, \psi) = E_\theta \log \left\{ \frac{f_\theta(Y)}{f_\psi(Y)} \right\} = E_\theta(\log f_\theta(Y)) - E_\theta(\log f_\psi(Y)). \qquad (3.9.1)$$

It follows from the discussion in Sec. 3.5 that $\mathrm{KL}(\theta, \psi)$ is zero for $\psi = \theta$ and is non-negative for all $\psi \neq \theta$. The Kullback-Leibler information measures the discrepancy between $f_\psi(Y)$ and $f_\theta(Y)$. Akaike's approach was to select the parameter vector so as to minimize this discrepancy. Since the term $E_\theta(\log f_\theta(Y))$ is fixed, this is equivalent to maximizing $E_\theta(\log f_\psi(Y))$. For i.i.d. measurements and under the regularity conditions specified in Sec. 3.5, $E_\theta(\log f_\psi(Y))$ can be consistently estimated by $N^{-1} \sum_{n=1}^{N} \log f_\psi(y_n)$, so the latter can be maximized instead. Thus, as long as the dimension of the parameter space is fixed, minimization of the Kullback-Leibler information is equivalent to maximum likelihood.

Akaike's main observation was that $N^{-1} \sum_{n=1}^{N} \log f_{\hat\theta}(y_n)$ is a biased estimate of $\log f_\theta(Y)$, and the bias is approximately M/N. The bias expression was derived using a small-error assumption and Taylor series approximations. Akaike therefore suggested the use of an *unbiased estimate* of the minimum Kullback-Leibler information, which is given [for the parameter space $\Theta(M)$] by $-N^{-1} \sum_{n=1}^{N} \log f_{\hat\theta(M)}(y_n) + N^{-1}M$. He finally multiplied this expression by $2N$ (which is independent of M) and defined

$$\mathrm{AIC}(M) = -2 \log f_{\hat\theta(M)}(Y) + 2M. \qquad (3.9.2)$$

The AIC estimate is obtained by finding the ML estimate $\hat\theta(M)$ for each $M_{\min} \leq M \leq M_{\max}$ and choosing M and $\hat\theta(M)$ that minimize $\mathrm{AIC}(M)$.

When increasing M, the term $-2 \log f_{\hat\theta(M)}(Y)$ will usually decrease monotonically. However, the *penalty term* $2M$ will increase. At some point the decrease in the first term will start being negligible, while the penalty term will continue to increase linearly. Therefore, the AIC is expected to reach a global minimum. In principle, we can start with some $M_{\min}$ that represents the smallest dimension of interest for the given problem and increase M ad infinitum. In practice, the search may be stopped when the AIC starts increasing at an approximately linear rate.

We remark that, despite the fact that the above argument depends on the i.i.d. assumption, the AIC has been successfully applied to non-i.i.d. models. We will encounter such applications in Chapters 6 and 7.

Around 1978, three independent investigators, using different approaches, derived similar modifications of Akaike's criterion. Schwarz has modified Akaike's arguments [Schwarz, 1978]. Akaike applied his principle to ARMA processes assuming prior (Bayesian) distribution of the model parameters [Akaike, 1978]. Rissanen advocated the principle of describing the observed data by the smallest number of binary digits [Rissanen, 1978]. The asymptotic form of all these criteria is the same and is given by

$$\mathrm{MDL}(M) = -2 \log f_{\hat\theta(M)}(Y) + M \log N. \qquad (3.9.3)$$

The notation MDL stands for "minimum description length," which is Rissanen's terminology. Akaike's notation [for the ARMA case, in which (3.9.3) assumes a slightly different form] is BIC (for "Bayesian information criterion"). The MDL criterion has been popularized in signal-processing applications in recent years, mainly by M. Wax and his collaborators (e.g., Wax and Kailath [1985]). In the statistics literature, the reference to Akaike's BIC criterion is more common.

Comparing (3.9.3) to (3.9.2), we see that the MDL uses a larger penalty term than the AIC, so the estimated MDL model order will usually be smaller than the estimated AIC order. The theoretical asymptotic properties of these methods will be described in Chapters 6 and 7 for AR and ARMA processes, respectively.

REFERENCES

Akaike, H., "A New Look at Statistical Model Identification," *IEEE Trans. Automatic Control*, AC-19, pp. 716–723, December 1974.

Akaike, H., "Time Series Analysis and Control Through Parametric Models," in *Applied Time Series Analysis*, D. F. Findley (ed.), Academic Press, New York, 1978.

Apostol, T. M., *Mathematical Analysis*, 2nd ed., Addison-Wesley, Reading, MA, 1974.

Bahadur, R. R., "On the Asymptotic Efficiency of Tests and Estimators," *Sankhya*, 22, pp. 229–252, 1960.

Barankin, E. W., "Locally Best Unbiased Estimates," *Ann. Math. Statist.*, 20, pp. 477-501, 1949.

Bhattacharyya, A., "On Some Analogues to the Amount of Information and Their Uses in Statistical Estimation," *Sankhya*, 8, pp. 1–14, 1946.

Billingsley, P., *Probability and Measure*, 2nd ed., Wiley, New York, 1986.

Chapman, D. G., and Robbins, H., "Minimum Variance Estimation without Regularity Assumptions," *Ann. Math. Statist.*, 22, pp. 581–586, 1951.

Chung, K. L., *A Course in Probability Theory*, Academic Press, New York, 1974.

Cramér, H., "A Contribution to the Theory of Statistical Estimation," *Aktuariestidskrift*, 29, pp. 458–463, 1946.

Feller, W., *An Introduction to Probability Theory and Its Applications, Volume I*, Wiley, New York, 1950.

Fletcher, R., *Practical Methods of Optimization*, 2nd ed., Wiley, New York, 1987.

Golub, G. H., and Van Loan, C. F., *Matrix Computations*, Johns Hopkins University Press, Baltimore, MD, 1983.

Kullback, S., *Information Theory and Statistics*, Wiley, New York, 1959.

Lehmann, E. L., *Theory of Point Estimation,* Wiley, New York, 1983.

Lehmann, E. L., *Testing Statistical Hypotheses,* 2nd ed., Wiley, New York, 1986.

Rao, C. R., "Information and the Accuracy Attainable in the Estimation of Statistical Parameters," *Bull. Calcutta Math. Soc.,* 37, pp. 81–91, 1945.

Rissanen, J., "Modeling by Shortest Data Description," *Automatica,* 14, pp. 465–471, 1978.

Schwarz, G., "Estimating the Dimension of a Model," *Ann. Statistics,* 6, No. 2, 1978.

Wax, M., and Kailath, T., "Detection of Signals by Information Theoretic Criteria," *IEEE Trans. Acoustics, Speech, Signal Processing,* ASSP-33, pp. 387–392, 1985.

Zacks, S., *The Theory of Statistical Inference,* Wiley, New York, 1971.

PROBLEMS

3.1. Completeness of families of distributions.
  (a) Let $\{F_\theta(Y)\}$ be the family of uniform distributions on the interval $[-0.5\theta, 0.5\theta]$, where $\theta > 0$. Prove that this family is not complete.
  (b) Investigate the completeness of the family of Laplace distributions with parameter $\theta > 0$. The p.d.f. of the Laplace distribution is given by $f_\theta(Y) = 0.5\theta \exp(-\theta|Y|)$.

3.2. Sufficient statistics.
  (a) Let $\{F_\theta(Y)\}$ be the family of uniform distributions on the interval $[0, \theta]$, where $\theta > 0$. Let $[Y_1, Y_2, \ldots, Y_N]^T$ be a vector of N independent measurements from $F_\theta(Y)$. Prove that $\max_{1 \leq n \leq N}\{Y_n\}$ is a sufficient statistic for θ.
  (b) Find a sufficient statistic for the Laplace distribution with parameter θ (case of N independent measurements).

3.3. Show the equality of the two expressions in (3.1.2) under the assumption (3.1.3).

3.4. Compute the Fisher information of N independent measurements from an exponential distribution with parameter θ. The p.d.f. is $f_\theta(Y) = \theta \exp(-\theta Y), Y \geq 0$.

3.5. Let Y be a Gaussian vector with mean vector $\boldsymbol{\mu}(\theta)$ depending on a parameter vector θ and covariance matrix Γ independent of θ. Prove that the Fisher information matrix of Y is given by

$$J(\theta) = \left(\frac{\partial \boldsymbol{\mu}(\theta)}{\partial \theta}\right)^T \Gamma^{-1} \left(\frac{\partial \boldsymbol{\mu}(\theta)}{\partial \theta}\right).$$

3.6. Let $\Gamma = \sigma^2 \mathbf{1}$ in the previous problem, but assume that σ^2 is an unknown parameter. Show that the information matrix of Y is given by

$$J(\theta) = \frac{1}{\sigma^2} \begin{bmatrix} \frac{N}{2\sigma^2} & 0 \\ 0 & \left(\frac{\partial \boldsymbol{\mu}(\theta)}{\partial \theta}\right)^T \left(\frac{\partial \boldsymbol{\mu}(\theta)}{\partial \theta}\right) \end{bmatrix}.$$

3.7. Refer to Example 3.3.
   (a) Show that $T_1(Y)$ is an unbiased estimate of θ_1.
   (b) Show that $T_2(Y)$ is a biased estimate of θ_2.
   (c) Suggest a modification of $T_2(Y)$ that yields an unbiased estimate of θ_2.

3.8. Where is assumption (1) used in the proof of Theorem 3.4?

3.9. Suppose the parameter vector θ is partitioned into two subvectors θ_1 and θ_2, and we are only interested in estimating θ_1. Consider two cases: (a) θ_2 is known; (b) θ_2 is unknown (in this case it is called a *nuisance parameter*). Prove that the CRB for unbiased estimates of θ_1 is never smaller in case (b) than in case (a) (thus, the presence of nuisance parameters usually spoils the accuracy of the parameters of interest). Under what conditions will the two be equal?

3.10. Find a nonestimable function $g(\theta)$ under the exponential distribution with parameter θ.

3.11. Show that $f_\theta(Y)$ in Example 3.3 has the form (3.3.7), hence it is an exponential family. Give another example of a probability density function such that the corresponding joint p.d.f. of N i.i.d. random variables has the form (3.3.7).

3.12. Assume we are given N i.i.d. measurements from the Cauchy probability density $f_\theta(Y) = [\pi(1 + (x - \theta)^2)]^{-1}$.
   (a) Compute the CRB for unbiased estimates of θ.
   (b) Consider the estimate $\hat{\theta} = N^{-1} \sum_{n=1}^{N} Y_n$. What is the variance of $\hat{\theta}$? Does it attain the CRB?

3.13. Show that the supremum in the right side of (3.3.10) is greater or equal to the CRB under an appropriate regularity condition, and specify this condition. Hint: Consider the special choice $\theta_m(h) = \theta + he_m$, where e_m is the mth M-dimensional unit vector. Then let h tend to zero.

3.14. Consider the Chapman-Robbins bound for scalar θ and N i.i.d. measurements, using the supremum as in (3.3.10). Prove that the bound for N measurements is less or equal to the bound for one measurement divided by N. Remark: Note that in the case of the CRB there is strict equality, but here we only have inequality.

3.15. Maximum likelihood in the Gaussian case.
   (a) Compute the variances of the ML estimates in Example 3.6, and compare them with the CRB. Hint: The Fisher information was computed in Example 3.4. However, since $\hat{\theta}_{2,\text{ML}}$ is biased, the CRB is *not* the inverse of the information.

(b) Repeat for the modified unbiased estimate developed in Prob. 3.7(c).

3.16. Consider the following estimation problem. We are given M vectors $\{\boldsymbol{h}_m(\theta), 1 \leq m \leq M\}$ and M real numbers $\{a_m, 1 \leq m \leq M\}$. Let the measured vector be given by

$$Y = \sum_{m=1}^{M} a_m \boldsymbol{h}_m(\theta) + U,$$

where U is a zero mean Gaussian vector with $EUU^T = \sigma_u^2 \mathbf{1}$. All vectors are N-dimensional. In concise notation,

$$Y = H(\theta)\boldsymbol{a} + U,$$

where $\boldsymbol{a}$ is the vector whose elements are $\{a_m\}$. The parameters $\{\sigma_u^2, \boldsymbol{a}, \theta\}$ are unknown. This is a mixed linear/nonlinear model, where the dependence on $\boldsymbol{a}$ is linear, while the dependence on θ is nonlinear. We can think of the columns of $H(\theta)$ as (sampled) signal waveforms and of the components of $\boldsymbol{a}$ as amplitudes of the corresponding waveforms. The measurement consists of the sum of the signals and a Gaussian white noise. The problem is to develop the maximum likelihood estimate of the waveform parameters, the amplitudes, and the noise variance.

(a) Write the log likelihood function and maximize it with respect to σ_u^2. Show that

$$\hat{\sigma}_u^2 = \frac{1}{N}[Y - H(\theta)\boldsymbol{a}]^T[Y - H(\theta)\boldsymbol{a}].$$

(b) After substituting $\hat{\sigma}_u^2$ in the log likelihood, maximize with respect to $\boldsymbol{a}$. Show that this gives

$$\hat{\boldsymbol{a}} = [H(\theta)^T H(\theta)]^{-1} H(\theta)^T Y.$$

(c) After substituting $\hat{\boldsymbol{a}}$ in the log likelihood, show that we are led to the maximization of the cost function

$$C(\theta) = Y^T H(\theta)^T [H(\theta)^T H(\theta)]^{-1} H(\theta)^T Y.$$

This maximization has to be carried out numerically in general.

3.17. Prove formulas (3.6.6) and (3.6.8).

3.18. Prove that the partial derivatives of first and second orders of the function in (3.6.9) at $\psi = \theta$, $\eta = \eta_0$ are equal to those of the function in (3.6.3).

3.19. The following special case of the method of moments is of great practical importance. Assume that the parameter vector θ and the statistic $s(\theta)$ satisfy a linear constraint

$$A(s)\theta = \boldsymbol{b}(s)$$

for some $L \times M$ matrix $A(s)$ and L-dimensional vector $\boldsymbol{b}(s)$, where $L \geq M$. Let W be an $L \times L$ positive definite matrix. Define the *weighted least-squares estimate* of θ, corresponding to the weight matrix W, as

$$\hat{\theta}_N = [A^T(s_N(Y_N))WA(s_N(Y_N))]^{-1}A^T(s_N(Y_N))W\boldsymbol{b}(s_N(Y_N)).$$

Let $D(s)$ be the $L \times K$ matrix whose ith column is

$$D_i = \frac{\partial \boldsymbol{b}}{\partial s_i} - \frac{\partial A}{\partial s_i} \theta \bigg|_{s=s(\theta)}$$

Using Theorem 3.16, prove that the normalized asymptotic variance of the weighted least-square estimate is given by

$$V = [A^T(s)WA(s)]^{-1}A^T(s)WD(s)\Sigma(\theta)D^T(s)WA(s)[A^T(s)WA(s)]^{-1}.$$

3.20. Using the method of the proof of Lemma 3.1, prove that the weight matrix

$$W = [D(s)\Sigma(\theta)D^T(s)]^{-1}$$

minimizes V in Prob. 3.19, and the minimum V is given by

$$V = \{A^T(s)[D(s)\Sigma(\theta)D^T(s)]^{-1}A(s)\}^{-1}$$

The weighted least-square estimate based on this W is not a valid estimate (explain why). Nevertheless, a valid estimate having the same asymptotic properties can be constructed by replacing $W(\theta)$ with a consistent estimate of it, say $\hat{W}$. This is similar to the replacement of $\Sigma(\theta)$ by $\Sigma_N(Y_N)$ in (3.6.9).

3.21. Consider a special case of the linear model (3.7.4), in which $\boldsymbol{h}(Y) = Y$, $H(Y) = H$ is independent of Y, and $\boldsymbol{e}$ is Gaussian with zero mean and covariance matrix Γ. Show that the maximum likelihood estimate of θ coincides with a certain weighted least-squares estimate, and find the corresponding weight matrix.

CHAPTER 4

Nonparametric Spectrum Estimation

4.1. INTRODUCTION

This chapter deals with the problem of estimating the power spectral density of a wide sense stationary regular process. For the spectral density to exist, we must assume that the process under consideration is purely indeterministic. Then the spectral density is the Fourier transform of the covariance sequence. The subjects of covariance estimation and spectrum estimation are closely related. We will therefore discuss the estimation of the covariances first and then turn our attention to the spectral density. We will present several types of spectrum estimates and analyze their statistical properties.

The analysis of some of the statistical properties of spectral estimates requires that the process be stationary not only in the wide sense, but in the moments of fourth order as well. When this requirement is satisfied, the fourth-order cumulants are independent of the time index and are given by [cf. (2.4.6)]

$$\kappa_{k,\ell,m} = E y_{t+k} y_{t+\ell} y_{t+m} y_t - r_{k-\ell} r_m - r_{k-m} r_\ell - r_k r_{\ell-m}. \qquad (4.1.1)$$

As long as we deal with estimation from a finite number of measurements, there is no need for further assumptions. However, in dealing with the asymptotic properties of sequential estimates, we will have to impose additional conditions. Specifically, some or all of the following conditions will be assumed to hold at various parts of the analysis.

Condition R0. The covariances of the process are absolutely summable; that is,

$$\sum_{n=-\infty}^{\infty} |r_n| \le R_0 < \infty. \qquad (4.1.2)$$

Condition R1. The covariances satisfy

$$\sum_{n=-\infty}^{\infty} |n|\,|r_n| \le R_1 < \infty. \tag{4.1.3}$$

Condition R1 obviously implies Condition R0, but not vice versa. While Condition R0 implies the continuity of $S(\omega)$, Condition R1 implies existence and continuity of $dS(\omega)/d\omega$.

Condition K0. The fourth-order cumulants of the process satisfy

$$\sum_{n=-\infty}^{\infty} |\kappa_{n+k,n,\ell}| \le K_0 < \infty. \tag{4.1.4}$$

Condition K1. The fourth-order cumulants satisfy

$$\sum_{n=-\infty}^{\infty} |n|\,|\kappa_{n+k,n,\ell}| \le K_1 < \infty. \tag{4.1.5}$$

Note that the bounds K_0 and K_1 should be uniform (independent of k and ℓ). Condition K1 obviously implies Condition K0, but not vice versa. Conditions K0 and K1 neither imply R0 and R1 nor are implied by them.

Of special interest to us is the class of linear processes defined by

$$y_t = \sum_{n=-\infty}^{\infty} \gamma_n w_{t-n}, \tag{4.1.6}$$

where $\{w_t\}$ are i.i.d. and the sequence $\{\gamma_n\}$ satisfies the following condition.

Condition G0.

$$\sum_{n=-\infty}^{\infty} |\gamma_n| \le G_0 < \infty. \tag{4.1.7}$$

Every linear process is purely indeterministic, subject only to the condition $Ew_t^2 = \sigma_w^2 < \infty$. Therefore, it yields the Wold representation

$$y_t = \sum_{n=0}^{\infty} \beta_n u_{t-n} \tag{4.1.8}$$

in addition to (4.1.6). Note, however, that the innovation sequence $\{u_t\}$ is not i.i.d., but only uncorrelated. The covariances of a linear process are given by

$$r_k = \sigma_w^2 \sum_{n=-\infty}^{\infty} \gamma_{n+k}\gamma_n = \sigma_u^2 \sum_{n=0}^{\infty} \beta_{n+k}\beta_n. \tag{4.1.9}$$

The fourth-order cumulants are given by

$$\kappa_{k,\ell,m} = \kappa_4(w) \sum_{t=-\infty}^{\infty} \gamma_t \gamma_{t+k} \gamma_{t+\ell} \gamma_{t+m}, \tag{4.1.10}$$

where $\kappa_4(w) = Ew_t^4 - 3(Ew_t^2)^2$ is the fourth-order cumulant of w_t, assumed to be finite. Equation (4.1.10) will be proved in Chapter 10.

It is easy to show (Prob. 4.1) that linear processes satisfy both conditions R0 and K0. A condition stronger than G0 is:

Condition G1.

$$\sum_{n=-\infty}^{\infty} |n|\,|\gamma_n| \leq G_1 < \infty. \qquad (4.1.11)$$

Condition G1 implies both R1 and K1 (Prob. 4.2).

4.2. ESTIMATION OF THE MEAN AND THE COVARIANCES

A stationary process must have a constant mean, which may or may not be equal to zero. If the mean μ is unknown, its standard estimate from a set of N consecutive measurements is

$$\hat{\mu} = \frac{1}{N} \sum_{t=0}^{N-1} y_t. \qquad (4.2.1)$$

This estimate is clearly unbiased. Its variance is given as follows [A17].

$$\text{var}\{\hat{\mu}\} = E(\hat{\mu})^2 - \mu^2 = \frac{1}{N^2} \sum_{t=0}^{N-1} \sum_{s=0}^{N-1} E(y_t y_s) - \mu^2 = \frac{1}{N^2} \sum_{t=0}^{N-1} \sum_{s=0}^{N-1} r_{t-s}$$

$$= \frac{1}{N} \sum_{n=-(N-1)}^{N-1} \left(1 - \frac{|n|}{N}\right) r_n.$$

If the process satisfies Condition R0, we get from [A10]

$$\lim_{N\to\infty} N\text{var}\{\hat{\mu}\} = \sum_{n=-\infty}^{\infty} r_n = S(0). \qquad (4.2.2)$$

Once the mean has been estimated, we can subtract it from the measurements to get the sequence $\{\tilde{y}_n = y_n - \hat{\mu}, 0 \leq n \leq N-1\}$. In the sequel, we will assume that the mean is known to be zero, so this subtraction is unnecessary. All the asymptotic results derived in this chapter can be shown to be unaffected by an unknown mean, but we will skip these details (see Probs. 4.6, 4.7, and 4.11 for some related results).

The two standard estimates of the covariances of a zero mean stationary process are

$$\tilde{r}_k = \tilde{r}_{-k} = \frac{1}{N-k} \sum_{t=0}^{N-k-1} y_{k+t} y_t, \quad \hat{r}_k = \hat{r}_{-k} = \frac{1}{N} \sum_{t=0}^{N-k-1} y_{k+t} y_t. \qquad (4.2.3)$$

The $\{\tilde{r}_k\}$ are called the *unbiased sample covariances*, while $\{\hat{r}_k\}$ are the *biased sample covariances*. Indeed, it is easy to check that the former are unbiased while the bias of the latter is $-kr_k/N$. The definitions in (4.2.3) hold for all $0 \le k \le N-1$. The biased sample covariances will turn out to be the more useful of the two definitions.

Next we wish to compute the variances and covariances of the sample covariances. Assume, without loss of generality, that $k \ge \ell \ge 0$. Then

$$\hat{r}_k = \frac{1}{N} \sum_{t=0}^{N-k-1} y_{t+k}y_t, \quad \hat{r}_\ell = \frac{1}{N} \sum_{s=0}^{N-\ell-1} y_{s+l}y_s.$$

Therefore,

$$E(\hat{r}_k\hat{r}_\ell) = \frac{1}{N^2} \sum_{t=0}^{N-k-1} \sum_{s=0}^{N-\ell-1} E(y_{t+k}y_t y_{s+\ell}y_s)$$

$$= \frac{1}{N^2} \sum_{t=0}^{N-k-1} \sum_{s=0}^{N-\ell-1} (r_k r_\ell + r_{t-s} r_{t-s+k-\ell} + r_{t-s+k} r_{t-s-\ell} + \kappa_{t-s+k,t-s,\ell})$$

$$= \left(1 - \frac{k}{N}\right)\left(1 - \frac{\ell}{N}\right) r_k r_\ell$$

$$+ \frac{1}{N^2} \sum_{t=0}^{N-k-1} \sum_{s=0}^{N-\ell-1} (r_{t-s} r_{t-s+k-\ell} + r_{t-s+k} r_{t-s-\ell} + \kappa_{t-s+k,t-s,\ell}).$$

So

$$\mathrm{cov}\{\hat{r}_k, \hat{r}_\ell\} = \frac{1}{N^2} \sum_{t=0}^{N-k-1} \sum_{s=0}^{N-\ell-1} (r_{t-s} r_{t-s+k-\ell} + r_{t-s+k} r_{t-s-\ell} + \kappa_{t-s+k,t-s,\ell}).$$

$$(4.2.4)$$

To derive the limit of this expression, we will need the following result.

Lemma 4.1.

(a) Let $k \ge \ell \ge 0$ and assume

$$\sum_{n=-\infty}^{\infty} |a_n| = A_0 < \infty, \quad \sum_{n=-\infty}^{\infty} a_n = A. \qquad (4.2.5)$$

Then

$$\lim_{N \to \infty} \frac{1}{N} \sum_{t=0}^{N-k-1} \sum_{s=0}^{N-\ell-1} a_{t-s} = A. \qquad (4.2.6)$$

(b) If

$$\sum_{n=-\infty}^{\infty} |n| |a_n| = A_1 < \infty, \qquad (4.2.7)$$

then

$$\frac{1}{N}\sum_{t=0}^{N-k-1}\sum_{s=0}^{N-\ell-1}a_{t-s}=\sum_{n=-\infty}^{-(k-\ell+1)}\left(1-\frac{\ell}{N}\right)a_n$$

$$+\sum_{n=-(k-\ell)}^{\infty}\left(1-\frac{k}{N}\right)a_n+O(N^{-1}),\qquad(4.2.8)$$

where the term $O(N^{-1})$ is uniformly bounded by $N^{-1}A_1$ (i.e., the bound is independent of k and ℓ).

Proof.
(a)

$$\sum_{t=0}^{N-k-1}\sum_{s=0}^{N-\ell-1}a_{t-s}=\sum_{n=-(N-\ell-1)}^{-(k-\ell+1)}(N-\ell+n)a_n$$

$$+(N-k)\sum_{n=-(k-\ell)}^{-1}a_n+\sum_{n=0}^{N-k-1}(N-k-n)a_n.$$

So

$$\frac{1}{N}\sum_{t=0}^{N-k-1}\sum_{s=0}^{N-\ell-1}a_{t-s}=\left(1-\frac{\ell}{N}\right)\sum_{n=-(N-\ell-1)}^{-(k-\ell+1)}\left(1-\frac{|n|}{N-\ell}\right)a_n$$

$$+\left(1-\frac{k}{N}\right)\sum_{n=-(k-\ell)}^{-1}a_n+\left(1-\frac{k}{N}\right)\sum_{n=0}^{N-k-1}\left(1-\frac{|n|}{N-k}\right)a_n.$$

Hence, by the Cesàro summability property [A10],

$$\lim_{N\to\infty}\frac{1}{N}\sum_{t=0}^{N-k-1}\sum_{s=0}^{N-\ell-1}a_{t-s}=\sum_{n=-\infty}^{-(k-\ell+1)}a_n+\sum_{n=-(k-\ell)}^{-1}a_n+\sum_{n=0}^{\infty}a_n=A.$$

(b)

$$\left|\sum_{n=-\infty}^{-(k-\ell+1)}(N-\ell)a_n+\sum_{n=-(k-\ell)}^{\infty}(N-k)a_n-\sum_{t=0}^{N-k-1}\sum_{s=0}^{N-\ell-1}a_{t-s}\right|=$$

$$\left|\sum_{n=-\infty}^{-(N-\ell)}(N-\ell)a_n+\sum_{n=-(N-\ell-1)}^{-(k-\ell+1)}|n|a_n+\sum_{n=0}^{N-k-1}|n|a_n+\sum_{n=N-k}^{\infty}(N-k)a_n\right|\le$$

$$\sum_{n=-\infty}^{-(N-\ell)}|n|\,|a_n|+\sum_{n=-(N-\ell-1)}^{-(k-\ell+1)}|n|\,|a_n|+\sum_{n=0}^{N-k-1}|n|\,|a_n|+\sum_{n=N-k}^{\infty}|n|\,|a_n|\le A_1.$$

Dividing by N gives (4.2.8).   ∎

Let us apply this lemma to

$$a_n(k, \ell) = r_n r_{n+k-\ell} + r_{n+k} r_{n-\ell} + \kappa_{n+k,n,\ell}. \tag{4.2.9}$$

Under conditions R0 and K0, we have

$$\sum_{n=-\infty}^{\infty} |a_n| \leq r_0 R_0 + r_0 R_0 + K_0.$$

Therefore, we get from (4.2.4) and Lemma 4.1(a):

Theorem 4.1. The asymptotic covariances of the sample covariances of a process satisfying R0 and K0 are given by

$$\lim_{N \to \infty} N\text{cov}\{\hat{r}_k, \hat{r}_\ell\} = \sum_{n=-\infty}^{\infty} (r_n r_{n+k-\ell} + r_{n+k} r_{n-\ell} + \kappa_{n+k,n,\ell}). \tag{4.2.10}$$

∎

Equation (4.2.10) is known as *Bartlett's formula*. Note that if the given process is Gaussian the fourth-order cumulants are identically zero, and then the third term in Bartlett's formula vanishes. Using Parseval's theorem and the Fourier relationship between the covariances and the spectral density, we get the following alternative form for Bartlett's formula (Prob. 4.3)

$$\lim_{N \to \infty} N\text{cov}\{\hat{r}_k, \hat{r}_\ell\} = \frac{1}{2\pi} \int_{-\pi}^{\pi} S^2(\omega)(e^{j(k-\ell)\omega} + e^{j(k+\ell)\omega})d\omega + \sum_{n=-\infty}^{\infty} \kappa_{n+k,n,\ell}$$

$$= \overline{r}_{k-\ell} + \overline{r}_{k+\ell} + \sum_{n=-\infty}^{\infty} \kappa_{n+k,n,\ell}, \tag{4.2.11}$$

where $\{\overline{r}_k\}$ is the inverse Fourier transform of $S^2(\omega)$.

If the given process is linear, conditions R0 and K0 hold automatically. Then it can be shown that (Prob. 4.4)

$$\sum_{n=-\infty}^{\infty} \kappa_{n+k,n,\ell} = \frac{\kappa_4(w)}{\sigma_w^4} r_k r_\ell. \tag{4.2.12}$$

Therefore, Bartlett's formula assumes the forms

$$\lim_{N \to \infty} N\text{cov}\{\hat{r}_k, \hat{r}_\ell\} = \sum_{n=-\infty}^{\infty} (r_n r_{n+k-\ell} + r_{n+k} r_{n-\ell}) + \frac{\kappa_4(w)}{\sigma_w^4} r_k r_\ell$$

$$= \frac{1}{2\pi} \int_{-\pi}^{\pi} S^2(\omega)(e^{j(k-\ell)\omega} + e^{j(k+\ell)\omega})d\omega + \frac{\kappa_4(w)}{\sigma_w^4} r_k r_\ell$$

$$= \overline{r}_{k-\ell} + \overline{r}_{k+\ell} + \frac{\kappa_4(w)}{\sigma_w^4} r_k r_\ell. \tag{4.2.13}$$

When conditions R1 and K1 hold, we can strengthen (4.2.10), using Lemma 4.1(b).

Theorem 4.2. Assume that conditions R1 and K1 hold for the given process. Then

$$\text{cov}\{\hat{r}_k, \hat{r}_\ell\} = \frac{1}{N} \sum_{n=-\infty}^{-(k-\ell+1)} \left(1 - \frac{\ell}{N}\right) a_n(k, \ell)$$

$$+ \frac{1}{N} \sum_{n=-(k-\ell)}^{\infty} \left(1 - \frac{k}{N}\right) a_n(k, \ell) + O(N^{-2}), \qquad (4.2.14)$$

where $a_n(k, \ell)$ is given in (4.2.9). The proof follows directly from Lemma 4.1 and Prob. 4.5.   ∎

Example 4.1. Let us compute the matrix $\Sigma(a)$ used in Example 3.7. The spectrum of an AR(1) process with unit innovation variance is given by

$$S(\omega) = \frac{1}{(1 + ae^{-j\omega})(1 + ae^{j\omega})}$$

so

$$S^2(\omega) = \frac{1}{(1 + 2ae^{-j\omega} + a^2 e^{-j2\omega})(1 + 2ae^{j\omega} + a^2 e^{j2\omega})}.$$

The first three terms of the inverse Fourier transform of $S^2(\omega)$ can be computed by solving the linear equations (2.12.7). Here these equations take the form

$$\begin{bmatrix} 1 & 2a & a^2 \\ 2a & 1+a^2 & 0 \\ a^2 & 2a & 1 \end{bmatrix} \begin{bmatrix} \bar{r}_0 \\ \bar{r}_1 \\ \bar{r}_2 \end{bmatrix} = \begin{bmatrix} 1 \\ 0 \\ 0 \end{bmatrix}.$$

The solution is

$$\begin{bmatrix} \bar{r}_0 \\ \bar{r}_1 \\ \bar{r}_2 \end{bmatrix} = \frac{1}{(1-a^2)^3} \begin{bmatrix} 1+a^2 \\ -2a \\ 3a^2 - a^4 \end{bmatrix}.$$

According to (4.2.11), when the process is Gaussian, the matrix $\Sigma(a)$ is given by

$$\Sigma(a) = \begin{bmatrix} 2\bar{r}_0 & 2\bar{r}_1 \\ 2\bar{r}_1 & \bar{r}_0 + \bar{r}_2 \end{bmatrix} = \frac{1}{(1-a^2)^3} \begin{bmatrix} 2(1+a^2) & -4a \\ -4a & 1 + 4a^2 - a^4 \end{bmatrix}.$$

This is the result used in Example 3.7.   ∎

We conclude this section by discussing the asymptotic normality of the sample covariances. We limit ourselves to the case of Gaussian processes, because their treatment is based on relatively elementary tools. The asymptotic normality has been established under more general conditions (see, e.g., Rosenblatt [1985, Ch. 3]).

Theorem 4.3. Let $\{y_t\}$ be a zero mean, purely indeterministic Gaussian process, whose covariance sequence is absolutely summable, and let K be a fixed positive integer. Then $\{N^{1/2}(\hat{r}_k - r_k), 0 \leq k \leq K - 1\}$ are jointly asymptotically normal with asymptotic mean zero and asymptotic covariance as given by Bartlett's formula (4.2.10).

Proof. By the Cramér-Wold theorem, a vector random variable is jointly asymptotically normal if and only if every linear combination of its components is asymptotically normal (see App. C). It is therefore sufficient to prove the asymptotic normality of $N^{1/2}X_N$, where $X_N = \sum_{k=0}^{K-1} a_k(\hat{r}_k - r_k)$, for arbitrary scalars $\{a_k\}$, not all zero. This will be done by showing that X_N and $d_N = N^{1/2}$ satisfy assumptions (1) to (3) in Lemma C.3. Asymptotic normality of X_N will then follow from the lemma.

(1) We have by Bartlett's formula (4.2.10) (with the third term omitted since the process is Gaussian)

$$\lim_{N \to \infty} d_N^2 \text{var}\{X_N\} = \sum_{k=0}^{K-1} \sum_{\ell=0}^{K-1} \sum_{n=-\infty}^{\infty} a_k a_\ell (r_n r_{n+k-\ell} + r_{n+k} r_{n-\ell}) < \infty.$$

(2) We have

$$\lim_{N \to \infty} d_N E(X_N) = - \lim_{N \to \infty} N^{1/2} N^{-1} \sum_{k=0}^{K-1} k a_k r_k = 0.$$

(3) This part will be shown using Theorem C.6, after some preparatory definitions. Let Y_N be the vector of $\{y_t, 0 \leq t \leq N - 1\}$ and $Z_{N,k}$ the $N \times N$ shift matrix $[Z_{N,k}]_{i,j} = \delta(i - j - k)$, where $-N + 1 \leq k \leq N - 1$. Then the sample covariance $\hat{r}_k$ can be expressed as

$$\hat{r}_k = N^{-1} Y_N^T Z_{N,k} Y_N = N^{-1} Y_N^T Z_{N,-k} Y_N = 0.5 N^{-1} Y_N^T (Z_{N,k} + Z_{N,-k}) Y_N.$$

Let $U_N = R_N^{-1/2} Y_N$, where $R_N^{1/2}$ is the lower triangular square root of R_N. Then U_N is a zero mean Gaussian vector, and $E U_N U_N^T = \mathbf{1}_N$, so the elements of U_N are i.i.d. We can write $\hat{r}_k$ in terms of U_N as

$$\hat{r}_k = 0.5 N^{-1} U_N^T R_N^{T/2} (Z_{N,k} + Z_{N,-k}) R_N^{1/2} U_N.$$

Therefore,

$$X_N = 0.5 N^{-1} U_N^T R_N^{T/2} A_N R_N^{1/2} U_N - \sum_{k=0}^{K-1} a_k r_k,$$

where A_N is the symmetric banded Toeplitz matrix constructed from the $\{a_k\}$ (but with $2a_0$ on the main diagonal). Let us also denote $C_N = R_N^{T/2} A_N R_N^{1/2}$, so

$$X_N - E(X_N) = 0.5 N^{-1} (U_N^T C_N U_N - \text{tr}\{C_N\}).$$

By Theorem C.6, $(X_N - E(X_N))/(\text{var}\{X_N\})^{1/2}$ is asymptotically normal $(0,1)$ if

$$\lim_{N \to \infty} \frac{\sum_{i=1}^{N} |\lambda_{N,i}|^3}{[\text{tr}\{C_N^2\}]^{3/2}} = 0,$$

where $\{\lambda_{N,i}\}$ are the eigenvalues of C_N. By Prob. 2.13, these eigenvalues are bounded in magnitude, so $N^{-1}\sum_{i=1}^{N} |\lambda_{N,i}|^3$ is bounded from above. Also, $\text{tr}\{C_N^2\} = \text{tr}\{R_N A_N R_N A_N\}$, so, by Prob. 2.15, $\lim_{N \to \infty} N^{-1}\text{tr}\{[C_N]^2\}$ exists and is strictly positive. Therefore,

$$\lim_{N \to \infty} \frac{\sum_{i=1}^{N} |\lambda_{N,i}|^3}{[\text{tr}\{[C_N]^2\}]^{3/2}} = \lim_{N \to \infty} \frac{N^{-1}\sum_{i=1}^{N} |\lambda_{N,i}|^3}{N^{1/2}[\text{tr}\{N^{-1}[C_N]^2\}]^{3/2}} = 0.$$

We have established assumptions (1) to (3) in Lemma C.3, hence the asymptotic normality of $N^{1/2}X_N$ and hence that of $\{N^{1/2}(\hat{r}_k - r_k), 0 \le k \le K - 1\}$. ∎

4.3. THE PERIODOGRAM

In looking for potential estimates of the spectral density, it is most natural perhaps to attempt an estimate based on the Fourier transform of the observed sequence, that is,

$$Y(\omega) = \sum_{t=0}^{N-1} y_t e^{-j\omega t}.$$

Since $S(\omega)$ is real and nonnegative, the square magnitude of $Y(\omega)$ is a reasonable candidate estimate of the power spectrum. As will soon become clear, the square magnitude should be normalized by the number of measurements N. We thus define the *periodogram* of $\{y_t, 0 \le t \le N - 1\}$ as

$$I(\omega) = \frac{1}{N} \left| \sum_{t=0}^{N-1} y_t e^{-j\omega t} \right|^2. \tag{4.3.1}$$

Like the spectral density $S(\omega)$, the periodogram is periodic with period 2π and is an even function of ω. Therefore, the values of the periodogram on the interval $[0, \pi]$ define it for all ω.

The periodogram was originally introduced to power spectrum estimation by Schuster [1898]. As we will presently see, the periodogram has a major shortcoming as an estimate of the spectral density. Despite this shortcoming, the periodogram is important both from a historical point of view and as a basis for other estimates of the spectral density.

The periodogram can be computed at frequencies $\{\omega_m = 2\pi m/N, -N/2+1 \le m \le N/2\}$ by means of the fast Fourier transform (FFT) [Oppenheim and Schafer, 1975]. Due to the symmetry of the periodogram, the values at negative frequencies are redundant. For other values of ω in the range $[0, \pi]$, we can either use the technique of zero padding [Oppenheim and Schafer, 1975] or resort to some form of interpolation. If the periodogram is needed only at one or a few values of ω, it is perhaps best to compute it directly from the definition.

An alternative expression of the periodogram can be given as follows.

$$I(\omega) = \frac{1}{N} \sum_{t=0}^{N-1} \sum_{s=0}^{N-1} y_t y_s e^{-j\omega(t-s)} = \frac{1}{N} \sum_{n=-(N-1)}^{N-1} \sum_{t=0}^{N-1-|n|} y_{t+|n|} y_t e^{-j\omega n}$$

$$= \sum_{n=-(N-1)}^{N-1} \hat{r}_n e^{-j\omega n}. \tag{4.3.2}$$

The periodogram is therefore the Fourier transform of the biased sample covariances taken up to the maximum order permitted by the number of measurements. This intuitively appealing expression can serve as justification for the use of the periodogram as an estimate of the spectral density and explains the normalization term $1/N$.

We now turn our attention to the statistical properties of the periodogram. We will assume that the process is linear and zero mean. Define for convenience the Cesàro spectrum

$$S_N(\omega) = \sum_{n=-(N-1)}^{N-1} \left(1 - \frac{|n|}{N}\right) r_n e^{-j\omega n}. \tag{4.3.3}$$

Theorem 4.4. The periodogram is an asymptotically unbiased estimate of the spectral density.

Proof. The mean of the periodogram is given by

$$E\, I(\omega) = \sum_{n=-(N-1)}^{N-1} \left(1 - \frac{|n|}{N}\right) r_n e^{-j\omega n} = S_N(\omega). \tag{4.3.4}$$

Therefore, by [A10],

$$\lim_{N\to\infty} E\, I(\omega) = S(\omega). \tag{4.3.5}$$

∎

Corollary. If Condition R1 holds, then

$$E\, I(\omega) = S(\omega) + O(N^{-1}) \tag{4.3.6}$$

uniformly in ω. This follows immediately from [A10].

∎

Problem 4.11 discusses the case of nonzero mean. It turns out that in this case the periodogram is not asymptotically unbiased at $\omega = 0$.

Next consider the variance of the periodogram.

Theorem 4.5. The asymptotic variance of the periodogram is given by

$$\lim_{N\to\infty} \text{var}\{I(\omega)\} = \begin{cases} 2S^2(\omega), & \omega = 0 \text{ or } \pi \\ S^2(\omega), & \text{otherwise.} \end{cases} \tag{4.3.7}$$

Proof. We have

$$I^2(\omega) = \frac{1}{N^2} \sum_{s=0}^{N-1} \sum_{t=0}^{N-1} \sum_{u=0}^{N-1} \sum_{v=0}^{N-1} y_s y_t y_u y_v e^{-j\omega(s-t+u-v)}.$$

Therefore,

$$E\,I^2(\omega) = \frac{1}{N^2} \sum_{s=0}^{N-1} \sum_{t=0}^{N-1} \sum_{u=0}^{N-1} \sum_{v=0}^{N-1} [r_{s-t}r_{u-v} + r_{s-v}r_{t-u} + r_{s-u}r_{t-v} + \kappa_{t-s,u-s,v-s}]$$
$$e^{-j\omega(s-t+u-v)}.$$

Now

$$\frac{1}{N} \sum_{s=0}^{N-1} \sum_{t=0}^{N-1} r_{s-t} e^{-j\omega(s-t)} = \sum_{n=-(N-1)}^{N-1} \left(1 - \frac{|n|}{N}\right) r_n e^{-j\omega n} = S_N(\omega) = E\,I(\omega).$$

Hence

$$\text{var}\{I(\omega)\} = \frac{1}{N^2} \sum_{s=0}^{N-1} \sum_{t=0}^{N-1} \sum_{u=0}^{N-1} \sum_{v=0}^{N-1} [r_{s-v}r_{t-u} + r_{s-u}r_{t-v} + \kappa_{t-s,u-s,v-s}]$$
$$e^{-j\omega(s-t+u-v)}. \tag{4.3.8}$$

Let us denote the three terms in (4.3.8) by A, B, and C in their order of appearance. Then

$$A = \frac{1}{N^2} \left[\sum_{s=0}^{N-1} \sum_{v=0}^{N-1} r_{s-v} e^{-j\omega(s-v)}\right] \left[\sum_{t=0}^{N-1} \sum_{u=0}^{N-1} r_{t-u} e^{j\omega(t-u)}\right]$$
$$= \left[\sum_{n=-(N-1)}^{N-1} \left(1 - \frac{|n|}{N}\right) r_n e^{-j\omega n}\right]^2 = S_N^2(\omega).$$

Therefore,

$$\lim_{N\to\infty} A = S^2(\omega).$$

For the second term we get

$$B = \frac{1}{N^2} \left[\sum_{s=0}^{N-1} \sum_{u=0}^{N-1} r_{s-u} e^{-j\omega(s+u)}\right] \left[\sum_{t=0}^{N-1} \sum_{v=0}^{N-1} r_{t-v} e^{j\omega(t+v)}\right] = B_1 B_1^*,$$

where

$$B_1 = \frac{1}{N} \sum_{n=-(N-1)}^{N-1} r_n e^{-j\omega n} \sum_{m=0}^{N-|n|-1} e^{-j2\omega m}$$

Note that

$$\sum_{m=0}^{N-|n|-1} e^{-j2\omega m} = \begin{cases} N - |n|, & \omega = 0 \text{ or } \pi \\ \frac{e^{-j2\omega(N-|n|)}-1}{e^{-j2\omega}-1}, & \text{otherwise.} \end{cases}$$

But

$$\left| \frac{e^{-j2\omega(N-|n|)} - 1}{e^{-j2\omega} - 1} \right| < \frac{1}{|\sin\omega|}$$

unless ω is an integer multiple of π. Therefore,

$$\lim_{N\to\infty} B = \begin{cases} S^2(\omega), & \omega = 0 \text{ or } \pi \\ 0, & \text{otherwise.} \end{cases}$$

For the third term we get

$$C = \frac{1}{N^2}\kappa_4(w) \sum_{i=-\infty}^{\infty} \gamma_i \sum_{s=0}^{N-1} C_1^2 C_1^* e^{-j\omega s},$$

where

$$C_1 = \sum_{t=0}^{N-1} \gamma_{i+t-s} e^{j\omega t}.$$

The term C_1 is bounded in magnitude by $\sum_{t=-\infty}^{\infty} |\gamma_t|$. Therefore,

$$\lim_{N\to\infty} C = 0.$$

Collecting the asymptotic expressions for A, B, and C yields (4.3.7).   ∎

This result implies that the periodogram, while asymptotically unbiased, is not a consistent estimate of the spectral density. Its asymptotic standard deviation at frequency ω is equal to $S(\omega)$ itself, except at $\omega = 0$ or π. This is the main shortcoming of the periodogram and the reason for the need for alternate estimates. The asymptotic distribution of the periodogram is exponential, except at $\omega = 0$ or π. See Prob. 4.12 for a proof of this property in the special case of a Gaussian process.

The asymptotic result (4.3.7) can be strengthened when Condition G1 holds and ω is one of the frequencies $\{\omega_k = 2\pi k/N, 0 \le k \le N/2\}$. We then have

Theorem 4.6.

$$\text{var}\{I(\omega_k)\} = \begin{cases} 2S^2(\omega_k) + O(N^{-1}), & k = 0 \text{ or } N/2 \\ S^2(\omega_k) + O(N^{-1}), & \text{otherwise.} \end{cases} \tag{4.3.9}$$

Proof. From $A = S_N^2(\omega)$ it follows that $A = S^2(\omega) + O(N^{-1})$. For the term B note that

$$\sum_{m=0}^{N-|n|-1} e^{-j4\pi km/N} = \begin{cases} N - |n|, & k = 0 \text{ or } N/2 \\ \frac{e^{j4\pi k|n|/N} - 1}{e^{-j4\pi k/N} - 1}, & \text{otherwise,} \end{cases}$$

and

$$\left| \frac{e^{j4\pi k|n|/N} - 1}{e^{-j4\pi k/N} - 1} \right| \leq |n|.$$

Therefore, since Condition G1 implies R1,

$$B = \begin{cases} S^2(\omega_k) + O(N^{-1}), & k = 0 \text{ or } N/2 \\ O(N^{-1}), & \text{otherwise.} \end{cases}$$

The term C can be easily verified to be $O(N^{-1})$, so in conclusion we get (4.3.9). ∎

Finally, we consider the covariance of the periodogram at different frequencies $\omega, \nu \in [0, \pi]$.

Theorem 4.7.

$$\lim_{N\to\infty} \text{cov}\{I(\omega), I(\nu)\} = 0, \quad \omega \neq \nu. \tag{4.3.10}$$

When Condition G1 holds,

$$\text{cov}\{I(\omega_k), I(\omega_\ell)\} = O(N^{-1}), \quad k \neq \ell. \tag{4.3.11}$$

Proof. We have, similarly to (4.3.8),

$$\text{cov}\{I(\omega), I(\nu)\} = \frac{1}{N^2} \sum_{s=0}^{N-1} \sum_{t=0}^{N-1} \sum_{u=0}^{N-1} \sum_{v=0}^{N-1} [r_{s-v}r_{t-u} + r_{s-u}r_{t-v} + \kappa_{t-s,u-s,v-s}]$$
$$e^{-j\omega(s-t)} e^{-j\nu(u-v)}. \tag{4.3.12}$$

Let us denote the three terms in (4.3.12) by A, B, and C in their order of appearance. Then

$$A = \frac{1}{N^2} \left[\sum_{s=0}^{N-1} \sum_{v=0}^{N-1} r_{s-v} e^{-j(\omega s - \nu v)} \right] \left[\sum_{t=0}^{N-1} \sum_{u=0}^{N-1} r_{t-u} e^{j(\omega t - \nu u)} \right] = A_1 A_1^*,$$

where

$$A_1 = \frac{1}{N} \sum_{n=-(N-1)}^{-1} r_n e^{-j\nu n} \sum_{m=0}^{N-|n|-1} e^{-j(\omega-\nu)m} + \frac{1}{N} \sum_{n=0}^{N-1} r_n e^{-j\omega n} \sum_{m=0}^{N-|n|-1} e^{-j(\omega-\nu)m}.$$

The sum $\sum_{m=0}^{N-|n|-1} e^{-j(\omega-\nu)m}$ is bounded in magnitude by $1/|\sin(0.5(\omega-\nu))|$ when $\omega \neq \nu$. Therefore, $\lim_{N\to\infty} A_1 = \lim_{N\to\infty} A = 0$. In exactly the same manner, it is shown that $\lim_{N\to\infty} B = 0$. That $\lim_{N\to\infty} C = 0$ is shown exactly as before. Collecting the three asymptotic expressions together gives (4.3.10). Equation (4.3.11) is shown in the same manner as (4.3.9). ∎

The conclusion drawn from (4.3.11) is that the sequence $\{I(\omega_k), 0 \le k \le N/2\}$ is approximately uncorrelated for large N. This will serve as a basis for the spectrum estimates introduced in Sec. 4.5.

Example 4.2. Let $\{y_t\}$ be the Gaussian ARMA(2,2) process

$$y_t = 1.3435y_{t-1} - 0.9025y_{t-2} + u_t + 1.3435u_{t-1} + 0.9025u_{t-2}.$$

Figure 4.1 shows the true spectral density of the process, as well as three periodograms, for 128, 512, and 2048 data points, respectively (the vertical axes are in decibels in all cases). As can be seen, the noise level in all three periodograms is about the same, in agreement with Theorem 4.5. ∎

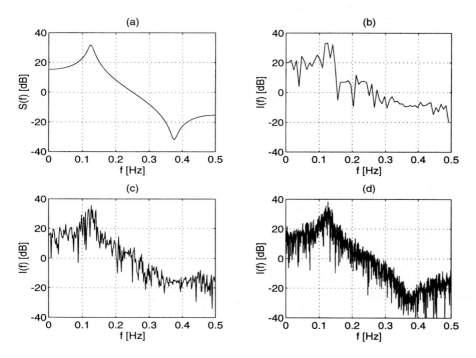

Figure 4.1. (a) The true spectral density of the process in Example 4.2; (b) the 128-point periodogram; (c) the 512-point periodogram; (d) the 2048-point periodogram (vertical axes in decibels).

4.4. PERIODOGRAM AVERAGING

To make the periodogram a useful spectrum estimate, it is necessary to reduce its variance, preferably to a small fraction of $S^2(\omega)$. A simple way of accomplishing this is to break the measurements into several segments, compute a separate periodogram for each segment, and finally compute the average of the separate periodograms. This method is called *periodogram averaging*.

The simplest way to perform periodogram averaging, due to Bartlett [1950], is as follows. Factor the number of measurements as $N = N_1 \times N_2$. Let the nth periodogram be

$$I_n(\omega) = \frac{1}{N_1} \left| \sum_{t=0}^{N_1-1} y_{t+nN_1} e^{-j\omega t} \right|^2, \quad 0 \leq n \leq N_2 - 1. \tag{4.4.1}$$

Then let the averaged periodogram be

$$\hat{I}(\omega) = \frac{1}{N_2} \sum_{n=0}^{N_2-1} I_n(\omega). \tag{4.4.2}$$

It can be shown that the individual periodograms $I_n(\omega)$ are approximately independent and identically distributed for large N_1. Therefore, the bias of $\hat{I}(\omega)$ is the same as that of a single periodogram based on N_1 measurements, while the variance is reduced by a factor of about N_2. The price paid for the reduced variance is the decrease of resolution, since $\hat{I}(\omega)$ has only N_1 approximately independent points.

A more sophisticated method, due to Welch [1970], includes both overlapping and windowing of the individual periodograms. Let N_1 be the desired length of each periodogram, as in the Bartlett method. Let L be an integer *overlap parameter* in the range $1 \leq L < N_1$. Let $N_2 = \lfloor (N - L)/(N_1 - L) \rfloor$. Let $\{w_t, 0 \leq t \leq N_1 - 1\}$ be some window (see, e.g., Oppenheim and Schafer [1975], Harris [1978], or Sec. 4.6 for common windows). Normalize the window weights such that $N^{-1} \sum_{t=0}^{N_1-1} w_t^2 = 1$ (Prob. 4.14). Then let

$$I_n(\omega) = \frac{1}{N_1} \left| \sum_{t=0}^{N_1-1} y_{t+n(N_1-L)} w_t e^{-j\omega t} \right|^2, \quad 0 \leq n \leq N_2 - 1. \tag{4.4.3}$$

The averaged periodogram is defined as in (4.4.2), using $I_n(\omega)$ as defined in (4.4.3).

The overlap parameter L is usually chosen as a fraction of N_1; a common choice is $L = 0.5N_1$. A common window for use in (4.4.3) is the Hann window, defined by

$$w_t = 0.5(1 - \cos(2\pi t/N_1)), \quad 0 \leq t \leq N_1 - 1. \tag{4.4.4}$$

The method of periodogram averaging is useful only if the number of measurements is fairly large—in the range of a thousand or more. The choice of the parameters N_1 and L (as well as the choice of window) is arbitrary to a large degree. From an implementation viewpoint it is desirable to choose N_1 as a power of 2 and use FFT for the individual periodograms. Periodogram averaging has been very popular due to its simplicity of implementation. However, the methods of periodogram smoothing and periodogram windowing, discussed in the subsequent sections, typically yield better spectral estimates than the averaged periodogram when the number of measurements is relatively small.

Example 4.3. Let $\{y_t\}$ be the Gaussian ARMA(2,2) process given in Example 4.2. The number of measurements is 2048. Figure 4.2 shows four Bartlett periodograms, for $N_1 = 128, 256, 512,$ and 1024 (the vertical axes are in decibels in all cases). As expected, the noise level increases with N_1, since the number of periodograms N_2 is in inverse proportion to N_1. It should be noted that the averaged periodogram is biased at frequencies where the true spectral density is small [cf. Figure 4.1(a)], and the bias decreases as N_1 increases. Figure 4.3 shows four Welch periodograms, for the values of N_1 as above, using a Hann window with 50% overlap. The Welch periodograms look better than the corresponding Bartlett periodograms; in particular, they are less biased at the frequencies of small spectral density. ∎

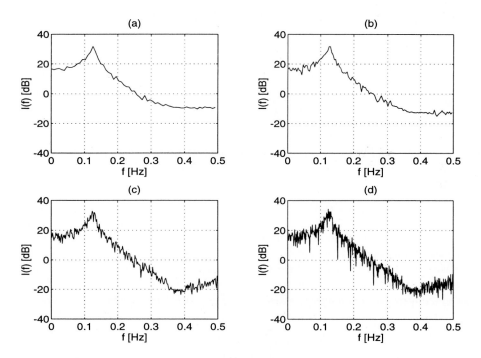

Figure 4.2. Bartlett periodograms of the process in Example 4.3: (a) $N_1 = 128$; (b) $N_1 = 256$; (c) $N_1 = 512$; (d) $N_1 = 1024$ (vertical axes in decibels).

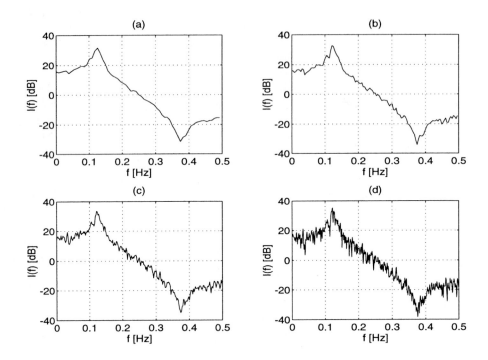

Figure 4.3. Welch periodograms of the process in Example 4.3: (a) $N_1 = 128$; (b) $N_1 = 256$; (c) $N_1 = 512$; (d) $N_1 = 1024$ (vertical axes in decibels).

4.5. SMOOTHED PERIODOGRAMS

The method of periodogram averaging allows a reduction of the variance at the expense of decreased resolution. A similar effect can be achieved by averaging neighboring frequencies of a single periodogram. Let $I(\omega)$ be the periodogram computed from the complete measurements, and let $\{W(k), -K \leq k \leq K\}$ be a set of nonnegative real numbers satisfying $\sum_{k=-K}^{K} W(k) = 1$. Define

$$\hat{S}(\omega_n) = \sum_{k=-K}^{K} W(k) I(\omega_{n+k}), \qquad (4.5.1)$$

where $\{\omega_n = 2\pi n/N, 0 \leq n \leq N/2\}$. We will call $\hat{S}(\omega_n)$ a *smoothed periodogram* and $\{W(k)\}$ its *weights*. Note that the symmetry and periodicity of $I(\cdot)$ are used to define the values needed in the right side of (4.5.1). The smoothed periodogram is obviously nonunique; each choice of K and a set of weights yields a different $\hat{S}(\omega_n)$.

It is common, though not necessary, to require that the weights be symmetric; that is, $W(k) = W(-k)$.

A minor difficulty occurs when the sum on the right side of (4.5.1) includes the frequencies $\omega = 0, \pi$. For one thing, the variance of $I(0)$ [or $I(\pi)$] is approximately twice the variance at neighboring frequencies [cf. (4.3.9)], so its inclusion in the sum may slightly degrade the quality of the result. Second, $I(0)$ is not even asymptotically unbiased if the given process has unknown mean, as is demonstrated in Prob. 4.11. It is therefore preferable to modify (4.5.1) by omitting the frequencies $\omega = 0, \pi$. The modified smoothed periodogram is defined by

$$\hat{S}(\omega_n) = \frac{\sum_{k=-K}^{K} W(k)(1 - \delta(n+k))(1 - \delta(n+k-N/2))I(\omega_{n+k})}{\sum_{k=-K}^{K} W(k)(1 - \delta(n+k))(1 - \delta(n+k-N/2))}. \tag{4.5.2}$$

The Kronecker deltas guarantee exclusion of the undesired frequencies, and the denominator renormalizes the sum of the effective weights to 1.

The smoothed periodogram is defined on the same discrete set of frequencies as the periodogram. The central questions are (a) how to extend the definition to any frequency ω and (b) how to achieve consistency of $\hat{S}(\omega)$ as $N \to \infty$. We now discuss these two questions.

Let $\omega \in [0, \pi]$. For every positive integer N, let $n(N) = \lceil \omega N/2\pi \rceil$; that is, $n(N)$ is the smallest integer greater than $\omega N/2\pi$. Then $n(N)$ is in the range $[0, N/2]$. Let $K(N)$ be an integer function of N, which

(1) is positive and monotone nondecreasing;
(2) satisfies $\lim_{N \to \infty} K(N) = \infty$; and
(3) satisfies $\lim_{N \to \infty} K(N)/N = 0$.

For every N, let $\{W_N(k), -K(N) \le k \le K(N)\}$ be a set of nonnegative numbers satisfying

(4) $W_N(k) = W_N(-k)$;
(5) $\sum_{k=-K(N)}^{K(N)} W_N(k) = 1$; and
(6) $\lim_{N \to \infty} \sum_{k=-K(N)}^{K(N)} W_N^2(k) = 0$.

We define the smoothed periodogram at the frequency ω by

$$\hat{S}(\omega) = \frac{\sum_{k=-K(N)}^{K(N)} W_N(k)(1 - \delta(n(N)+k))(1 - \delta(n(N)+k-N/2))I(\omega_{n(N)+k})}{\sum_{k=-K(N)}^{K(N)} W_N(k)(1 - \delta(n(N)+k))(1 - \delta(n(N)+k-N/2))}. \tag{4.5.3}$$

Note that this definition, like (4.5.2), uses only the periodogram values at the discrete set of frequencies. The above definition of $n(N)$ guarantees that the term $\omega_{n(N)}$ in the sum becomes closer to ω as N increases. The purpose of the various conditions is as follows. Condition (2) ensures averaging over an increasing number of periodogram values, thus reducing the variance of $\hat{S}(\omega)$ as the number of measurements increases. This condition is necessary for (6) to hold. Condition (6), in turn, ensures that all weights go to zero approximately uniformly and that no finite number of weights will dominate the average in the limit. As we will see below, this condition is necessary to reduce the variance to zero in the limit. Condition (3) provides localization of the averaging in the vicinity of the desired frequency; note that all the terms in (4.5.3) are distant from the center frequency by no more than $2\pi K(N)/N$. Condition (4) is for convenience only. Condition (5) is obvious.

The asymptotic properties of the smoothed periodogram will be derived under condition G1 and assumptions (1) to (6) above.

Theorem 4.8. The smoothed periodogram is asymptotically unbiased.

Proof. Let us assume first that $\omega \in (0, \pi)$. By assumption (3), there exists N_1 such that

$$K(N) < \min\left\{ \left[\frac{\omega N}{2\pi}\right], \frac{N}{2} - \left[\frac{\omega N}{2\pi}\right] \right\}, \quad \forall N > N_1.$$

Since we are interested in the limiting behavior as $N \to \infty$, we can confine ourselves to values of N greater than N_1. This simplifies (4.5.3) to

$$\hat{S}(\omega) = \sum_{k=-K(N)}^{K(N)} W_N(k) I(\omega_{n(N)+k}).$$

Consider the asymptotic mean of $\hat{S}(\omega)$. By (4.3.6), we have

$$E\,\hat{S}(\omega) = \sum_{k=-K(N)}^{K(N)} W_N(k) S(\omega_{n(N)+k}) + O(N^{-1});$$

hence

$$|E\,\hat{S}(\omega) - S(\omega)| \le \sum_{k=-K(N)}^{K(N)} W_N(k)|S(\omega_{n(N)+k}) - S(\omega)| + O(N^{-1}).$$

Choose $\epsilon > 0$. Due to the continuity of $S(\omega)$, there exists δ such that $|S(\nu) - S(\omega)| < \epsilon$ whenever $|\nu - \omega| < \delta$. By assumption (3), there exists N_2 such that $2\pi(K(N) + 1)/N < \delta$ for all $N > N_2$. Taking $N_0 = \max\{N_1, N_2\}$ guarantees that $|\omega_{n(N)+k} - \omega| < \delta$ for all $N > N_0$ and $|k| \le K(N)$. Therefore,

$$|E\,\hat{S}(\omega) - S(\omega)| < \epsilon + O(N^{-1}), \quad \forall N > N_0.$$

This proves that

$$\lim_{N\to\infty} E\,\hat{S}(\omega) = S(\omega). \tag{4.5.4}$$

The cases $\omega = 0, \pi$ are treated in a similar manner. For example, if $\omega = 0$, we get by the symmetry of $I(\cdot)$ and $W_N(k)$

$$\hat{S}(0) = 2(1 - W_N(0))^{-1} \sum_{k=1}^{K(N)} W_N(k) I(\omega_k)$$

and continue the proof exactly as before. The case $\omega = \pi$ is similarly handled. ∎

Next consider the asymptotic variance of $\hat{S}(\omega)$. Denote for convenience

$$C(N) = \sum_{k=-K(N)}^{K(N)} W_N^2(k).$$

Theorem 4.9.

$$\lim_{N\to\infty} C^{-1}(N)\operatorname{var}\{\hat{S}(\omega)\} = \begin{cases} S^2(\omega), & \text{if } \omega \in (0,\pi) \\ 2S^2(\omega), & \omega = 0, \pi. \end{cases} \tag{4.5.5}$$

Proof. Assume first, as in the proof of the previous theorem, that $\omega \in (0,\pi)$. By (4.3.9) and (4.3.11), we have

$$\operatorname{var}\{\hat{S}(\omega)\} = \sum_{k=-K(N)}^{K(N)} \sum_{\ell=-K(N)}^{K(N)} W_N(k)W_N(\ell)\operatorname{cov}\{I(\omega_{n(N)+k}), I(\omega_{n(N)+\ell})\}$$

$$= \sum_{k=-K(N)}^{K(N)} W_N^2(k)S^2(\omega_{n(N)+k}) + O(N^{-1}).$$

In exactly the same manner as in the proof of Theorem 4.8, it can be shown that for all $\epsilon > 0$ there exists N_0 such that

$$|\operatorname{var}\{\hat{S}(\omega)\} - C(N)S^2(\omega)| < C(N)\epsilon + O(N^{-1}), \quad \forall N > N_0.$$

Assumption (5) implies that (Prob. 4.16)

$$C(N) \geq \frac{1}{2K(N)+1}. \tag{4.5.6}$$

Now (4.5.6) and assumption (3) imply that $\lim_{N\to\infty} C^{-1}(N)O(N^{-1}) = 0$. This proves the first case in (4.5.5). The cases $\omega = 0, \pi$ are treated in a similar manner. For example, if $\omega = 0$, we have, as in the proof of Theorem 4.8,

$$\hat{S}(0) = 2(1 - W_N(0))^{-1} \sum_{k=1}^{K(N)} W_N(k)I(\omega_k).$$

Therefore,

$$\operatorname{var}\{\hat{S}(0)\} = 4(1 - W_N(0))^{-2} \sum_{k=1}^{K(N)} \sum_{\ell=1}^{K(N)} W_N(k)W_N(\ell)\operatorname{cov}\{I(\omega_k), I(\omega_\ell)\}$$

$$= 4(1 - W_N(0))^{-2} \sum_{k=1}^{K(N)} W_N^2(k)S^2(\omega_k) + O(N^{-1}).$$

Since $\lim_{N\to\infty} W_N(0) = 0$, the sum of the weights on the right side approaches $0.5C(N)$. This, together with the factor 4, yields the second case of (4.5.5). The case $\omega = \pi$ is similarly handled. ∎

Since $\lim_{N \to \infty} C(N) = 0$ by assumption (6), the limiting variance of $\hat{S}(\omega)$ is zero. This, together with (4.5.4), proves the mean-square consistency of $\hat{S}(\omega)$.

Example 4.4. Let $\{y_t\}$ be the Gaussian ARMA(2,2) process given in Example 4.2. The number of measurements is 2048. Figure 4.4 shows four smoothed periodograms, corresponding to $K = 2, 4, 8, 16$. Triangular weights were used in all cases. As expected, the noise level decreases with K. Similarly to the averaged periodogram, the smoothed periodogram is biased at frequencies where the true spectral density is small. ∎

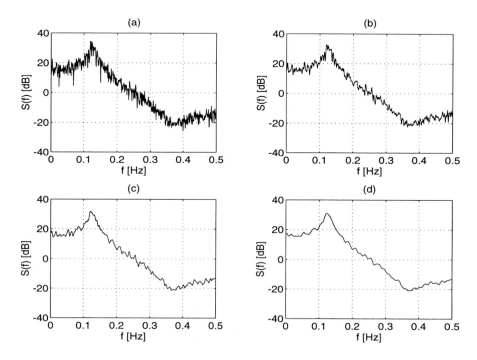

Figure 4.4. Smoothed periodograms of the process in Example 4.4: (a) $K = 2$; (b) $K = 4$; (c) $K = 8$; (d) $K = 16$ (vertical axes in decibels).

We conclude this section with a remark about the implementation of the smoothed periodogram. Equation (4.5.1) represents a circular convolution, so it can be computed by multiplying the inverse FFT of $I(\omega_k)$ by the inverse FFT of $W(k)$ and taking the FFT of the product. However, if we wish to use (4.5.2), the use of FFT is inconvenient and it is probably best to use this formula directly. If the number of measurements is large (in the range of several hundreds or more), the difference between (4.5.1) and (4.5.2) is small, so it is recommended to use (4.5.1) and do the computations with FFT. If the number of measurements is small, statistical accuracy calls for using (4.5.2).

4.6. WINDOWED PERIODOGRAMS

As a motivation for the estimates to be discussed next, recall the expression for the periodogram obtained in Sec. 4.3,

$$I(\omega) = \sum_{n=-(N-1)}^{N-1} \hat{r}_n e^{-j\omega n}. \tag{4.6.1}$$

The lack of consistency of the periodogram can be qualitatively explained by the use of the high-lag sample covariances, the covariances of lags close to $N-1$. These sample covariances involve very little data averaging; thus their variances are large. A possible remedy is therefore to reduce the number of terms in (4.6.1), making it small relative to N. This will ensure that only "sufficiently stable" covariance terms will enter the sum and will help to reduce its variance.

The *Gibbs phenomenon* is a well-known undesirable effect of Fourier series truncation [Bracewell, 1986, p. 209]. The common method of combating the Gibbs phenomenon is the use of windows. A *window* is a function $w_K(k)$, defined for all $K \geq 1$ and for all $-K \leq k \leq K$, having the following properties:
(1) Symmetry: $w_K(k) = w_K(-k)$.
(2) $w_K(0) = 1$.
(3) $0 \leq w_K(k) \leq 1$.
The *windowed periodogram* corresponding to the window $w_K(k)$ (where $K < N$ by assumption) is defined as

$$\hat{S}(\omega) = \sum_{k=-K}^{K} w_K(k)\hat{r}_k e^{-j\omega k}. \tag{4.6.2}$$

Note the differences between this definition and (4.6.1): the truncation from $N-1$ to K and the term-by-term multiplication by the window coefficients. The windowed periodogram is attributed to Blackman and Tukey [1958].

To understand the windowing operation better, consider (4.6.2) in the frequency domain. Denote the Fourier transform of the window by

$$W_K(\omega) = \sum_{k=-K}^{K} w_K(k)e^{-j\omega k}. \tag{4.6.3}$$

$W_K(\omega)$ is called the *smoothing kernel*. The symmetry of the window implies that the smoothing kernel is real and symmetric. Since multiplication of sequences translates to convolution of their transforms, (4.6.2) is equivalent to

$$\hat{S}(\omega) = \frac{1}{2\pi} \int_{-\pi}^{\pi} W_K(\nu)I(\omega - \nu)d\nu = \frac{1}{2\pi} \int_{-\pi}^{\pi} W_K(\nu)I(\nu + \omega)d\nu. \tag{4.6.4}$$

This formula shows the close relationship between the windowed periodogram and the smoothed periodogram defined in (4.5.1). Equation (4.6.4) can be viewed as a "refined" version of (4.5.1), in the sense that the summation is replaced by integration over a continuous range of frequencies. In contrast with the smoothed periodogram, the windowed periodogram is immediately defined for all ω.

The windowed periodogram is real and symmetric. Unlike the periodogram, it is not guaranteed to be nonnegative, since the smoothing kernel is not necessarily nonnegative for all ω. If, however, the window is chosen so that $W_K(\omega) \geq 0$, the windowed periodogram will also be nonnegative. Smoothing kernels having this property are called positive. A simple procedure for generating positive kernels is by convolving an arbitrary window with itself and normalizing $w_K(0)$ to 1. For example, a triangular window is a convolution of a rectangular window with itself; hence it is positive.

We now describe a general method of constructing windows. Let $v(x)$ be a symmetric continuous function on $[-1, 1]$, satisfying $v(0) = 1$ and $0 \leq v(x) \leq 1$ for all x. Then let $w_K(k) = v(k/K)$. Properties (1) to (3) are obviously satisfied for $w_K(k)$. The following lemma establishes an important property of windows thus constructed.

Lemma 4.2. Let $\{a_n\}$ be an absolutely summable sequence, with $\sum_{n=-\infty}^{\infty} a_n = A$ and $\sum_{n=-\infty}^{\infty} |a_n| = B$. Let $w_N(n)$ be derived from $v(x)$ as above. Then

$$\lim_{N \to \infty} \sum_{n=-N}^{N} w_N(n)a_n = A.$$

Proof. For all $1 \leq N_1 < N$, we have

$$A - \sum_{n=-N}^{N} w_N(n)a_n = \sum_{n=-N_1}^{N_1} [1 - w_N(n)]a_n + \sum_{|n|=N_1+1}^{N} [1 - w_N(n)]a_n + \sum_{|n|=N+1}^{\infty} a_n.$$

Now

$$\left| \sum_{|n|=N_1+1}^{N} [1 - w_N(n)]a_n + \sum_{|n|=N+1}^{\infty} a_n \right| \leq \sum_{|n|=N_1+1}^{\infty} |a_n|,$$

so there exists N_1 such that this term is less than $\epsilon/2$. To handle the first term, observe that

$$\left| \sum_{n=-N_1}^{N_1} [1 - w_N(n)]a_n \right| \leq \sup_{0 \leq n \leq N_1} |1 - w_N(n)| \sum_{n=-N_1}^{N_1} |a_n| \leq \sup_{0 \leq n \leq N_1} |1 - v(n/N)|B.$$

By the continuity of $v(x)$, there exists N_2 such that $\sup_{0 \leq n \leq N_1} |1 - v(n/N)| < \epsilon/2B$ for all $N \geq N_2$. This will make the first term less than $\epsilon/2$; hence $|A - \sum_{n=-N}^{N} w_N(n)a_n| < \epsilon$.   ∎

For reasons to be clarified later, we further require the window function $v(x)$ to satisfy a Lipschitz condition of order α; that is [A18],

$$|v(x_1) - v(x_2)| \leq C|x_1 - x_2|^\alpha, \quad \forall x_1, x_2 \in [-1, 1],$$

for some $\alpha > 0$ and $C > 0$. The following consequence of the Lipschitz condition will be needed to prove Theorem 4.11.

Lemma 4.3. Let $v(x)$ satisfy the Lipschitz condition of order α and let $\omega \in (0, 2\pi)$. Then

$$\lim_{K \to \infty} \frac{1}{K} \sum_{k=0}^{K-1} v(k/K)e^{-j\omega k} = 0.$$

Proof. Let $L = \lfloor K^{1/2} \rfloor$ and $K = LM + P$, where $0 \leq P < L \leq K^{1/2}$. The three integers L, M, and P depend on K and $L(K) \to \infty$, $M(K) \to \infty$ as $K \to \infty$. Then

$$\frac{1}{K} \sum_{k=0}^{LM-1} v\left(\frac{k}{K}\right)e^{-j\omega k} = \frac{1}{K} \sum_{\ell=0}^{L-1} \sum_{m=0}^{M-1} v\left(\frac{\ell M}{K} + \frac{m}{K}\right)e^{-j\omega(\ell M + m)}$$

$$= \frac{1}{K} \sum_{\ell=0}^{L-1} e^{-j\omega \ell M} \sum_{m=0}^{M-1} e^{-j\omega m}\left[v\left(\frac{\ell M}{K}\right) + O\left(\left(\frac{m}{K}\right)^\alpha\right)\right]$$

$$= \frac{1}{K} \sum_{\ell=0}^{L-1} e^{-j\omega \ell M} v\left(\frac{\ell M}{K}\right) \sum_{m=0}^{M-1} e^{-j\omega m} + \frac{1}{K} \sum_{\ell=0}^{L-1} e^{-j\omega \ell M} \sum_{m=0}^{M-1} O\left(\left(\frac{m}{K}\right)^\alpha\right)$$

$$= \frac{1}{K} \sum_{\ell=0}^{L-1} e^{-j\omega \ell M} v\left(\frac{\ell M}{K}\right) \frac{e^{-j\omega M} - 1}{e^{-j\omega} - 1} + \frac{1}{K} \sum_{\ell=0}^{L-1} O\left(\frac{M^{1+\alpha}}{K^\alpha}\right).$$

Therefore,

$$\left|\frac{1}{K} \sum_{k=0}^{LM-1} v\left(\frac{k}{K}\right)e^{-j\omega k}\right| \leq \frac{1}{\sin 0.5\omega} \frac{1}{K} \sum_{\ell=0}^{L-1} \left|v\left(\frac{\ell M}{K}\right)\right| + O\left(\frac{LM^{1+\alpha}}{K^{1+\alpha}}\right)$$

$$= O(M^{-1}) + O(L^{-\alpha}).$$

Also,

$$\left|\frac{1}{K} \sum_{k=LM}^{K-1} v\left(\frac{k}{K}\right)e^{-j\omega k}\right| \leq K^{-1/2},$$

so in summary

$$\left|\frac{1}{K} \sum_{k=0}^{K-1} v\left(\frac{k}{K}\right)e^{-j\omega k}\right| \leq O(M^{-1}) + O(L^{-\alpha}) + K^{-1/2}.$$

Taking the limit $K \to \infty$ yields the asserted result. ∎

To establish the mean-square consistency of the windowed periodogram, we need to make the window length K dependent on the number of measurements. Let us therefore assume, similarly to the case of a smoothed periodogram, that

(1) $K(N)$ is a monotone nondecreasing function of N;
(2) $\lim_{N \to \infty} K(N) = \infty$; and
(3) $\lim_{N \to \infty} K(N)/N = 0$.

In addition, we will assume that the process is linear and satisfies Condition G1 (hence R1 and K1).

Theorem 4.10. The windowed periodogram corresponding to the window $w_K(k)$ $= v(k/K(N))$ is asymptotically unbiased.

Proof.

$$E\hat{S}(\omega) = \sum_{k=-K(N)}^{K(N)} \left(1 - \frac{|k|}{N}\right) v\left(\frac{k}{K(N)}\right) r_k e^{-j\omega k}$$

$$= \sum_{k=-K(N)}^{K(N)} v\left(\frac{k}{K(N)}\right) r_k e^{-j\omega k} - \sum_{k=-K(N)}^{K(N)} \frac{|k|}{N} v\left(\frac{k}{K(N)}\right) r_k e^{-j\omega k}.$$

The first term on the right side approaches $S(\omega)$ by Lemma 4.2. The magnitude of the second term is bounded by $(K(N)/N) \sum_{k=-\infty}^{\infty} |r_k|$, so it goes to zero as $N \to \infty$. Therefore,

$$\lim_{N \to \infty} E\hat{S}(\omega) = S(\omega). \tag{4.6.5}$$

∎

The asymptotic variance of the windowed periodogram requires considerably more work. We will need the following lemma.

Lemma 4.4. Under condition R1,

$$\sum_{k=0}^{\infty} \sum_{\ell=0}^{\infty} \sum_{n=-\infty}^{\infty} |r_n||r_{n+k+\ell}| < \infty.$$

Proof.

$$\sum_{k=0}^{\infty} \sum_{\ell=0}^{\infty} \sum_{n=-\infty}^{\infty} |r_n||r_{n+k+\ell}| = \sum_{m=0}^{\infty} \sum_{n=-\infty}^{\infty} (m+1)|r_n||r_{n+m}|$$

$$= \sum_{n=-\infty}^{\infty} \sum_{p=n}^{\infty} (p-n+1)|r_n||r_p| \leq \sum_{n=-\infty}^{\infty} \sum_{p=-\infty}^{\infty} (|p| + |n| + 1)|r_n||r_p|$$

$$= 2R_0 R_1 + R_0^2 < \infty.$$

∎

We are now ready to prove:

Theorem 4.11. The asymptotic variance of the windowed periodogram corresponding to the window $w_{K(N)}(k) = v(k/K(N))$ is

$$\lim_{N \to \infty} \frac{N}{K(N)} \text{var}\{\hat{S}(\omega)\} = \begin{cases} VS^2(\omega), & \text{if } \omega \in (0, \pi) \\ 2VS^2(\omega), & \text{if } \omega = 0, \pi, \end{cases} \qquad (4.6.6)$$

where

$$V = \int_{-1}^{1} v^2(x)\, dx.$$

Proof. Observe that $\hat{S}(\omega)$ can be written as

$$\hat{S}(\omega) = 2 \sum_{k=0}^{K} [1 - 0.5\delta(k)] w_K(k) \hat{r}_k \cos \omega k.$$

Therefore, the variance of $\hat{S}(\omega)$ is

$$\text{var}\{\hat{S}(\omega)\} = 4 \sum_{k=0}^{K} \sum_{\ell=0}^{K}$$

$$[1 - 0.5\delta(k)][1 - 0.5\delta(\ell)] w_K(k) w_K(\ell) \text{cov}\{\hat{r}_k, \hat{r}_\ell\} \cos \omega k \cos \omega \ell = 8 \sum_{k=0}^{K} \sum_{\ell=0}^{k}$$

$$[1 - 0.5\delta(k)][1 - 0.5\delta(\ell)][1 - 0.5\delta(k - \ell)] w_K(k) w_K(\ell) \text{cov}\{\hat{r}_k, \hat{r}_\ell\} \cos \omega k \cos \omega \ell.$$

Recall formulas (4.2.9) and (4.2.14) for $\text{cov}\{\hat{r}_k, \hat{r}_\ell\}$ under conditions R1 and K1. Substitution in the above gives four terms, to be denoted by $A, B, C,$ and D [$A, B,$ and C correspond to the three terms of a_n, while D corresponds to $O(N^{-2})$]. We will show that only term A affects the asymptotic result. For term D, we have $|D| \leq 8K^2(N)O(N^{-2})$; hence

$$\lim_{N \to \infty} \frac{N}{K(N)} |D| = 0.$$

For the term C we have

$$|C| \leq \frac{8}{N} \sum_{k=0}^{\infty} \sum_{\ell=0}^{\infty} \sum_{n=-\infty}^{\infty} |\kappa_{n+k,n,\ell}| \leq \frac{8}{N} |\kappa_4(w)| \left[\sum_{s=-\infty}^{\infty} |\gamma_s| \right]^4,$$

so

$$\lim_{N \to \infty} \frac{N}{K(N)} |C| = 0.$$

For the term B we have, by Lemma 4.4,

$$|B| \leq \frac{8}{N} \sum_{k=0}^{\infty} \sum_{\ell=0}^{\infty} \sum_{n=-\infty}^{\infty} |r_n||r_{n+k+\ell}| \leq \frac{8(2R_0 R_1 + R_0^2)}{N},$$

so

$$\lim_{N \to \infty} \frac{N}{K(N)} |B| = 0.$$

Let us summarize what we have so far:

$$\frac{N}{K} \text{var}\{\hat{S}(\omega)\} = \frac{8}{K} \sum_{k=0}^{K} \sum_{\ell=0}^{k} f_K(k,\ell) \cos \omega k \cos \omega \ell$$

$$\left[\left(1 - \frac{\ell}{N}\right) \sum_{n=-\infty}^{-(k-\ell+1)} r_n r_{n+k-\ell} + \left(1 - \frac{k}{N}\right) \sum_{n=-(k-\ell)}^{\infty} r_n r_{n+k-\ell} \right] + o(1),$$

where

$$f_K(k,\ell) = [1 - 0.5\delta(k)][1 - 0.5\delta(\ell)][1 - 0.5\delta(k-\ell)] w_K(k) w_K(\ell).$$

Now let $\eta_{k-\ell} = \sum_{n=-\infty}^{-(k-\ell+1)} |r_n r_{n+k-\ell}|$. Then

$$\frac{1}{KN} \sum_{k=0}^{K} \sum_{\ell=0}^{k} \ell \eta_{k-\ell} = \frac{1}{KN} \sum_{m=0}^{K} \eta_m \sum_{\ell=0}^{K-m} \ell = \frac{1}{2KN} \sum_{m=0}^{K} (K - m + 1)(K - m) \eta_m$$

$$\leq \frac{K}{N} \sum_{m=0}^{\infty} \eta_m.$$

It is easy to show that η_m is summable, so the above is $O(K/N) = o(1)$. A similar argument holds for the term involving k/N, so

$$\frac{N}{K} \text{var}\{\hat{S}(\omega)\} = \frac{8}{K} \sum_{k=0}^{K} \sum_{\ell=0}^{k} f_K(k,\ell) \cos \omega k \cos \omega \ell \sum_{n=-\infty}^{\infty} r_n r_{n+k-\ell} + o(1)$$

$$= \frac{8}{K} \sum_{k=0}^{K} \sum_{\ell=0}^{k} f_K(k,\ell) \bar{r}_{k-\ell} \cos \omega k \cos \omega \ell + o(1)$$

$$= \frac{8}{K} \sum_{m=0}^{K} \sum_{\ell=0}^{K-m} f_K(\ell + m, \ell) \bar{r}_m \cos \omega(\ell + m) \cos \omega \ell + o(1),$$

where $\bar{r}_m$ was defined in (4.2.11). Let us further denote

$$a_K(m) = \frac{2}{K} \sum_{\ell=0}^{K-m} [1 - 0.5\delta(\ell + m)][1 - 0.5\delta(\ell)] v\left(\frac{\ell + m}{K}\right) v\left(\frac{\ell}{K}\right)$$

$$b_K(m, \omega) = \frac{2}{K} \sum_{\ell=0}^{K-m} [1 - 0.5\delta(\ell + m)][1 - 0.5\delta(\ell)] v\left(\frac{\ell + m}{K}\right) v\left(\frac{\ell}{K}\right) \cos 2\omega \ell$$

$$c_K(m, \omega) = \frac{2}{K} \sum_{\ell=0}^{K-m} [1 - 0.5\delta(\ell + m)][1 - 0.5\delta(\ell)] v\left(\frac{\ell + m}{K}\right) v\left(\frac{\ell}{K}\right) \sin 2\omega \ell.$$

Then we get, by applying some trigonometric identities,

$$\frac{N}{K}\text{var}\{\hat{S}(\omega)\} = 2\sum_{m=0}^{K}[1 - 0.5\delta(m)]$$

$$[(a_K(m) + b_K(m,\omega))\cos\omega m - c_K(m,\omega)\sin\omega m]\bar{r}_m + o(1). \qquad (4.6.7)$$

It follows as in the proof of Lemma 4.3 that, for any *fixed* m,

$$\lim_{K\to\infty} b_K(m,\omega) = \lim_{K\to\infty} c_K(m,\omega) = 0, \quad \omega \in (0,\pi).$$

The convergence is, however, not uniform in m or ω. For $\omega = 0, \pi$ we clearly have $b_K(m,\omega) = a_K(m)$ and $c_K(m,\omega) = 0$. Also, it is easy to verify that $|a_K(m)| \le 2$, $|b_K(m,\omega)| \le 2$, $|c_K(m,\omega)| \le 2$, and

$$\lim_{K\to\infty} a_K(m) = 2\int_0^1 v^2(x)dx = V$$

for any fixed m.

To complete the proof, we need to show that the limit of (4.6.7) is equal to $VS^2(\omega)$ for $\omega \in (0,\pi)$ and to $2VS^2(\omega)$ for $\omega = 0, \pi$. Consider first the case $\omega \in (0,\pi)$ and denote $\sum_{m=0}^{\infty} |\bar{r}_m| = G$. Note that $S^2(\omega) = 2\sum_{m=0}^{\infty}[1 - 0.5\delta(m)]\bar{r}_m \cos\omega m$. Let M be an integer smaller than K. Then

$$\left| V\sum_{m=0}^{\infty}[1 - 0.5\delta(m)]\bar{r}_m \cos\omega m \right.$$

$$\left. - \sum_{m=0}^{K}[1 - 0.5\delta(m)][(a_K(m) + b_K(m,\omega))\cos\omega m - c_K(m,\omega)\sin\omega m]\bar{r}_m \right| =$$

$$\left| V\sum_{m=M+1}^{\infty}\bar{r}_m \cos\omega m - \sum_{m=M+1}^{K}[(a_K(m) + b_K(m,\omega))\cos\omega m - c_K(m,\omega)\sin\omega m]\bar{r}_m \right.$$

$$\left. + \sum_{m=0}^{M}[V - a_K(m) - b_K(m,\omega)][1 - 0.5\delta(m)]\bar{r}_m \cos\omega m - \sum_{m=1}^{M}c_K(m,\omega)\bar{r}_m \sin\omega m \right|$$

$$\le V\sum_{m=M+1}^{\infty}|\bar{r}_m| + 6\sum_{m=M+1}^{K}|\bar{r}_m| + \sup_{0\le m\le M}|V - a_K(m) - b_K(m,\omega)|\sum_{m=0}^{M}|\bar{r}_m|$$

$$+ \sup_{1\le m\le M}|c_K(m,\omega)|\sum_{m=1}^{M}|\bar{r}_m|.$$

Let $\epsilon > 0$. Choose M such that the sum of the first two terms is less than 0.5ϵ. Then choose K_1 such that both $\sup_{0\le m\le M}|V - a_K(m) - b_K(m,\omega)|$ and $\sup_{1\le m\le M}|c_K(m,\omega)|$ are less than $0.5\epsilon/G$ for all $K > K_1$. This is possible since M is finite and fixed (i.e., independent of K). Then the above sum will be less than

ϵ for all $K > K_1$. Therefore, we have proved that

$$\lim_{K\to\infty} \frac{N}{K}\mathrm{var}\{\hat{S}(\omega)\} = \lim_{N\to\infty} \frac{N}{K}\mathrm{var}\{\hat{S}(\omega)\} = VS^2(\omega)$$

for all $\omega \in (0,\pi)$. To prove the case $\omega = 0, \pi$ recall that in this case $b_K(m,\omega) = a_K(m)$, and repeat the same argument.　　　　　　　■

From Theorems 4.10 and 4.11 we conclude that the windowed periodogram corresponding to the window $w_{K(N)}(k) = v(k/K(N))$ is mean-square consistent.

For a given N, the choice of K is a matter of trade-off between bias and variance. As K increases, the variance of the periodogram generally increases, while the bias decreases. Ideally, we would like to choose K so as to minimize the mean-square error (the square of the bias plus the variance), either at a given frequency or on the average over all frequencies. However, there is no easy way to accomplish this, since the optimum K depends on the true spectral density of the process, which is unknown (and the optimum is probably computationally intractable anyway). Theoretical analysis shows that, under some smoothness assumptions on the spectral density and on the window function, K should be proportional to $N^{1/5}$. With this choice, the bias and the standard deviation tend to zero at the same rate (i.e., neither dominates the other asymptotically), and the asymptotic mean-square error is proportional to $N^{-4/5}$ [Anderson, 1971, Sec. 9.3.4]. However, this rule says nothing about the proportionality constant. At the bottom line, the choice of K is up to the user and is often based on engineering judgment.

The choice of window function is another matter of engineering judgment. As is known from window theory [Oppenheim and Schafer, 1975], the kernel function $W(\omega)$ (the Fourier transform of the window) has a main lobe and side lobes. In general, a narrow main lobe is accompanied by relatively high side lobes, and vice versa. Thus, the window is chosen as a trade-off between the width of the main lobe (which should be as narrow as possible) and the magnitude of the highest side lobe (which should be as small as possible). Common windows include the following [the definitions are in terms of the function $v(x)$]

$$v(x) = 1 - |x| \quad \text{(triangular)} \tag{4.6.8a}$$

$$v(x) = 0.5 + 0.5\cos(\pi x) \quad \text{(Hann)} \tag{4.6.8b}$$

$$v(x) = 0.54 + 0.46\cos(\pi x) \quad \text{(Hamming)} \tag{4.6.8c}$$

$$v(x) = 0.42 + 0.5\cos(\pi x) + 0.08\cos(2\pi x) \quad \text{(Blackman)} \tag{4.6.8d}$$

$$v(x) = \begin{cases} 1 - 6x^2(1-|x|), & 0 \le |x| \le 0.5 \\ 2(1-|x|)^3, & 0.5 < |x| \le 1 \end{cases} \quad \text{(de la Vallé-Poussin)} \tag{4.6.8e}$$

$$v(x) = \frac{I_0(\pi\beta\sqrt{1-x^2})}{I_0(\pi\beta)} \quad \text{(Kaiser)} \tag{4.6.8f}$$

The function $I_0(\cdot)$ in the Kaiser window is the modified Bessel function of order zero, given by the infinite series $I_0(y) = \sum_{n=0}^{\infty}[(0.5y)^n/n!]^2$. The user-chosen parameter β controls the width of the main lobe and the magnitude of the highest side lobe. Of these windows, only the triangular and the de la Vallé-Poussin are positive (Prob. 4.20). However, as we have already said, a positive window can be constructed from any window by convolving it with itself.

Example 4.5. Let $\{y_t\}$ be the Gaussian ARMA(2,2) process given in Example 4.2. The number of measurements is 2048. Figure 4.5 shows four windowed periodograms, corresponding to $K = 32, 64, 128, 256$. The window was obtained by convolving a Hamming window with itself (to make it positive). As expected, the noise level increases with K. Similarly to the other spectrum estimates, the windowed periodogram is biased at frequencies where the true spectral density is small, and the bias decreases when K increases.   ■

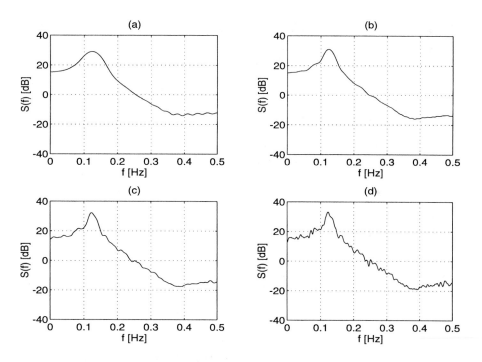

Figure 4.5. Windowed periodograms of the process in Example 4.5: (a) $K = 32$; (b) $K = 64$; (c) $K = 128$; (d) $K = 256$ (vertical axes in decibels).

4.7. MATHEMATICA PACKAGES

The procedure `CovSeq[data_, nr_]` in the package `ArmaData.m` generates the biased sample covariances of the sequence `data` of orders 0 through `nr`. An alternative procedure for this purpose is developed in Prob. 4.13.

The package `Period.m` implements the spectral estimates described in this chapter, as well as some common windows. Here is a brief description of the functions in this package.

`Periodogram[data_]` computes the periodogram of the sequence `data`.

`AveragedPeriodogram[data_, window_, overlap_]` computes the averaged periodogram of the sequence `data`. `window` is the desired window (e.g., a Hann window for the standard Welch method). The length of the window determines the length of the periodogram (see Prob. 4.15 for relaxation of this restriction). `overlap` is the overlap parameter L (an integer), which should be less than the length of the periodogram. A rectangular window with zero overlap gives the Bartlett method.

`SmoothedPeriodogram[data_, window_]` computes the smoothed periodogram of the sequence `data`. `window` is the set of weights, assumed to be symmetric. The computation of the circular convolution is done by FFT, as explained in Sec. 4.5.

`WindowedPeriodogram[data_, window_, npoint_]` computes the windowed periodogram of the sequence `data`. `window` is the desired window. `npoint` is the desired number of frequency points.

`RectangularWindow[n_]` returns a rectangular window of length n.

`TriangularWindow[n_]` returns a triangular window of length n.

`HannWindow[n_]` returns a Hann (cosine) window of length n.

`HammingWindow[n_]` returns a Hamming (raised cosine) window of length n.

`BlackmanWindow[n_]` returns a Blackman window of length n.

`KaiserWindow[n_, beta_]` returns a Kaiser window of length n and parameter beta.

`ConvWindow[window_]` convolves a window with itself to yield a positive window of twice the length.

REFERENCES

Anderson, T. W., *The Statistical Analysis of Time Series,* Wiley, New York, 1971.

Bartlett, M. S., "Periodogram Analysis and Continuous Spectra," *Biometrika,* 37, pp. 1–16, 1950.

Blackman, R. B., and Tukey, J. W., *The Measurement of Power Spectra from the Point of View of Communications Engineering*, Dover Publications, New York, 1958.

Bracewell, R. N., *The Fourier Transform and Its Applications,* 2nd ed., McGraw-Hill, New York, 1986.

Harris, F. J., "On the Use of Windows for Harmonic Analysis with the Discrete Fourier Transform," *Proc. IEEE,* 66, pp. 51–83, 1978.

Oppenheim, A. V., and Schafer, R. W., *Digital Signal Processing,* Prentice Hall, Englewood Cliffs, NJ, 1975.

Rosen, Y., and Porat, B., "The Second Order Moments of the Sample Covariances for Time Series with Missing Observations," *IEEE Trans. Information Theory,* IT-35, pp. 334–341, 1989.

Rosenblatt, M., *Stationary Sequences and Random Fields,* Birkhäuser, Boston, 1985.

Schuster, A., "On the Investigation of Hidden Periodicities with Application to a Supposed 26 Day Period of Meteorological Phenomena," *Terr. Magn.,* 3, pp. 13–41, 1898.

Welch, P. D., "The Use of Fast Fourier Transform for the Estimation of Power Spectra," *IEEE Trans. Audio and Electroacoustics,* AU-15, pp. 70–73, 1970.

PROBLEMS

4.1. Show that Condition G0 for a linear process implies Conditions R0 and K0, and determine the corresponding bounds R_0 and K_0.

4.2. Show that Condition G1 for a linear process implies Conditions R1 and K1, and determine the corresponding bounds R_1 and K_1.

4.3. Establish the equality between (4.2.10) and the first form in the right side of (4.2.11).

4.4. Prove formula (4.2.12).

4.5. If Condition R1 holds, prove that

$$\sum_{n=-\infty}^{\infty} |n|\,|r_n r_{n+k-\ell}| \le r_0 R_1, \qquad \sum_{n=-\infty}^{\infty} |n|\,|r_{n+k} r_{n-\ell}| \le r_0 R_1.$$

Remark: Take care to properly eliminate the apparent dependence on k, ℓ and obtain *uniform* bounds.

4.6. Suppose the process $\{y_t\}$ has an unknown mean μ. Let $\hat{\mu}$ be the estimated mean as in (4.2.1) and define the biased sample covariances by

$$\hat{r}_k = \frac{1}{N} \sum_{t=0}^{N-k-1} (y_{k+t} - \hat{\mu})(y_t - \hat{\mu}).$$

Derive the equivalent of formula (4.2.4) for this case.

4.7. Show that the asymptotic Bartlett formula (4.2.10) holds for the sample covariances defined in Prob. 4.6.

4.8. Define the correlation coefficients and the sample correlation coefficients

$$\rho_k = \frac{r_k}{r_0}, \qquad \hat{\rho}_k = \frac{\hat{r}_k}{\hat{r}_0}.$$

Use (4.2.13) and Theorem 3.15 to derive the following asymptotic formula for the covariances of the sample correlation coefficients in case of a linear process

$$\lim_{N\to\infty} N\mathrm{cov}\{\hat{\rho}_k, \hat{\rho}_\ell\} =$$

$$\sum_{n=-\infty}^{\infty} (\rho_n\rho_{n+k-\ell} + \rho_{n+k}\rho_{n-\ell} + 2\rho_k\rho_\ell\rho_n^2 - 2\rho_k\rho_n\rho_{n+\ell} - 2\rho_\ell\rho_n\rho_{n+k}).$$

This formula is due to Bartlett. Note that the terms involving $\kappa_4(w)$ cancel out.

4.9. Repeat the computation in Example 4.1 for an ARMA(1,1) process with AR and MA coefficients a and b, respectively.

4.10. Prove that, under the conditions of Theorem 4.3, the sample covariance $\hat{r}_k$ converges to the true covariance r_k almost surely. Hint: Use Theorem C.5 and Prob. 2.15.

4.11. Explore the asymptotic bias of the periodogram in the case of unknown mean. Consider the following cases: (a) when the definition (4.3.1) remains unchanged; (b) when the definition is modified to

$$I(\omega) = \frac{1}{N}\left|\sum_{t=0}^{N-1}(y_t - \hat{\mu})e^{-j\omega t}\right|^2.$$

Distinguish between the bias at $\omega = 0$ and at $\omega \neq 0$.

4.12. Assume that $\{y_t\}$ is Gaussian. Prove that $I(\omega)$ converges in distribution to an exponential random variable with parameter $S(\omega)$, except at $\omega = 0$ or π. Hint: Consider the joint distribution of the real and imaginary parts of $N^{-1/2}Y(\omega)$.

4.13. Develop a procedure for computing the sample covariances $\{\hat{r}_k, 0 \leq k \leq N-1\}$ from the data $\{y_n, 0 \leq n \leq N-1\}$ in a number of operations in the order of $N \log_2 N$. Program this procedure in Mathematica. Assume for convenience that N is an integer power of 2. Hint: Zero-pad the data to $2N$ and use the two forms of the periodogram (4.3.1) and (4.3.2).

4.14. Explain the normalization $N^{-1}\sum_{t=0}^{N_1-1} w_t^2 = 1$ in the Welch periodogram (4.4.3).

4.15. Modify the procedure `AveragedPeriodogram` so as to enable computation of the averaged periodogram at more frequency points than the length of the window (use zero padding).

4.16. Prove the inequality (4.5.6), and show that equality holds if and only if all the weights are equal.

4.17. The following situation is an example of the so-called *missing data* problem. Suppose that the stationary Gaussian process $\{y_t\}$ is measured at time t with probability p and missed (not measured) with probability $1-p$. The misses are i.i.d. and independent of the process $\{y_t\}$. Define $w_t = 1$ if the process is measured and $w_t = 0$ if it is missed at time t. Thus $\{w_t\}$ is a

Bernoulli process, independent of $\{y_t\}$. A natural definition of the sample covariances for this situation is

$$\tilde{r}_k = \frac{\sum_{t=0}^{N-k-1} y_{t+k} w_{t+k} y_t w_t}{\sum_{t=0}^{N-k-1} w_{t+k} w_t}.$$

(a) Show that the sample covariances thus defined are unbiased.

(b) Develop a Bartlett-like formula for the asymptotic covariances of the sample covariances, that is, for $\lim_{N\to\infty} N\text{cov}\{\tilde{r}_k, \tilde{r}_\ell\}$. See Rosen and Porat [1989] for more details.

4.18. This problem introduces the notion of *spectral resolution* and illustrates it in the context of nonparametric spectrum estimation. Let $\{x_t^{(1)}\}$ and $\{x_t^{(2)}\}$ be two AR(2) processes,

$$x_t^{(i)} = 2\rho_i \cos\omega_i x_{t-1}^{(i)} - \rho_i^2 x_{t-2}^{(i)} + u_t^{(i)}, \quad i = 1, 2,$$

where $\{u_t^{(1)}\}$ and $\{u_t^{(2)}\}$ are mutually independent unit-variance white Gaussian processes, and let $y_t = x_t^{(1)} + x_t^{(2)}$.

(a) Write a Mathematica procedure that computes and plots the spectral density of $\{y_t\}$. The input parameters to the procedure are ρ_1, ρ_2, ω_1, and ω_2. Use the procedure `Spectrum` as necessary.

(b) If ρ_1, ρ_2 are close to 1 (say 0.9 and above), and ω_1, ω_2 are not too close, you should be able to see two peaks in the spectral density at the frequencies ω_1, ω_2. As the frequencies become closer, the two peaks will eventually merge. Experiment with various values of the input parameters and find values for which the two peaks are close, but still distinguishable.

(c) Write a Mathematica procedure that generates the process $\{y_t\}$. Use the procedure `ArmaData` as necessary.

(d) The two components $\{x_t^{(1)}\}$ and $\{x_t^{(2)}\}$ are said to be *resolved* if the estimated spectral density of $\{y_t\}$ exhibits two peaks. If there is only one peak in the estimated spectral density, we say that the two components are not resolved. Thus, resolution is the ability to infer about the existence of two (or more in the general case) narrow-band components in the observed signal from the shape of the estimated spectral density. Run the procedure that generates $\{y_t\}$ and compute the corresponding spectral density estimates. Experiment with different values of N and with different estimators, and observe when the two components are resolved. Try to deduce a rule of thumb for the relation between N and $|\omega_1 - \omega_2|$ that provides resolution. Do so for $\rho_1 = \rho_2 = 0.95$, as well as for 0.99.

4.19. Let Y_N have its usual meaning (see Sec. 4.2) and let $M_N(\omega)$ be a $N \times N$ matrix function of ω. Consider spectral density estimates of the form

$$\hat{S}_N(\omega) = N^{-1} Y_N^T M_N(\omega) Y_N.$$

(a) Show that all spectral density estimates discussed in this chapter are of this form.

(b) Show that no estimate of this form can be unbiased in general (they may be asymptotically unbiased, however).

(c) Find an estimate of this form that is unbiased in the special case of $MA(q)$ processes.

4.20. Show that the de la Vallé-Poussin window is positive. Hint: Show that it is obtained as a convolution of a triangular window with itself.

CHAPTER 5

Parameter Estimation Theory
for Gaussian Processes

5.1. INTRODUCTION

This chapter serves as a general introduction to the parametric estimation of stationary Gaussian processes. As we saw in Chapter 4, the main issue in nonparametric estimation is that of consistency. When dealing with parametric estimation, consistency is still important, but it is not sufficient to describe the relative merits of candidate estimation algorithms. The Cramér-Rao lower bound provides a convenient and objective benchmark for the achievable accuracy of consistent estimators. Algorithms that are asymptotically efficient are the best from a performance point of view, but may be difficult to implement in practice. Algorithms that are convenient to implement may not be asymptotically efficient, so it is of interest to explore their *relative efficiency*, that is, their asymptotic variance relative to the CRB.

We will begin our discussion by deriving the information matrix (hence the CRB) for Gaussian processes. Both finite-sample and asymptotic formulas will be given. Then we consider maximum likelihood estimation for Gaussian processes and prove its asymptotic efficiency and normality (under certain regularity conditions). Following that, we discuss estimation of the process parameters from a finite set of sample covariances. This is a special case of the method of moments presented in Chapter 3.

The following assumptions will be made throughout this chapter.

(1) The process $\{y_t\}$ is Gaussian, zero-mean, and purely indeterministic, and its covariance sequence $\{r_n\}$ depends on a parameter vector $\theta \in \Theta$, where Θ is open in $\mathbf{R}^M$.

(2) The $\{r_n\}$ are differentiable with respect to θ.

(3) The sequences $\{r_n\}$ and $\{\partial r_n / \partial \theta_k\}$ are absolutely summable for all θ.

(4) The spectral density $S_\theta(\omega)$, which is continuous by assumption (3), is strictly positive for all ω and all θ.

In this chapter we will denote the partial derivative of any function (scalar, vector, or matrix) $G(\theta)$ with respect to θ_k by $G^{(k)}$.

5.2. THE FISHER INFORMATION OF STATIONARY GAUSSIAN PROCESSES

In this section we will derive exact and asymptotic formulas for the Fisher information of stationary, zero-mean, purely indeterministic Gaussian processes, characterized by finite parametric models. We begin by deriving the general form of the information of a zero-mean Gaussian vector.

Theorem 5.1. Let Y be a zero-mean Gaussian random vector of dimension N, whose covariance matrix Γ depends on a parameter vector θ. Assume that Γ is positive definite and differentiable with respect to θ for all θ. Then the information matrix of Y is given by

$$J_{k,\ell}(\theta) = \frac{1}{2}\mathrm{tr}\{\Gamma^{-1}\Gamma^{(k)}\Gamma^{-1}\Gamma^{(\ell)}\}. \tag{5.2.1}$$

Proof. The derivation uses the following formula, whose proof is left as an exercise to the reader (Prob. 5.1)

$$\frac{\partial \log |\Gamma|}{\partial \theta_k} = \mathrm{tr}\{\Gamma^{-1}\Gamma^{(k)}\}. \tag{5.2.2}$$

We have

$$\log f_\theta(Y) = -\frac{N}{2}\log(2\pi) - \frac{1}{2}\log|\Gamma| - \frac{1}{2}Y^T\Gamma^{-1}Y,$$

so

$$\frac{\partial \log f_\theta(Y)}{\partial \theta_k} = -\frac{1}{2}\mathrm{tr}\{\Gamma^{-1}\Gamma^{(k)}\} + \frac{1}{2}Y^T\Gamma^{-1}\Gamma^{(k)}\Gamma^{-1}Y.$$

The expectation of the second term in the right side is clearly equal to the first term (as it must be, since the expectation of the left side must be zero). Therefore,

$$E_\theta\left\{\frac{\partial \log f_\theta(Y)}{\partial \theta_k} \cdot \frac{\partial \log f_\theta(Y)}{\partial \theta_\ell}\right\} = \frac{1}{4}E_\theta Y^T\Gamma^{-1}\Gamma^{(k)}\Gamma^{-1}YY^T\Gamma^{-1}\Gamma^{(\ell)}\Gamma^{-1}Y$$

$$-\frac{1}{4}\mathrm{tr}\{\Gamma^{-1}\Gamma^{(k)}\}\mathrm{tr}\{\Gamma^{-1}\Gamma^{(\ell)}\}. \tag{5.2.3}$$

Let us denote $Z = \Gamma^{-1}Y$. Then $E_\theta ZZ^T = \Gamma^{-1}$ and

$$E_\theta Z^T\Gamma^{(k)}ZZ^T\Gamma^{(\ell)}Z = \sum_g\sum_h\sum_p\sum_q E_\theta(Z_gZ_hZ_pZ_q)\Gamma^{(k)}_{g,h}\Gamma^{(\ell)}_{p,q}$$

$$= \sum_{g} \sum_{h} \sum_{p} \sum_{q} [(\Gamma^{-1})_{g,h}(\Gamma^{-1})_{p,q} + (\Gamma^{-1})_{g,p}(\Gamma^{-1})_{h,q} + (\Gamma^{-1})_{g,q}(\Gamma^{-1})_{h,p}]\Gamma_{g,h}^{(k)}\Gamma_{p,q}^{(\ell)}$$

$$= \mathrm{tr}\{\Gamma^{-1}\Gamma^{(k)}\}\mathrm{tr}\{\Gamma^{-1}\Gamma^{(\ell)}\} + 2\,\mathrm{tr}\{\Gamma^{-1}\Gamma^{(k)}\Gamma^{-1}\Gamma^{(\ell)}\}. \qquad (5.2.4)$$

Finally, substitution of (5.2.4) in (5.2.3) yields (5.2.2). ∎

Next we derive the Fisher information for stationary zero-mean Gaussian processes. Let Y_N be the column vector of $\{y_n, 0 \le n \le N-1\}$. Let $\{\alpha_{n,0} = 1, \alpha_{n,1}, \ldots, \alpha_{n,n}\}$ be the coefficients of the nth-order best linear predictor of y_n and $u_{n,n}$ the nth-order partial innovation $u_{n,n} = \sum_{i=0}^{n} \alpha_{n,i} y_{n-i}$ (cf. Sec. 2.6). Denote $d_n = E u_{n,n}^2$ and $\alpha_n(e^{j\omega}) = \sum_{i=0}^{n} \alpha_{n,i} e^{-j\omega i}$. Also, let

$$S_n(\omega) = \frac{d_n}{\alpha_n(e^{j\omega})\alpha_n(e^{-j\omega})}. \qquad (5.2.5)$$

Observe that $S_n(\omega)$ is the spectral density of an AR(n) process whose covariances of orders 0 through n coincide with those of y_n. This follows from the corollary to Theorem 2.21 and formula (2.12.7).

Theorem 5.2.

$$[J_N(\theta)]_{k,\ell} = \sum_{n=0}^{N-1} \frac{1}{4\pi} \int_{-\pi}^{\pi} \frac{S_n^{(k)}(\omega)S_n^{(\ell)}(\omega)}{S_n^2(\omega)} d\omega. \qquad (5.2.6)$$

Proof. Let U_N be the column vector of $\{u_{n,n}, 0 \le n \le N-1\}$. Let A_N be the $N \times N$ lower triangular matrix $A_N = [\alpha_{n,n-m}]_{n \ge m}$ and D_N the $N \times N$ diagonal matrix of the $\{d_n\}$ [cf. (2.10.3)]. Finally, let $R_N = E Y_N Y_N^T = [r_{n-m}]_{n,m=0}^{N-1}$. Then we have

$$Y_N = A_N^{-1} U_N, \quad E U_N U_N^T = D_N, \quad E Y_N U_N^T = A_N^{-1} D_N, \quad R_N^{-1} = A_N^T D_N^{-1} A_N.$$

The log likelihood of Y_N is given by

$$\log f_\theta(Y_N) = -\frac{N}{2}\log(2\pi) - \frac{1}{2}\log|R_N| - \frac{1}{2}Y_N^T R_N^{-1} Y_N$$

$$= -\frac{N}{2}\log(2\pi) - \frac{1}{2}\sum_{n=0}^{N-1}\log d_n - \frac{1}{2}Y_N^T A_N^T D_N^{-1} A_N Y_N.$$

Therefore,

$$\frac{\partial \log f_\theta(Y_N)}{\partial \theta_k} = -\frac{1}{2}\sum_{n=0}^{N-1}\frac{d_n^{(k)}}{d_n}$$

$$- \frac{1}{2}[Y_N^T (A_N^{(k)})^T D_N^{-1} A_N Y_N + Y_N^T A_N^T D_N^{-1} A_N^{(k)} Y_N - Y_N^T A_N^T D_N^{-1} D_N^{(k)} D_N^{-1} A_N Y_N]$$

$$= -\frac{1}{2}\sum_{n=0}^{N-1}\frac{d_n^{(k)}}{d_n} - U_N^T D_N^{-1} A_N^{(k)} Y_N + \frac{1}{2}U_N^T D_N^{-1} D_N^{(k)} D_N^{-1} U_N.$$

We get, after some lengthy but straightforward calculations (Prob. 5.2),

$$[J_N(\theta)]_{k,\ell} = \frac{1}{2} \sum_{n=0}^{N-1} \frac{d_n^{(k)} d_n^{(\ell)}}{d_n^2} + \mathrm{tr}\{D_N^{-1} A_N^{(k)} R_N (A_N^{(\ell)})^T\}$$

$$= \frac{1}{2} \sum_{n=0}^{N-1} \frac{d_n^{(k)} d_n^{(\ell)}}{d_n^2} + \sum_{n=0}^{N-1} \frac{1}{d_n} \left(\sum_{p=0}^{n} \sum_{q=0}^{n} \alpha_{n,n-p}^{(k)} r_{p-q} \alpha_{n,n-q}^{(\ell)} \right). \quad (5.2.7)$$

By (5.2.5),

$$r_m = \frac{1}{2\pi} \int_{-\pi}^{\pi} \frac{d_n e^{j\omega m}}{\alpha_n(e^{j\omega}) \alpha_n(e^{-j\omega})} d\omega, \quad 0 \le m \le n. \quad (5.2.8)$$

Substitute (5.2.8) in (5.2.7) to get

$$[J_N(\theta)]_{k,\ell} = \frac{1}{2} \sum_{n=0}^{N-1} \frac{d_n^{(k)} d_n^{(\ell)}}{d_n^2} + \sum_{n=0}^{N-1} \frac{1}{2\pi} \int_{-\pi}^{\pi} \left(\sum_{p=0}^{n} \sum_{q=0}^{n} \frac{\alpha_{n,n-p}^{(k)} \alpha_{n,n-q}^{(\ell)} e^{j\omega(p-q)}}{\alpha_n(e^{j\omega}) \alpha_n(e^{-j\omega})} \right) d\omega$$

$$= \frac{1}{2} \sum_{n=0}^{N-1} \frac{d_n^{(k)} d_n^{(\ell)}}{d_n^2} + \sum_{n=0}^{N-1} \frac{1}{2\pi} \int_{-\pi}^{\pi} \frac{\alpha_n^{(k)}(e^{j\omega}) \alpha_n^{(\ell)}(e^{-j\omega})}{\alpha_n(e^{j\omega}) \alpha_n(e^{-j\omega})} d\omega. \quad (5.2.9)$$

Next note that

$$\frac{S_n^{(k)}(\omega)}{S_n(\omega)} = \frac{d_n^{(k)}}{d_n} - \frac{\alpha_n^{(k)}(e^{j\omega})}{\alpha_n(e^{j\omega})} - \frac{\alpha_n^{(k)}(e^{-j\omega})}{\alpha_n(e^{-j\omega})}.$$

This gives, after some more calculations (Prob. 5.3),

$$\frac{1}{2\pi} \int_{-\pi}^{\pi} \frac{S_n^{(k)}(\omega) S_n^{(\ell)}(\omega)}{S_n^2(\omega)} d\omega = \frac{d_n^{(k)} d_n^{(\ell)}}{d_n^2} + \frac{2}{2\pi} \int_{-\pi}^{\pi} \frac{\alpha_n^{(k)}(e^{j\omega}) \alpha_n^{(\ell)}(e^{-j\omega})}{\alpha_n(e^{j\omega}) \alpha_n(e^{-j\omega})} d\omega. \quad (5.2.10)$$

Finally, substitution of (5.2.10) in (5.2.9) yields (5.2.6). ∎

The limit of (5.2.6) yields the asymptotic information matrix as given by the following theorem, due to Whittle [1953].

Theorem 5.3.

$$[J_0(\theta)]_{k,\ell} = \lim_{N \to \infty} N^{-1} [J_N(\theta)]_{k,\ell} = \frac{1}{4\pi} \int_{-\pi}^{\pi} \frac{S^{(k)}(\omega) S^{(\ell)}(\omega)}{S^2(\omega)} d\omega. \quad (5.2.11)$$

Partial Proof. Under assumptions (1) to (4), the sequence $\{S_n(\omega)\}$ converges to $S(\omega)$ uniformly on $[-\pi, \pi]$, and the sequence $\{S_n^{(k)}(\omega)\}$ converges to $S^{(k)}(\omega)$ uniformly on this interval. The proof of these statements is not difficult, but lengthy and technical, so it will be skipped. Since $\{S_n(\omega)\}$ and $S(\omega)$ are strictly positive, the sequence $\{S_n^{(k)}(\omega) S_n^{(\ell)}(\omega)/S_n^2(\omega)\}$ converges uniformly to $S^{(k)}(\omega) S^{(\ell)}(\omega)/S^2(\omega)$

on $[-\pi, \pi]$. Since uniform convergence on a closed interval implies convergence of the integrals to the integral of the limit [Rudin, 1964, p. 137], we get

$$\lim_{n \to \infty} \int_{-\pi}^{\pi} \frac{S_n^{(k)}(\omega)S_n^{(\ell)}(\omega)}{S_n^2(\omega)} d\omega = \int_{-\pi}^{\pi} \frac{S^{(k)}(\omega)S^{(\ell)}(\omega)}{S^2(\omega)} d\omega.$$

Therefore, (5.2.11) follows from (5.2.6) by the convergence theorem of arithmetic means [A10].

Remark. Whittle's formula (5.2.11) can be written in the alternative form

$$[J_0(\theta)]_{k,\ell} = \frac{1}{4\pi} \int_{-\pi}^{\pi} [\log S(\omega)]^{(k)} [\log S(\omega)]^{(\ell)} d\omega. \qquad (5.2.12)$$

∎

We now use the result (5.2.12) to derive the asymptotic information matrix for Gaussian ARMA processes. This will serve as an example of the application of Whittle's formula and will turn out to be a highly useful result in its own merit.

The spectral density of an ARMA(p, q) process is

$$S(\omega) = \frac{\sigma_u^2 b(e^{j\omega})b(e^{-j\omega})}{a(e^{j\omega})a(e^{-j\omega})},$$

where

$$a(e^{j\omega}) = 1 + \sum_{k=1}^{p} a_k e^{-j\omega k}, \quad b(e^{j\omega}) = 1 + \sum_{k=1}^{q} b_k e^{-j\omega k}.$$

The polynomials $z^p a(z)$ and $z^q b(z)$ are assumed to be coprime and nonzero on and outside the unit circle. The coefficients a_p and b_q are assumed to be nonzero. The parameter vector is $\theta = [\sigma_u^2, a_1, \ldots, a_p, b_1, \ldots, b_q]^T$. We have

$$\frac{\partial \log S(\omega)}{\partial \sigma_u^2} = \frac{1}{\sigma_u^2}$$

$$\frac{\partial \log S(\omega)}{\partial a_k} = -\frac{e^{-j\omega k}}{a(e^{j\omega})} - \frac{e^{j\omega k}}{a(e^{-j\omega})}$$

$$\frac{\partial \log S(\omega)}{\partial b_k} = \frac{e^{-j\omega k}}{b(e^{j\omega})} + \frac{e^{j\omega k}}{b(e^{-j\omega})}.$$

Therefore,

$$\frac{1}{4\pi} \int_{-\pi}^{\pi} \left(\frac{\partial \log S(\omega)}{\partial \sigma_u^2} \right)^2 d\omega = \frac{1}{2\sigma_u^4}$$

$$\frac{1}{4\pi} \int_{-\pi}^{\pi} \frac{\partial \log S(\omega)}{\partial \sigma_u^2} \cdot \frac{\partial \log S(\omega)}{\partial a_k} d\omega = 0$$

$$\frac{1}{4\pi} \int_{-\pi}^{\pi} \frac{\partial \log S(\omega)}{\partial \sigma_u^2} \cdot \frac{\partial \log S(\omega)}{\partial b_k} d\omega = 0$$

$$\frac{1}{4\pi}\int_{-\pi}^{\pi}\frac{\partial \log S(\omega)}{\partial a_k}\cdot\frac{\partial \log S(\omega)}{\partial a_\ell}d\omega = \frac{1}{2\pi}\int_{-\pi}^{\pi}\frac{e^{j\omega(\ell-k)}}{a(e^{j\omega})a(e^{-j\omega})}d\omega$$

$$\frac{1}{4\pi}\int_{-\pi}^{\pi}\frac{\partial \log S(\omega)}{\partial a_k}\cdot\frac{\partial \log S(\omega)}{\partial b_\ell}d\omega = -\frac{1}{2\pi}\int_{-\pi}^{\pi}\frac{e^{j\omega(\ell-k)}}{a(e^{j\omega})b(e^{-j\omega})}d\omega$$

$$\frac{1}{4\pi}\int_{-\pi}^{\pi}\frac{\partial \log S(\omega)}{\partial b_k}\cdot\frac{\partial S(\omega)}{\partial b_\ell}d\omega = \frac{1}{2\pi}\int_{-\pi}^{\pi}\frac{e^{j\omega(\ell-k)}}{b(e^{j\omega})b(e^{-j\omega})}d\omega.$$

Therefore, the asymptotic information matrix is given by

$$J_0(\theta) = \begin{bmatrix} 0.5\sigma_u^{-4} & 0 & 0 \\ 0 & R_{xx} & -R_{xv} \\ 0 & -R_{xv}^T & R_{vv} \end{bmatrix}, \qquad (5.2.13)$$

where:

- R_{xx} is the $p \times p$ covariance matrix of an AR(p) process $\{x_t\}$ whose characteristic polynomial is $a(z)$ and whose innovation variance is 1. It is a Toeplitz matrix whose entries are given by

$$[R_{xx}]_{k,\ell} = \frac{1}{2\pi}\int_{-\pi}^{\pi}\frac{e^{j\omega(\ell-k)}}{a(e^{j\omega})a(e^{-j\omega})}d\omega. \qquad (5.2.14)$$

- R_{vv} is the $q \times q$ covariance matrix of an AR(q) process $\{v_t\}$ whose characteristic polynomial is $b(z)$ and whose innovation variance is 1. It is a Toeplitz matrix whose entries are given by

$$[R_{vv}]_{k,\ell} = \frac{1}{2\pi}\int_{-\pi}^{\pi}\frac{e^{j\omega(\ell-k)}}{b(e^{j\omega})b(e^{-j\omega})}d\omega. \qquad (5.2.15)$$

- R_{xv} is the $p \times q$ cross-covariance matrix of the processes $\{x_t\}$ and $\{v_t\}$. It is a Toeplitz matrix whose entries are given by

$$[R_{xv}]_{k,\ell} = \frac{1}{2\pi}\int_{-\pi}^{\pi}\frac{e^{j\omega(\ell-k)}}{a(e^{j\omega})b(e^{-j\omega})}d\omega. \qquad (5.2.16)$$

The entries of the matrix R_{xx} can be computed by solving the set of equations (2.12.7) (with $\sigma_u^2 = 1$). The entries of R_{vv} can be computed in a similar manner. A simple procedure for computing the entries of R_{xv} is developed in Prob. 5.4.

The asymptotic information matrices of pure AR and pure MA processes are obtained as special cases of (5.2.13). For the former, only R_{xx} appears, while for the latter, only R_{vv}. For all three types of processes, the information of σ_u^2 is asymptotically decoupled from that of the other parameters, as is seen from the block-diagonal structure of (5.2.13). In the case of ARMA processes, there is coupling between the numerator and the denominator parameters. We should emphasize again that the above result is valid only when all the zeros of $b(z)$ are inside the unit circle.

5.3. MAXIMUM LIKELIHOOD PARAMETER ESTIMATION

In this section we explore the asymptotic properties of the maximum likelihood estimates for parametric Gaussian processes.

Let us first study the limiting behavior of the log likelihood function. As in Chapter 3, we denote the true value of the parameter by θ and any other value by ψ. The true value is, by definition, the one governing the distribution of the process $\{y_t\}$.

Theorem 5.4. Under the above assumptions,

$$\lim_{N\to\infty} -\frac{2}{N}\log f_\psi(Y_N) = \log(2\pi) + \frac{1}{2\pi}\int_{-\pi}^{\pi}\left[\frac{S_\theta(\omega)}{S_\psi(\omega)} + \log S_\psi(\omega)\right]d\omega \quad \text{a. s. } (\theta).$$

$$(5.3.1)$$

Proof. We have

$$-\frac{2}{N}\log f_\psi(Y_N) = \log(2\pi) + \frac{1}{N}\sum_{n=0}^{N-1}\log d_{\psi,n} + \frac{1}{N}Y_N^T R_{\psi,N}^{-1} Y_N.$$

The proof will be in two parts. First, we will show that

$$\lim_{N\to\infty}\frac{1}{N}[Y_N^T R_{\psi,N}^{-1} Y_N - \text{tr}\{R_{\psi,N}^{-1} R_{\theta,N}\}] = 0 \quad \text{a. s. } (\theta),\qquad (5.3.2)$$

and then that

$$\lim_{N\to\infty}\frac{1}{N}\left[\text{tr}\{R_{\psi,N}^{-1} R_{\theta,N}\} + \sum_{n=0}^{N-1}\log d_{\psi,n}\right] = \frac{1}{2\pi}\int_{-\pi}^{\pi}\left[\frac{S_\theta(\omega)}{S_\psi(\omega)} + \log S_\psi(\omega)\right]d\omega. \quad (5.3.3)$$

Letting $U_N = R_{\theta,N}^{-1/2}Y_N$, we get[†]

$$\frac{1}{N}[Y_N^T R_{\psi,N}^{-1} Y_N - \text{tr}\{R_{\psi,N}^{-1} R_{\theta,N}\}] = \frac{1}{N}[U_N^T R_{\theta,N}^{T/2} R_{\psi,N}^{-1} R_{\theta,N}^{1/2} U_N - \text{tr}\{R_{\psi,N}^{-1} R_{\theta,N}\}].$$

The components of U_N are Gaussian i.i.d. (cf. Theorem 4.3), so we can use Theorem C.5 to prove the almost sure convergence to zero. All we need to show is that the eigenvalues of $R_{\theta,N}^{T/2} R_{\psi,N}^{-1} R_{\theta,N}^{1/2}$ are uniformly bounded in magnitude. These are equal to the eigenvalues of $R_{\psi,N}^{-1/2} R_{\theta,N} R_{\psi,N}^{-T/2}$ (Prob. 5.5), which are bounded in magnitude by Prob. 2.13. This proves (5.3.3).

To prove (5.3.3), write $\text{tr}\{R_{\psi,N}^{-1} R_{\theta,N}\}$ as in the proof of Theorem 5.2:

$$\text{tr}\{R_{\psi,N}^{-1} R_{\theta,N}\} = \text{tr}\{D_{\psi,N}^{-1} A_{\psi,N} R_{\theta,N} A_{\psi,N}^T\}$$

$$= \sum_{n=0}^{N-1}\frac{1}{2\pi}\int_{-\pi}^{\pi}\frac{d_{\theta,n}\alpha_{\psi,n}(e^{j\omega})\alpha_{\psi,n}(e^{-j\omega})}{d_{\psi,n}\alpha_{\theta,n}(e^{j\omega})\alpha_{\theta,n}(e^{-j\omega})}d\omega,$$

[†] Note that U_N is different from the one used in the proof of Theorem 5.2. How are they related?

so, by the same argument as in the proof of Theorem 5.3,

$$\lim_{N \to \infty} \frac{1}{N} \text{tr}\{R_{\psi,N}^{-1} R_{\theta,N}\} = \frac{1}{2\pi} \int_{-\pi}^{\pi} \frac{S_\theta(\omega)}{S_\psi(\omega)} d\omega.$$

Also, by Kolmogorov's formula (2.8.6),

$$\lim_{N \to \infty} \frac{1}{N} \sum_{n=0}^{N-1} \log d_{\psi,n} = \log \sigma_{u,\psi}^2 = \frac{1}{2\pi} \int_{-\pi}^{\pi} \log S_\psi(\omega) d\omega.$$

Adding these two equalities yields (5.3.3).   ∎

Corollary. The value $\psi = \theta$ is the global maximizer of the almost sure limit of the log likelihood.

Proof. It is easy to check that $x - 1 - \log x \geq 0$ for all $x > 0$, with equality for $x = 1$ only. Substitute $x = S_\theta(\omega)/S_\psi(\omega)$ to get

$$\frac{S_\theta(\omega)}{S_\psi(\omega)} + \log S_\psi(\omega) \geq 1 + \log S_\theta(\omega),$$

so

$$\frac{1}{2\pi} \int_{-\pi}^{\pi} \left[\frac{S_\theta(\omega)}{S_\psi(\omega)} + \log S_\psi(\omega) \right] d\omega \geq \frac{1}{2\pi} \int_{-\pi}^{\pi} [1 + \log S_\theta(\omega)] d\omega$$

with equality only if $S_\psi(\omega) = S_\theta(\omega)$ almost everywhere on $[-\pi, \pi]$.   ∎

An interesting property of Gaussian ARMA models is that their asymptotic likelihood function has a *unique* stationary point (i.e., a point where the gradient with respect to θ is zero), which is the point of global maximum. This was proved in Åström and Söderström [1974], and later in Stoica and Söderström [1982] using a different technique.

Theorem 5.4 and its conclusion suggest an interpretation similar to the one in Sec. 3.5 for the i.i.d. case. The normalized log likelihood converges to a deterministic function that has a global maximum at the true value of the parameter vector. This, by itself, does not prove the consistency of the maximum likelihood estimate, since it says nothing about the convergence of the global maxima of the finite-sample log likelihood functions. However, the equivalent of Theorem 3.12 holds for the present case, subject to the following additional assumptions.

(5) The r_n are continuously differentiable with respect to θ up to a third-order and the sequences $\{r_n\}$, $\{\partial r_n/\partial\theta_k\}$, $\{\partial^2 r_n/\partial\theta_k\partial\theta_\ell\}$, and $\{\partial^3 r_n/\partial\theta_k\partial\theta_\ell\partial\theta_m\}$ are absolutely summable for all θ.

(6) The asymptotic Fisher information matrix $J_0(\theta)$ is positive definite.

Theorem 5.5. Under assumptions (1) to (6), there exists a sequence of local maxima of the log likelihood function that converges to θ almost surely.

Partial Proof. The proof is very similar to that of Theorem 3.12, so we only sketch it. The first step is to verify that condition (3.5.2) holds. Since the parameter set is open, there is an open sphere $\{\psi : \|\psi - \theta\| < \rho\}$ such that $R_{\psi,N}$ is positive definite for all N and all ψ in the sphere. In this sphere, the third derivative of the log likelihood is a quadratic form in Y_N, so the expectation of its magnitude is finite and bounded. The rest of the proof follows as in Theorem 3.12. We only need to show the almost sure convergence of the first three derivatives of the log likelihood. For the first derivative, we have

$$\frac{2}{N}\frac{\partial \log f_\theta(Y_N)}{\partial \theta_k} = -\frac{1}{N}\sum_{n=0}^{N-1}\frac{d_{\theta,n}^{(k)}}{d_{\theta,n}} + \frac{1}{N}Y_N^T R_{\theta,N}^{-1} R_{\theta,N}^{(k)} R_{\theta,N}^{-1} Y_N$$

$$= -\frac{1}{N}\sum_{n=0}^{N-1}\frac{d_{\theta,n}^{(k)}}{d_{\theta,n}} + \frac{1}{N}U_N^T R_{\theta,N}^{-1/2} R_{\theta,N}^{(k)} R_{\theta,N}^{-T/2} U_N.$$

The almost sure convergence of the right side to zero is established exactly as in the proof of Theorem 5.3. The second and third derivatives are handled in a similar manner. In each case we get a quadratic form in U_N, the almost sure convergence of which follows from Theorem C.5.   ∎

As in the i.i.d. case, the sequence of local maxima can be used to construct a strongly consistent estimate $\hat{\theta}_N(Y_N)$, using some consistent initial estimate $\bar{\theta}_N(Y_N)$. The asymptotic normality and efficiency of $\hat{\theta}_N(Y_N)$ can be shown as in Theorem 3.13.

Theorem 5.6. Under the assumptions (1) to (6), the estimate $\hat{\theta}_N(Y_N)$ is best asymptotically normal.

Partial Proof. We only need to show the asymptotic normality of $\partial \log f_\theta(Y_N)/\partial \theta$. The rest of the proof is then identical to that of Theorem 3.13. The asymptotic normality of the derivative of the log likelihood will be proved similarly to Theorem 4.3. Let

$$\xi = 2\sum_{m=1}^{M} a_m \frac{\partial \log f_\theta(Y_N)}{\partial \theta_m}$$

be some linear combination of the partial derivatives. Then $\xi = U_N^T C_N U_N$, where

$$C_N = \sum_{m=1}^{M} a_m R_{\theta,N}^{-1/2} R_{\theta,N}^{(m)} R_{\theta,N}^{-T/2} = R_{\theta,N}^{-1/2}\left[\sum_{m=1}^{M} a_m R_{\theta,N}^{(m)}\right] R_{\theta,N}^{-T/2}$$

$$= R_{\theta,N}^{-1/2} B_N R_{\theta,N}^{-T/2}.$$

By Prob. 2.13, the eigenvalues of C_N are uniformly bounded in magnitude. Also $\text{tr}\{[C_N]^2\} = \text{tr}\{R_{\theta,N}^{-1} B_N R_{\theta,N}^{-1} B_N\}$, so, by Prob. 2.16, $N^{-1}\text{tr}\{[C_N]^2\}$ is bounded from below for sufficiently large N. The conclusion follows as in Theorem 4.3.   ∎

5.4. THE RELATIVE EFFICIENCY OF THE SAMPLE COVARIANCES

In Chapter 4 we discussed the problem of estimating the covariance sequence of general stationary processes and, in particular, analyzed the asymptotic properties of the sample covariances $\hat{r}_k$. Now we specialize the discussion to Gaussian processes with finite parametric models. We compare the asymptotic covariance of the sample covariances with the asymptotic Cramér-Rao bound and draw conclusions about the relative efficiency of the sample covariances. The main reason for our interest in this question is as an introduction and preparation for the next topic, the estimation of the process parameters from the sample covariances by the method of moments.

Let $L_2^e[-\pi, \pi]$ be the subspace of all real even functions in $L_2[-\pi, \pi]$. This subspace is easily verified to be closed; hence it is a Hilbert space. The functions $\{\cos \omega k, 0 \leq k < \infty\}$ constitute a basis for $L_2^e[-\pi, \pi]$ (this is a consequence of the fact that $\{e^{j\omega k}, -\infty < k < \infty\}$ form a basis for $L_2[-\pi, \pi]$). Define

$$\phi_k(\omega) = \sqrt{2} S(\omega) \cos \omega k, \quad 0 \leq k < \infty, \tag{5.4.1a}$$

$$\psi_m(\omega) = \frac{S^{(m)}(\omega)}{\sqrt{2} S(\omega)}, \quad 1 \leq m \leq M. \tag{5.4.1b}$$

The functions $\{\phi_k(\omega), 0 \leq k < \infty\}$ are also a basis for $L_2^e[-\pi, \pi]$, albeit not an orthonormal one (Prob. 5.13). Let $\underline{\phi}(\omega)$ be the column vector of $\{\phi_k(\omega), 0 \leq k \leq K - 1\}$ and $\underline{\psi}(\omega)$ the column vector of $\{\psi_m(\omega), 1 \leq m \leq M\}$.

Let $\Sigma(\underline{\theta})$ be the normalized asymptotic covariance of $\{\hat{r}_k, 0 \leq k \leq K - 1\}$ and $F(\theta)$ the Jacobian matrix of the functions $r_k(\theta)$.[†] Let $J_0(\theta)$ be the normalized asymptotic information matrix given in (5.2.11). We will show that these three matrices can be expressed as inner products of the functions defined in (5.4.1).

Recall formula (4.2.11) for the asymptotic covariances of the sample covariances. If the process is Gaussian, then the third term (the one involving the fourth-order cumulants) vanishes. By expressing the complex exponentials in terms of sines and cosines, we get

$$\Sigma_{k,\ell}(\theta) = \lim_{N \to \infty} N \operatorname{cov}\{\hat{r}_k, \hat{r}_\ell\} = \frac{1}{2\pi} \int_{-\pi}^{\pi} 2S^2(\omega) \cos \omega k \cos \omega \ell \, d\omega = \langle \phi_k(\omega), \phi_\ell(\omega) \rangle. \tag{5.4.2}$$

Also (Prob. 5.14),

$$F_{k,m}(\theta) = \frac{\partial r_k}{\partial \theta_m} = \frac{\partial}{\partial \theta_m} \frac{1}{2\pi} \int_{-\pi}^{\pi} S(\omega) \cos \omega k \, d\omega = \frac{1}{2\pi} \int_{-\pi}^{\pi} \frac{\partial S(\omega)}{\partial \theta_m} \cos \omega k \, d\omega$$
$$= \langle \phi_k(\omega), \psi_m(\omega) \rangle, \tag{5.4.3}$$

and by (5.2.11)

$$[J_0(\theta)]_{k,\ell} = \langle \psi_k(\omega), \psi_\ell(\omega) \rangle. \tag{5.4.4}$$

[†] This is the matrix denoted by $S(\theta)$ in Sec. 3.6; we change notation to avoid conflict with the notation $S(\omega)$ for the spectral density.

Formulas (5.4.2) through (5.4.4) immediately imply the following result.

Theorem 5.7. $\{\hat{r}_k, 0 \leq k \leq K-1\}$ are asymptotically efficient estimates of $\{r_k, 0 \leq k \leq K-1\}$ if and only if the functions in $\underline{\phi}(\omega)$ are linear combinations of the functions in $\underline{\psi}(\omega)$.

Proof. Let $\underline{\epsilon}(\omega)$ be the residual from the projection of $\underline{\phi}(\omega)$ on the space spanned by the elements of $\underline{\psi}(\omega)$. Then (Prob. 5.15)

$$\underline{\epsilon}(\omega) = \underline{\phi}(\omega) - F(\theta)J_0^{-1}(\theta)\underline{\psi}(\omega).$$

The Gram matrix of the components of $\underline{\epsilon}(\omega)$ is nonnegative; that is,

$$\Sigma(\theta) - F(\theta)J_0^{-1}(\theta)F^T(\theta) \geq 0.$$

This inequality simply shows that the asymptotic covariance of the sample covariances is greater than or equal to the asymptotic Cramér-Rao bound, which is not at all surprising. However, it also shows that equality holds if and only if $\underline{\epsilon}(\omega)$ is identically zero, that is, if each of the elements of $\underline{\phi}(\omega)$ is a linear combination of the elements of $\underline{\psi}(\omega)$. ∎

To illustrate the use of this theorem, consider the case of ARMA processes. The functions of $\underline{\psi}(\omega)$ are given by

$$\psi_0(\omega) = \frac{1}{\sqrt{2}S(\omega)} \cdot \frac{\partial S(\omega)}{\partial \sigma_u^2} = \frac{1}{\sqrt{2}\sigma_u^2} = \frac{a(e^{j\omega})a(e^{-j\omega})}{\sqrt{2}\sigma_u^2 a(e^{j\omega})a(e^{-j\omega})}$$

$$\psi_m(\omega) = \frac{1}{\sqrt{2}S(\omega)} \cdot \frac{\partial S(\omega)}{\partial a_m} = -\frac{e^{-j\omega m}}{\sqrt{2}a(e^{j\omega})} - \frac{e^{j\omega m}}{\sqrt{2}a(e^{-j\omega})}$$

$$= -\frac{e^{-j\omega m}a(e^{-j\omega}) + e^{j\omega m}a(e^{j\omega})}{\sqrt{2}a(e^{j\omega})a(e^{-j\omega})}, \quad 1 \leq m \leq p$$

$$\psi_{m+p}(\omega) = \frac{1}{\sqrt{2}S(\omega)} \cdot \frac{\partial S(\omega)}{\partial b_m} = \frac{e^{-j\omega m}}{\sqrt{2}b(e^{j\omega})} + \frac{e^{j\omega m}}{\sqrt{2}b(e^{-j\omega})}$$

$$= \frac{e^{-j\omega m}b(e^{-j\omega}) + e^{j\omega m}b(e^{j\omega})}{\sqrt{2}b(e^{j\omega})b(e^{-j\omega})}, \quad 1 \leq m \leq q.$$

As we see, the first $p+1$ entries of $\underline{\psi}(\omega)$ have the form

$$\psi_m(\omega) = \frac{d_m(e^{j\omega})}{a(e^{j\omega})a(e^{-j\omega})},$$

where $d_m(e^{j\omega})$ is a symmetric polynomial in $e^{j\omega}$ of degree $\max\{m, p-m\}$. These functions are linearly independent, so together they span the same $(p+1)$-dimensional subspace as the functions $\{(e^{j\omega m} + e^{-j\omega m})/a(e^{j\omega})a(e^{-j\omega}), 0 \leq m \leq p\}$. Returning to the function $\phi_k(\omega)$, we observe that this function has the form $g_k(e^{j\omega})/a(e^{j\omega})a(e^{-j\omega})$, where $g_k(e^{j\omega})$ is a symmetric polynomial of degree $k+q$. Therefore,

$\phi_k(\omega)$ is a linear combination of the elements of $\underline{\psi}(\omega)$ if $0 \leq k+q \leq p$. We therefore get the following interesting result.

Theorem 5.8. The sample covariances $\{\hat{r}_k, 0 \leq k \leq p-q\}$ of a Gaussian ARMA(p, q) process are asymptotically efficient. $\hat{r}_k$ is not asymptotically efficient for any $k > p - q$.

Proof. The first part was proved above. The second part follows upon observing that (a) $e^{j\omega k} + e^{-j\omega k}$ is linearly independent of all $\{e^{j\omega m} + e^{-j\omega m}, m < k\}$, so $\phi_k(\omega)$ is not a linear combination of the first $p+1$ elements of $\underline{\psi}(\omega)$ if $k > p - q$; (b) the last q elements of $\underline{\psi}(\omega)$ are of no help, since the denominator polynomials $b(e^{j\omega})b(e^{-j\omega})$ and $a(e^{j\omega})a(e^{-j\omega})$ are coprime. ∎

Corollaries.

(a) The first $p+1$ sample covariances of an AR(p) process are asymptotically efficient.

(b) None of the sample covariances of a MA(q) process is asymptotically efficient.

(c) The sample variance $\hat{r}_0$ of an ARMA(p, p) process is asymptotically efficient, but none of the other sample covariances is. ∎

5.5. PARAMETER ESTIMATION FROM THE SAMPLE COVARIANCES

In this section we explore the method of moments of Sec. 3.6 in the context of stationary Gaussian processes. In particular, we examine the potential accuracy of estimates based on the sample covariance statistics $\{\hat{r}_k, 0 \leq k \leq K-1\}$ in comparison to the asymptotic Cramér-Rao bound. We use the notations and definitions introduced in the previous section.

Theorem 5.9.

$$[F^T(\theta)\Sigma^{-1}(\theta)F(\theta)]^{-1} \geq J_0^{-1}(\theta) \tag{5.5.1}$$

with equality if and only if all the elements of $\underline{\psi}(\omega)$ are in the space spanned by the elements of $\underline{\phi}(\omega)$.

Proof. The proof is similar to that of Theorem 5.7. Let $\underline{\epsilon}(\omega)$ be the residual from the projection of $\underline{\psi}(\omega)$ on the space spanned by the elements of $\underline{\phi}(\omega)$. Then

$$\underline{\epsilon}(\omega) = \underline{\psi}(\omega) - F^T(\theta)\Sigma^{-1}(\theta)\underline{\phi}(\omega).$$

The Gram matrix of the components of $\underline{\epsilon}(\omega)$ is nonnegative; that is,

$$J_0(\theta) - F^T(\theta)\Sigma^{-1}(\theta)F(\theta) \geq 0.$$

Therefore [A14],

$$[F^T(\theta)\Sigma^{-1}(\theta)F(\theta)]^{-1} \geq J_0^{-1}(\theta).$$

Equality holds if and only if $\underline{\epsilon}(\omega) = 0$, that is, if and only if all the components of $\psi(\omega)$ are in the space spanned by the components of $\phi(\omega)$.   ∎

Recall that the left side of (5.5.1) is the lower bound on the asymptotic covariance matrix of any estimate of θ obtained from $\{\hat{r}_k, 0 \le k \le K - 1\}$. Theorem 5.9 says, in essence, that estimates based on the sample covariances are generally not asymptotically efficient. The only exception is given by the following theorem.

Theorem 5.10. The parameters of a Gaussian process satisfying assumptions (1) to (4) can be estimated from the sample covariances $\{\hat{r}_k, 0 \le k \le K - 1\}$ in an asymptotically efficient manner if and only if the spectral density of the process has the functional form

$$S_\theta(\omega) = \frac{1}{d_\theta(e^{j\omega}) + d_0(\omega)}, \tag{5.5.2}$$

where (i) $d_\theta(e^{j\omega})$ is a symmetric polynomial of degree $K-1$ in $e^{j\omega}$, whose coefficients are functions of θ; (ii) $d_0(\omega)$ is a real and symmetric function of ω, independent of θ and subject to the regularity assumptions (1) to (4), but arbitrary otherwise.

Proof. Let $U_\theta(\omega) = 1/S_\theta(\omega)$. Then we can express the condition $\underline{\epsilon}(\omega) = 0$ as

$$-\frac{1}{U_\theta(\omega)} \cdot \frac{\partial U_\theta(\omega)}{\partial \theta_m} = \frac{1}{U_\theta(\omega)} \sum_{k=0}^{K-1} f_k(\theta) \cos \omega k$$

for some set of coefficients $f_k(\theta)$. This is equivalent to saying that $\partial U_\theta(\omega)/\partial \theta_m$ is a symmetric polynomial in $e^{j\omega}$ for every θ; hence

$$U_\theta(\omega) = d_\theta(e^{j\omega}) + d_0(\omega)$$

for some symmetric polynomial $d_\theta(e^{j\omega})$ of degree $K - 1$ and an arbitrary function $d_0(\omega)$ independent of θ.   ∎

Autoregressive processes provide an example for Theorem 5.10. In the case of AR(p) processes we have $d_\theta(e^{j\omega}) = a(e^{j\omega})a(e^{-j\omega})/\sigma_u^2$ and $d_0(\omega) = 0$. The immediate conclusion is that the parameters of an AR(p) process can be estimated from $\{\hat{r}_k, 0 \le k \le p\}$ in an asymptotically efficient manner. On the other hand, an ARMA(p, q) spectrum is not of the form (5.5.2) for any finite K. This implies that the parameters of a Gaussian ARMA process cannot be efficiently estimated from any finite number of sample covariances.

By Theorem 5.9, the difference between the bounds $[F^T(\theta)\Sigma^{-1}(\theta)F(\theta)]^{-1}$ and $J_0^{-1}(\theta)$ is roughly proportional to the norm of $\underline{\epsilon}(\omega)$. By Theorem 5.10, this is related to the achievable error in approximating $U_\theta(\omega)$ by a symmetric polynomial in $e^{j\omega}$. In the case of an ARMA process, $U_\theta(\omega) = a(e^{j\omega})a(e^{-j\omega})/\sigma_u^2 b(e^{j\omega})b(e^{-j\omega})$. This function has poles at the zeros of $b(z)$. The closer the poles to the unit circle, the harder it is to approximate $U_\theta(\omega)$ by a finite-order polynomial. We can therefore make the following qualitative statement: the closer the zeros of an ARMA process to the unit circle, the greater the distance between the CRB and the asymptotic variance of the best estimate based on a fixed number of sample covariances.

When the number of sample covariances K approaches infinity, the inequality in Theorem 5.9 becomes an equality, as we will now prove.

Theorem 5.11.
$$\lim_{K\to\infty}[F^T(\theta)\Sigma^{-1}(\theta)F(\theta)]^{-1} = J_0^{-1}(\theta). \tag{5.5.3}$$

Proof. From the fact that $\{\phi_k(\omega), 0 \le k < \infty\}$ form a basis for $L_2^e[-\pi, \pi]$, it follows that
$$\underset{K\to\infty}{\text{l.i.m.}}\ \epsilon(\omega) = 0.$$
This implies that
$$J_0(\theta) - \lim_{K\to\infty} F^T(\theta)\Sigma^{-1}(\theta)F(\theta) = 0,$$
whence (5.5.2) follows. ∎

Recall that in Chapter 4 we used an increasing sequence of sample covariances $\{\hat{r}_k, 0 \le k \le K(N)\}$ to obtain consistent estimates of the spectral density, by imposing the conditions $K(N) \to \infty$ and $K(N)/N \to 0$. By Theorem 5.11, a similar idea can be used here to obtain asymptotically efficient estimates of θ from an increasing sequence of sample covariances. Let $\hat{\theta}_N(Y_N)$ be the estimate obtained through global minimization of the cost function (3.6.3), using the statistics $\{\hat{r}_k, 0 \le k \le K(N)\}$. The condition $\lim_{N\to\infty} K(N) = \infty$ enables us to use Theorem 5.11. A condition of the form $\lim_{N\to\infty} K^m(N)/N = 0$, for some positive m, is necessary to guarantee the desired asymptotic properties of $\hat{\theta}_N(Y_N)$. The proper value of m has not been explored, but the author believes it to be $m = 3$.

We conclude our discussion with a remark about non-Gaussian linear processes. The Cramér-Rao bound derived in this chapter does not apply to non-Gaussian processes, so any comparison to this bound would be meaningless. However, the asymptotic variance of estimates based on the sample covariances can still be analyzed using the same techniques. In fact, the only difference in the analysis for the non-Gaussian case is the appearance of the fourth-order cumulant term in the matrix $\Sigma(\theta)$ [cf. (4.2.11)]. For convenience, we continue to denote the Gaussian expression (i.e., without the cumulant term) by $\Sigma(\theta)$, while denoting the non-Gaussian expression by $\overline{\Sigma}(\theta)$. Assuming that the process is linear, it follows from (4.2.13) that
$$\overline{\Sigma}(\theta) = \Sigma(\theta) + \delta rr^T, \tag{5.5.4}$$
where r is the vector of true covariances of orders 0 through $K-1$ and $\delta = \kappa_4(w)/\sigma_w^4$. By the matrix identity [A19],
$$F^T\overline{\Sigma}^{-1}F = F^T\Sigma^{-1}F - \frac{\delta F^T\Sigma^{-1}rr^T\Sigma^{-1}F}{1+\delta r^T\Sigma^{-1}r} = F^T\Sigma^{-1}F - \bar{\delta}F^T\Sigma^{-1}rr^T\Sigma^{-1}F, \tag{5.5.5}$$

where

$$\overline{\delta} = \frac{\delta}{1 + \delta r^T \Sigma^{-1} r}. \tag{5.5.6}$$

Applying [A19] again to (5.5.5) gives

$$[F^T \overline{\Sigma}^{-1} F]^{-1} = [F^T \Sigma^{-1} F]^{-1} + \frac{\overline{\delta}[F^T \Sigma^{-1} F]^{-1} F^T \Sigma^{-1} r r^T \Sigma^{-1} F [F^T \Sigma^{-1} F]^{-1}}{1 - \overline{\delta} r^T \Sigma^{-1} F [F^T \Sigma^{-1} F]^{-1} F^T \Sigma^{-1} r}. \tag{5.5.7}$$

Let us now assume that the parametric model in question is an innovation model and that the innovation variance σ_u^2 is one of the parameters, say the first one. In this case, σ_u^2 explicitly multiplies each of the components of r, so $\partial r / \partial \sigma_u^2 = (1/\sigma_u^2) r$. This means that r is proportional to the first column of F, or $r = \sigma_u^2 F e_1$, where e_1 is the first standard unit vector. Substitution of this fact in (5.5.7) yields (see also Prob. 5.17)

$$[F^T \overline{\Sigma}^{-1} F]^{-1} = [F^T \Sigma^{-1} F]^{-1} + \frac{\overline{\delta} \sigma_u^4 e_1 e_1^T}{1 - \overline{\delta} \sigma_u^4 e_1^T [F^T \Sigma^{-1} F] e_1}$$

$$= [F^T \Sigma^{-1} F]^{-1} + \kappa_4(u) e_1 e_1^T. \tag{5.5.8}$$

Note that the right side of (5.5.8) differs from $[F^T \Sigma^{-1} F]^{-1}$ only in the element in position (1,1). The conclusion from this result is that, in the special case of non-Gaussian innovation models, the asymptotic variance of the estimates based on the sample covariances is almost identical to that for Gaussian processes. The only parameter affected by the non-Gaussian property is the innovation variance, while the asymptotic variances of all other parameters remain unchanged. This result holds for AR, MA, and ARMA processes as a special case.

5.6. MATHEMATICA PACKAGES

The package `Bounds.m` and the auxiliary packages `Rarmad.m` and `CRarma.m` implement the bounds presented in this chapter for Gaussian ARMA processes (AR and MA as special cases). Here is a brief description of the functions in these packages.

`BndCov[a_, b_, sigu2_, m_]` computes the lower bound on the left side of Eq. (5.5.1), that is, the bound on the normalized asymptotic variance of estimates based on the sample covariances of orders 0 through `m`. The parameters of the ARMA process are `a`, `b` and `sigu2`. The returned values are the square roots of the diagonal elements of the lower-bound matrix.

`BndCovList[a_, b_, sigu2_, m_]` is identical to `BndCov` with the same parameters, except that it gives a list of the bounds for covariances of orders 0 through i, for all $p + q \leq i \leq m$.

`CRB[a_, b_, sigu2_]` computes the normalized asymptotic Cramér-Rao bound for the ARMA process with parameters `a`, `b` and `sigu2`; see Eqs. (5.2.13) through (5.2.16). The internal function `Rarcross` implements the part R_{xv} as discussed in Prob. 5.4. The returned values are the square roots of the diagonal elements of the CRB matrix.

`Rarmad[a_, b_, sigu2_, m_]` computes the partial derivatives of the covariances of an ARMA process of orders 0 through m with respect to the ARMA parameters. It uses the internal functions `Rarmada` and `Rarmadb`, which implement the computation as discussed in Prob. 5.16.

`Crarma[a_, b_, sigu2_, m_]` computes the normalized asymptotic variances of the sample covariances of an ARMA process, of orders 0 through m. The computation is based on (4.2.13); that is, the covariances $\bar{r}_k$ of the ARMA process with parameters $a^2(z)$, $b^2(z)$, and σ_u^4 are computed first and used to build the matrix Σ.

REFERENCES

Åström, K. J., and Söderström, T., "Uniqueness of the Maximum Likelihood Estimates of the Parameters of an ARMA Model," *IEEE Trans. Automatic Control*, AC-19, pp. 769–773, 1974.

Box, G. E. P., and Jenkins, G. M., *Time Series Analysis: Forecasting and Control*, rev. ed., Holden-Day, San Francisco, 1976.

Friedlander, B., and Porat, B., "On the Computation of an Asymptotic Bound for Estimating Autoregressive Signals in White Noise," *Signal Processing*, 8, pp. 291–302, 1985.

Porat, B., and Friedlander, B., "The Exact Cramer-Rao Bound for Gaussian Autoregressive Processes," *IEEE Trans. Aerospace Electronic Systems*, AES-23, pp. 537–542, 1987.

Rudin, W., *Principles of Mathematical Analysis*, McGraw-Hill, New York, 1964.

Stoica, P., and Söderström, T., "Uniqueness of the Maximum Likelihood Estimates of ARMA Model Parameters—An Elementary Proof," *IEEE Trans. Automatic Control*, AC-27, pp. 736–738, 1982.

Whittle, P., "The Analysis of Multiple Stationary Time Series," *J. Roy. Statist. Soc.*, 15, pp. 125–139, 1953.

PROBLEMS

5.1. Let $A(x)$ be a square matrix of differentiable functions of x. Prove that

$$\frac{d}{dx}|A(x)| = \text{tr}\left\{\text{adj}(A(x))\frac{dA(x)}{dx}\right\},$$

where adj$(A(x))$ is the adjoint matrix of $A(x)$ (the matrix of cofactors). Hence, if $A(x)$ is nonsingular at x,

$$\frac{d}{dx}\log|A(x)| = \text{tr}\left\{A^{-1}(x)\frac{dA(x)}{dx}\right\}.$$

This result is used in (5.2.2). Hint: Use the chain rule to write

$$\frac{d}{dx}|A(x)| = \sum_i\sum_k \frac{\partial|A(x)|}{\partial a_{ik}}\frac{da_{ik}}{dx};$$

then use the Laplace expansion of $|A(x)|$.

5.2. Complete the missing steps in the derivation of (5.2.7).

5.3. Let $p(z) = \sum_{i=0}^{n} p_i z^{-i}$ be nonzero on and outside the unit circle. Show that

$$\int_{-\pi}^{\pi}\frac{e^{-j\omega k}}{p(e^{j\omega})}d\omega = \int_{-\pi}^{\pi}\frac{e^{j\omega k}}{p(e^{-j\omega})}d\omega = 0$$

for all integers $k \geq 1$. This fact was used in the derivation of (5.2.10), as well as in the derivation of the information matrix of ARMA processes. Explain where exactly it was needed.

5.4. Develop a procedure for computing the entries of the matrix R_{xv} defined in (5.2.16). Hint: Let $1/a(z)b(z^{-1}) = \sum_{i=-\infty}^{\infty} d_i z^{-i}$. Then $\{d_i, -q \leq i \leq p\}$ appear as entries of R_{xv}. Let $n = \max\{p, q\}$. From

$$\frac{1}{a(z)} = \sum_{k=0}^{q}\sum_{i=-\infty}^{\infty} b_k d_i z^{-(i-k)}$$

it is seen that the right side must contain only nonpositive powers of z. Use this to get $n+1$ equations for $\{d_i, -n \leq i \leq n\}$. Similarly, from

$$\frac{1}{b(z^{-1})} = \sum_{k=0}^{p}\sum_{i=-\infty}^{\infty} a_k d_i z^{-(i+k)}$$

it is seen that the right side must contain only nonnegative powers of z. Use this to get n additional equations for $\{d_i, -n \leq i \leq n\}$. The $2n+1$ equations can be uniquely solved for the d_i.

An alternative procedure is based on the observation that

$$\frac{1}{a(e^{j\omega})b(e^{-j\omega})} = \frac{a(e^{-j\omega})b(e^{j\omega})}{a(e^{j\omega})a(e^{-j\omega})b(e^{j\omega})b(e^{-j\omega})}.$$

Therefore, the entries of R_{xv} can be obtained as certain linear combinations of the covariances of an AR process with the characteristic polynomial $a(z)b(z)$. Write down the details of this procedure.

5.5. Prove that the matrices $R_{\theta,N}^{T/2}R_{\psi,N}^{-1}R_{\theta,N}^{1/2}$ and $R_{\psi,N}^{-1/2}R_{\theta,N}R_{\psi,N}^{-T/2}$ have the same set of eigenvalues (this is used in the proof of Theorem 5.4).

5.6. Use the proof of Theorem 5.2 to derive an *exact* expression for the information matrix of an $AR(p)$ process when the number of measurements N is finite. Show that the information is given by

$$J_N(\theta) = (N - p)J_0(\theta) + J(\theta),$$

where $J_0(\theta)$ is the asymptotic information and $J(\theta)$ is a constant matrix (independent of N). Derive an expression for this matrix. See Porat and Friedlander [1987] for further details on this problem and the next.

5.7. Use the result of Prob. 5.6 to write the exact information matrix for an $AR(1)$ process. Explore the magnitude of the exact information relative to the asymptotic one as a function of a_1.

5.8. Derive an explicit expression for the asymptotic information matrix of an $ARMA(1,1)$ process, and use it to derive the asymptotic Cramér-Rao lower bounds on the variances of $\hat{a}_1$ and $\hat{b}_1$.

5.9. Let $\{y_t\}$ be a Gaussian AR process in white Gaussian noise, as defined in Prob. 2.19; that is, $y_t = x_t + v_t$, where $\{x_t\}$ is Gaussian $AR(p)$ and $\{v_t\}$ white Gaussian noise.

    (a) Use the spectral density formula derived in Prob. 2.19 and Whittle's formula (5.2.11) to write the asymptotic Fisher information of the parameters $\{\sigma_v^2, \sigma_u^2, a_1, \ldots, a_p\}$. Hint: See Friedlander and Porat [1985] for further details.

    (b) Develop a computational procedure for the information matrix and program it in Mathematica.

5.10. Suppose we try to fit an $AR(2)$ model to a given process, the true model of which is $AR(1)$ with parameters $\{\sigma_u^2, a_1\}$. Since the process can be equally well described as an $AR(2)$ model with parameters $\{\sigma_u^2, a_1, 0\}$, it is conceivable that $\hat{a}_1$ and $\hat{a}_2$ of the estimated model will approach a_1 and 0, respectively, as $N \to \infty$. However, the asymptotic variance of $\hat{a}_1$ may be adversely affected by this procedure. Compute the CRB on $\hat{a}_1$ in this case and compare it with the CRB in the case where the true order is used for estimation.

5.11. Repeat Prob. 5.10 in the case where the process in question is $ARMA(1,1)$ and we use an $ARMA(2,2)$ model for estimation. How would the estimates of a_1 and b_1 be affected? Hint: The analytic CRB formula is rather complicated and you are not advised to attempt it. However, you can try some numerical tests using the Mathematica program for the CRB, observe the results, and try to explain them.

5.12. Consider an $ARMA(p, q)$ process parameterized in terms of its poles and zeros:

$$\frac{b(z)}{a(z)} = \frac{\prod_{k=1}^{q}(1 - \nu_k z^{-1})}{\prod_{k=1}^{p}(1 - \mu_k z^{-1})}.$$

Assume that all poles and zeros are simple. Compute the asymptotic Fisher information matrix for the parameters $\{\sigma_u^2, \mu_1, \ldots, \mu_p, \nu_1, \ldots, \nu_q\}$. Hint: Use

Whittle's asymptotic formula. Take care to interpret complex poles and zeros correctly. You can find the answer in Box and Jenkins [1976, p. 242].

5.13. Prove that, if $S(\omega) > 0$ for all ω, the functions $\{\phi_k(\omega), 0 \le k < \infty\}$ form a basis for $L_2^e[-\pi, \pi]$.

5.14. Show that the condition in [A15] holds for (5.4.3), so differentiation under the integral sign is permitted.

5.15. Let $\{x_1, \ldots, x_n\}$ be a finite set of linearly independent vectors in a Hilbert space $\mathcal{H}$ and y be another vector in this space. Prove that the projection of y on the space spanned by $\{x_k\}$ is given by

$$\hat{y} = [\, \langle y, x_1 \rangle \quad \cdots \quad \langle y, x_n \rangle \,] \begin{bmatrix} \langle x_1, x_1 \rangle & \cdots & \langle x_1, x_n \rangle \\ \vdots & & \vdots \\ \langle x_n, x_1 \rangle & \cdots & \langle x_n, x_n \rangle \end{bmatrix}^{-1} \begin{bmatrix} x_1 \\ \vdots \\ x_n \end{bmatrix}.$$

Remark: The matrix whose inverse appears on the right side is called the *Gram matrix* of $\{x_1, \ldots, x_n\}$. This matrix is always positive semidefinite; it is nonsingular if and only if the vectors are linearly independent.

5.16. Develop a procedure for computing the partial derivatives $\partial r_m / \partial a_k$ and $\partial r_m / \partial b_k$ for ARMA(p, q) processes to be used in the matrix $F(\theta)$. Hint: Observe that

$$\frac{\partial r_m}{\partial a_k} = -\frac{1}{2\pi} \int_{-\pi}^{\pi} \frac{\sigma_u^2 b(e^{j\omega}) b(e^{-j\omega})}{a^2(e^{j\omega}) a^2(e^{-j\omega})} [a(e^{-j\omega}) e^{-j\omega k} + a(e^{j\omega}) e^{j\omega k}] e^{j\omega m} d\omega,$$

$$\frac{\partial r_m}{\partial b_k} = \frac{1}{2\pi} \int_{-\pi}^{\pi} \frac{\sigma_u^2}{a(e^{j\omega}) a(e^{-j\omega})} [b(e^{-j\omega}) e^{-j\omega k} + b(e^{j\omega}) e^{j\omega k}] e^{j\omega m} d\omega.$$

Therefore, $\partial r_m / \partial a_k$ can be expressed as a finite linear combination of the covariances of the ARMA$(2p, q)$ process corresponding to the transfer function $\sigma_u b(e^{j\omega})/a^2(e^{j\omega})$, and $\partial r_m / \partial a_b$ can be expressed as a finite linear combination of the covariances of the AR(p) process corresponding to the transfer function $\sigma_u / a(e^{j\omega})$. These covariances can be computed using (2.12.7), (2.12.8), and (2.12.11). Write down the details of the computation.

5.17. Prove the second equality in (5.5.8).

CHAPTER 6

Autoregressive Parameter Estimation

6.1. THE YULE-WALKER ESTIMATE

This chapter is devoted to the subject of estimation of autoregressive processes. The autoregressive model, introduced in Sec. 2.11, has a unique place among the parametric models for discrete-time stationary processes. In the linear prediction framework, it represents the special case where the best linear predictor of y_t from *all* past measurements is identical to the best linear predictor from a finite number of them. As we will see, the theory of AR processes is rich and elegant. As a result, AR modeling has become one of the most popular techniques in applied signal processing and time series analysis. We begin our discussion of AR parameter estimation with the classical Yule-Walker method. Yule [1927] was the first to propose the use of autoregressive models for spectral analysis and applied it to the estimation of the period of sunspot numbers (this period is about 11 years).

From the definition of the AR(p) model (2.12.5), we have that the predicted value of y_t is given by

$$\hat{y}_t = -\sum_{k=1}^{p} a_k y_{t-k}. \tag{6.1.1}$$

This implies that the coefficients of the process satisfy the pth-order Yule-Walker equation [cf. (2.10.2) and use the symmetry of the covariance sequence]

$$
\begin{bmatrix}
r_0 & r_1 & \cdots & r_p \\
r_1 & r_0 & \cdots & r_{p-1} \\
\vdots & \vdots & \ddots & \vdots \\
r_p & r_{p-1} & \cdots & r_0
\end{bmatrix}
\begin{bmatrix}
1 \\
a_1 \\
\vdots \\
a_p
\end{bmatrix}
=
\begin{bmatrix}
\sigma_u^2 \\
0 \\
\vdots \\
0
\end{bmatrix}. \tag{6.1.2}
$$

If the covariances $\{r_0, \ldots, r_p\}$ were known, this equation could be solved for the unknowns $\{\sigma_u^2, a_1, \ldots, a_p\}$. The Yule-Walker estimate of the AR parameters is simply the solution of (6.1.2) obtained upon replacing the true unknown covariances by their estimates—the sample covariances defined in (4.2.3) [Yule, 1927; Walker, 1931]. Explicitly,

$$
\begin{bmatrix} \hat{a}_1 \\ \hat{a}_2 \\ \vdots \\ \hat{a}_p \end{bmatrix} = - \begin{bmatrix} \hat{r}_0 & \hat{r}_1 & \cdots & \hat{r}_{p-1} \\ \hat{r}_1 & \hat{r}_0 & \cdots & \hat{r}_{p-2} \\ \vdots & \vdots & \ddots & \vdots \\ \hat{r}_{p-1} & \hat{r}_{p-2} & \cdots & \hat{r}_0 \end{bmatrix}^{-1} \begin{bmatrix} \hat{r}_1 \\ \hat{r}_2 \\ \vdots \\ \hat{r}_p \end{bmatrix}, \tag{6.1.3a}
$$

$$
\hat{\sigma}_u^2 = \hat{r}_0 + \sum_{k=1}^{p} \hat{r}_k \hat{a}_k. \tag{6.1.3b}
$$

This is also called the *autocorrelation method*, especially in the signal-processing literature. We will continue to refer to it as the Yule-Walker estimate.

Another estimate can be obtained by using the unbiased sample covariances $\tilde{r}_k$ in place of the biased ones. When the number of measurements N increases, the difference between the two estimates becomes smaller, and asymptotically they are equivalent. When the number of measurements is relatively small, the difference may not be negligible. In particular, using the biased sample covariances guarantees the stability of the estimated AR polynomial, as we will now prove.

Theorem 6.1. Assume that $N \geq p+1$ and that the measurements $\{y_t, 0 \leq t \leq N-1\}$ are not identically zero. Then:

(a) The Toeplitz matrix $\hat{R}_{p+1}$ built from the biased sample covariances is positive definite.

(b) The polynomial $z^p \hat{a}(z) = \sum_{k=0}^{p} \hat{a}_k z^{p-k}$ built from the Yule-Walker estimates is stable, that is, all its zeros are inside the unit circle.

Proof. Let $\boldsymbol{Y}^T$ be the $(p+1) \times (N+p)$-dimensional matrix

$$
\boldsymbol{Y}^T = \begin{bmatrix} y_0 & y_1 & \cdots & y_p & \cdots & y_{N-1} & 0 & \cdots & 0 \\ 0 & y_0 & \cdots & y_{p-1} & \cdots & y_{N-2} & y_{N-1} & \cdots & 0 \\ \vdots & \vdots & \ddots & \vdots & & \vdots & \vdots & \ddots & \vdots \\ 0 & 0 & \cdots & y_0 & \cdots & y_{N-p-1} & y_{N-p} & \cdots & y_{N-1} \end{bmatrix}. \tag{6.1.4}
$$

Then it is easy to verify that

$$
\hat{R}_{p+1} = \frac{1}{N} \boldsymbol{Y}^T \boldsymbol{Y};
$$

hence this matrix is always positive semidefinite. Moreover, $\boldsymbol{Y}$ is full-rank, except when all the y_t are zero. To see this, let t_0 be the smallest value of t for which $y_t \neq 0$. Then the determinant of the submatrix of $\boldsymbol{Y}^T$ consisting of the columns t_0 through $t_0 + p$ is $y_{t_0}^{p+1}$, which is nonzero. Therefore, the rank of $\boldsymbol{Y}^T$, hence of

$\hat{R}_{p+1}$, is $p+1$. This proves that $\hat{R}_{p+1}$ is positive definite. The stability of $z^p\hat{a}(z)$ now follows from Theorem 2.15. ∎

We remark that the Toeplitz matrix $\tilde{R}_{p+1}$ constructed from the unbiased sample covariances is *not* guaranteed to be positive definite (Prob. 6.1). Consequently, the AR polynomial estimate obtained from $\tilde{R}_{p+1}$ is not guaranteed to be stable. Due to this reason, the biased sample covariances are the ones commonly employed in the Yule-Walker estimate.

The main properties of the Yule-Walker estimate are given in the following theorem.

Theorem 6.2. If the AR process is Gaussian, the Yule-Walker estimate is consistent, asymptotically normal, and asymptotically efficient.

Proof. The Yule-Walker estimate is a special case of the method of moments presented in Sec. 3.6. The sample covariances of an AR process are consistent asymptotically normal estimates of the true covariances (Theorem 4.3). The transformation from the covariances to the parameters $\{\sigma_u^2, a_1, \ldots, a_p\}$ (i.e., the solution of the Yule-Walker equation) is continuously differentiable and one to one. By Theorem 3.16, the estimate is consistent and asymptotically normal. Let $G(\theta)$ be the Jacobian of the transformation from the covariances to the parameters and $S(\theta)$ the Jacobian of the inverse transformation. Both matrices are $(p+1) \times (p+1)$, and $S(\theta) = G^{-1}(\theta)$. The matrix $\Sigma(\theta)$, the asymptotic covariance of the sample covariances, is also $(p+1) \times (p+1)$. Therefore, we have in this case (see the corollary to Lemma 3.1)

$$G(\theta)\Sigma(\theta)G^T(\theta) = (S^T(\theta)\Sigma^{-1}(\theta)S(\theta))^{-1},$$

so the Yule-Walker estimate achieves the smallest possible asymptotic variance among all estimates that are functions of $\{\hat{r}_k, 0 \le k \le p\}$. By Theorem 5.10, this is equal to the asymptotic Cramér-Rao bound; hence the estimate is asymptotically efficient. ∎

Example 6.1. As was mentioned before, Yule was the first to suggest the autoregressive model for spectral analysis and applied it to the determination of the period of sunspot numbers. Yule used annual averages of observed sunspot numbers over the years 1749–1924. Those numbers are tabulated in Yule [1927] and are plotted in Figure 6.1(a). A windowed periodogram of the sunspot numbers, using 25 covariance terms and Hamming window, is shown in Figure 6.1(b). The peak value of the windowed periodogram is at frequency 0.0943 year^{-1}, corresponding to a period of 10.6 years. Solution of the Yule-Walker equations with $p=2$ yields $\hat{a} = \{1.0, -1.3266, 0.6424\}$ and $\hat{\sigma}_u^2 = 245.4$. The period is obtained as $[(\arccos(-0.5\hat{a}_1/(\hat{a}_2)^{1/2}))/(2\pi)]^{-1}$ [see Prob. 6.5(a)], which is 10.54 years in this case. Figure 6.1(d) shows the spectral density obtained from the second-order autoregressive model, while Figure 6.1(c) shows the estimated innovation sequence obtained by convolving the data (after removing the sample mean) with $\hat{a}(z)$. As can be seen, the estimated innovation does not quite look like stationary white noise, indicating that the AR(2) model is not quite satisfactory for this data set. However, this example is still of historical interest. ∎

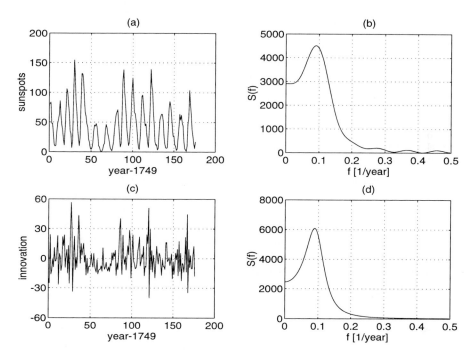

Figure 6.1. (a) Wolfer's sunspot data for the years 1749–1924; (b) the windowed periodogram; (c) the estimated innovation for an AR(2) model; (d) the spectral density of the AR(2) model.

Example 6.2. This example illustrates the technique of *Monte-Carlo simulation* for statistical analysis of estimators. The AR(4) model with $\sigma_u^2 = 1$ and $a(z) = 1 - 0.6z^{-1} + 0.1z^{-2} - 0.38z^{-3} + 0.72z^{-4}$ was used to generate 1000 statistically independent trials of $N = 256$ data points each. Each sample was fed to the Yule-Walker algorithm to yield estimates $\hat{\sigma}_u^2$ and $\hat{a}(z)$. The mean and standard deviations of the 1000 independent estimates were then computed. The mean was subtracted from the true value to yield the *empirical bias* (which should be close to the theoretical bias if the number of experiments is large enough). The standard deviation was compared with the corresponding (asymptotic) Cramér-Rao bound. By Theorem 6.2, the two should be close if N and the number of experiments are large enough.

Table 6.1 summarizes the results of the simulation. As can be seen, the bias is a fraction of the standard deviation in all cases, and the standard deviation is very close to the CRB. These results are in good agreement with the theory.   ∎

Parameter	σ_u^2	a_1	a_2	a_3	a_4
True value	1.0	-0.6	0.1	-0.38	0.72
Bias	0.04956	0.007718	-0.004867	0.01474	-0.02600
St. dev.	0.09871	0.04595	0.05010	0.05262	0.04383
CRB	0.08839	0.04337	0.05219	0.05219	0.04337

Table 6.1. Results of Monte-Carlo simulation of Example 6.1.

6.2. THE LEVINSON-DURBIN ALGORITHM

A direct solution of the Yule-Walker equations requires a number of operations proportional to p^3. It frequently happens that the user is interested in examining models of different orders for a given set of sample covariances. The computation of the predictor coefficients of all orders up to p requires a number of operations proportional to p^4. The algorithm of Levinson and Durbin takes advantage of the Toeplitz structure of the covariance matrix; it computes the predictor coefficients of all orders up to p in a number of operations proportional to p^2. Besides its practical advantages, the algorithm is a starting point for a rich theory of fast algorithms in linear algebra, some of which will be mentioned later.

We will derive the Levinson-Durbin algorithm using a stochastic approach. An equivalent algebraic approach is developed in Prob. 6.2. As in Chapters 2 and 5, we denote the coefficients of the nth-order best linear predictor by $\{-\alpha_{n,k}, 1 \leq k \leq n\}$, the nth-order prediction error at time t by $u_{t,n}$, and the variance of $u_{t,n}$ by d_n. The prediction error is given by

$$u_{t,n} = \sum_{k=0}^{n} \alpha_{n,k} y_{t-k}, \quad \alpha_{n,0} = 1.$$

The predictor coefficients satisfy [cf. (2.6.9)]

$$r_\ell + \sum_{k=1}^{n} \alpha_{n,k} r_{\ell-k} = 0, \quad 1 \leq \ell \leq n. \tag{6.2.1}$$

The equations of backward prediction can be derived analogously to the forward prediction equations. Let $\{-\tilde{\alpha}_{n,k}, 1 \leq k \leq n\}$ be the coefficients of the best linear predictor of y_t from $\{y_{t+k}, 1 \leq k \leq n\}$, and let $\tilde{u}_{t,n}$ be the corresponding prediction error. Then

$$\tilde{u}_{t,n} = \sum_{k=0}^{n} \tilde{\alpha}_{n,k} y_{t+k}, \quad \tilde{\alpha}_{n,0} = 1.$$

By the orthogonality property of the best linear predictor (Theorem B.1),

$$E\tilde{u}_{t,n} y_{t+\ell} = r_{-\ell} + \sum_{k=1}^{n} \tilde{\alpha}_{n,k} r_{k-\ell} = 0, \quad 1 \leq \ell \leq n. \tag{6.2.2}$$

Due to the symmetry of the covariances $\{r_k\}$, this set of equations is identical to (6.2.1). We conclude that, for real scalar stationary processes, the best linear backward predictor is identical to the best linear forward predictor; that is, $\tilde{\alpha}_{n,k} = \alpha_{n,k}$ for all n and k. We remark that stationary *vector* processes do not share this property (see Prob. 6.4).

Let X, Y, and Z be zero mean random variables, scalar or vector. Assume that EZZ^T is nonsingular (nonzero in the scalar case). Let $\hat{X}$ and $\hat{Y}$ be the projections of X and Y on the Hilbert space spanned by Z. Let $\tilde{X}$ and $\tilde{Y}$ be the corresponding errors. We have (Prob. 5.15)

$$\tilde{X} = X - \hat{X} = X - (EXZ^T)(EZZ^T)^{-1}Z, \quad \tilde{Y} = Y - \hat{Y} = Y - (EYZ^T)(EZZ^T)^{-1}Z.$$

Therefore,

$$E\tilde{X}\tilde{Y}^T = EX\tilde{Y}^T = E\tilde{X}Y^T = EXY^T - (EXZ^T)(EZZ^T)^{-1}(EZY^T).$$

The quantity $E\tilde{X}\tilde{Y}^T$ is called the *partial correlation* of X and Y modulo Z. The partial correlation is thus the correlation between the errors from projecting both X and Y on a common space—the one spanned by the third random variable Z. If both X and Y are scalars, the partial correlation is also a scalar; otherwise, it is a matrix.

Returning to the case of stationary processes, let y_t and y_{t-n-1} take the roles of X and Y above, and let $Z = [y_{t-1}, \ldots, y_{t-n}]^T$. The corresponding errors are then $u_{t,n}$ and $\tilde{u}_{t-n-1,n}$. Let Δ_{n+1} be the partial correlation of y_t and y_{t-n-1} modulo Z (note that the index $n+1$ corresponds to the difference between the two time points). Then Δ_{n+1} is given by

$$\Delta_{n+1} = Eu_{t,n}\tilde{u}_{t-n-1,n} = Ey_t\tilde{u}_{t-n-1,n} = Eu_{t,n}y_{t-n-1}.$$

Δ_n is called the nth-order partial correlation of the process and is defined for all $n \geq 1$. The partial correlation is independent of t due to the stationarity of the process. An explicit expression for the partial correlation is given as follows.

$$\Delta_{n+1} = Eu_{t,n}y_{t-n-1} = \sum_{k=0}^{n} \alpha_{n,k} Ey_{t-k}y_{t-n-1} = \sum_{k=0}^{n} \alpha_{n,k} r_{n+1-k}. \tag{6.2.3}$$

The *partial correlation coefficient* is defined by

$$K_{n+1} = \frac{E u_{t,n} \tilde{u}_{t-n-1,n}}{\sqrt{E u_{t,n}^2} \sqrt{E \tilde{u}_{t-n-1,n}^2}}. \tag{6.2.4}$$

It is easy to verify that $u_{t,n}$ and $\tilde{u}_{t-n-1,n}$ have the same variance (Prob. 6.3), denoted by d_n. Therefore,

$$K_{n+1} = \frac{\Delta_{n+1}}{d_n}. \tag{6.2.5}$$

The partial correlation coefficients satisfy $|K_n| \leq 1$ (this follows directly from their definition and the Cauchy-Schwarz inequality).

From the orthogonality of $\tilde{u}_{t-n-1,n}$ and Z, it follows that the projection of y_t on the space spanned by both random variables together is the sum of the individual projections. Also, Z and $\tilde{u}_{t-n-1,n}$ together clearly span the same space as $[y_{t-1}, \ldots, y_{t-n-1}]^T$. Evaluating the combined projection yields

$$\hat{y}_{t,n+1} = \hat{y}_{t,n} + \frac{E y_t \tilde{u}_{t-n-1,n}}{E \tilde{u}_{t-n-1,n}^2} \tilde{u}_{t-n-1,n} = \hat{y}_{t,n} + K_{n+1} \tilde{u}_{t-n-1,n},$$

so

$$u_{t,n+1} = y_t - \hat{y}_{t,n+1} = u_{t,n} - K_{n+1} \tilde{u}_{t-n-1,n}. \tag{6.2.6}$$

Expressing the two prediction errors in terms of the prediction coefficients gives

$$\sum_{k=0}^{n+1} \alpha_{n+1,k} y_{t-k} = \sum_{k=0}^{n} \alpha_{n,k} y_{t-k} - K_{n+1} \sum_{k=0}^{n} \alpha_{n,k} y_{t-n-1+k},$$

or, in vector notation,

$$[\alpha_{n+1,0}, \ldots, \alpha_{n+1,n+1}][y_t, \ldots, y_{t-n-1}]^T$$
$$= \{[\alpha_{n,0}, \ldots, \alpha_{n,n}, 0] - K_{n+1}[0, \alpha_{n,n}, \ldots, \alpha_{n,0}]\}[y_t, \ldots, y_{t-n-1}]^T.$$

Since this holds identically for all values of $\{y_t\}$, we get

$$[\alpha_{n+1,0}, \ldots, \alpha_{n+1,n+1}] = [\alpha_{n,0}, \ldots, \alpha_{n,n}, 0] - K_{n+1}[0, \alpha_{n,n}, \ldots, \alpha_{n,0}]. \tag{6.2.7}$$

This formula expresses the predictor coefficients of order $n+1$ in terms of those of order n and the partial correlation coefficient of order $n+1$.

Taking the expectation of the square of (6.2.6) gives

$$d_{n+1} = d_n - 2K_{n+1}\Delta_{n+1} + K_{n+1}^2 d_n = d_n(1 - K_{n+1}^2). \tag{6.2.8}$$

Equations (6.2.3), (6.2.5), (6.2.7), and (6.2.8) together form the Levinson-Durbin algorithm [Levinson, 1947; Durbin, 1960]. The initial conditions are

$$d_0 = r_0, \quad \alpha_{0,0} = 1. \tag{6.2.9}$$

The complete algorithm is summarized in Table 6.2. It takes as input the covariance sequence of $\{y_t\}$ (or an estimate thereof) and delivers the coefficients of the best linear predictors of orders 1 through p, the partial correlation coefficients, and the partial innovation variances.

$$
\begin{aligned}
&\text{Set } d_0 = r_0,\ \alpha_{0,0} = 1 \\
&\text{For } n = 0 \text{ to } p - 1 \text{ do} \\
&\qquad K_{n+1} = d_n^{-1} \sum_{k=0}^{n} \alpha_{n,k} r_{n+1-k} \\
&\qquad d_{n+1} = d_n(1 - K_{n+1}^2) \\
&\qquad \alpha_{n+1,0} = 1 \\
&\qquad \text{For } k = 1 \text{ to } n \text{ do} \\
&\qquad\qquad \alpha_{n+1,k} = \alpha_{n,k} - K_{n+1}\alpha_{n,n+1-k} \\
&\qquad \alpha_{n+1,n+1} = -K_{n+1}
\end{aligned}
$$

Table 6.2. A summary of the Levinson-Durbin algorithm.

Each step of the algorithm computes the predictor coefficients of order $n + 1$ in terms of those of order n and the elements of the covariance sequence. The algorithm terminates at $n = p$, the order of the AR process. The nth step involves about $2n$ multiplications [n for (6.2.3) and n for (6.2.7)] and a similar number of additions. Thus, the complete algorithm requires about p^2 multiplications and additions.

A multichannel version of the algorithm is developed in Prob. 6.4. This version will be used in Chapter 9, so the reader is advised to work it out in detail.

To carry out the Levinson-Durbin algorithm, we do not have to assume that the process is true autoregressive, only that its covariance sequence is positive definite. It follows from (6.2.8) that a regular process must satisfy $|K_n| < 1$ for all n. This is because $|K_n| = 1$ for some n would imply $d_{n+1} = 0$, hence $\sigma_u^2 = 0$, in contradiction to the regularity assumption.

If the given process is AR(p), then necessarily

$$
\alpha_{p+\ell,k} = a_k = \begin{cases} \alpha_{p,k}, & \text{if } 0 \leq k \leq p \\ 0, & \text{if } k > p \end{cases}, \quad \forall \ell \geq 0.
$$

In this case it follows from (6.2.7) that $K_{p+\ell} = 0$ for all $\ell > 0$. Thus, a process is AR(p) if and only if its partial correlation coefficients of orders larger than p are identically zero. Any stationary regular process that is not autoregressive is characterized by a nonterminating sequence of partial correlation coefficients.

Another interesting consequence of the Levinson-Durbin algorithm is the following relationship, obtained as the limit of (6.2.8).

$$
\sigma_u^2 = r_0 \prod_{n=1}^{\infty} (1 - K_n^2). \tag{6.2.10}
$$

This, in turn, implies the following result.

Theorem 6.3. A stationary process is regular if and only if $r_0 > 0$ and its sequence of partial correlation coefficients is square summable and each of its terms is strictly less than 1 in magnitude.

Proof.

(a) Necessity: The necessity of $r_0 > 0$ is obvious. The necessity of the condition $|K_n| < 1$ has been shown above. Since $x \leq -\log(1-x)$ for all $x \geq 0$, we get from (6.2.10)

$$\sum_{n=1}^{\infty} K_n^2 \leq \sum_{n=1}^{\infty} -\log(1-K_n^2) = \log r_0 - \log \sigma_u^2.$$

If the process is regular, $\sigma_u^2 > 0$; hence the right side is finite. Hence $\{K_n\}$ is square summable.

(b) Sufficiency: If $\sum_{n=1}^{\infty} K_n^2 < \infty$, then there exists N such that $K_n^2 < 0.5$ for all $n > N$. It is easy to verify that $-\log(1-x) < 2x$ if $0 \leq x \leq 0.5$. Therefore,

$$-\log \sigma_u^2 = -\log r_0 + \sum_{n=1}^{N} -\log(1-K_n^2) + \sum_{n=N+1}^{\infty} -\log(1-K_n^2)$$

$$< -\log r_0 + \sum_{n=1}^{N} -\log(1-K_n^2) + 2\sum_{n=N+1}^{\infty} K_n^2 < \infty,$$

so $\sigma_u^2 > 0$. ∎

Example 6.3. Application of the Levinson-Durbin algorithm to the sunspot data in Example 6.1, up to order $p = 10$, yields the following sequence of partial correlation coefficients: $\{0.8077, -0.6424, -0.0960, -0.0084, -0.0430, 0.1383, 0.1144, 0.2127, 0.0346, 0.0960\}$. Note that K_3 and K_4 are rather small in magnitude, which explains why Yule did not get an improvement upon increasing the order of the AR model from 2 to 4. However, the partial correlation coefficients 6 through 8 are rather large. Thus, using an AR(8) model would reveal additional periodicities and would give the estimated innovation a "whiter" appearance (Prob. 6.6). ∎

Example 6.4. This example illustrates the use of AR modeling for speech signals. In this context, AR modeling is better known as *linear predictive coding* (LPC). A short segment of a speech signal (on the order of a few tens of milliseconds) can be regarded as nearly stationary. Speech signals are commonly sampled at 8 kHz (more if very high fidelity is required) and collected in batches of a few hundreds of samples each. AR parameter estimation is then performed on each batch separately. Basic units of speech (over which stationarity holds approximately) are called *phonemes*. The phonemes of the English language are listed in Rabiner and Schafer [1978, p. 43]. They can be classified into two groups: the voiced and the unvoiced phonemes. A voiced sound is generated by vibrating the vocal cords. The fundamental vibration frequency is called the *pitch* and is in the range of 100 to 150 Hz for males and about twice this for females. The vibration excites the vocal tract, which acts approximately as a linear time-invariant filter. The vocal tract has several resonant frequencies (typically three), which are called *formants*. The values of the formant frequencies depend on the individual speaker and the sounded phoneme. An unvoiced sound does not involve the vocal cords, but is generated by the lips, teeth, and other parts. Correspondingly, the spectrum of an unvoiced sound typically does not exhibit formants.

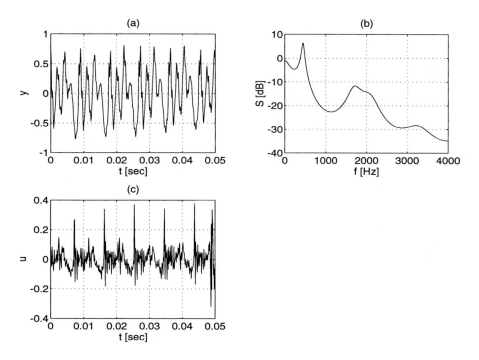

Figure 6.2. (a) Waveform of the sound "ae"; (b) the estimated spectral density; (c) the estimated innovation.

Figure 6.2(a) shows 50 milliseconds (400 samples) of the voiced phoneme "ae" expressed by an adult male. The signal was used for estimating an AR(10) model through the Levinson-Durbin algorithm. The resulting partial correlation coefficients are 0.900, −0.504, 0.301, −0.578, −0.319, −0.092, 0.139, 0.112, 0.340, and 0.100. The estimated spectral density is shown in Figure 6.2(b). The three formants are at frequencies of 450, 1750, and 2100 Hz. Figure 6.2(c) shows the estimated innovation. The impulse-like periodic component is the pitch frequency characteristic to voiced sounds. In this case it is about 110 Hz. As we see, voiced speech is non-Gaussian.

Figure 6.3(a) shows 50 milliseconds (400 samples) of the unvoiced phoneme "sh" expressed by an adult male. The signal was used for estimating an AR(10) model through the Levinson-Durbin algorithm. The resulting partial correlation coefficients are 0.321, −0.125, 0.551, 0.307, 0.323, 0.005, −0.119, −0.298, −0.238, and −0.178. The estimated spectral density is shown in Figure 6.3(b). Figure 6.3(c) shows the estimated innovation. As we see, the innovation has a Gaussian white-noise appearance in this case.

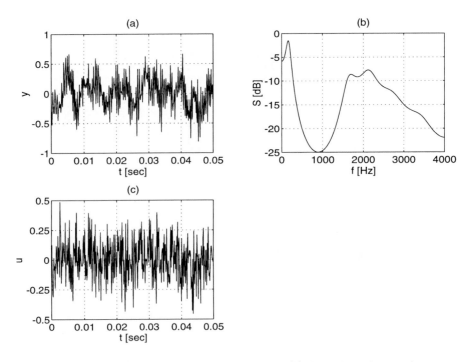

Figure 6.3. (a) Waveform of the sound "sh"; (b) the estimated spectral density; (c) the estimated innovation.

LPC is mainly used for speech compression. The compressed information consists of the AR parameters, a gain parameter (proportional to σ_u), voiced/unvoiced flag, and (in the case of a voiced sound) the pitch frequency. The reconstructed speech is generated by exciting an all-pole filter whose transfer function is $1/\hat{a}(z)$ with a synthetic innovation sequence. For voiced sound, the synthetic innovation is a periodic impulse train at the pitch frequency. For unvoiced sound, it is white Gaussian noise. ∎

6.3. ALGORITHMS RELATED TO LEVINSON-DURBIN

The Levinson-Durbin algorithm computes both the predictor polynomials and the partial correlation coefficients. In some cases we do not need the predictor polynomials, only the partial correlation coefficients. For example, suppose we are interested in retrieving the innovation sequence u_t. This can be done by iterating the equations

$$u_{t,n+1} = u_{t,n} - K_{n+1}\tilde{u}_{t-n-1,n}, \qquad (6.3.1a)$$

$$\tilde{u}_{t-n-1,n+1} = \tilde{u}_{t-n-1,n} - K_{n+1}u_{t,n} \qquad (6.3.1b)$$

for $n = 0$ to $p - 1$, starting with the initial condition

$$u_{t,0} = y_t, \quad \tilde{u}_{t-1,0} = y_{t-1}. \qquad (6.3.2)$$

Equation (6.3.1a) was derived above [cf. (6.2.6)], while Eq. (6.3.1b) is derived in a dual manner. The forward and backward innovations are obtained at the last step. To carry out this computation, we only need to know the partial correlation coefficients of orders 1 through p.

The classical Schur algorithm computes the partial correlation coefficients directly from the covariance sequence, without explicitly computing the predictor coefficients along the way. The Schur algorithm was originally devised as a test for positive definiteness of the covariance sequence (equivalently, the positive definiteness of the Toeplitz matrix constructed from the covariances). As was shown above, the covariance matrix is positive definite if and only if the partial correlation coefficients are all less than 1 in magnitude. The Schur algorithm produces the $\{K_n\}$ sequentially, starting at $n = 1$. The test consists of verifying that $|K_n| < 1$ for each new n. If this condition is violated for some n, the given sequence is not positive definite and the procedure is terminated. Otherwise, the sequence is positive definite.

To derive the Schur algorithm, substitute $t - k$ for t in Eqs. (6.3.1a) and (6.3.1b), multiply by y_t, and take the expected values to get

$$Ey_t u_{t-k,n+1} = Ey_t u_{t-k,n} - K_{n+1}Ey_t \tilde{u}_{t-k-n-1,n}, \qquad (6.3.3)$$

$$Ey_t \tilde{u}_{t-k-n-1,n+1} = Ey_t \tilde{u}_{t-k-n-1,n} - K_{n+1}Ey_t u_{t-k,n}. \qquad (6.3.4)$$

These equations hold for all $k \geq 0$. In particular, for $k = 0$ we have $Ey_t \tilde{u}_{t-n-1,n+1} = 0$ by definition. Therefore, we can solve (6.3.4) for K_{n+1}, obtaining

$$K_{n+1} = \frac{Ey_t \tilde{u}_{t-n-1,n}}{Ey_t u_{t,n}}. \qquad (6.3.5)$$

The initial conditions for $n = 0$ are obtained from the definitions

$$Ey_t u_{t-k,0} = Ey_t y_{t-k} = r_k, \quad Ey_t \tilde{u}_{t-k-1,0} = Ey_t y_{t-k-1} = r_{k+1}, \quad k \geq 0. \quad (6.3.6)$$

The complete algorithm consists of Eqs. (6.3.3) through (6.3.6) and is summarized in Table 6.3. Its computational complexity is the same as that of the Levinson-Durbin algorithm.

For $k = 0$ to $p - 1$ do
$\quad Ey_t u_{t-k,0} = r_k$
$\quad Ey_t \tilde{u}_{t-k-1,0} = r_{k+1}$
For $n = 0$ to $p - 1$ do
$\quad K_{n+1} = \dfrac{Ey_t \tilde{u}_{t-n-1,n}}{Ey_t u_{t,n}}$
$\quad$ For $k = 0$ to $p - n - 1$ do
$\quad\quad Ey_t u_{t-k,n+1} = Ey_t u_{t-k,n} - K_{n+1} Ey_t \tilde{u}_{t-k-n-1,n}$
$\quad\quad Ey_t \tilde{u}_{t-k-n-1,n+1} = Ey_t \tilde{u}_{t-k-n-1,n} - K_{n+1} Ey_t u_{t-k,n}$

Table 6.3. A summary of the Schur algorithm.

The Schur algorithm has another interesting application. Recall the factorization (2.10.4)

$$R_{p+1} = A_{p+1}^{-1} D_{p+1} A_{p+1}^{-T},$$

where A_{p+1} and D_{p+1} were defined in Eq. (2.10.3). From this we get

$$A_{p+1} R_{p+1} = D_{p+1} (A_{p+1}^T)^{-1}, \tag{6.3.7}$$

where the right side is an upper triangular matrix with the d_k along the main diagonal. The (n, ℓ)th element of the left side is given by

$$\sum_{k=0}^{n} \alpha_{n,k} r_{\ell-n+k} = \sum_{k=0}^{n} \alpha_{n,k} Ey_t y_{t+n-\ell-k} = Ey_t u_{t+n-\ell,n}. \tag{6.3.8}$$

The right side of (6.3.8) is zero for $0 \leq \ell < n$, as it should be by (6.3.7). The values for $n = 0$ are simply the elements of the given covariance sequence. The values for $1 \leq n \leq p$ and $n \leq \ell \leq p$ are computed by the Schur algorithm, as is seen from Table 6.3. Therefore, both A_{p+1}^{-T} and D_{p+1} can be computed, so the algorithm provides the lower-diagonal-upper factorization of R_{p+1} (2.10.4) (see Prob. 6.7). The Levinson-Durbin algorithm, on the other hand, provides the upper-diagonal-lower factorization of R_{p+1}^{-1} (2.10.5).

In recent years, both Levinson and Schur algorithms were modified by taking advantage of certain symmetries in Yule-Walker equations. The resulting modifications are known as the *split Levinson/Schur algorithms*. These algorithms offer computational advantages over the ones described above (they require about half the number of operations), but they do not have obvious stochastic interpretations. We will not discuss the split algorithms in this book, but refer the interested reader to Therrien [1992, Ch. 8] and the references within.

As is clear from the discussion of the Levinson-Durbin algorithm, the parameter vectors $\theta_a = [\sigma_u^2, a_1, \ldots, a_p]$, $\theta_k = [\sigma_u^2, K_1, \ldots, K_p]$, and $\theta_r = [r_0, \ldots, r_p]$ are related by one-one transformations, so they provide equivalent representations of the AR process. The Levinson-Durbin algorithm computes θ_a and θ_k from θ_r. The same algorithm can be used to compute θ_a from θ_k, by iterating (6.2.7) for n from 1 to p. The Schur algorithm provides an alternative means for computing θ_k from θ_r. The linear equations (2.12.7) give the transformation from θ_a to θ_r. The transformation from θ_a to θ_k can be done by carrying out the Levinson-Durbin algorithm in reverse order, starting from the given pth order predictor. This is known as the *inverse Levinson algorithm* and is summarized in Table 6.4. Its derivation is left to the reader (Prob. 6.8). The inverse Levinson algorithm can be used to test the stability of a given polynomial $z^p a(z)$. If the magnitude of all the K_n is less than 1, the polynomial is stable. If $|K_n| \geq 1$ is encountered, the procedure is terminated and the polynomial is declared unstable. The inverse Levinson algorithm is very similar to Jury's test for the location of the zeros of a polynomial with respect to the unit circle [Phillips and Nagle, 1984, p. 198].

$$
\begin{array}{l}
\text{For } k = 0 \text{ to } p \text{ do} \\
\quad \alpha_{p,k} = a_k \\
\text{For } n = p \text{ to } 1 \text{ do} \\
\quad K_n = -\alpha_{n,n} \\
\quad \text{For } k = 1 \text{ to } n - 1 \text{ do} \\
\qquad \alpha_{n-1,k} = \dfrac{\alpha_{n,k} + K_n \alpha_{n,n-k}}{1 - K_n^2}
\end{array}
$$

Table 6.4. A summary of the inverse Levinson algorithm.

The transformation from θ_k to θ_r can be done by an intermediate computation of θ_a, but for completeness we will show how to do it directly. Substitute $t + k$ for t and $n - 1$ for n in (6.3.1a) and (6.3.1b), multiply by y_t, and take expected values to get

$$E y_t u_{t+k,n} = E y_t u_{t+k,n-1} - K_n E y_t \tilde{u}_{t+k-n,n-1},$$
$$E y_t \tilde{u}_{t+k-n,n} = E y_t \tilde{u}_{t+k-n,n-1} - K_n E y_t u_{t+k,n-1}.$$

These are equivalent to

$$E y_t u_{t+k,n-1} = E y_t u_{t+k,n} + K_n E y_t \tilde{u}_{t+k-n,n-1}, \qquad (6.3.9)$$
$$E y_t \tilde{u}_{t+k-n,n} = -K_n E y_t u_{t+k,n} + (1 - K_n^2) E y_t \tilde{u}_{t+k-n,n-1}. \qquad (6.3.10)$$

The initial conditions are

$$E y_t u_{t,0} = E y_t \tilde{u}_{t,0} = r_0 = \frac{\sigma_u^2}{\prod_{n=1}^{p}(1 - K_n^2)}, \qquad (6.3.11)$$

$$Ey_t u_{t+k,k} = 0, \quad k > 0. \tag{6.3.12}$$

The complete algorithm is summarized in Table 6.5. This algorithm is, in a sense, a reversed-order Schur algorithm. We will therefore call it the *inverse Schur algorithm*, although this name is not standard.

Set $Ey_t u_{t,0} = Ey_t \tilde{u}_{t,0} = r_0 = \sigma_u^2 \prod_{n=1}^p (1 - K_n^2)^{-1}$
For $k = 1$ to p do
$\qquad Ey_t u_{t+k,k} = 0$
For $k = 1$ to p do
$\qquad$ For $n = k$ down to 1 do
$\qquad\qquad Ey_t u_{t+k,n-1} = Ey_t u_{t+k,n} + K_n Ey_t \tilde{u}_{t+k-n,n-1}$
$\qquad\qquad Ey_t \tilde{u}_{t+k-n,n} = -K_n Ey_t u_{t+k,n} + (1-K_n^2) Ey_t \tilde{u}_{t+k-n,n-1}$
$\qquad$ Set $r_k = Ey_t \tilde{u}_{t+k,0} = Ey_t u_{t+k,0}$

Table 6.5. A summary of the inverse Schur algorithm.

The inverse Levinson and the inverse Schur algorithms can be used together to compute the covariance sequence from the AR parameters. The first algorithm computes θ_k from θ_a, while the second uses θ_a to compute θ_r. This procedure is more efficient than a direct solution of (2.12.7), because it avoids the explicit matrix inversion and requires a number of operations proportional to p^2 instead of p^3. This method of computing the covariances of an autoregressive process is used in the Mathematica procedure `Rar` mentioned in Sec. 2.13.

We finally mention another useful relation between the parameter vectors θ_r and θ_a, the *Gohberg-Semencul formula*. Let R_p have its usual meaning and let A_1 and A_2 be the $p \times p$ lower triangular Toeplitz matrices $[A_1]_{i,j} = a_{i-j}$, $[A_2]_{i,j} = a_{p-i+j}$. The Gohberg-Semencul formula is then

$$R_p^{-1} = \sigma_u^{-2}(A_1 A_1^T - A_2 A_2^T). \tag{6.3.13}$$

The proof is given in App. D, Theorem D.9. This formula is interesting since it provides an explicit expression for the inverse covariance matrix in terms of the AR parameters. Moreover, the formula can be easily extended to yield R_N^{-1} for any $N \geq p$. Simply observe that an AR(p) process can also be viewed as an AR(N) process, where $N > p$ and $a_k = 0$ for all $k > p$ (σ_u^2 remains unchanged). Redefine A_1 and A_2 as the $N \times N$ lower triangular Toeplitz matrices $[A_1]_{i,j} = a_{i-j}$, $[A_2]_{i,j} = a_{N-i+j}$. Then

$$R_N^{-1} = \sigma_u^{-2}(A_1 A_1^T - A_2 A_2^T). \tag{6.3.14}$$

Since the likelihood function of a Gaussian process includes the matrix R_N^{-1}, formula (6.3.14) is useful for maximum likelihood estimation of the AR parameters. This subject will be brought up in Sec. 6.7.

6.4. LATTICE FILTERS

We now digress from estimation theory for a while and discuss some system theoretical aspects of the Levinson-Durbin algorithm. The results of this chapter are of interest to digital signal processing as a whole and will also be used to motivate some of the material of Chapter 8.

Let us write Eq. (6.2.7) and its reversed-order version in the form

$$
\begin{bmatrix} \alpha_{n+1,0} & \alpha_{n+1,1} & \cdots & \alpha_{n+1,n} & \alpha_{n+1,n+1} \\ \alpha_{n+1,n+1} & \alpha_{n+1,n} & \cdots & \alpha_{n+1,1} & \alpha_{n+1,0} \end{bmatrix}
$$
$$
= \begin{bmatrix} 1 & -K_{n+1} \\ -K_{n+1} & 1 \end{bmatrix} \begin{bmatrix} \alpha_{n,0} & \alpha_{n,1} & \cdots & \alpha_{n,n} & 0 \\ 0 & \alpha_{n,n} & \cdots & \alpha_{n,1} & \alpha_{n,0} \end{bmatrix}.
$$

Multiply both sides by the column vector $[1, z^{-1}, \ldots, z^{-n}, z^{-(n+1)}]^T$ to get

$$
\begin{bmatrix} \alpha_{n+1}(z) \\ \tilde{\alpha}_{n+1}(z) \end{bmatrix} = \begin{bmatrix} 1 & -K_{n+1} \\ -K_{n+1} & 1 \end{bmatrix} \begin{bmatrix} \alpha_n(z) \\ z^{-1}\tilde{\alpha}_n(z) \end{bmatrix}
$$
$$
= \begin{bmatrix} 1 & -K_{n+1} \\ -K_{n+1} & 1 \end{bmatrix} \begin{bmatrix} 1 & 0 \\ 0 & z^{-1} \end{bmatrix} \begin{bmatrix} \alpha_n(z) \\ \tilde{\alpha}_n(z) \end{bmatrix}, \qquad (6.4.1)
$$

where $\alpha_n(z)$ is the nth-order predictor polynomial defined in (2.10.6), and $\tilde{\alpha}_n(z) = z^{-n}\alpha_n(z^{-1})$ is the reversed-order polynomial, also called the backward predictor polynomial. Now define

$$
T_{n+1}(z) = \begin{bmatrix} 1 & -K_{n+1} \\ -K_{n+1} & 1 \end{bmatrix} \begin{bmatrix} 1 & 0 \\ 0 & z^{-1} \end{bmatrix}, \qquad (6.4.2)
$$

and then

$$
\begin{bmatrix} \alpha_{n+1}(z) \\ \tilde{\alpha}_{n+1}(z) \end{bmatrix} = T_{n+1}(z) \begin{bmatrix} \alpha_n(z) \\ \tilde{\alpha}_n(z) \end{bmatrix}.
$$

Also, $\alpha_0(z) = \tilde{\alpha}_0(z) = 1$, so we get by induction

$$
\begin{bmatrix} a_p(z) \\ \tilde{a}_p(z) \end{bmatrix} = T_p(z)T_{p-1}(z)\ldots T_1(z) \begin{bmatrix} 1 \\ 1 \end{bmatrix}. \qquad (6.4.3)
$$

The 2×1 vector of $a_p(z)$ and $\tilde{a}_p(z)$ can be regarded as the transfer matrix of a one-input, two-output discrete-time linear system. The two channels represent the forward and backward pth-order predictors, respectively. Both are finite-impulse-response (FIR) filters. Each of the $T_n(z)$ represents a two-input, two-output system, and the product in (6.4.3) represents a cascade connection. Therefore, (6.4.3) represents a cascade realization of the two filters, as shown in Figure 6.4. Each block in the diagram has the internal structure shown in Figure 6.5. The block z^{-1} represents one-sample delay. This structure is known in the signal-processing literature as a *lattice filter*. The lattice filter has been used in the seismic signal-processing literature as a model for acoustic reflections from earth layers. The parameters $\{K_n\}$ have become known in the signal-processing literature as the *reflection coefficients*.

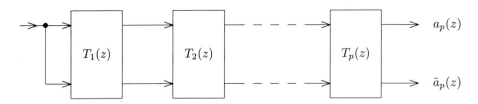

Figure 6.4. The structure of a lattice filter.

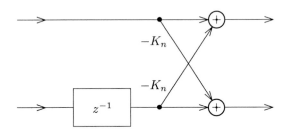

Figure 6.5. The structure of the lattice section $T_n(z)$.

The principal use of the lattice filter is to compute the sequence of prediction errors (innovations) $u_{t,p}$ from the given data sequence. In this application it is commonly called the *whitening filter*, because the sequence $\{u_{t,p}\}$ is white noise if the input process is AR(p) and the partial correlation coefficients used in the filter are the true ones. Some applications require the reverse operation: the synthesis of an AR(p) process from a white-noise sequence. We know from linear system theory that the filter for this purpose must have the transfer function $1/a_p(z)$. Mason's celebrated rule of network inversion [Mason, 1956] provides a lattice filter with this transfer function. Mason's rule applies to networks with a direct forward path from the input to the output (a direct path is characterized by a constant nonzero gain, without delay elements). According to that rule, the network's transfer function will be inverted if:

1. The direction of the flow of information along the direct path is reversed (i.e., the roles of input and output are interchanged).

2. The gains along the direct path are replaced by their reciprocals.

3. The polarities of the all the signals entering the summing junctions along the direct path are reversed, except the polarities along the direct path itself.

Figures 6.6 and 6.7 show the effect of applying Mason's rule to the lattice filter. Figure 6.6 shows the global structure and Figure 6.7 the details of a typical section. Note the flow of information: from right to left in the top line (which is the direct path) and from left to right in the bottom line. The transfer matrix of the nth section is given by (Prob. 6.10)

$$U_n(z) = \begin{bmatrix} 1 & K_n \\ -K_n & 1 - K_n^2 \end{bmatrix} \begin{bmatrix} 1 & 0 \\ 0 & z^{-1} \end{bmatrix}. \tag{6.4.4}$$

Figure 6.8 shows a rearrangement of the section, which preserves its transfer function, but is free of the "local feedback" seen in Figure 6.7.

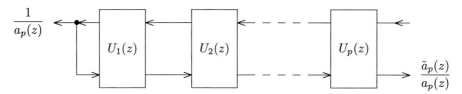

Figure 6.6. The structure of a feedback lattice filter.

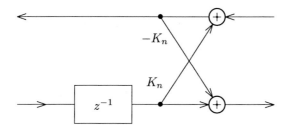

Figure 6.7. The structure of the lattice section $U_n(z)$.

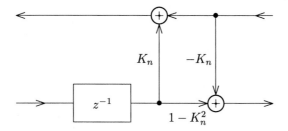

Figure 6.8. An alternative structure of the lattice section $U_n(z)$.

Lattice filters have several useful features, as follows.

1. The forward lattice is *nested*, in the sense that all intermediate-order filters are embedded in the pth-order filter. By feeding the data sequence $\{y_t\}$ at the input, we get the nth-order (forward and backward) prediction errors at the outputs of the nth section for all $1 \leq n \leq p$. We can therefore change the desired filter's order simply by switching, and there is no need to modify the

filter structure or its coefficients. Note, however, that the nesting property does not hold for the feedback filter in the same manner. To decrease the order of the feedback lattice we must disable the higher-order sections, for example, by resetting the corresponding K_n to zero.

2. The internal gains are less than 1 in magnitude, so the filter can be conveniently implemented in fixed-point arithmetic.

3. The stability of the feedback lattice is guaranteed as long as the K_n are less than 1 in magnitude (the stability of the forward lattice is guaranteed in any case because it is FIR, so it is not special in this regard).

4. The lattice structure, particularly the feedback one, is known to have good round-off and limit-cycle behavior.

In Chapter 8 we will further discuss lattice filters in the context of adaptive AR estimation.

6.5. MAXIMUM ENTROPY ESTIMATION

In Sec. 3.9 we introduced the concept of entropy and outlined the principle of maximum entropy estimation. We will now show how the Yule-Walker estimate can be viewed as a special case of the maximum entropy principle. This connection has been established by Burg [1967, 1975].

We start by deriving the entropy of an N-dimensional Gaussian vector Y with zero mean and covariance matrix Γ. We have, by the definition of entropy,

$$H(\Gamma) = -E \log f(Y) = \frac{N}{2} \log 2\pi + \frac{1}{2} \log |\Gamma| + \frac{1}{2} E(Y^T \Gamma^{-1} Y)$$

$$= \frac{N}{2} \log 2\pi + \frac{1}{2} \log |\Gamma| + \frac{N}{2}. \tag{6.5.1}$$

Let us specialize this to the case where Y is a vector of N consecutive measurements from a purely indeterministic Gaussian process. In this case $\Gamma = R_N$ (the Toeplitz covariance of the process), and we have by the lower-diagonal-upper decomposition of the covariance (2.10.4)

$$|R_N| = \prod_{n=0}^{N-1} d_n \quad \Longrightarrow \quad \log |R_N| = \sum_{n=0}^{N-1} \log d_n.$$

We also have, by (6.2.8),

$$d_n = r_0 \prod_{k=1}^{n} (1 - K_k^2) \quad \Longrightarrow \quad \log d_n = \log r_0 + \sum_{k=1}^{n} \log(1 - K_k^2).$$

Therefore,

$$\frac{2}{N}H(R_N) = \log 2\pi + 1 + \log r_0 + \sum_{n=1}^{N-1}\left(1 - \frac{n}{N}\right)\log(1 - K_n^2). \qquad (6.5.2)$$

A purely indeterministic zero-mean Gaussian process is fully characterized by its covariance sequences. Given N measurements, we clearly cannot estimate any of the covariances beyond r_{N-1}. Moreover, the estimation of all first N covariances is impractical both computationally and from a statistical accuracy viewpoint. Suppose, therefore, that we decide to estimate the first $p+1$ covariances and choose the remaining ones by some other criterion. From the theory of the Levinson-Durbin algorithm, we know that the first $p+1$ covariances are in one-one correspondence with r_0 and the first p partial correlation coefficients. Therefore, the estimates $\{\hat{r}_n, 0 \le n \le p\}$ uniquely determine $\{\hat{r}_0, \hat{K}_n, 1 \le n \le p\}$. From (6.5.2) we see that $\{K_n, p+1 \le n \le N-1\}$ provide the extra degrees of freedom for controlling the entropy. It is also clear from (6.5.2) that the entropy is a monotone decreasing function of each of the $|K_n|$ individually. The entropy will be maximized if we choose all these partial correlation coefficients to be zero. This is equivalent to choosing the process as an autoregressive of order p. In summary of all purely indeterministic Gaussian processes whose first $p+1$ covariances are given, the one whose entropy is maximum is the autoregressive process of order p possessing the given covariances. This is exactly the process provided by the Yule-Walker estimate (or its Levinson-Durbin implementation).

The above argument also solves the problem of *covariance extension*, which is: given a finite sequence $\{r_n, 0 \le n \le p\}$ such that the corresponding R_p is positive definite, how can we extend the sequence infinitely while preserving its positive definiteness? The obvious answer is: choose $\{K_n, n > p\}$ arbitrarily, subject only to the constraint $|K_n| < 1$. The set of all such K_n fully parameterizes the set of all covariance extensions. Moreover, among all covariance extensions there is a unique extension whose corresponding entropy is maximum: that of a pth-order autoregressive process.

The principle of maximum entropy can be applied to other estimation problems besides the one mentioned here, which is probably the simplest one encountered in the realm of discrete-time random processes. In most other cases it leads to functional maximization problems that can sometimes be solved by variational methods. The maximum entropy method will not be further treated in this book.

6.6. LEAST-SQUARES ESTIMATION

The two prediction equations

$$y_t + \sum_{k=1}^{p} a_k y_{t-k} = u_{t,p}, \quad y_t + \sum_{k=1}^{p} a_k y_{t+k} = \tilde{u}_{t,p}$$

can be regarded as special cases of the estimation problem (3.7.1), in which θ is the set of AR coefficients and $u_{t,p}, \tilde{u}_{t,p}$ are the modeling errors. Given the measurements $\{y_t, 0 \leq t \leq N-1\}$, we can form the two systems of equations

$$
\begin{bmatrix} y_p \\ y_{p+1} \\ \vdots \\ y_{N-1} \end{bmatrix} + \begin{bmatrix} y_{p-1} & y_{p-2} & \cdots & y_0 \\ y_p & y_{p-1} & \cdots & y_1 \\ \vdots & \vdots & & \vdots \\ y_{N-2} & y_{N-3} & \cdots & y_{N-p-1} \end{bmatrix} \begin{bmatrix} a_1 \\ a_2 \\ \vdots \\ a_p \end{bmatrix} = \begin{bmatrix} u_{p,p} \\ u_{p+1,p} \\ \vdots \\ u_{N-1,p} \end{bmatrix}, \tag{6.6.1}
$$

$$
\begin{bmatrix} y_0 \\ y_1 \\ \vdots \\ y_{N-p-1} \end{bmatrix} + \begin{bmatrix} y_1 & y_2 & \cdots & y_p \\ y_2 & y_3 & \cdots & y_{p+1} \\ \vdots & \vdots & & \vdots \\ y_{N-p} & y_{N-p+1} & \cdots & y_{N-1} \end{bmatrix} \begin{bmatrix} a_1 \\ a_2 \\ \vdots \\ a_p \end{bmatrix} = \begin{bmatrix} \tilde{u}_{0,p} \\ \tilde{u}_{1,p} \\ \vdots \\ \tilde{u}_{N-p-1,p} \end{bmatrix}, \tag{6.6.2}
$$

or, in concise notation,

$$
\boldsymbol{y}_f + \boldsymbol{Y}_f \boldsymbol{a} = \boldsymbol{e}_f, \quad \boldsymbol{y}_b + \boldsymbol{Y}_b \boldsymbol{a} = \boldsymbol{e}_b. \tag{6.6.3}
$$

These are special cases of the linear least-squares problem (3.7.6). The corresponding least-square solutions are [cf. (3.7.8)]

$$
\hat{\boldsymbol{a}}_f = -(\boldsymbol{Y}_f^T \boldsymbol{Y}_f)^{-1} \boldsymbol{Y}_f^T \boldsymbol{y}_f, \quad \hat{\boldsymbol{a}}_b = -(\boldsymbol{Y}_b^T \boldsymbol{Y}_b)^{-1} \boldsymbol{Y}_b^T \boldsymbol{y}_b. \tag{6.6.4}
$$

$\hat{\boldsymbol{a}}_f$ and $\hat{\boldsymbol{a}}_b$ are called the *forward* and *backward least-square estimates*, respectively. In the signal-processing literature, the names *forward* and *backward covariance method* are common. The two sets of equations can also be combined to yield

$$
\begin{bmatrix} \boldsymbol{y}_f \\ \boldsymbol{y}_b \end{bmatrix} + \begin{bmatrix} \boldsymbol{Y}_f \\ \boldsymbol{Y}_b \end{bmatrix} \boldsymbol{a} = \begin{bmatrix} \boldsymbol{e}_f \\ \boldsymbol{e}_b \end{bmatrix} \implies \hat{\boldsymbol{a}}_{fb} = -(\boldsymbol{Y}_f^T \boldsymbol{Y}_f + \boldsymbol{Y}_b^T \boldsymbol{Y}_b)^{-1}(\boldsymbol{Y}_f^T \boldsymbol{y}_f + \boldsymbol{Y}_b^T \boldsymbol{y}_b). \tag{6.6.5}
$$

This is known as the *forward-backward covariance method* in the signal-processing literature.

An estimate closely related to the forward covariance is the *prewindowed estimate*. This is defined as the least-squares solution of the system of equations

$$
\begin{bmatrix} y_0 \\ y_1 \\ \vdots \\ y_{N-1} \end{bmatrix} + \begin{bmatrix} 0 & 0 & \cdots & 0 \\ y_0 & 0 & \cdots & 0 \\ \vdots & \vdots & & \vdots \\ y_{N-2} & y_{N-3} & \cdots & y_{N-p-1} \end{bmatrix} \begin{bmatrix} a_1 \\ a_2 \\ \vdots \\ a_p \end{bmatrix} = \begin{bmatrix} u_{0,p} \\ u_{1,p} \\ \vdots \\ u_{N-1,p} \end{bmatrix}. \tag{6.6.6}
$$

Comparing (6.6.6) to (6.6.1), we see that the prewindowed estimate is obtained by adding p equations to the forward covariance equations, while forcing $\{y_t, -p \leq t \leq -1\}$ to zero (i.e., prewindowing the available data by zeros). The prewindowed method plays an important role in adaptive AR estimation, as will be discussed in Chapter 8.

The relationship among the four least-square estimates presented in this section and the Yule-Walker estimate is explored in Prob. 6.13. Asymptotically, all these methods are equivalent. For short data records they may yield substantially different results though. Neither of the least-squares estimates is guaranteed to yield a stable predictor polynomial. However, these estimates have advantages that compensate for this drawback. Unlike the Yule-Walker estimate, they lend themselves to adaptive implementations, as will be discussed in the Chapter 8. In addition, they have been observed to yield better spectral estimates than the Yule-Walker estimate for short data records. The forward-backward covariance method is generally regarded as the best in this respect, but its implementation requires about twice as many calculations as any of the other methods.

6.7. MAXIMUM LIKELIHOOD ESTIMATION

The likelihood function of a Gaussian AR process can be expressed as a relatively simple function of the process parameters. In fact, a Gaussian AR process admits a nontrivial sufficient statistic, as given by the following theorem.

Theorem 6.4. If the number of measurements from a Gaussian AR(p) process satisfies $N \geq 3p + 1$, there exists a sufficient statistic of dimension $3p + 1$.

Proof. Recalling the general form of the Gaussian density function (2.2.1), we see that Y_N appears only in the quadratic form in the exponent. By the factorization theorem 3.1, it is sufficient to show that the quadratic form

$$Q(Y_N, \theta_a) = \sigma_u^2 Y_N^T R_N^{-1} Y_N$$

can be expressed in terms of $3p + 1$ functions of the data vector Y_N. By the Gohberg-Semencul formula (6.3.14), we have

$$Q(Y_N, \theta_a) = Y_N^T (A_1 A_1^T - A_2 A_2^T) Y_N. \tag{6.7.1}$$

Let $\boldsymbol{Y}^T$ be the matrix defined in (6.1.4) and define the $(p+1) \times p$ matrices

$$\boldsymbol{Y}_1 = \begin{bmatrix} y_0 & y_1 & \cdots & y_{p-1} \\ 0 & y_0 & \cdots & y_{p-2} \\ \vdots & \vdots & \ddots & \vdots \\ 0 & 0 & \cdots & y_0 \\ 0 & 0 & \cdots & 0 \end{bmatrix}, \quad \boldsymbol{Y}_2 = \begin{bmatrix} y_{N-1} & y_{N-2} & \cdots & y_{N-p} \\ 0 & y_{N-1} & \cdots & y_{N-p-1} \\ \vdots & \vdots & \ddots & \vdots \\ 0 & 0 & \cdots & y_{N-1} \\ 0 & 0 & \cdots & 0 \end{bmatrix}.$$

Define also $\boldsymbol{a} = [a_p, \ldots, a_1, 1]^T$. Then it can be verified that (Prob. 6.14)

$$\begin{aligned} Q(Y_N, \theta_a) &= \boldsymbol{a}^T (\boldsymbol{Y}^T \boldsymbol{Y} - \boldsymbol{Y}_1^T \boldsymbol{Y}_1 - \boldsymbol{Y}_2^T \boldsymbol{Y}_2) \boldsymbol{a} \\ &= \boldsymbol{a}^T (N \hat{R}_{p+1} - \boldsymbol{Y}_1^T \boldsymbol{Y}_1 - \boldsymbol{Y}_2^T \boldsymbol{Y}_2) \boldsymbol{a}. \end{aligned} \tag{6.7.2}$$

Now the matrix $\hat{R}_{p+1}$ depends only on $\{\hat{r}_n, 0 \leq n \leq p\}$, the matrix $\boldsymbol{Y}_1$ depends only on $\{y_n, 0 \leq n \leq p-1\}$, and the matrix $\boldsymbol{Y}_2$ depends only on $\{y_n, N-p \leq n \leq N-1\}$. In summary, $Q(Y_N, \theta_a)$ depends on Y_N only through the statistic

$$T(Y_N) = [y_0, \ldots, y_{p-1}, y_{N-p}, \ldots, y_{N-1}, \hat{r}_0, \ldots, \hat{r}_p], \tag{6.7.3}$$

whose dimension is $3p + 1$.   ∎

The existence of a low dimension sufficient statistic is another unique property of AR processes, not shared by other common parametric models (such as MA and ARMA). We now apply the result of Theorem 6.4 to the problem of maximum likelihood estimation of the AR parameters. We have

$$-2 \log f_\theta(Y_N) = N \log 2\pi + \log |R_N| + \sigma_u^{-2} Q(Y_N, \theta_a).$$

We also have, from Sec. 6.5,

$$\log |R_N| = \sum_{n=0}^{N-1} \log d_n = \sum_{n=0}^{p-1} \log d_n + (N - p) \log \sigma_u^2 = N \log \sigma_u^2 - \sum_{n=1}^{p} n \log(1 - K_n^2).$$

Therefore,

$$-2 \log f_\theta(Y_N) = N \log 2\pi + N \log \sigma_u^2 - \sum_{n=1}^{p} n \log(1 - K_n^2) + \sigma_u^{-2} Q(Y_N, \theta_a). \quad (6.7.4)$$

Minimizing (6.7.4) with respect to σ_u^2 gives

$$\hat{\sigma}_u^2 = N^{-1} Q(Y_N, \theta_a).$$

Substitution of this estimate in (6.7.4) yields

$$-\frac{2}{N} \log f_\theta(Y_N) \Big|_{\hat{\sigma}_u^2} = \log 2\pi + \log N^{-1} Q(Y_N, \theta_a) - \frac{1}{N} \sum_{n=1}^{p} n \log(1 - K_n^2) + 1$$

$$= \log 2\pi + 1 + \log[a^T (\hat{R}_{p+1} - N^{-1} Y_1^T Y_1 - N^{-1} Y_2^T Y_2) a]$$

$$- N^{-1} \sum_{n=1}^{p} n \log(1 - K_n^2). \quad (6.7.5)$$

An alternative procedure for computing the likelihood function will be presented in Sec. 7.7.

The maximum likelihood estimate of the AR parameters is obtained by minimizing the right side of (6.7.5) with respect to the $\{a_n\}$. Recall that the $\{K_n\}$ are functions of the $\{a_n\}$ and can be computed using the inverse Levinson algorithm. Alternatively, (6.7.5) can be minimized with respect to the $\{K_n\}$, using the Levinson-Durbin algorithm to compute the $\{a_n\}$. In any case, the minimization involves nonlinear operations and needs to be carried out numerically. The Yule-Walker estimate can be used as initial condition for the numerical minimization. Problem 6.15 deals with the special case of the AR(1) process, in which there is no need for numerical minimization.

When the number of measurements N approaches infinity, the terms proportional to N^{-1} become negligible. Minimization of (6.7.5) then becomes equivalent to minimization of $a^T \hat{R}_{p+1} a$. The solution of this minimization problem is precisely the Yule-Walker estimate (Prob. 6.16). The maximum likelihood estimate is therefore asymptotically equivalent to the Yule-Walker estimate, as well as to the least-squares estimates of Sec. 6.6. For short data records, it can yield significantly different results. Due to its computational complexity, the maximum likelihood estimate is seldom used in practice.

6.8. ESTIMATION OF THE PARTIAL CORRELATION COEFFICIENTS

All the algorithms discussed so far estimate the AR parameters and, if necessary, compute the partial correlation coefficients from the estimated parameters. An alternative approach, which estimates the partial correlation coefficients directly from the measurements, is the subject of this section.

As is clear from Eqs. (6.3.1a) and (6.3.1b), the partial correlation coefficients $\{K_n, 1 \leq n \leq p\}$ can be used to compute the forward and backward partial innovations of orders 1 through p, starting with the initial conditions $u_{t,0} = \tilde{u}_{t,0} = y_t$. More precisely, given $\{y_t, 0 \leq t \leq N-1\}$, we can compute $\{u_{t,n}, \tilde{u}_{t-n,n}, 1 \leq n \leq p, n \leq t \leq N-1\}$. When only estimates of the partial correlation coefficients are given, we can compute the *forward and backward residual errors*, defined recursively by

$$e_{t,n+1} = e_{t,n} - \hat{K}_{n+1}\tilde{e}_{t-n-1,n}, \tag{6.8.1}$$

$$\tilde{e}_{t-n-1,n+1} = \tilde{e}_{t-n-1,n} - \hat{K}_{n+1}e_{t,n}. \tag{6.8.2}$$

The algorithm of Burg is based on the following idea [Burg, 1967, 1975]. Suppose we have estimated the partial reflection coefficients of orders up to n and used them to compute the forward and backward residual errors of order n. Let $\hat{K}_{n+1}$ be chosen to minimize the total energy of the forward and backward residual errors of order $n+1$; that is, $\sum_{t=n+1}^{N-1}[e_{t,n+1}^2 + \tilde{e}_{t-n-1,n+1}^2]$. Using (6.8.1) and (6.8.2), we get

$$\sum_{t=n+1}^{N-1}[e_{t,n+1}^2 + \tilde{e}_{t-n-1,n+1}^2] =$$

$$\sum_{t=n+1}^{N-1}[(e_{t,n} - \hat{K}_{n+1}\tilde{e}_{t-n-1,n})^2 + (\tilde{e}_{t-n-1,n} - \hat{K}_{n+1}e_{t,n})^2] =$$

$$\hat{K}_{n+1}^2\sum_{t=n+1}^{N-1}[e_{t,n}^2 + \tilde{e}_{t-n-1,n}^2] - 4\hat{K}_{n+1}\sum_{t=n+1}^{N-1}e_{t,n}\tilde{e}_{t-n-1,n} + \sum_{t=n+1}^{N-1}[e_{t,n}^2 + \tilde{e}_{t-n-1,n}^2].$$

The minimizer of this expression is

$$\hat{K}_{n+1} = \frac{2\sum_{t=n+1}^{N-1}e_{t,n}\tilde{e}_{t-n-1,n}}{\sum_{t=n+1}^{N-1}[e_{t,n}^2 + \tilde{e}_{t-n-1,n}^2]}. \tag{6.8.3}$$

Burg's algorithm consists of (6.8.3), (6.8.1), and (6.8.2), computed iteratively for $0 \leq n \leq p-1$.

A similar algorithm, due to Itakura and Saito [1971], estimates the partial correlation coefficients from

$$\hat{K}_{n+1} = \frac{\sum_{t=n+1}^{N-1} e_{t,n}\tilde{e}_{t-n-1,n}}{\left[\sum_{t=n+1}^{N-1} e_{t,n}^2\right]^{1/2} \left[\sum_{t=n+1}^{N-1} \tilde{e}_{t-n-1,n}^2\right]^{1/2}}. \qquad (6.8.4)$$

This estimate, while not based on a minimum criterion like (6.8.3), is intuitively appealing, since it is obtained from the definition

$$K_{n+1} = \frac{E u_{t,n}\tilde{u}_{t-n-1,n}}{\left[E u_{t,n}^2\right]^{1/2} \left[E \tilde{u}_{t-n-1,n}^2\right]^{1/2}}$$

upon replacing the expectations by the sample averages.

For $t = 0$ to $N - 1$ do
$\quad e_{t,0} = \tilde{e}_{t,0} = y_t$
For $n = 0$ to $p - 1$ do
$\quad A = \sum_{t=n+1}^{N-1} e_{t,n}\tilde{e}_{t-n-1,n}$
$\quad B = \sum_{t=n+1}^{N-1} e_{t,n}^2$
$\quad C = \sum_{t=n+1}^{N-1} \tilde{e}_{t-n-1,n}^2$
$\quad \hat{K}_{n+1} = \dfrac{2A}{B+C} \quad$ (Burg)
$\qquad$ or
$\quad \hat{K}_{n+1} = \dfrac{A}{\sqrt{BC}} \quad$ (Itakura-Saito)
$\quad$ For $t = n + 1$ to $N - 1$ do
$\qquad e_{t,n+1} = e_{t,n} - \hat{K}_{n+1}\tilde{e}_{t-n-1,n}$
$\qquad \tilde{e}_{t-n-1,n+1} = \tilde{e}_{t-n-1,n} - \hat{K}_{n+1}e_{t,n}$

Table 6.6. A summary of Burg and Itakura-Saito algorithms.

Table 6.6 summarizes the Burg and Itakura-Saito algorithms. In both cases, the AR parameters can be computed from the $\{\hat{K}_n\}$ using the Levinson-Durbin algorithm. We remark that both algorithms guarantee that the estimated partial correlation coefficients are less than 1 in magnitude, hence the stability of the estimated AR polynomial. The right side of (6.8.4) is less than 1 in magnitude by the Cauchy-Schwarz inequality. The right side of (6.8.3) is less than (6.8.4) in magnitude (hence less than 1) since the arithmetic mean of two positive numbers is larger than their geometric mean. Asymptotically, both algorithms are equivalent and are also equivalent to the Yule-Walker estimate. The algorithms of Burg and Itakura-Saito used to be very popular during the 1970s. In the 1980s they were accompanied by the adaptive lattice algorithms to be studied in Chapter 8.

Example 6.5. This example illustrates the least-squares algorithms described in Sec. 6.6 and the partial correlation algorithms described in this section. The simulated AR process has a characteristic polynomial

$$a(z) = 1 - 2.8z^{-1} + 3.82z^{-2} - 2.66z^{-3} + 0.9025z^{-4}$$

and unit innovation variance. The spectral density of this process exhibits two closely spaced peaks, as shown in Figure 6.9 (cf. the discussion on spectral resolution in Prob. 4.18). The number of data points was taken as $N = 64$, and 1000 Monte-Carlo simulations were performed. The asymptotic analysis does not apply to such a small number of data, so simulation is the only means of performance analysis in this case. Table 6.7 shows the empirical biases and standard deviations of the parameters for the five algorithms. As can be seen, the algorithms perform almost identically, but the forward-backward covariance method has a slightly smaller bias. The Yule-Walker method was also tested, but failed completely. The Yule-Walker estimate was unable to resolve the two spectral peaks in most of the 1000 runs. The conclusion (which has been verified by many researchers) is that both the least-squares methods and the partial correlation methods are preferable to the Yule-Walker method when the number of data is small.   ∎

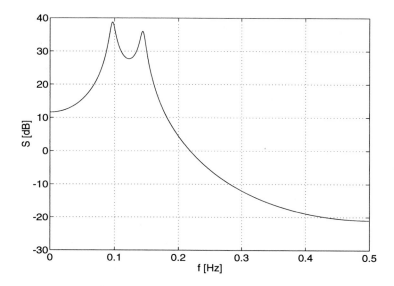

Figure 6.9. The spectral density of the signal in Example 6.5.

Parameter	σ_u^2	a_1	a_2	a_3	a_4
True value	1.0	−2.8	3.82	−2.66	0.9025
CRB	0.1768	0.0538	0.1218	0.1218	0.0538
Forward covariance	−0.07631 0.1711	0.03245 0.06872	−0.08453 0.1563	0.08586 0.1573	−0.03774 0.07153
Backward covariance	−0.06627 0.1748	0.02947 0.07386	−0.06606 0.1720	0.06261 0.1740	−0.02817 0.07811
Forward-backward	−0.05720 0.1711	0.03150 0.06555	−0.07667 0.1460	0.07611 0.1460	−0.03385 0.0660
Burg	−0.03692 0.1749	0.03551 0.06885	−0.08245 0.1481	0.08052 0.1457	−0.03431 0.06559
Itakura-Saito	−0.04880 0.1726	0.03407 0.06863	−0.07965 0.1478	0.07835 0.1454	−0.03374 0.06557

Table 6.7. Results of Monte-Carlo simulation of Example 6.5. Upper entry: empirical bias; lower entry: empirical standard deviation.

6.9. MODEL ORDER SELECTION

In Sec. 3.9 we introduced two approaches to model order selection. We now describe the application of these methods to order selection for autoregressive processes. It is clear from the discussion in Sec. 6.7 that for large N the first term in the AIC (3.9.2) or the MDL (3.9.3) criteria is given by $N \log \hat{\sigma}_u^2$ up to an additive constant. Also, for either the Levinson-Durbin algorithm or any of the methods of Sec. 6.8, $\hat{\sigma}_u^2$ is obtained as $\hat{r}_0 \prod_{n=1}^{p}(1 - \hat{K}_n^2)$. Therefore,

$$N^{-1}\text{AIC}(p) = \log \hat{r}_0 + \sum_{n=1}^{p} \log(1 - \hat{K}_n^2) + \frac{2(p+1)}{N}. \qquad (6.9.1)$$

The MDL expression is similar, except that the last term is $(p+1)\log N/N$.

The AIC (or the MDL) criterion can be combined with the order-recursive algorithms as follows. We set $\text{AIC}(0) = 0$ (the constant additive term can be neglected). Then, for each $n \geq 1$, we set

$$\text{AIC}(n) = \text{AIC}(n-1) + \log(1 - \hat{K}_n^2) + 2/N. \qquad (6.9.2)$$

After the maximum order is reached, the minimum value of the AIC is found and used to select the final model order p.

The AIC is not a consistent estimate of the model order. If the process is truly AR(p), there is a nonzero probability that the order selected by the AIC will be larger than p, even when N tends to infinity. This was shown by Shibata [1976], Söderström [1977], and Kashyap [1980] and is explained by the following argument.

It can be shown, using the methods of Sec. 3.6, that the asymptotic mean of $\hat{K}_{p+1}$ of an AR(p) process is zero, and its asymptotic variance is $\lim_{N \to \infty} N \hat{K}_{p+1} = 1$. Also, $\hat{K}_{p+1}$ is asymptotically normal. Therefore, there is a finite probability (about 0.15) that $|\hat{K}_{p+1}| > \sqrt{2/N}$. But, since $\log(1 - \hat{K}_{p+1}^2) < -\hat{K}_{p+1}^2$, it follows that there is a finite probability that $\log(1 - \hat{K}_{p+1}^2) + 2/N < 0$ and hence that $\text{AIC}(p+1) < \text{AIC}(p)$. But then the AIC cannot reach its minimum at the true order p.

In contrast to the AIC, the MDL (or Akaike's BIC, see Sec. 3.9) is consistent: the probability that the estimated order be equal to the true order approaches 1 as N tends to infinity. This was shown by Hannan [1980] for ARMA processes, and it follows for AR processes as a special case. An intuitive explanation is that the probability of $|\hat{K}_{p+1}| > \sqrt{\log N / N}$ tends to zero as $N \to \infty$.

6.10. ESTIMATION OF THE SPECTRAL DENSITY

One of the most common uses of autoregressive estimation is for estimating the spectral density of the measured process. The pth-order AR spectral estimator is given by

$$\hat{S}_{\text{AR}}^{(p)}(\omega) = \frac{\hat{\sigma}_u^2}{|\hat{a}(e^{j\omega})|^2} = \frac{\hat{\sigma}_u^2}{\left| \sum_{k=0}^{p} \hat{a}_k e^{-j\omega k} \right|^2}. \tag{6.10.1}$$

The AR(p) spectral estimator is not consistent, unless the given process is truly autoregressive, of order p or less. If the given process is Gaussian autoregressive of order p, the AR(p) spectral estimator is asymptotically efficient, and its normalized asymptotic variance is given by [cf. (3.3.3) and (5.2.13)]

$$\lim_{N \to \infty} N \operatorname{var}\{\hat{S}_{\text{AR}}^{(p)}(\omega)\} = \left(\frac{\partial S(\omega)}{\partial \theta_a} \right) \begin{bmatrix} 2\sigma_u^4 & 0 \\ 0 & R_p^{-1} \end{bmatrix} \left(\frac{\partial S(\omega)}{\partial \theta_a} \right)^T. \tag{6.10.2}$$

The individual components of the Jacobian matrix in (6.10.2) are given by

$$\frac{\partial S(\omega)}{\partial \sigma_u^2} = \frac{S(\omega)}{\sigma_u^2} \tag{6.10.3a}$$

$$\frac{\partial S(\omega)}{\partial a_k} = -S(\omega) \left[\frac{e^{-j\omega k}}{a(e^{j\omega})} + \frac{e^{j\omega k}}{a(e^{-j\omega})} \right]. \tag{6.10.3b}$$

The computation of the AR spectral density estimator requires a number of operations proportional to $(p+1)L$, where L is the desired number of frequency points, in addition to the operations needed to estimate the AR parameters. If p is greater than $\log_2 L$, it is more efficient to compute (6.10.1) by padding the vector $[1, \hat{a}_1, \ldots, \hat{a}_p]$ with $L - p - 1$ zeros and obtain $\hat{a}(e^{j\omega})$ by FFT. This procedure takes about $L \log_2 L$ operations.

As we have said, a fixed-order AR model does not yield a consistent spectrum estimate if the given process is not autoregressive. Nevertheless, it is possible to obtain a consistent estimate by increasing the order p as a function of the number of data points N. As was shown by Berk [1974], the choice $\lim_{N\to\infty} p(N) = \infty$ and $\lim_{N\to\infty} p^3(N)/N = 0$ guarantees consistency of the spectral estimate, provided the true spectral density is strictly positive and subject to some additional conditions. For the exact conditions and the proof, see Berk [1974].

Another spectral estimator, closely related to the AR estimator, is the *minimum variance spectral estimator* (MV), or the *Capon estimator* [Capon, 1969]. Denote

$$\boldsymbol{v}(\omega) = [1, e^{-j\omega}, \ldots, e^{-j\omega p}]^H.$$

Then the minimum variance spectral estimator is defined as

$$\hat{S}_{\text{MV}}^{(p)}(\omega) = \frac{p+1}{\boldsymbol{v}^H(\omega)\hat{R}_{p+1}^{-1}\boldsymbol{v}(\omega)}. \tag{6.10.4}$$

The derivation of the minimum variance estimator is based on the following argument. We have already seen, in developing the Yule-Walker estimate, that the output variance of the whitening filter $a(z)$ is given by $\boldsymbol{a}^T R_{p+1}\boldsymbol{a}$, where $\boldsymbol{a} = [a_0, a_1, \ldots, a_p]^T$. In the Yule-Walker estimate, the output variance is minimized subject to the constraint $a_0 = 1$ (see Prob. 6.16). Suppose that we remove this constraint, but instead impose the constraint $\sum_{k=0}^{p} a_k e^{-j\omega_0 k} = \boldsymbol{v}^H(\omega_0)\boldsymbol{a} = 1$ at a given frequency ω_0. This means that we wish to minimize the output variance of the whitening filter while maintaining a unit response to a complex sinusoid at a given frequency. The solution to this constrained minimization problem is (Prob. 6.21)

$$\boldsymbol{a} = \frac{R_{p+1}^{-1}\boldsymbol{v}(\omega_0)}{\boldsymbol{v}(\omega_0)^H R_{p+1}^{-1}\boldsymbol{v}(\omega_0)} \tag{6.10.5}$$

(note that $\boldsymbol{a}$ is a complex vector). Correspondingly, the minimum output variance at the frequency ω_0 is

$$(\boldsymbol{a}^H R_{p+1}\boldsymbol{a})_{\min} = \frac{1}{\boldsymbol{v}(\omega_0)^H R_{p+1}^{-1}\boldsymbol{v}(\omega_0)}. \tag{6.10.6}$$

Comparing (6.10.6) to (6.10.4), we see that the minimum variance spectral estimator at any frequency ω is (up to a constant factor) the minimum output variance of the whitening filter at that frequency, subject to the constraint that the response to a complex sinusoid at that frequency is unity.

To see the connection to AR estimation, recall the upper-diagonal-lower decomposition of R_{p+1}^{-1} in (2.10.5),

$$R_{p+1}^{-1} = A_{p+1}^T D_{p+1}^{-1} A_{p+1}.$$

Note that

$$[A_{p+1}\boldsymbol{v}]_k = \sum_{i=0}^{k} \alpha_{k,i} e^{j\omega(k-i)} = e^{j\omega k}\sum_{i=0}^{k} \alpha_{k,i} e^{-j\omega i} = e^{j\omega k}\alpha_k(e^{j\omega}).$$

Therefore,

$$\hat{S}_{\text{MV}}^{(p)}(\omega) = \frac{p+1}{\sum_{k=0}^{p} \hat{d}_k^{-1} |\hat{\alpha}_k(e^{j\omega})|^2}.$$

But

$$\hat{d}_k^{-1} |\hat{\alpha}_k(e^{j\omega})|^2 = \frac{1}{\hat{S}_{\text{AR}}^{(k)}(\omega)},$$

where $\hat{S}_{\text{AR}}^{(k)}(\omega)$ is the kth-order AR spectral estimate obtained from the covariance sequence $\{\hat{r}_0, \hat{r}_1, \ldots\}$. Therefore,

$$\hat{S}_{\text{MV}}^{(p)}(\omega) = \left[\frac{1}{p+1} \sum_{k=0}^{p} \frac{1}{\hat{S}_{\text{AR}}^{(k)}(\omega)} \right]^{-1}. \qquad (6.10.7)$$

The minimum variance spectral estimator is thus the *harmonic mean* of the AR spectral estimators of orders 0 through p.

The minimum variance estimator is not consistent even for a true AR(p) process, as is clear from (6.10.7). It is sometimes advocated on the basis of providing an estimate smoother than the corresponding AR(p) estimate (Prob. 6.17). Its usefulness for spectral analysis is limited. However, it is quite popular in a related field: that of direction finding of point sources by sensor arrays, the *array processing problem*. Musicus [1985] has developed a computationally efficient algorithm for the minimum variance spectral estimator based on the AR parameters and using FFT.

6.11. MATHEMATICA PACKAGES

The package `ArAlg.m` implements the algorithms presented in this chapter, as follows.

`YuleWalker[cov_]` solves the Yule-Walker equations for the given covariance sequence by direct matrix inversion. The returned values are the AR coefficients `a` and the innovation variance `sigu2`.

`Levinson[cov_]` implements the Levinson-Durbin algorithm for the given covariance sequence. The returned values are as for `YuleWalker`.

`KtoA[K_]` implements the part of the Levinson-Durbin algorithm that computes the AR coefficients from the partial correlation coefficients. The returned values are the AR coefficients `a`.

`AtoK[a_]` implements the inverse Levinson algorithm, computing the sequence of partial correlation coefficients from the given AR parameters.

`KtoR[K_, sigu2_, m_]` implements the inverse Schur algorithm, computing the covariance sequence from the given partial correlation coefficients of an AR(p) process and the innovation variance. It returns `r` in the range 0 through `m`.

`RtoK[r_]` implements the Schur algorithm, computing the partial correlation coefficients and the innovation variance from the given covariance sequence. The returned values are `sigu2` and `K`.

`LstSqrAr[data_, p_, opts___]` implements the four least-square algorithms described in Sec. 6.6. `data` is the data sequence, and `p` is the order of the AR model. The options `opts` are `Method->FCov` for the forward covariance method (this is the default), `Method->BCov` for the backward covariance method, `Method->FBCov` for the forward-backward covariance method, and `Method->PWin` for the prewindowed method. The returned values are `a` and `sigu2`.

`ParCor[data_, p_, opts___]` implements the two methods for direct computation of the partial correlation coefficients described in Sec. 6.8. `data` is the data sequence, and `p` is the order of the AR model. The options `opts` are `Method->ItakuraSaito` for the Itakura-Saito method (this is the default) and `Method->Burg` for the Burg method. The function then calls `KtoA` to compute the AR coefficients. The returned values are `a`, `K`, and `sigu2`.

`MvSpec[cov_, omega_]` computes the minimum-variance spectral estimator corresponding to the given covariance sequence using formula (6.10.5). The parameter `omega` can be either an integer or a list of frequency values (in radians). In the former case, the function returns a list of spectrum values at the given number of frequency points, equally spaced in the range $[0, \pi]$. In the latter case, the function returns the spectrum values at the given frequencies.

REFERENCES

Berk, K. N., "Consistent Autoregressive Spectral Estimates," *Ann. Statistics,* 2, pp. 489–502, 1974.

Burg, J. P., "Maximum Entropy Spectral Analysis," *37th Ann. Int. S.E.G. Meeting,* Oklahoma, 1967.

Burg, J. P., *Maximum Entropy Spectral Analysis,* Ph.D. dissertation, Stanford University, CA, 1975.

Capon, J., "High-resolution Frequency-wavenumber Spectrum Analysis," *Proc. IEEE,* 57, pp. 1408–1418, 1969.

Durbin, J., "The Fitting of Time-series Models," *Rev. Inst. Int. Statist.,* 28, pp. 233–243, 1960.

Hannan, E. J., "The Estimation of the Order of an ARMA Process," *Ann. Statistics,* 8, pp. 1071–1081, 1980.

Itakura, F., and Saito, S., "Digital Filtering Techniques for Speech Analysis and Synthesis," *7th Int. Cong. Acoust.,* Budapest, Paper 25-C-1, pp. 261–264, 1971.

Kashyap, R. L., "Inconsistency of the AIC Rule for Estimating the Order of Autoregressive Models," *IEEE Trans. Automatic Control,* AC-25, pp. 996–998, 1980.

Levinson, N., "The Wiener RMS (Root Mean Square) Error Criterion in Filter Design and Prediction," *J. Math. Phys.,* 25, pp. 261–278, 1947.

Mason, S. L., "Feedback Theory: Further Properties of Signal Flow Graphs," *Proc. IRE,* 44, pp. 920–926, 1956.

Musicus, B., "Fast MLM Power Spectrum Estimation from Uniformly Spaced Correlations," *IEEE Trans. Acoustics, Speech, Signal Processing,* ASSP-33, pp. 1333–1335, 1985.

Phillips, C. L., and Nagle, H. T., *Digital Control System Analysis and Design,* Prentice Hall, Englewood Cliffs, NJ, 1984.

Rabiner, L. R., and Schafer, R. W., *Digital Processing of Speech Signals,* Prentice Hall, Englewood Cliffs, NJ, 1978.

Shibata, R., "Selection of the Order of an Autoregressive Model by Akaike's Information Criterion," *Biometrika,* 63, pp. 117–126, 1976.

Söderström, T., "On Model Structure Testing in System Identification," *Int. J. Control,* 26, pp. 1–18, 1977.

Therrien, C. W., *Discrete Random Signals and Statistical Signal Processing,*, Prentice Hall, Englewood Cliffs, NJ, 1992.

Walker, G., "On periodicity in Series of Related Terms," *Proc. Roy. Soc. London Ser. A,* 131, pp. 518–532, 1931.

Yule, G. U., "On a Method for Investigating Periodicities in Disturbed Series with Special Reference to Wolfer's Sunspot Numbers," *Philos. Trans. Roy. Soc. London Ser. A,* 226, pp. 267–298, 1927.

PROBLEMS

6.1. Show by an example that the matrix $\tilde{R}_{p+1}$ constructed from the unbiased sample covariances may not be positive definite. Hint: Find an example with $p = 1$, $N = 2$, in which this matrix is positive semidefinite, and an example with $p = 2$, $N = 3$, in which this matrix is indefinite.

6.2. Derive the Levinson-Durbin algorithm by algebraic means, without relying on the stochastic interpretation of the quantities involved. Hint: Assume we have obtained a solution to the Yule-Walker equation of order n and we wish to use it to get a solution of order $n + 1$. Consider the expression

$$
\begin{bmatrix}
\alpha_{n,0} & \alpha_{n,1} & \cdots & \alpha_{n,n} & 0 \\
0 & \alpha_{n,n} & \cdots & \alpha_{n,1} & \alpha_{n,0}
\end{bmatrix}
\begin{bmatrix}
r_0 & r_1 & \cdots & r_n & r_{n+1} \\
r_1 & r_0 & \cdots & r_{n-1} & r_n \\
\vdots & \vdots & \ddots & \vdots & \vdots \\
r_{n+1} & r_n & \cdots & r_1 & r_0
\end{bmatrix}.
$$

Show that the result contains only four nonzero terms (at the four corners of the matrix) and write the expressions for these terms. Interpret them in the context of Sec. 6.2. Then find a 2×2 matrix that forces two of these four terms to zero when it premultiplies the above expression. This will yield the update formulas of the algorithm.

6.3. Prove that $u_{t,n}$ and $\tilde{u}_{t-n-1,n}$ have the same variance.

6.4. Derive a multichannel version of the Levinson-Durbin algorithm. Proceed through the following steps.

(a) Let $\{y_t\}$ be a stationary vector process, where the dimension of y_t is m. The forward and backward prediction errors are given by

$$
u_{t,n} = \sum_{k=0}^{n} A_{n,k} y_{t-k}, \quad A_{n,0} = \mathbf{1}_m,
$$

$$
\tilde{u}_{t,n} = \sum_{k=0}^{n} \tilde{A}_{n,k} y_{t+k}, \quad \tilde{A}_{n,0} = \mathbf{1}_m,
$$

where $\{A_{n,k}\}$ and $\{\tilde{A}_{n,k}\}$, the predictor coefficients, are $m \times m$ matrices. Show that the predictor coefficients satisfy the forward and backward Yule-Walker equations

$$
r_\ell + \sum_{k=1}^{n} A_{n,k} r_{\ell-k} = 0, \quad r_{-\ell} + \sum_{k=1}^{n} \tilde{A}_{n,k} r_{k-\ell} = 0
$$

for $1 \le \ell \le n$. Note that the $\{r_k\}$ are $m \times m$ matrices.

(b) Define

$$
d_n = u_{t,n} u_{t,n}^T, \quad \tilde{d}_n = \tilde{u}_{t,n} \tilde{u}_{t,n}^T,
$$

$$
\Delta_{n+1} = E u_{t,n} \tilde{u}_{t-n-1,n}^T, \quad K_{n+1} = \Delta_{n+1}^T d_n^{-1}, \quad \tilde{K}_{n+1} = \Delta_{n+1} \tilde{d}_n^{-1}.
$$

Show that

$$
\Delta_{n+1} = \sum_{k=0}^{n} A_{n,k} r_{n+1-k}.
$$

Then show that

$$
u_{t,n+1} = u_{t,n} - \tilde{K}_{n+1} \tilde{u}_{t-n-1,n}
$$

$$
\tilde{u}_{t-n-1,n+1} = \tilde{u}_{t-n-1,n} - K_{n+1} u_{t,n}.
$$

(c) From the above deduce that

$$
\begin{bmatrix} \mathbf{1}_m & A_{n+1,1} & \cdots & A_{n+1,n} & A_{n+1,n+1} \\ \tilde{A}_{n+1,n+1} & \tilde{A}_{n+1,n} & \cdots & \tilde{A}_{n+1,1} & \mathbf{1}_m \end{bmatrix}
$$
$$
= \begin{bmatrix} \mathbf{1}_m & -\tilde{K}_{n+1} \\ -K_{n+1} & \mathbf{1}_m \end{bmatrix} \begin{bmatrix} \mathbf{1}_m & A_{n,1} & \cdots & A_{n,n} & 0 \\ \tilde{0} & \tilde{A}_{n,n} & \cdots & \tilde{A}_{n,1} & \mathbf{1}_m \end{bmatrix}.
$$

$$
d_{n+1} = d_n - \tilde{K}_{n+1}\Delta_{n+1}^T, \quad \tilde{d}_{n+1} = \tilde{d}_n - K_{n+1}\Delta_{n+1}.
$$

You now have the complete Levinson-Durbin algorithm for vector processes. Write down the equations in their proper order of evaluation and initial conditions.

6.5. AR modeling is often used to identify spectral peaks and estimate their frequencies. Consider the special case of AR(2) process. Assuming that the poles are complex, there is a single peak near the frequency $\omega_0 = \arccos(-a_1/2\sqrt{a_2})$. It is therefore reasonable to estimate the frequency by $\hat{\omega}_0 = \arccos(-\hat{a}_1/2\sqrt{\hat{a}_2})$.

   (a) What exactly is ω_0? Why is it close to the frequency of the spectral peak?

   (b) Find an expression for the asymptotic CRB on the variance of $\hat{\omega}_0$ as a function of a_1, a_2.

6.6. Use the Levinson-Durbin algorithm to construct an AR(8) model for the sunspot data (the file `sunspot.dat` on the disk). Plot the corresponding spectral density and the estimated innovation sequence, and compare them with the results of the AR(2) model shown in Example 6.1.

6.7. Modify the Mathematica procedure `RtoK` so as to produce the lower-diagonal-upper decomposition of the Toeplitz covariance matrix R_{p+1}.

6.8. Derive the inverse Levinson algorithm given in Table 6.4.

6.9. Show that each of the algorithms in Sec. 6.3 can be implemented with an amount of storage proportional to p (rather than p^2). Rewrite the algorithms using the explicit storage variables.

6.10. Derive (6.4.4) from Figure 6.7.

6.11. Consider the state-space realization of an AR(p) model in controller form, as in (2.12.12) and (2.12.13). Show that the state vector x_t is equal to $[y_t, y_{t-1}, \ldots, y_{t-p+1}]^T$ in this case.

6.12. Consider the lattice filter realization of an AR(p) model depicted in Figures 6.6, 6.7, and 6.8.

   (a) Write the expressions for the state-space matrices of this realization in terms of the partial correlation coefficients $\{K_n\}$.

   (b) Let the innovation process $\{u_t\}$ be the input to the lattice filter. Use the relationships in Sec. 6.3 to show that the upper inputs to the blocks $\{U_n(z), 1 \le n \le p\}$ are $\{u_{t,n}\}$, while the lower inputs are $\{\tilde{u}_{t-n+1,n-1}\}$. Hence conclude that the state vector of this realization is $[\tilde{u}_{t,0}, \tilde{u}_{t-1,1}, \ldots, \tilde{u}_{t-p+1,p-1}]^T$.

(c) Use Prob. 6.11 and part (b) to show that the state vectors of the controller form realization and the lattice realization are related through the similarity transformation matrix A_p defined in (2.10.3).

(d) Use part (c) to deduce the algebraic relationship between the top-row companion matrix of $a(z)$ and the state-space matrix A of the lattice realization.

6.13. Show how the Yule-Walker estimate can be derived as the solution to a certain least-squares problem, similar to the ones presented in Sec. 6.6.

6.14. Show that the expressions (6.7.1) and (6.7.2) for $Q(Y_N, \theta_a)$ are equivalent by examining the individual entries of the corresponding matrices.

6.15. Show that the maximum likelihood estimate of the parameter a_1 of an AR(1) process can be obtained by solving a third-order polynomial equation, and write this equation explicitly.

6.16. Show that the minimizer of $a^T \hat{R}_{p+1} a$ [cf. (6.7.5)] is the Yule-Walker estimate.

6.17. Explain, in view of (6.10.5), why $\hat{S}_{MV}^{(p)}(\omega)$ is smoother (as a function of ω) than $\hat{S}_{AR}^{(p)}(\omega)$.

6.18. Use (6.5.2) to show that the limiting normalized entropy of a purely indeterministic stationary process is given by

$$\lim_{N\to\infty} \frac{2}{N} H(R_N) = 1 + \log 2\pi + \log \sigma_u^2 = 1 + \log 2\pi + \frac{1}{2\pi} \int_{-\pi}^{\pi} \log S(\omega) d\omega.$$

Remark: Burg originally derived his maximum entropy solution from this form of the entropy [Burg, 1967, 1975].

6.19. Add the computation of the AIC and MDL criteria to the procedure Levinson.

6.20. When the Levinson-Durbin (or Schur's) algorithm is used with the sample covariances $\{\hat{r}_n\}$, it produces estimates of the partial correlation coefficients $\{\hat{K}_n\}$. Develop a procedure for computing the asymptotic covariance of the $\{\hat{K}_n\}$, implement it in Mathematica, and test it. Hint: There are at least two ways to perform this computation. One is to compute the Jacobian matrix of the $\{K_n\}$ with respect to the $\{r_n\}$ and use it with Bartlett's formula for the asymptotic covariances of the $\{\hat{r}_n\}$. Another way is to compute the Jacobian matrix of the $\{K_n\}$ with respect to the $\{a_n\}$ and use it with the formula for the asymptotic covariances of the $\{\hat{a}_n\}$. Choose one of these ways and develop a procedure for computing the corresponding Jacobian.

6.21. Establish (6.10.5), proceeding as follows. Define $\lambda = (v^H(\omega_0) R_{p+1}^{-1} v(\omega_0))^{-1}$. Express the output variance as

$$a^H R_{p+1} a =$$

$$(a - \lambda R_{p+1}^{-1} v(\omega_0) + \lambda R_{p+1}^{-1} v(\omega_0))^H R_{p+1} (a - \lambda R_{p+1}^{-1} v(\omega_0) + \lambda R_{p+1}^{-1} v(\omega_0)).$$

Expand this expression and use the constraint $v^H(\omega_0) a = 1$ to show that $a^H R_{p+1} a \geq \lambda$ with equality if and only if $a = \lambda R_{p+1}^{-1} v(\omega_0)$.

CHAPTER 7

Moving Average and ARMA Parameter Estimation

7.1. INTRODUCTION

This chapter is devoted to estimation of the parameters of moving average and autoregressive moving average processes. MA and ARMA estimation is problematic due to several reasons. Contrary to AR, neither of these models admits a sufficient statistic of a fixed dimension, so any data reduction invariably leads to a loss of information. In particular, any estimate based on a fixed number of sample covariances cannot be asymptotically efficient, as was shown in Chapter 5. Linear estimation methods exist for MA and ARMA models, but their performance is often poor. The better estimation methods are nonlinear, so they require iterative procedures and are computation intensive. The presence of zeros near the unit circle implies frequency regions of low energy density. This offers an intuitive explanation to a well-known phenomenon: zeros near the unit circle are hard to estimate accurately. The asymptotic Cramér-Rao bound has a symmetric dependence on the numerator and the denominator polynomials [cf. (5.2.13)]. This symmetry is misleading, however; when the zeros are close to the unit circle, the number of measurements for which (5.2.13) applies is very large.

An "ideal" algorithm for MA and ARMA estimation does not exist. This may explain the large number of different algorithms proposed in the literature over the years. The selection of algorithms to be included in this book was based on their performance on the one hand, and on their computational complexity on the other. Performance is measured relative to the asymptotic Cramér-Rao bound, although this has its limitations, especially for short data records. Computational complexity cannot be judged against a standard measure, so algorithms with good performance were included in this book even though they may require a lot of computations. Algorithms with moderate accuracy and moderate computational

187

cost are important, too, at least for two reasons. First, they are desirable in cases where the number of data is large and processing speed is a major factor (mainly real-time applications). Second, they are useful for providing initial conditions for the more sophisticated algorithms.

7.2. MOVING AVERAGE PARAMETER ESTIMATION: ELEMENTARY METHODS

The moving average model to be treated in this chapter is

$$y_t = \sum_{k=0}^{q} b_k u_{t-k}, \quad b_0 = 1, \tag{7.2.1}$$

where $\{u_t\}$ is the innovation of $\{y_t\}$, assumed to be Gaussian with zero mean and variance σ_u^2. As was explained in Sec. 2.12, this amounts to assuming that the zeros of $b(z)$ are inside or on the unit circle. Noninnovation MA models are indistinguishable from innovation models when the process in question is Gaussian and only output measurements are available. In this section we present some elementary methods for estimating the MA parameters, while in the next section we discuss more advanced methods.

Recall the relationship between the MA parameters $\{b_k\}$, the covariances $\{r_k\}$, and the spectral density $S(\omega)$:

$$S(\omega) = \sum_{k=-q}^{q} r_k e^{-j\omega k} = \sigma_u^2 b(e^{j\omega}) b(e^{-j\omega}). \tag{7.2.2}$$

This relationship enables the computation of the MA parameters from the covariance sequence $\{r_k\}$ through the following procedure.

1. Form the polynomial

$$\overline{S}(z) = z^q \sum_{k=-q}^{q} r_k z^{-k}$$

and compute it roots, say $\{\mu_k, 1 \leq k \leq 2q\}$. Since the coefficients of this polynomial are symmetric, its $2q$ roots can be grouped in two sets of q elements each: the first set contains all the roots inside the unit circle and the second, all the roots outside it. Roots of $\overline{S}(z)$ on the unit circle must have even multiplicity (Prob. 2.19), so they are divided equally between the two sets. The elements of the second set are the reciprocals of the elements of the first.

2. Use the roots in the first set to build $b(z)$ as

$$b(z) = \prod_{k=1}^{q} (1 - \mu_k z^{-1}). \tag{7.2.3}$$

The coefficients of $b(z)$ are then the MA parameters of the process.

3. The innovation variance can be obtained upon equating the free terms in both sides of (7.2.2). This gives

$$\sigma_u^2 = \frac{r_0}{1 + \sum_{k=1}^{q} b_k^2}.$$

This procedure can be applied, in principle, to the sample covariances $\{\hat{r}_k\}$, thus yielding estimates of the MA parameters $\{\hat{\sigma}_u^2, \hat{b}_1, \ldots, \hat{b}_q\}$. In practice, the factorization of $\hat{\overline{S}}(z)$ may fail to yield two separable sets, as in the case of the true spectral density $\overline{S}(z)$. This happens whenever $\sum_{k=-q}^{q} \hat{r}_k e^{-j\omega k}$ takes negative values on the interval $[-\pi, \pi]$ (Prob. 7.1). A simple remedy is given as follows. Recall that the Toeplitz matrix $\hat{R}_{q+1}$ constructed from the sample covariances is guaranteed to be positive definite. Therefore, the windowed periodogram

$$\hat{S}_q(\omega) = \sum_{k=-q}^{q} \left(1 - \frac{|k|}{q+1}\right) \hat{r}_k e^{-j\omega k}$$

$$= \frac{1}{q+1}[1, e^{-j\omega}, \ldots, e^{-j\omega q}]\hat{R}_{q+1}[1, e^{-j\omega}, \ldots, e^{-j\omega q}]^H$$

is positive for all ω. Applying the above factorization procedure to $\hat{S}_q(\omega)$ is therefore guaranteed to succeed. Note, however, that this estimate is not consistent. Better windowing procedures exist that impose positive semidefiniteness of the estimated spectrum, but we will not discuss them here.

Several MA estimation methods are based on the idea of precomputing an AR estimate from the given data and using the estimated AR parameters to estimate the MA parameters. The order ℓ of the intermediate AR estimate should be much higher than the order q of the desired MA model. Assume we have computed the AR estimates $\{\hat{a}_k, 1 \le k \le \ell\}$ by one of the methods described in Chapter 6. Methods that compute $\{\hat{b}_k, 1 \le k \le q\}$ as a function of $\{\hat{a}_k, 1 \le k \le \ell\}$ are sometimes called "MA by long AR" methods. All these methods are necessarily inconsistent, since no MA process can be exactly described by a finite-order AR model (except in the trivial case $q = 0$). However, their bias can be made arbitrarily small by sufficiently increasing the AR order ℓ. The best known of these methods is perhaps the one due to Durbin [1959]. Durbin's method is usually presented as an approximate maximum likelihood (relying on the asymptotic normality of the AR estimates) [Anderson, 1971; Kay, 1988]. However, it can also be derived using a simple least-squares argument, as follows.

Define

$$\left(\sum_{k=0}^{q} \hat{b}_k z^{-k}\right)\left(\sum_{k=0}^{\ell} \hat{a}_k z^{-k}\right) = \sum_{k=0}^{\ell+q} \epsilon_k z^{-k}.$$

Then $\epsilon_0 = 1$ and

$$
\begin{bmatrix} \hat{a}_1 \\ \vdots \\ \vdots \\ 0 \\ \vdots \\ 0 \end{bmatrix} + \begin{bmatrix} 1 & \cdots & & & 0 \\ \vdots & \ddots & & & \vdots \\ \vdots & & \ddots & 1 & \\ 0 & & \ddots & & \vdots \\ \hat{a}_\ell & \ddots & & & \vdots \\ \vdots & & \ddots & & \\ 0 & \cdots & & & \hat{a}_\ell \end{bmatrix} \begin{bmatrix} \hat{b}_1 \\ \vdots \\ \hat{b}_q \end{bmatrix} = \begin{bmatrix} \epsilon_1 \\ \vdots \\ \vdots \\ \epsilon_{\ell+1} \\ \vdots \\ \epsilon_{\ell+q} \end{bmatrix}, \tag{7.2.4}
$$

or, in a compact notation,

$$\hat{a} + \hat{A}\hat{b} = e.$$

Suppose we replace $\{\hat{b}_k\}$ in (7.2.4) by $\{b_k\}$ and $\{\hat{a}_k\}$ by $\{\alpha_k\}$ of (2.7.8) and let ℓ tend to infinity. Then we know, by Theorem 2.7, that the $\{\epsilon_k\}$ will be identically zero. Since the $\{\hat{a}_k\}$ are presumably good estimates of the $\{\alpha_k\}$, we can take $\{\hat{b}_k\}$ in (7.2.4) as the minimizers of the sum of squares of the $\{\epsilon_k\}$. By (3.7.8) we know that the solution of this least-squares problem is

$$\hat{b} = -(\hat{A}^T \hat{A})^{-1} \hat{A}^T \hat{a}. \tag{7.2.5}$$

This is Durbin's estimate of the MA parameters. We observe that (7.2.5) has the form of the Yule-Walker equation, with the "sample covariances" $\{\sum_{i=0}^{\ell-k} \hat{a}_{i+k}\hat{a}_i, 0 \le k \le q\}$. Durbin's method can therefore be summarized by the following two-step procedure.

1. Pick $q \ll \ell \ll N$ and obtain an AR(ℓ) estimate from the data (by any of the methods presented in Chapter 6).
2. Using the AR coefficients as data, solve the qth-order Yule-Walker equations built from the sample covariances of these "data." The solution of the Yule-Walker equations is the vector of MA parameter estimates.

The main advantage of Durbin's method is that it uses only simple operations, of the type encountered in AR estimation. In particular, there is no need for polynomial root-finding or iterative calculations. The estimated MA polynomial is guaranteed to have all its roots inside the unit circle. On the negative side, the estimate lacks consistency and its bias is heavily dependent on the choice of the AR order ℓ. When the MA polynomial of the given process has zeros near the unit circle, a very high ℓ may be needed to reduce the bias to an acceptable level. If there are zeros on the unit circle, the estimates remain biased no matter how large ℓ is. Durbin's method is a good choice for obtaining initial conditions for the more sophisticated algorithms described in the next section.

Example 7.1. This example illustrates the performance of Durbin's MA algorithm for various values of the AR order ℓ. The MA polynomial was taken as $b(z) = 1 - z^{-1} + 0.8z^{-2}$ and the number of measurements was $N = 256$. Table 7.1 shows the empirical biases and standard deviations obtained from 1000 Monte-Carlo runs, for $\ell = 5, 10, 15, 20$, and 25. Also shown in the table is the asymptotic CRB (standard deviation). As can be seen, the bias of the estimates decreases as ℓ increases, but is not negligible even for $\ell = 25$. The standard deviation does not vary much with ℓ and is slightly larger than the CRB.   ∎

Parameter	σ_u^2	b_1	b_2
True value	1.0	−1.0	0.8
CRB	0.08839	0.0375	0.0375
$\ell = 5$	0.2243 0.1322	0.1001 0.04104	−0.2148 0.03200
$\ell = 10$	0.1254 0.1259	0.06954 0.04164	−0.1110 0.03703
$\ell = 15$	0.07463 0.1214	0.04020 0.04312	−0.07166 0.04254
$\ell = 20$	0.06516 0.1218	0.03742 0.04263	−0.06021 0.04471
$\ell = 25$	0.06165 0.1223	0.03494 0.04452	−0.05776 0.04628

Table 7.1. Results of Monte-Carlo simulation of Example 7.1. Upper entry: empirical bias; lower entry: empirical standard deviation.

7.3. MOVING AVERAGE PARAMETER ESTIMATION: ADVANCED METHODS

In Sec. 3.6 we presented two general schemes for parameter estimation based on sample moments: one that minimizes the cost function (3.6.3) and the other that minimizes the function (3.6.9). In the context of MA estimation, the statistic $s_N(Y_N)$ represents the set of sample covariances $\{\hat{r}_k, 0 \leq k \leq K\}$, the vector $s(\psi)$ represents the corresponding covariances $\{r_k, 0 \leq k \leq K\}$ as functions of the MA parameters, the matrix $\Sigma(\psi)$ is the covariance matrix of the $\hat{r}_k$, and $\hat{\Sigma}_N(Y_N)$ is any consistent estimate of $\Sigma(\theta)$ (θ being the vector of true MA parameters).

The explicit expressions for $r_k(\psi)$ and the entries of $\Sigma(\psi)$ are as follows.

$$r_k(\psi) = \begin{cases} \sigma_u^2 \sum_{i=0}^{q-k} b_i b_{i+k}, & 0 \leq k \leq q \\ 0, & \text{otherwise,} \end{cases} \tag{7.3.1}$$

$$\lim_{N \to \infty} N\text{cov}\{\hat{r}_k, \hat{r}_\ell\} = \sum_{i=-q}^{q-|k-\ell|} r_i r_{i+|k-\ell|} + \sum_{i=-q}^{q-k-\ell} r_i r_{i+k+\ell}. \tag{7.3.2}$$

Equation (7.3.2) is a special case of Bartlett's formula (4.2.12).

Minimization of (3.6.3) with respect to $\psi = \{\sigma_u^2, b_1, \ldots, b_q\}$ can be carried out by a numerical procedure. The estimates of Durbin's method can be used as initial conditions. Since most good numerical procedures rely on gradients, we need expressions for the partial derivatives of $C(\psi)$. We have

$$\frac{\partial r_k}{\partial \sigma_u^2} = \frac{r_k}{\sigma_u^2}, \quad \frac{\partial r_k}{\partial b_m} = \sigma_u^2 (b_{m+k} + b_{m-k}). \tag{7.3.3}$$

The partial derivatives of $\Sigma(\psi)$ and $C(\psi)$ can be expressed in terms of (7.3.3) as discussed in Prob. 7.2.

The estimate $\hat{\Sigma}$ needed for the cost function (3.6.9) can be obtained by substituting $\{\hat{r}_k\}$ for $\{r_k\}$ in (7.3.2). However, this is not guaranteed to yield a positive definite $\hat{\Sigma}$, since the sample covariances themselves are not guaranteed to be positive definite (this is the same problem encountered in Sec. 7.2 in estimating the MA parameters from the sample covariances). Failure of $\hat{\Sigma}$ to be positive definite may lead to divergence of the minimization algorithm for $\hat{C}(\psi)$. An alternative method is to compute $\hat{\Sigma}$ as a function of the Durbin MA estimates, which are also used as initial conditions for the nonlinear minimization procedure. This will always give a positive definite $\hat{\Sigma}$. Practice has shown that the accuracy of the MA estimates obtained from (3.6.9) is typically only slightly inferior to that of the estimates obtained from (3.6.3). Since (3.6.9) is much simpler to minimize than (3.6.3), it is the preferred procedure when using the method of moments for MA processes.

Another moment-based method uses the *sample correlation coefficients* in place of the sample covariances. These are defined by

$$\hat{\rho}_k = \frac{\hat{r}_k}{\hat{r}_0}, \quad k \geq 1. \tag{7.3.4}$$

The correlation coefficients are scale invariant; that is, they are independent of the innovation variance σ_u^2. This saves one parameter in the nonlinear minimization and slightly improves the numerical stability. The normalized asymptotic covariances of the sample correlation coefficients are given by

$$\Sigma_{k,\ell} = \lim_{N \to \infty} N \text{cov}\{\hat{\rho}_k, \hat{\rho}_\ell\} = \sum_{i=-q}^{q-|k-\ell|} \rho_i \rho_{i+|k-\ell|} + \sum_{i=-q}^{q-k-\ell} \rho_i \rho_{i+k+\ell} + 2\rho_k \rho_\ell \sum_{i=-q}^{q} \rho_i^2$$

$$- 2\rho_k \sum_{i=-q}^{q-\ell} \rho_i \rho_{i+\ell} - 2\rho_\ell \sum_{i=-q}^{q-k} \rho_i \rho_{i+k}. \tag{7.3.5}$$

This is a special case of Bartlett's formula for the correlation coefficients; see Prob. 4.8. Estimation based on the sample correlation coefficients is further developed in Prob. 7.4.

We finally remind the reader that the method of moments is not efficient, as was explained in Chapter 5. The relative efficiency improves with K, at the price of higher computational complexity. We also note that the method is not limited to Gaussian processes. As was proved in Chapter 5, Eq. (5.5.8), the method will have the same asymptotic performance for the MA coefficients if the process is non-Gaussian, and only the asymptotic variance of $\hat{\sigma}_u^2$ will change. The same remarks apply to ARMA processes, to be discussed in Sec. 7.5.

Example 7.2. The asymptotic covariance of the above estimates (based on either the sample covariances or the sample correlations) can be computed from the left side of (5.5.1) since, as we have shown in Sec. 3.6, these estimates achieve the lower bound asymptotically. Figure 7.1 shows the theoretical standard deviations of $\hat{b}_1$ and $\hat{b}_2$ for the process in Example 7.1 (with $N = 256$ measurements). We see that the graphs flatten out at about $K = 20$. In order to verify the theoretical results, we also ran 100 Monte-Carlo simulations of the algorithm, for $K = 5$ and 15. Table 7.2 shows the simulation results. The empirical biases of $\hat{b}_1$ and $\hat{b}_2$ are negligible, and the empirical standard deviations are in good agreement with the theoretical values. The last row of the table involves the maximum likelihood estimates, to be discussed later. ∎

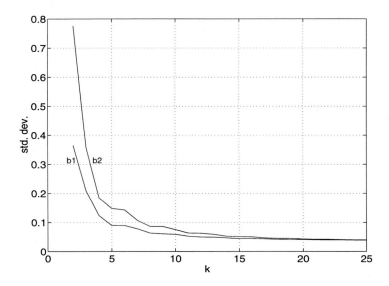

Figure 7.1. The theoretical standard deviations of $\hat{b}_1$ and $\hat{b}_2$ in Example 7.2.

To achieve full asymptotic efficiency, we must resort to maximum likelihood estimation. Exact maximum likelihood estimation of MA processes is deferred to Sec. 7.7, where it will be treated as a special case of maximum likelihood ARMA estimation. For the time being, we will settle for approximate maximum likelihood MA estimation. Recall that all the AR methods presented in Chapter 6 could be interpreted as approximate maximum likelihood techniques. As such, they all achieved asymptotic efficiency, even though they were not strictly maximum likelihood. In a similar vein, the approximate maximum likelihood MA estimate presented next is asymptotically efficient.

Parameter	σ_u^2	b_1	b_2
True value	1.0	-1.0	0.8
$K = 5$	0.01725 0.1337 0.1345	-0.00762 0.08347 0.08982	0.01656 0.1321 0.1480
$K = 15$	0.1277 0.1093 0.08965	0.00211 0.05375 0.04475	-0.00714 0.06625 0.05159
Maximum likelihood	0.02085 0.09337 0.08339	0.00804 0.03740 0.03750	-0.00860 0.04460 0.03750

Table 7.2. Results of Monte-Carlo simulation of Example 7.2. Upper entry: empirical bias; middle entry: empirical standard deviation; lower entry: theoretical standard deviation.

As usual, the measurements are $\{y_t, 0 \leq t \leq N-1\}$, and the vector of measurements is denoted by Y_N. Let U_N denote the vector of $\{u_t, 0 \leq t \leq N-1\}$ and let $\boldsymbol{u}$ be the vector of $\{u_t, -q \leq t \leq -1\}$. Then we can express Y_N as

$$Y_N = LU_N + B\boldsymbol{u}, \tag{7.3.6}$$

where

$$L_{k,\ell} = \begin{cases} b_{k-\ell}, & 0 \leq k-\ell \leq q \\ 0, & \text{otherwise} \end{cases}, \quad B_{k,\ell} = \begin{cases} b_{q+k-\ell}, & -q+1 \leq k-\ell \leq 0 \\ 0, & \text{otherwise.} \end{cases} \tag{7.3.7}$$

We now approximate Y_N by omitting the term $B\boldsymbol{u}$ from (7.3.6) or, equivalently, by assuming that $\boldsymbol{u} = 0$. With this approximation we have $R_N = EY_N Y_N^T \approx \sigma_u^2 LL^T$, so we can approximate the log likelihood function as

$$\log f_\theta(Y_N) \approx -\frac{N}{2}\log 2\pi - \frac{N}{2}\log \sigma_u^2 - \frac{1}{2\sigma_u^2}U_N^T U_N. \tag{7.3.8}$$

Maximizing (7.3.8) first with respect to σ_u^2 gives

$$\hat{\sigma}_u^2 = \frac{U_N^T U_N}{N}. \tag{7.3.9}$$

Substitution of (7.3.9) in (7.3.8) leads to the minimization of the cost function

$$C(\psi) = U_N^T U_N = Y_N^T L^{-T} L^{-1} Y_N, \tag{7.3.10}$$

where $\psi = [b_1, \ldots, b_q]^T$. Note that $C(\psi)$ depends on ψ through L [cf. (7.3.7)].

To compute the $\{u_t\}$, we use the recursion

$$u_t = y_t - \sum_{k=1}^{q} b_k u_{t-k}, \tag{7.3.11}$$

with the initial conditions $u_t = 0, t < 0$. The partial derivatives of $C(\psi)$ are given by

$$\frac{\partial C(\psi)}{\partial b_\ell} = 2U_N^T \frac{\partial U_N}{\partial b_\ell}, \tag{7.3.12}$$

where $\{\partial u_t / \partial b_\ell\}$ are computed from the recursion

$$\frac{\partial u_t}{\partial b_\ell} = -u_{t-\ell} - \sum_{k=1}^{q} b_k \frac{\partial u_{t-k}}{\partial b_\ell} \tag{7.3.13}$$

with zero initial conditions.

The approximate maximum likelihood algorithm can be initialized with Durbin's estimate. As was said before, the algorithm is asymptotically efficient, since the term Bu becomes negligible with respect to LU_N when N tends to infinity. However, when N is relatively small, the algorithm is biased due to the transient phenomenon in (7.3.11) and (7.3.13) caused by the zero initial conditions. This effect is stronger when the MA polynomial has roots near the unit circle, because it then takes longer for the filter $1/b(z)$ to recover from the zero initial conditions. Therefore, the closer the roots of $b(z)$ to the unit circle, the larger the number N required to eliminate the bias of the estimate.

The computational complexity of the approximate maximum likelihood algorithm is discussed in Prob. 7.5. The conclusion from Probs. 7.3 and 7.5 is that the maximum likelihood is more computationally intensive than the algorithms based on the sample covariances if the number of data points is large. It is therefore reasonable to use the latter in most cases and reserve the maximum likelihood approach for applications in which the number of data points is relatively small.

Example 7.2 (continued). The Monte-Carlo results of 100 simulations with approximate maximum likelihood estimation are shown in the last row of Table 7.2. Also shown in the table are the corresponding CRBs. As can be seen, the empirical biases of the ML estimates are negligible in this case, and the standard deviations are close to the CRB. ∎

7.4. ARMA ESTIMATION: THE MODIFIED YULE-WALKER METHOD

We begin our discussion of ARMA parameter estimation with a simple linear algorithm for estimation of the AR parameters, the modified Yule-Walker method. Consider the ARMA(p,q) process

$$y_t = -\sum_{k=1}^{p} a_k y_{t-k} + \sum_{k=0}^{q} b_k u_{t-k}. \tag{7.4.1}$$

Assume that all the roots of the polynomial $b(z)$ are inside (or on) the unit circle, so $\{u_t\}$ is the innovation process of $\{y_t\}$, as explained in Sec. 2.12. Multiply (7.4.1) by $y_{t-\ell}$, where $\ell > q$. Since $E u_{t-k} y_{t-\ell} = 0$ for $\ell > k$, we get upon taking expected values

$$r_\ell = -\sum_{k=1}^{p} a_k r_{\ell-k}, \quad \ell \geq q+1. \tag{7.4.2}$$

These are known as the *modified Yule-Walker equations*, in view of their similarity to (2.6.9). By collecting p such equations, we get the system

$$\begin{bmatrix} r_q & r_{q-1} & \cdots & r_{q-p+1} \\ r_{q+1} & r_q & \cdots & r_{q-p+2} \\ \vdots & \vdots & \ddots & \vdots \\ r_{q+p-1} & r_{q+p-2} & \cdots & r_q \end{bmatrix} \begin{bmatrix} a_1 \\ a_2 \\ \vdots \\ a_p \end{bmatrix} = -\begin{bmatrix} r_{q+1} \\ r_{q+2} \\ \vdots \\ r_{q+p} \end{bmatrix}. \tag{7.4.3}$$

If the covariances $\{r_{q-p+1}, \ldots, r_{p+q}\}$ were known, these equations could be solved for $\{a_1, \ldots, a_p\}$. In practice, the covariances are replaced by their sample counterparts, thus yielding estimates $\{\hat{a}_1, \ldots, \hat{a}_p\}$. These are the modified Yule-Walker estimates of the AR parameters of the ARMA process.

The modified Yule-Walker estimates are consistent, but their relative efficiency is known to be poor (more on this later). It has been observed (by Mehra [1971] and Cadzow [1980], among others) that better estimates can sometimes be obtained by using a larger number of equations from (7.4.2) and forming the system

$$\begin{bmatrix} r_q & r_{q-1} & \cdots & r_{q-p+1} \\ r_{q+1} & r_q & \cdots & r_{q-p+2} \\ \vdots & \vdots & \ddots & \vdots \\ r_{m-1} & r_{m-2} & \cdots & r_{m-p} \end{bmatrix} \begin{bmatrix} a_1 \\ a_2 \\ \vdots \\ a_p \end{bmatrix} = -\begin{bmatrix} r_{q+1} \\ r_{q+2} \\ \vdots \\ r_m \end{bmatrix} \tag{7.4.4}$$

for some $m > p+q$. These equations hold exactly for the true covariances and AR parameters. When the true covariances are replaced by the sample covariances, the system is *overdetermined*, and equality does not hold in general. However, we can *define* the estimates $\{\hat{a}_1, \ldots, \hat{a}_p\}$ as the least-squares solution of (7.4.4); that is,

$$\hat{a} = -(\hat{R}^T \hat{R})^{-1} \hat{R}^T \hat{r}, \tag{7.4.5}$$

where $\hat{R}, \hat{r}$, and $\hat{a}$ are the estimated values of the corresponding entries of (7.4.4). This is known as the *overdetermined Yule-Walker estimate* of the AR parameters.

Contrary to the Yule-Walker estimate for AR processes, the modified Yule-Walker estimate is not guaranteed to yield a stable polynomial. An unstable denominator polynomial is not acceptable from either a theoretical or practical point of view. An ad hoc solution to this problem is to compute the roots of the estimate $\hat{a}(z)$ and reflect those outside the unit circles into the unit circle by taking their reciprocals. This procedure can be justified by observing that the spectral density of the estimated model is invariant under reflection of the poles.

Both the modified and the overdetermined Yule-Walker estimates belong to the class of weighted least-squares estimates introduced in Prob. 3.19. In this case the weighting matrix W is the identity matrix. Therefore, the normalized asymptotic variances of these estimates can be analyzed using the formula given in Prob. 3.19. The statistic $s_N(Y_N)$ is $\{\hat{r}_{q-p+1}, \ldots, \hat{r}_m\}$ and its dimension is $K = m+p-q$. The matrix D in Prob. 3.19 is easily computed to be the $(m-q) \times (m+p-q)$ Toeplitz matrix

$$D_{k,\ell} = \begin{cases} a_{p+k-\ell}, & -p \leq k - \ell \leq 0 \\ 0, & \text{otherwise.} \end{cases} \tag{7.4.6}$$

Therefore, the normalized asymptotic variance of the estimate is

$$V = (R^T R)^{-1} R^T D \Sigma D^T R (R^T R)^{-1}, \tag{7.4.7}$$

where R is the coefficient matrix in (7.4.4) and Σ is the normalized asymptotic covariance of $\{\hat{r}_{q-p+1}, \ldots, \hat{r}_m\}$. This formula holds equally well for the modified and the overdetermined Yule-Walker estimates, and it depends on the number m. When $m = p + q$, it simplifies to

$$V = R^{-1} D \Sigma D^T R^{-1}. \tag{7.4.8}$$

Problem 7.6 shows how to compute the matrix $D\Sigma D^T$ without explicitly computing Σ first.

The overdetermined Yule-Walker estimate is asymptotically unbiased, but it can be considerably biased for a finite sample size. The bias typically increases with m, which puts a practical upper limit on m of about three to four times p. This phenomenon has been analyzed in Porat and Friedlander [1985].

As was discussed in Prob. 3.20, the normalized asymptotic variance of any weighted least-squares estimate is bounded from below by $[R^T (D\Sigma D^T)^{-1} R]^{-1}$, and the lower bound is achieved by the weighting matrix $W = (D\Sigma D^T)^{-1}$. The use of this weighting matrix yields the *optimal modified Yule-Walker estimate*, or *optimal instrumental variable estimate* [Stoica, Söderström, and Friedlander, 1985]. Strictly speaking, the weight matrix $W = (D\Sigma D^T)^{-1}$ cannot be used in the algorithm, since it depends on the unknown ARMA parameters. However, an asymptotically equivalent estimate can be obtained by replacing W with a consistent estimate thereof. A similar approach was taken in Sec. 3.6, when we replaced $\Sigma(\theta)$ by $\Sigma_N(Y_N)$ in (3.6.9) to obtain an asymptotically minimum variance estimate. Methods for obtaining consistent estimates for the matrix W were developed in Stoica, Friedlander, and Söderström [1987b].

Example 7.3. Let $\{y_t\}$ be the Gaussian ARMA$(4,4)$ process

$$y_t = 1.16098y_{t-1} - 1.94297y_{t-2} + 1.04778y_{t-3} - 0.81451y_{t-4}$$
$$+ u_t - 0.92878u_{t-1} + 1.31018u_{t-2} - 0.50293u_{t-3} + 0.29322u_{t-4}, \quad \sigma_u^2 = 2.0.$$

Table 7.3 shows the results of 500 Monte-Carlo simulations of the overdetermined Yule-Walker method, with $N = 1024$ and $m = 8, 12, 16$, and 20. Also shown in the table are the theoretical standard deviations, computed from (7.4.7). As we see, the estimation accuracy obtained with $m = 8$ (the minimum number) is poor. Increasing the number of Yule-Walker equations improves the accuracy, reaching an optimum at about $m = 16$. However, the bias is not negligible, being of the same order of magnitude as the standard deviation. The empirical standard deviations are reasonably close to the theoretical values. ∎

Parameter	a_1	a_2	a_3	a_4
True value	-1.16098	1.94297	-1.04778	0.81451
$m = 8$	0.03080 0.20743 0.18472	-0.25732 0.26997 0.32000	0.17895 0.21069 0.12785	-0.15606 0.15877 0.13344
$m = 12$	0.01460 0.03858 0.03478	-0.02452 0.05428 0.04633	0.01976 0.04134 0.03615	-0.01481 0.03393 0.03092
$m = 16$	0.02092 0.03309 0.02740	-0.03649 0.04580 0.03534	0.03018 0.04219 0.03550	-0.02058 0.03319 0.02981
$m = 20$	0.02634 0.03559 0.03040	-0.04454 0.04933 0.03858	0.03585 0.04386 0.03689	-0.02363 0.03307 0.02800

Table 7.3. Results of Monte-Carlo simulation of Example 7.3. Upper entry: empirical bias; middle entry: empirical standard deviation; lower entry: theoretical standard deviation.

The modified Yule-Walker estimate handles only the AR parameters of the ARMA process and leaves open the question of estimating the MA parameters and the innovation variance. A simple way of completing the estimation procedure is to use Durbin's method presented in Sec. 7.2. To this end, we form the intermediate sequence

$$e_t = \sum_{k=0}^{p} \hat{a}_k y_{t-k}, \quad p \le t \le N - 1. \tag{7.4.9}$$

If the estimated AR parameters are sufficiently accurate, the filtered sequence $\{e_t\}$ is a reasonable approximation of the MA process $\sum_{k=0}^{q} b_k u_{t-k}$. Durbin's algorithm can thus be applied to $\{e_t\}$, to estimate $\{\sigma_u^2, b_1, \ldots, b_q\}$. An even simpler method is to apply Durbin's idea directly to $\{y_t\}$. Let $\hat{c}(z)$ be the result of fitting an ℓth-order AR model to $\{y_t\}$. Define

$$\left(\sum_{k=0}^{q} \hat{b}_k z^{-k}\right)\left(\sum_{k=0}^{\ell} \hat{c}_k z^{-k}\right) = \sum_{k=0}^{\ell+q} \epsilon_k z^{-k}.$$

Then

$$\begin{bmatrix} \hat{c}_{p+1} \\ \vdots \\ 0 \\ \vdots \\ 0 \end{bmatrix} + \begin{bmatrix} \hat{c}_p & \cdots & \hat{c}_{p-q+1} \\ \vdots & \ddots & \vdots \\ \hat{c}_\ell & \ddots & \vdots \\ \vdots & \ddots & \vdots \\ 0 & \cdots & \hat{c}_\ell \end{bmatrix} \begin{bmatrix} \hat{b}_1 \\ \vdots \\ \hat{b}_q \end{bmatrix} = \begin{bmatrix} \epsilon_{p+1} \\ \vdots \\ \epsilon_{\ell+1} \\ \vdots \\ \epsilon_{\ell+q} \end{bmatrix}, \qquad (7.4.10)$$

or, in compact notation,

$$\hat{\boldsymbol{c}} + \hat{C}\hat{\boldsymbol{b}} = \boldsymbol{e}.$$

As in the case of Durbin's MA estimate, we can use this relationship to compute $\hat{\boldsymbol{b}}$ as

$$\hat{\boldsymbol{b}} = -(\hat{C}^T\hat{C})^{-1}\hat{C}^T\hat{\boldsymbol{c}}. \qquad (7.4.11)$$

Note that, contrary to (7.2.5), $\hat{C}^T\hat{C}$ is not a Toeplitz matrix, so (7.4.11) cannot be solved by the Levinson-Durbin algorithm (it can be solved efficiently using the generalized Levinson algorithm [Friedlander et. al., 1979], but this is outside the scope of this book).

Neither of the two methods for estimating the MA parameters is very accurate. However, both are often adequate for generating initial conditions for the nonlinear ARMA methods described in the subsequent sections.

Example 7.3 (Continued). The estimated AR parameters obtained from the overdetermined Yule-Walker equations (with $m = 16$) were used to compute the sequence $\{e_t\}$ as in (7.4.9), and the MA parameters were estimated from this sequence by Durbin's MA method. This was repeated for the 500 Monte-Carlo simulations. Table 7.4 shows the results, as well as the CRBs for the MA parameters (the theoretical standard deviations of this method have not been derived, so they are not shown). The MA estimates are biased, as expected, but their empirical standard deviations are close to the corresponding CRBs. The method given by (7.4.10) and (7.4.11) was also tested, but its accuracy was very poor in this example. ∎

7.5. ARMA ESTIMATION FROM THE SAMPLE COVARIANCES

The methods presented in Sec. 7.3 can be extended to ARMA processes. As before, the statistic $s_N(Y_N)$ represents the set of sample covariances $\{\hat{r}_k, 0 \le k \le K\}$, the vector $s(\psi)$ represents the corresponding covariances $\{r_k, 0 \le k \le K\}$ as functions of the ARMA parameters, and the matrix $\Sigma(\psi)$ is the covariance matrix of the $\hat{r}_k$. The cost function to be minimized is (3.6.3).

Parameter	b_1	b_2	b_3	b_4
True value	-0.92878	1.31018	-0.50293	0.29322
$\ell = 20$	0.03002 0.04732 0.04351	-0.06263 0.06051 0.05521	0.03512 0.05700 0.05513	-0.02311 0.05373 0.04336

Table 7.4. Results of Monte-Carlo simulation of Example 7.3 (continued). Upper entry: empirical bias; middle entry: empirical standard deviation; lower entry: Cramér-Rao bound (standard deviation).

The covariances $r_k(\psi)$ are computed by (2.12.11). The AR covariances $\{c_k\}$ appearing in that formula are best computed by the inverse Levinson and inverse Schur procedures described in Sec. 6.3. The entries of $\Sigma(\psi)$ are computed by the Bartlett formula

$$\lim_{N \to \infty} N \operatorname{cov}\{\hat{r}_k, \hat{r}_\ell\} = \bar{r}_{k+\ell}(\psi) + \bar{r}_{k-\ell}(\psi) \qquad (7.5.1)$$

[cf. (4.2.13)]. The $\bar{r}_k(\psi)$ are the covariances of the ARMA$(2p, 2q)$ process corresponding to the transfer function $b^2(z)/a^2(z)$, so they are computed similarly to the $r_k(\psi)$. The computation of the partial derivatives of $r_k(\psi)$ was discussed in Prob. 5.6. The computation of the partial derivatives of $\bar{r}_k(\psi)$ is discussed in Prob. 7.8. The methods presented in Sec. 7.4 can be used to provide initial conditions for the numerical minimization procedure. The normalized asymptotic variance of the estimate is

$$V(\theta) = \left[\left(\frac{\partial \boldsymbol{r}}{\partial \theta} \right)^T \Sigma^{-1}(\theta) \left(\frac{\partial \boldsymbol{r}}{\partial \theta} \right) \right]^{-1} . \qquad (7.5.2)$$

As in the case of MA processes, the following variations of the above method can be suggested.

1. Replacing $\Sigma(\psi)$ by its estimate $\hat{\Sigma}$, computed from the initial estimates;
2. Working with the sample correlation coefficients $\{\hat{\rho}_k\}$ (and their normalized asymptotic covariances) instead of the sample covariances;
3. A combination of 1 and 2.

A numerical procedure for computing the normalized asymptotic covariance of the sample correlation coefficients is developed in Prob. 7.9. Asymptotically, all four methods are equivalent; however, for finite data they usually give different answers. Typically, the bias of these methods is not negligible even when the number of measurements is on the order of thousands, especially when the process has zeros near the unit circle. The methods based on $\hat{\Sigma}$ are preferred to the ones based on $\Sigma(\psi)$ from a computational viewpoint.

Example 7.4. Let $\{y_t\}$ be the ARMA$(1,1)$ process

$$y_t = 0.5y_{t-1} + u_t + 0.8u_{t-1}, \quad \sigma_u^2 = 1.$$

Table 7.5 shows the results of 100 Monte-Carlo simulations of the four algorithms mentioned above. The number of measurements in each run is $N = 1024$, and the number of sample covariances (or correlation coefficients) used in the algorithms is $K = 10$. As we see, the four algorithms are similar in their performance, and the empirical standard deviations are close to the theoretical values. However, the bias is not negligible with respect to the standard deviation.   ∎

Parameter	σ_u^2	a_1	b_1
True value	1.0	−0.5	0.8
$\Sigma(\theta), \hat{r}$	0.02907 0.03888	0.00072 0.03129	−0.01315 0.02765
$\hat{\Sigma}, \hat{r}$	0.01560 0.03880	0.00328 0.03040	−0.01505 0.02767
$\Sigma(\theta), \hat{\rho}$	0.03143 0.04000	0.01012 0.03059	−0.01326 0.02773
$\hat{\Sigma}, \hat{\rho}$	0.01933 0.03800	0.00314 0.03041	−0.01511 0.02766
Theoretical	0.04440	0.02957	0.02417

Table 7.5. Results of Monte-Carlo simulation of Example 7.4. Upper entry: empirical bias; lower entry: empirical standard deviation.

The estimation of the ARMA parameters from the sample covariances (or correlation coefficients) has been the subject of extensive research. The linear methods (the modified Yule-Walker and its variations) were discussed in the previous section. The methods of Walker ([1961], [1962] for MA and ARMA, respectively) achieve the same relative asymptotic efficiency as the methods discussed above. A modification of Walker's ARMA method is given in Stoica, Friedlander, and Söderström [1987a]. A relatively simple procedure is based on the minimization of the cost function

$$C(\psi) = 0.5 \sum_{k=0}^{K} [r_k(\psi) - \hat{r}_k]^2. \tag{7.5.3}$$

This avoids the need for $\Sigma(\psi)$ and its derivatives and is therefore much simpler to implement than the methods presented above. However, the asymptotic performance of this method is inferior to the ones above, as analyzed in Prob. 7.11. Yet another method, based on a nonlinear cost function of the sample partial correlation coefficients, is explored in Prob. 7.12.

7.6. APPROXIMATE MAXIMUM LIKELIHOOD ARMA ESTIMATION

An approximate maximum likelihood algorithm for the ARMA problem can be developed similarly to the one described in Sec. 7.3. Let Y_{N-p} be the vector of $\{y_t, p \leq t \leq N-1\}$ and $\boldsymbol{y}$ the vector of $\{y_t, 0 \leq t \leq p-1\}$. Also, let U_{N-p} be the vector of $\{u_t, p \leq t \leq N-1\}$ and $\boldsymbol{u}$ the vector of $\{u_t, p-q \leq t \leq p-1\}$. Then

$$L_1 Y_{N-p} + A\boldsymbol{y} = L_2 U_{N-p} + B\boldsymbol{u}, \qquad (7.6.1)$$

where L_2 and B are as in (7.3.5), and

$$(L_1)_{k,\ell} = \begin{cases} a_{k-\ell}, & 0 \leq k-\ell \leq p \\ 0, & \text{otherwise} \end{cases}, \quad A_{k,\ell} = \begin{cases} a_{p+k-\ell}, & -p+1 \leq k-\ell \leq 0 \\ 0, & \text{otherwise}. \end{cases}$$
$$(7.6.2)$$

As in the MA case, we approximate Y_N by omitting the term $B\boldsymbol{u}$ or, equivalently, assuming that $\boldsymbol{u} = 0$. In addition, we make the approximate assumption that the vector $\boldsymbol{y}$ is nonrandom. There is no need to assume that $\boldsymbol{y} = 0$, since this vector is available from the measurements. With these approximations we have

$$R_{N-p} = E[Y_{N-p} + L_1^{-1}A\boldsymbol{y}][Y_{N-p} + L_1^{-1}A\boldsymbol{y}]^T \approx \sigma_u^2 L_1^{-1} L_2 L_2^T L_1^{-T},$$

so we can approximate the log likelihood function as

$$\log f_\theta(Y_{N-p}) \approx -\frac{N-p}{2}\log 2\pi - \frac{N-p}{2}\log \sigma_u^2 - \frac{1}{2\sigma_u^2}U_{N-p}^T U_{N-p}. \qquad (7.6.3)$$

Maximizing (7.6.3) first with respect to σ_u^2 gives

$$\hat{\sigma}_u^2 = \frac{U_{N-p}^T U_{N-p}}{N-p}. \qquad (7.6.4)$$

Substitution of (7.6.4) in (7.6.3) leads to minimization of the cost function

$$C(\psi) = U_{N-p}^T U_{N-p} = [L_1 Y_{N-p} + A\boldsymbol{y}]^T L_2^{-T} L_2^{-1} [L_1 Y_{N-p} + A\boldsymbol{y}]^T, \qquad (7.6.5)$$

where $\psi = [a_1, \ldots, a_p, b_1, \ldots, b_q]^T$. $C(\psi)$ depends on ψ through L_1, L_2, and A.

To compute the $\{u_t\}$, we use the recursion

$$u_t = \sum_{k=0}^{p} a_k y_{t-k} - \sum_{k=1}^{q} b_k u_{t-k}, \qquad (7.6.6)$$

with the initial conditions $u_t = 0, t < p$. The partial derivatives of $C(\psi)$ are given by

$$\frac{\partial C(\psi)}{\partial a_\ell} = 2U_{N-p}^T \frac{\partial U_{N-p}}{\partial a_\ell}, \quad \frac{\partial C(\psi)}{\partial b_\ell} = 2U_{N-p}^T \frac{\partial U_{N-p}}{\partial b_\ell}, \qquad (7.6.7)$$

where $\{\partial u_t / \partial a_\ell\}$ and $\{\partial u_t / \partial b_\ell\}$ are computed from the recursions

$$\frac{\partial u_t}{\partial a_\ell} = y_{t-\ell} - \sum_{k=1}^{q} b_k \frac{\partial u_{t-k}}{\partial a_\ell}, \quad \frac{\partial u_t}{\partial b_\ell} = -u_{t-\ell} - \sum_{k=1}^{q} b_k \frac{\partial u_{t-k}}{\partial b_\ell} \qquad (7.6.8)$$

with zero initial conditions.

Example 7.5. Let $\{y_t\}$ be the ARMA$(2, 2)$ process

$$y_t = 1.4y_{t-1} - 0.9y_{t-2} + u_t + u_{t-1} + 0.8u_{t-2}, \quad \sigma_u^2 = 1.$$

The first row of Table 7.6 shows the results of 100 Monte-Carlo simulations of the approximate maximum likelihood algorithm, with $N = 128$. Despite the small number of measurements, the estimates have a very small bias, and the empirical standard deviation is close to the (asymptotic) CRB. This excellent performance is not matched by any of the algorithms described in the previous sections. The second row of the table will be discussed later. ∎

Parameter	σ_u^2	a_1	a_2	b_1	b_2
True value	1.0	-1.4	0.9	1.0	0.8
Approx. ML	-0.00956	-0.00515	0.00048	0.00454	-0.00036
(time)	0.1356	0.04232	0.04081	0.07073	0.08995
$N = 128$	0.1250	0.03919	0.03914	0.05385	0.05380
Approx. ML	-0.00023	-0.00149	0.00014	0.00033	0.00125
(frequency)	0.04779	0.01383	0.01420	0.01829	0.02301
$N = 1024$	0.04419	0.01385	0.01384	0.01904	0.01902

Table 7.6. Results of Monte-Carlo simulation of Example 7.5. Upper entry: empirical bias; middle entry: empirical standard deviation; lower entry: CRB (standard deviation).

Box and Jenkins [1970] have proposed an improvement over the approximate log likelihood (7.6.3). They have shown that the log likelihood can be expressed as

$$\log f_\theta(Y_N) = g(\theta) - \frac{1}{2} \sum_{t=-\infty}^{N-1} \hat{u}_t^2, \tag{7.6.9}$$

where $\hat{u}_t$ denotes the best linear estimate of u_t, given the measurement vector Y_N and the parameter vector θ, and $g(\theta)$ is a function that does not depend on Y_N. Furthermore, $g(\theta)$ becomes negligible relative to the second term in (7.6.9) as N tends to infinity. Box and Jenkins then suggested an iterative approximate procedure for the computation of $\sum_{t=-\infty}^{N-1} \hat{u}_t^2$. The Box-Jenkins method has gained considerable popularity in the statistical community, but it has never become a favorite in engineering applications. The algorithm is awkward to implement and does not appear to offer an advantage over the exact maximum likelihood method discussed in the next section. We will therefore not discuss this approach further.

An alternative maximum likelihood approach, originally suggested by Hannan [1969] and elaborated by Akaike [1973], is based on frequency domain description of the data. Recall that the periodogram values $I(\omega_k)$ are asymptotically exponentially distributed, except when $\omega_k = 0$ or π (Prob. 4.11). The mean of $I(\omega_k)$ is asymptotically equal to $S(\omega_k)$, the true spectral density at the same frequency.

Furthermore, it can be shown (see, e.g., Anderson [1971, p. 484]) that $I(\omega_k)$ and $I(\omega_\ell)$ are asymptotically independent for $k \neq \ell$ (asymptotic lack of correlation was proved in Theorem 4.7). Assume for convenience that N is even and let $\boldsymbol{I}_N = \{I(\omega_k), \omega_k = 2\pi(k - 0.5)/N, 1 \leq k \leq N/2\}$ (note that the frequencies are chosen so as to avoid using 0 and π). Then we have

$$f_\theta(\boldsymbol{I}_N) \approx \prod_{k=1}^{N/2} \frac{1}{S(\omega_k)} \exp\left\{ -\frac{I(\omega_k)}{S(\omega_k)} \right\}. \tag{7.6.10}$$

Let us further denote $S_0(\omega) = b(e^{j\omega})b(e^{-j\omega})/a(e^{j\omega})a(e^{-j\omega})$, so that $S(\omega) = \sigma_u^2 S_0(\omega)$. Then we get form (7.6.10),

$$\log f_\theta(\boldsymbol{I}_N) \approx -\frac{N}{2} \log \sigma_u^2 - \sum_{k=1}^{N/2} \log S_0(\omega_k) - \frac{1}{\sigma_u^2} \sum_{k=1}^{N/2} \frac{I(\omega_k)}{S_0(\omega_k)}. \tag{7.6.11}$$

Maximizing (7.6.11) first with respect to σ_u^2 yields

$$\hat{\sigma}_u^2 = \frac{2}{N} \sum_{k=1}^{N/2} \frac{I(\omega_k)}{S_0(\omega_k)}. \tag{7.6.12}$$

Substituting back to (7.6.11) gives

$$\left.\frac{2}{N} \log f_\theta(\boldsymbol{I}_N)\right|_{\sigma_u^2 = \hat{\sigma}_u^2} \approx -\log\left[\frac{2}{N} \sum_{k=1}^{N/2} \frac{I(\omega_k)}{S_0(\omega_k)} \right] - \frac{2}{N} \sum_{k=1}^{N/2} \log S_0(\omega_k) - 1. \tag{7.6.13}$$

The third term in (7.6.13) is constant, while the second approaches zero as N tends to infinity, independently of θ (Prob. 7.13). The approximate frequency domain maximum likelihood algorithm is thus based on the minimization of the cost function

$$C(\psi) = \frac{2}{N} \sum_{k=1}^{N/2} \frac{I(\omega_k)}{S_0(\omega_k)} = \frac{2}{N} \sum_{k=1}^{N/2} I(\omega_k) \frac{a(e^{j\omega_k})a(e^{-j\omega_k})}{b(e^{j\omega_k})b(e^{-j\omega_k})} \tag{7.6.14}$$

with respect to $\psi = \{a_1, \ldots, a_p, b_1, \ldots, b_q\}$. The partial derivatives of $C(\psi)$ are obtained in a straightforward manner.

The performance of the frequency domain algorithm is not satisfactory when the ARMA process contains poles or zeros near the unit circle and the number of data is relatively small. The reason is that the bias of the periodogram is significant in such cases, causing a large bias in the estimates.

Example 7.5 (continued). The frequency domain algorithm fails completely in this example when $N = 128$ due to the large bias of the estimates. However, when N is increased to 1024, the bias almost disappears and the standard deviation becomes close to the CRB, as shown in the second row of Table 7.6. ∎

7.7. EXACT MAXIMUM LIKELIHOOD ARMA ESTIMATION

All the methods presented so far rely on the asymptotic properties of the given ARMA process. The number of measurements needed for the asymptotic approximations to be valid is roughly determined by the poles and the zeros of largest magnitudes. The closer these magnitudes are to 1, the larger the required number of measurements. When the number of measurements is too small, the methods described so far tend to be biased, and their performance may be far worse than the theoretical predictions. Exact maximum likelihood is often the only good estimation method in such situations. It should be noted, however, that even the exact maximum likelihood is not likely to achieve the accuracy predicted by the asymptotic CRB formula. A method for computing the exact CRB for a small number of measurements has been derived (see, e.g., Porat and Friedlander [1986]), but the exact maximum likelihood may not achieve this lower bound. In addition, the exact maximum likelihood is not free of bias for a small number of measurements. Despite these remarks, exact maximum likelihood usually yields the most accurate estimates when the number of measurements is small.

In this section we derive an algorithm for computing the exact likelihood function of an ARMA process and use it to develop an exact maximum likelihood algorithm for ARMA parameter estimation. This will be done using the *Kalman-Bucy* filter. The Kalman-Bucy filter is a recursive algorithm that computes (among other things) the partial innovation of a process obeying a state-space model such as (2.12.13). Since we do not wish to assume prior knowledge of this algorithm, we will derive it from basic principles. We will do so only for the special case appearing in the ARMA maximum likelihood algorithm. The interested reader is referred to Kailath [1981] or Maybeck [1979–1982] for the general Kalman-Bucy filter and discussion of its properties.

Our starting point will be the state-space model (2.12.13) of the ARMA process,

$$x_t = Ax_{t-1} + Bu_t, \quad y_t = Cx_t, \tag{7.7.1}$$

with A, B, and C as in (2.12.12). Recall that the dimension of the state vector is $\max\{p, q+1\}$. Let $\hat{x}_{t|t-1}$ denote the best linear estimate of x_t using the measurements in the range $[0, t-1]$, and $\hat{x}_{t|t}$, the best linear estimate of x_t using the measurements in the range $[0, t]$. We will also denote $\tilde{x}_{t|t-1} = x_t - \hat{x}_{t|t-1}$, the prediction error of the state vector x_t. The quantities $\hat{y}_{t,t}$ and $u_{t,t}$ are as in Chapter 6; that is, the former is the best linear predictor of y_t based on the measurements in the range $[0, t-1]$, and the latter is the partial innovation $u_{t,t} = y_t - \hat{y}_{t,t}$. The variance of $u_{t,t}$ is denoted by d_t.

Since u_t is uncorrelated with x_{t-1},

$$\hat{x}_{t|t-1} = A\hat{x}_{t-1|t-1}. \tag{7.7.2}$$

Also, by Prob. 7.14,

$$\hat{x}_{t|t} = \hat{x}_{t|t-1} + K_t u_{t,t}, \tag{7.7.3}$$

where

$$K_t = (Ex_t u_{t,t})(Eu_{t,t}^2)^{-1} = (Ex_t u_{t,t})d_t^{-1}. \tag{7.7.4}$$

Combining (7.7.2) and (7.7.3), we get

$$\hat{x}_{t+1|t} = A\hat{x}_{t|t-1} + AK_t u_{t,t}. \tag{7.7.5}$$

We also have

$$u_{t,t} = y_t - \hat{y}_{t,t} = y_t - C\hat{x}_{t|t-1} = Cx_t - C\hat{x}_{t|t-1} = C\tilde{x}_{t|t-1}. \tag{7.7.6}$$

To compute the gain vector K_t, denote

$$P_{t|t-1} = E\tilde{x}_{t|t-1}\tilde{x}_{t|t-1}^T = Ex_t \tilde{x}_{t|t-1}^T = Ex_t x_t^T - E\hat{x}_{t|t-1}\hat{x}_{t|t-1}^T \tag{7.7.7}$$

(see Prob. 7.15). Then we get, by (7.7.4) and (7.7.6),

$$K_t = P_{t|t-1}C^T d_t^{-1}. \tag{7.7.8}$$

We know by Prob. 2.9 that

$$Ex_{t+1}x_{t+1}^T = A(Ex_t x_t^T)A^T + \sigma_u^2 BB^T, \tag{7.7.9}$$

and similarly, from (7.7.5),

$$E\hat{x}_{t+1|t}\hat{x}_{t+1|t}^T = A(E\hat{x}_{t|t-1}\hat{x}_{t|t-1}^T)A^T + d_t AK_t K_t^T A^T. \tag{7.7.10}$$

Subtracting (7.7.10) from (7.7.9) and using (7.7.7), we get

$$\begin{aligned} P_{t+1|t} &= AP_{t|t-1}A^T + \sigma_u^2 BB^T - d_t AK_t K_t^T A^T \\ &= AP_{t|t-1}A^T + \sigma_u^2 BB^T - d_t^{-1} AP_{t|t-1}C^T CP_{t|t-1}A^T. \end{aligned} \tag{7.7.11}$$

Finally, the innovation variance is obtained from (7.7.6) and (7.7.7) as

$$d_t = CP_{t|t-1}C^T. \tag{7.7.12}$$

The initial condition for (7.7.5) is $\hat{x}_{0|-1} = 0$. Therefore, the initial condition for (7.7.11) is $P_{0|-1} = Ex_t x_t^T$. This can be obtained as the solution of the Liapunov equation

$$P_{0|-1} - AP_{0|-1}A^T = \sigma_u^2 BB^T. \tag{7.7.13}$$

By Prob. 2.9, the solution of this equation is precisely the Toeplitz covariance matrix of the AR process with parameters σ_u^2 and $a(z)$ of dimension $\max\{p, q+1\}$. Table 7.7 summarizes the Kalman-Bucy filter in the proper order of evaluation.

Set $x_{0|-1} = 0$
Solve $P_{0|-1} - AP_{0|-1}A^T = \sigma_u^2 BB^T$ for $P_{0|-1}$
For $t = 0$ to $N - 1$ do
$$u_{t,t} = y_t - C\hat{x}_{t|t-1}$$
$$d_t = CP_{t|t-1}C^T$$
$$K_t = P_{t|t-1}C^T d_t^{-1}$$
$$\hat{x}_{t+1|t} = A\hat{x}_{t|t-1} + AK_t u_{t,t}$$
$$P_{t+1|t} = AP_{t|t-1}A^T + \sigma_u^2 BB^T - d_t AK_t K_t^T A^T$$

Table 7.7. A summary of Kalman-Bucy filter algorithm.

Let us now apply the Kalman-Bucy algorithm to the computation of the likelihood function. We have already seen (e.g., in the proof of Theorem 5.2) that the log likelihood function can be written as

$$\log f_\theta(Y_N) = -\frac{N}{2}\log 2\pi - \frac{1}{2}\sum_{t=0}^{N-1}\log d_t - \frac{1}{2}\sum_{t=0}^{N-1}\frac{u_{t,t}^2}{d_t}. \qquad (7.7.14)$$

The Kalman-Bucy filter provides us with $\{u_{t,t}, d_t, 0 \leq t \leq N-1\}$, and these can be substituted in (7.7.14) to compute the exact likelihood function for a given value of θ. The dependence of the likelihood function on $\theta = \{\sigma_u^2, a_1, \ldots, a_p, b_1, \ldots, b_q\}$ is through the matrices A and C and the parameter σ_u^2 in the filter's equations.

The partial derivative of the log likelihood with respect to any of the parameters θ_m is given by

$$\frac{\partial \log f_\theta(Y_N)}{\partial \theta_m} = \frac{1}{2}\sum_{t=0}^{N-1}\frac{u_{t,t}^2 - d_t}{d_t^2}\frac{\partial d_t}{\partial \theta_m} - \sum_{t=0}^{N-1}\frac{u_{t,t}}{d_t}\frac{\partial u_{t,t}}{\partial \theta_m}. \qquad (7.7.15)$$

To obtain the partial derivatives of $u_{t,t}$ and d_t, we need to differentiate the filter equations with respect to θ_m. This gives the algorithm summarized in Table 7.8. The superscript $(\cdot)^{(m)}$ denotes partial differentiation with respect to θ_m. The matrix B is independent of the ARMA parameters, so its partial derivatives are zero.

The exact maximum likelihood algorithm is considerably more time consuming than the one based on the approximate log likelihood (7.6.5). It is therefore recommended always to run the approximate maximum likelihood algorithm first, and use the resulting estimate as an initial condition for the exact maximum likelihood algorithm. Many iterations of the latter algorithm can be saved this way. We remark that the exact maximum likelihood algorithm presented here can handle AR and MA processes as special cases in a straightforward manner.

Set $x_{0|-1}^{(m)} = 0$

Solve $P_{0|-1}^{(m)} - A P_{0|-1}^{(m)} A^T = (\sigma_u^2)^{(m)} B B^T + A^{(m)} P_{0|-1} A^T + A P_{0|-1} (A^{(m)})^T$

  for $P_{0|-1}^{(m)}$

For $t = 0$ to $N - 1$ do

$$u_{t,t}^{(m)} = -C^{(m)} \hat{x}_{t|t-1} - C \hat{x}_{t|t-1}^{(m)}$$

$$d_t^{(m)} = C^{(m)} P_{t|t-1} C^T + C P_{t|t-1}^{(m)} C^T + C P_{t|t-1} (C^{(m)})^T$$

$$K_t^{(m)} = P_{t|t-1}^{(m)} C^T d_t^{-1} + P_{t|t-1} (C^{(m)})^T d_t^{-1} - P_{t|t-1} C^T d_t^{(m)} d_t^{-2}$$

$$\hat{x}_{t+1|t}^{(m)} = A^{(m)} \hat{x}_{t|t-1} + A \hat{x}_{t|t-1}^{(m)} + A^{(m)} K_t u_{t,t} + A K_t^{(m)} u_{t,t} + A K_t u_{t,t}^{(m)}$$

$$P_{t+1|t}^{(m)} = A^{(m)} P_{t|t-1} A^T + A P_{t|t-1}^{(m)} A^T + A P_{t|t-1} (A^{(m)})^T + (\sigma_u^2)^{(m)} B B^T$$
$$\qquad - d_t^{(m)} A K_t K_t^T A^T - d_t A^{(m)} K_t K_t^T A^T - d_t A K_t^{(m)} K_t^T A^T$$
$$\qquad - d_t A K_t (K_t^{(m)})^T A^T - d_t A K_t K_t^T (A^{(m)})^T$$

Table 7.8. A summary of the algorithm for the partial derivatives.

Example 7.6.   The exact maximum likelihood algorithm was applied to the sunspot data discussed in Example 6.1 using an ARMA$(2, 2)$ model. The resulting estimate is

$$y_t = 1.4212 y_{t-1} - 0.7234 y_{t-2} + u_t - 0.1562 u_{t-1} + 0.0270 u_{t-2}.$$

The estimated period, computed from the roots of $\hat{a}(z)$, is 10.8 years. This is closer to the commonly accepted value (which is about 11 years) than the period obtained from the AR(2) estimate in Example 6.1.   ∎

The selection of the model order parameters p and q of the ARMA process can be done by the AIC or the MDL criteria, similarly to the case of AR processes. For large N, the AIC is given by

$$N^{-1} \text{AIC}(p, q) = \hat{\sigma}_u^2 + \frac{2(p + q + 1)}{N}, \tag{7.7.16}$$

and similarly for the MDL. Unlike the AR case, there does not exist a convenient order-recursive procedure for ARMA estimation. Therefore, ML estimation (exact or approximate) has to be performed for each pair (p, q) in the desired range and used to compute the criterion. This procedure is very expensive in terms of computational load, so it is seldom performed in real-time applications. In off-line applications we can usually afford to do as many computations as necessary, so the use of order selection criteria is advisable.

7.8. AN EXAMPLE: EMG SIGNAL ANALYSIS

We conclude our presentation of MA and ARMA algorithms with an illustration of the application of ARMA modeling to biological signal analysis. The signal is a recording of muscle activity, known as an electromyogram (EMG). In this case it is surface EMG, taken from the biceps, filtered to a range of 10 to 250 Hz and sampled at 500 Hz. Surface EMG is known to consist of two components [Inbar and Noujaim, 1984]. The first component is the firing frequencies of the motor units, which typically includes one or more spectral peaks. The second component is wide band, resulting from the motor unit action potential. An EMG signal can be regarded as approximately stationary as long as the force exerted by the muscle is constant. In this example, 4000 data points were recorded spanning 8 seconds. The force exerted during the recording was nearly constant. Figure 7.2(a) shows the first 500 data points of the signal. Figure 7.2(b) shows the Welch periodogram of the complete record using 400 points for each segment and 50% overlap. One firing frequency is discerned in the periodogram, at about 20 Hz.

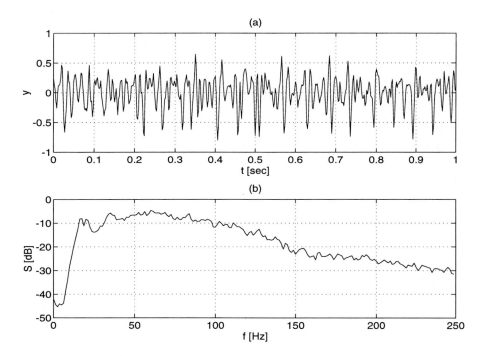

Figure 7.2. (a) One-second sample of an EMG signal; (b) the Welch periodogram of 8 seconds of an EMG signal.

An ARMA$(8,8)$ model was chosen for the analysis, and the minimum variance method described in Sec. 7.5 (with $\hat{r}$ and $\hat{\Sigma}$) was used to estimate the parameters. The complete data record was used to compute $\hat{r}$. Figure 7.3(a) shows the spectrum computed from the estimated parameters. The firing frequency is clearly visible, and its value can be determined with better resolution than from the periodogram (in this case the value is 16 Hz).

Next we took three different segments of 400 points each, starting at 0, 2, and 4 seconds, and used the time domain approximate maximum likelihood algorithm with each segment. Figures 7.3(b), (c), and (d) show the spectra computed from the estimated parameters. The firing frequency is visible in the first two cases, but not in the third. This type of analysis can be used to track time variations of the firing frequencies, for example, due to muscle fatigue. ∎

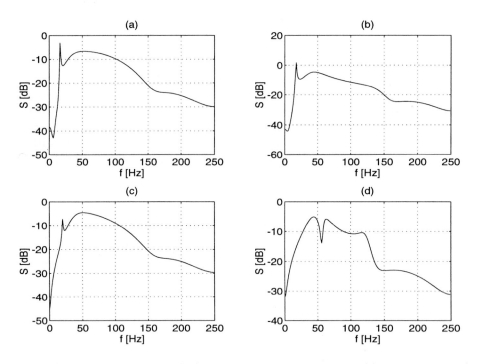

Figure 7.3. The ARMA (8,8) spectral density of an EMG signal: (a) estimate based on 8 seconds; (b), (c), (d) estimates based on nonoverlapping 0.8-second segments.

7.9. MATHEMATICA PACKAGES

The packages `MaAlg.m` and `ArmaAlg.m` and the auxiliary packages `Opt.m`, `Rarma.m`, `Rarmad.m`, `CRarma.m`, and `CRarmad.m` implement the algorithms presented in this chapter. Here is a brief description of the procedures in these packages.

`DurbinMa[rhat_, q_]` computes the Durbin $MA(q)$ estimate from the estimated covariance sequence `rhat` using an intermediate AR model whose order is determined by the length of `rhat`. `DurbinMa[data_, q_, l_]` computes the Durbin $MA(q)$ estimate for the data sequence `data` using an intermediate $AR(\ell)$ model. The returned values are the estimated `sigu2` and the list of coefficients `b`.

`MvrMa1[rhat_, init_]` computes the minimum variance $MA(q)$ estimate based on the sample covariances `rhat`. `init` is the initial condition of `sigu2` and `b`. This version uses $\Sigma(\psi)$ in the cost function. The returned values are the estimated `sigu2` and `b`. Options and default values: `RelAcc->10^(-6)` is the relative accuracy of the minimum value of the cost function. `MaxIter->200` is the maximum number of iterations in the nonlinear minimization. `Method->Gradient` is the nonlinear minimization method, either Davidon-Fletcher-Powell (`Gradient`) or conjugate directions without gradients (`ConjDir`).

`MvrMa2[rhat_, init_]` computes the minimum variance $MA(q)$ estimate based on the sample covariances `rhat`. `init` is the initial condition of `sigu2` and `b`. This version uses $\hat{\Sigma}$ in the cost function. The returned values are the estimated `sigu2` and `b`. Options and default values are as in `MaMvr1`.

`MvgMa1[rhat_, init_]` computes the minimum variance $MA(q)$ estimate based on the sample correlation coefficients `rhohat` computed internally from `rhat`. `init` is the initial condition of `b`. This version uses $\Sigma(\psi)$ in the cost function. The returned values are the estimated `sigu2` and `b`. Options and default values are as for `MvrMa1`.

`MvgMa2[rhat_, init_]` computes the minimum variance $MA(q)$ estimate based on the sample correlation coefficients `rhohat` computed internally from `rhat`. `init` is the initial condition of `b`. This version uses $\hat{\Sigma}$ in the cost function. The returned values are the estimated `sigu2` and `b`. Options and default values are as for `MvrMa1`.

`AmlMa[data_, init_]` computes the approximate maximum likelihood $MA(q)$ estimate for the data sequence `data`. `init` is the initial condition of `sigu2` and `b`. The returned values are the estimated `sigu2` and `b`. Options and default values are as for `MvrMa1`.

`ModYuleWalker[rhat_, p_, q_]` computes the modified or overdetermined Yule-Walker estimate of the AR parameters of an $ARMA(p, q)$ process from the given sample covariances `rhat`; see (7.4.4). The number of terms in `rhat` must be at least $p + q + 1$ (the minimum number gives the standard MYW estimate).

`PerfMyw[a_, b_, m_]` computes the normalized asymptotic variance of the modified or overdetermined Yule-Walker estimates of the AR parameters of an ARMA process, according to (7.4.6) and (7.4.7). `m` is the index of the highest-order covariance used in the estimate. It also computes the lower bound on this estimate when using the optimal weighting matrix; see Sec. 7.4.

DurbinArma[rhat_, p_, q_] estimates the MA parameters of an $ARMA(p, q)$ process by a variation of Durbin's MA method; see (7.4.10). rhat is the sample covariance sequence whose length determines the order of the long AR model. DurbinArma[data_, l_, p_, q_] estimates the MA parameters from data using a long AR model of order l. The returned values are the estimated sigu2 and b.

MvrArma1[rhat_, init_] computes the minimum variance $ARMA(p, q)$ estimate based on the sample covariances rhat. init is the initial condition of sigu2, a, and b. This version uses $\Sigma(\psi)$ in the cost function. The returned values are the estimated sigu2, a, and b. Options and default values are as for MvrMa1.

MvrArma2[rhat_, init_] computes the minimum variance $ARMA(p, q)$ estimate based on the sample covariances rhat. init is the initial condition of sigu2, a, and b. This version uses $\hat{\Sigma}$ in the cost function. The returned values are the estimated sigu2, a, and b. Options and default values are as for MvrMa1.

MvgArma1[rhat_, init_] computes the minimum variance $ARMA(p, q)$ estimate based on the sample correlation coefficients rhohat computed internally from rhat. init is the initial condition of sigu2, a, and b. This version uses $\Sigma(\psi)$ in the cost function. The returned values are the estimated sigu2, a, and b. Options and default values are as for MvrMa1.

MvgArma2[rhat_, init_] computes the minimum variance $ARMA(p, q)$ estimate based on the sample correlation coefficients rhohat computed internally from rhat. init is the initial condition of sigu2, a, and b. This version uses $\hat{\Sigma}$ in the cost function. The returned values are the estimated sigu2, a, and b. Options and default values are as for MvrMa1.

AmlArma1[data_, init_] computes the time domain approximate maximum likelihood $ARMA(p, q)$ estimate from the data sequence data. init is the initial condition of sigu2, a, and b. The returned values are the estimated sigu2, a, and b. Options and default values are as for MvrMa1.

AmlArma2[data_, init_] computes the frequency domain approximate maximum likelihood $ARMA(p, q)$ estimate from the data sequence data. init is the initial condition of sigu2, a, and b. The returned values are the estimated sigu2, a, and b. Options and default values are as for MvrMa1.

EmlArma[data_, init_] computes the exact maximum likelihood ARMA (p,q) estimate from the data sequence data. init is the initial condition of sigu2, a, and b. The returned values are the estimated sigu2, a, and b. Options and default values: RelAcc->10^(-6) is the relative accuracy of the minimum value of the cost function. MaxIter->200 is the maximum number of iterations in the nonlinear minimization. This procedure was only programmed to work with the conjugate directions method (i.e., without the gradient).

Rarma was described in Chapter 2. Garma[a_, b_, m_] uses Rarma to compute the correlation coefficients of an ARMA process for the MA and ARMA algorithms based on the sample correlation coefficients.

Rarmad was described in Chapter 5. Garmad[a_, b_, m_] uses Rarmad to compute the derivatives of the correlation coefficients of an ARMA process for the

MA and ARMA algorithms based on the sample correlation coefficients.

CRarma was described in Chapter 5. CGarma[a_, b_, m_] uses CRarma to compute the normalized asymptotic variances of the sample correlation coefficients of an ARMA process for the MA and ARMA algorithms based on the sample correlation coefficients. See Prob. 7.9 for details.

CRarmad[a_, b_, sigu2_, m_] computes the partial derivatives of $\Sigma(\psi)$ of the sample covariances. See Prob. 7.8 for details.

CGarmad[a_, b_, m_] computes the partial derivatives of $\Sigma(\psi)$ of the sample correlation coefficients. See Prob. 7.10 for details.

Minimize1[f_, g_, x0_, tol_] performs multidimensional numerical minimization by the Davidon-Fletcher-Powell method, which uses both the cost function and its gradient. This algorithm was adapted from Press et al. [1986]. The cost function f and its gradient g are supplied by the calling function. x0 is the initial value of the argument, tol is the stopping tolerance (relative decrease in the function value), and itmax is the limit on the number of iterations. The returned values are xmin, the minimum point, fmin, the minimum value of the cost function, and iter, the actual number of iterations.

Minimize2[f_, x0_, tol_] performs multidimensional numerical minimization by Powell's conjugate directions method, which uses only the cost function. This algorithm was adapted from Press et al. [1986]. It is typically much slower to converge than Minimize1, but is useful when the computation of the gradient is difficult. The cost function f is supplied by the calling function. x0 is the initial value of the argument, tol is the stopping tolerance (relative decrease in the function's value), and itmax is the limit on the number of iterations. The returned values are xmin, the minimum point, fmin, the minimum value of the cost function, and iter, the actual number of iterations.

These two functions use the internal functions LineMin, Bracket, and Brent. LineMin implements the line search. It first calls Bracket to do the initial bracketing of the minimum point along the direction of search and then calls Brent to find this minimum, using Brent's parabolic fit method.

REFERENCES

Akaike, H., "Maximum Likelihood Identification of Gaussian Autoregressive Moving Average Models," *Biometrika,* 60, pp. 255–265, 1973.

Anderson, T. W., *The Statistical Analysis of Time Series,* Wiley, New York, 1971.

Box, G. E. P., and Jenkins, G. M., *Time Series Analysis: Forecasting and Control,* Holden-Day, San Francisco, 1970.

Cadzow, J., "High Performance Spectral Estimation—A New ARMA Method," *IEEE Trans. Acoustics, Speech, Signal Processing,* ASSP-28, pp. 524–529, 1980.

Durbin, J., "Efficient Estimation of Parameters in Moving Average Models," *Biometrika,* 46, pp. 306–316, 1959.

Friedlander, B., Morf, M., Kailath, T., and Ljung, L., "New Inversion Formulas for Matrices Classified in Terms of Their Distance from Toeplitz Matrices," *Linear Algebra and Its Applications,* 27, pp. 31–60, 1979.

Hannan, E. J., "The Estimation of Mixed Moving Average Autoregressive Systems," *Biometrika,* 56, pp. 579–593, 1969.

Inbar, G. F., and Noujaim, A. E., "On Surface EMG Spectral Characterization and Its Application to Diagnostic Classification," *IEEE Trans. Biomedical Engineering,* BME-31, pp. 597–604, 1984.

Kailath, T., *Lectures on Wiener and Kalman Filtering,* 2nd printing, CISM Courses and Lectures No. 140, Springer-Verlag, New York, 1981.

Kay, S. M., *Modern Spectral Estimation, Theory and Applications,* Prentice Hall, Englewood Cliffs, NJ, 1988.

Maybeck, P. S., *Stochastic Models, Estimation and Control,* Academic Press, New York, 1979–1982.

Mehra, R. K., "On-line Identification of Linear Dynamic Systems with Applications to Kalman Filtering," *IEEE Trans. Automatic Control,* AC-16, pp. 12–21, 1971.

Press, W. H., Flannery, B. P., Teukolsky, S. A., and Vettering, W. T., *Numerical Recipes, the Art of Scientific Computing,* Cambridge University Press, Cambridge, 1986.

Porat, B., "ARMA Spectral Estimation Based on Partial Autocorrelations," *Circuits, Systems, Signal Processing,* 2, pp. 341–360, 1983.

Porat, B., and Friedlander, B., "Asymptotic Analysis of the Bias of the Modified Yule-Walker Estimator," *IEEE Trans. Automatic Control,* AC-30, pp. 765–767, 1985.

Porat, B., and Friedlander, B., "Computation of the Exact Information Matrix of Gaussian Time Series with Stationary Random Components," *IEEE Trans. Acoustics, Speech, Signal Processing,* ASSP-34, pp. 118–130, 1986.

Rosen, Y., and Porat, B., "ARMA Spectral Estimation Based on Partial Autocorrelations—Part II: Statistical Analysis," *Circuits, Systems, Signal Processing,* pp. 367–383, 1985.

Stoica, P., Friedlander, B., and Söderström, T., "Approximate Maximum Likelihood Approach to ARMA Spectral Estimation," *Int. J. Control,* 45, pp. 1281–
1310, 1987a.

Stoica, P., Friedlander, B., and Söderström, T., "Optimal Instrumental Variable
Multistep Algorithms for Estimation of the AR Parameters of an ARMA
Process," *Int. J. Control,* 45, pp. 2083–2107, 1987b.

Stoica, P., Söderström, T., and Friedlander, B., "Optimal Instrumental Variable Estimates of the AR Parameters of ARMA Processes,", *IEEE Trans. Automatic
Control,* AC-30, pp. 1066–1074, 1985.

Walker, A. M., "Large-sample Estimation of Parameters for Moving-average Models," *Biometrika,* 48, pp. 343–357, 1961.

Walker, A. M., "Large-sample Estimation of Parameters for Autoregressive Processes with Moving-average Residuals," *Biometrika,* 49, pp. 117–131, 1962.

PROBLEMS

7.1. Show that, if $\sum_{k=-q}^{q} \hat{r}_k e^{-j\omega k}$ takes negative values on the interval $[-\pi, \pi]$,
the roots of $\hat{\bar{S}}(z)$ will not consist of two separable sets, as needed for (7.2.3).

7.2. Use (7.3.3) to write the derivatives of $\Sigma(\psi)$ and $C(\psi)$.

7.3. Discuss the relative computational complexity of the algorithms based on
the cost functions (3.6.3) and (3.6.9). Give a rough estimate of the number
of operations for (a) computation of the cost function and (b) computation
of the gradient in each of the two methods.

7.4. Develop the algorithms based on the sample correlation coefficients, in analogy with the algorithms based on the sample covariances. Write down the
details of the computation of the cost functions and their gradients.

7.5. Discuss the computational complexity of the approximate maximum likelihood MA algorithm. Give a rough estimate of the number of operations for
(a) computation of the cost function and (b) computation of the gradient.

7.6. Show that $D\Sigma D^T$ appearing in (7.4.7) is a Toeplitz matrix whose entries are
the covariances of an ARMA process with numerator polynomial $b^2(e^{j\omega})$,
denominator polynomial $a(e^{j\omega})$, and innovation variance σ_u^4. Hint: Use
Bartlett's formula in Chapter 4 to express Σ as $\Sigma_1 + \Sigma_2$, where the first
term is Toeplitz and the second is Hankel (i.e., a matrix constant along
its antidiagonals). Express each entry in the two matrices as an inverse
Fourier transform. Then show that $D\Sigma_2 D^T = 0$ and $D\Sigma_1 D^T$ has the stated
structure.

7.7. The derivation in (7.4.6) and Prob. 7.6 appears to be incorrect if $q < p - 1$,
because it neglects the fact that negative-indexed sample covariances are
identical to the corresponding positive-indexed covariances. An alternative

approach in this case is to use the statistic $\{\hat{r}_0, \ldots, \hat{r}_m\}$ and modify (7.4.6) accordingly. Show that this approach yields the same $D\Sigma D^T$ as in Prob. 7.6.

7.8. Extend Prob. 5.6 to the computation of the partial derivatives of the covariances $\bar{r}_k$ used in (7.5.1). Hint: This can be done directly or by a two-stage procedure, using the chain rule for partial derivatives. Explore both methods.

7.9. Using Prob. 4.8, develop a computational procedure for the asymptotic covariance of the sample correlation coefficients $\lim_{N\to\infty} N\mathrm{cov}\{\hat{\rho}_k, \hat{\rho}_\ell\}$ in the case of an ARMA(p, q) process. Hint: Express all terms as functions of $\{r_k\}$ and $\{\bar{r}_k\}$.

7.10. Use the two previous problems to develop a computational procedure for the partial derivatives of $\Sigma(\psi)$ of the sample correlation coefficients.

7.11. Use the techniques of Sec. 3.6 to derive the following expression for the normalized asymptotic covariance of the estimate obtained from the minimization of (7.5.3).

$$V(\theta) = \left[\left(\frac{\partial r}{\partial \theta}\right)^T \left(\frac{\partial r}{\partial \theta}\right)\right]^{-1} \left(\frac{\partial r}{\partial \theta}\right)^T \Sigma(\theta) \left(\frac{\partial r}{\partial \theta}\right) \left[\left(\frac{\partial r}{\partial \theta}\right)^T \left(\frac{\partial r}{\partial \theta}\right)\right]^{-1}.$$

Write a Mathematica procedure that implements this expression for ARMA processes. Experiment with some test cases of your choice, and compare the numerical values with the ones corresponding to minimum-variance estimates (7.5.2). Report your results.

7.12. Let $\{\hat{K}_i, 1 \le i \le L\}$ be a set of sample partial correlation coefficients computed from the sample covariances by the Schur or Burg algorithms. Let $\{K_i(\theta), 1 \le i \le L\}$ be the corresponding true partial correlation coefficients of the ARMA process. Consider the cost function

$$C(\psi) = 0.5 \sum_{i=1}^{L} [K_i(\psi) - \hat{K}_i]^2.$$

Let $\hat{\theta}$ be the minimizer of $C(\psi)$. This estimate is based on the empirical observation that, unlike the sample covariances, the sample partial correlation coefficients are often weakly correlated, so the identity matrix is nearly the optimal weighting matrix for this statistic. Develop a working algorithm for this estimate and implement it in Mathematica. Explore its performance for some test cases and report the results. See Porat [1983] and Rosen and Porat [1985] for further details.

7.13. Explain why $(2/N) \sum_{k=1}^{N/2} \log S_0(\omega_k)$ in (7.6.13) tends to zero as N goes to infinity. Hint: Use Kolmogorov's formula (2.8.6).

7.14. Explain Eq. (7.7.3) in detail using the result of Prob. 5.15.

7.15. The first equality in (7.7.7) is the definition of $P_{t|t-1}$. The other two equalities follow from the definitions of $\hat{x}_{t|t-1}$ and $\tilde{x}_{t|t-1}$. Prove them.

7.16. Estimate the number of operations needed for the algorithm in Table 7.7. Remark: The sparseness of the matrices A and B can be used to save many computations.

7.17. The matrix A is independent of the parameters $\{\sigma_u^2, b_1, \ldots, b_q\}$, and its partial derivative with respect to a_m is a very simple matrix. Similarly, C is independent of the parameters $\{\sigma_u^2, a_1, \ldots, a_p\}$, and its partial derivative with respect to b_m is a very simple matrix. Use these facts to derive simplified versions of the formulas in Table 7.8, considering the partial derivatives with respect to σ_u^2, a_m, and b_m separately. Then estimate the number of operations needed to compute the partial derivatives.

7.18. Implement the result of the previous problem in a Mathematica procedure.

7.19. Suppose that the measurement vector Y_N contains gaps; that is, some data points are missing. The exact maximum likelihood algorithm can be easily modified to handle this case with only a minor change. Explain the necessary change. Hint: Substituting zeros for the missing data is *not* the correct answer!

7.20. Consider the model of AR process in noise described in Prob. 2.7. Propose two different algorithms for estimating the model parameters. Implement the algorithms in Mathematica, run them for several test cases, compare with the lower bound derived in Prob. 5.11, and report the results.

CHAPTER 8

Adaptive AR and ARMA Estimation

8.1. INTRODUCTION

Up to this point we were concerned with *batch estimation* problems, in which the entire data set is given to the estimator and the result is a single estimate based on this set. In engineering applications we frequently encounter situations in which data keep coming (sometimes at a very high rate), and it is required to update the estimate as new data arrive. There are three possible cases: (a) The underlying process is stationary, so we would like the estimation error to decrease as more data become available and to reach zero asymptotically (i.e., the estimate should be consistent). (b) The underlying process is nonstationary, but the time variation of the model is slow relative to the data rate. In this case we would like the estimation algorithm to adapt itself to the variations of the model and to provide estimates that track the true parameters. Thus the estimation error is not expected to decrease to zero, but to vary randomly within some range. (c) The underlying process is nonstationary, and the time variation of the model is not slow relative to the data rate.

Adaptive algorithms are designed to handle cases (a) and (b) above. For case (a) we have *growing memory* algorithms, while for case (b) there are *decaying memory* algorithms. This chapter is devoted to the study of such algorithms. Case (c) will be discussed in Chapters 11 and 12.

Adaptive estimation has been a subject of extensive research in recent years, and the amount of literature is vast. A recent book [Haykin, 1991] covers this subject in great detail. It is impossible to do justice to the entire field of adaptive estimation in a single chapter. Therefore, we will concentrate on the basics. Each of the main approaches to adaptive estimation is given a brief description, including the principles and at least one detailed algorithm. The main approaches are

218

recursive least-squares, stochastic gradient, and lattice filters. For each of these, we treat both the AR and the ARMA problems. The MA model is not given a separate treatment, but in most cases it can be handled as a special case of an ARMA process.

8.2. THE RECURSIVE LEAST-SQUARES ALGORITHM

Consider the weighted least-squares problem (3.7.3), and omit the dependence of H and $\boldsymbol{h}$ on Y for convenience. Let the matrices corresponding to time t be denoted by H_t and $\boldsymbol{h}_t$, respectively. Assume that the dimensionality of the problem grows with t, such that at each new time point a row η_t^T is added to the bottom of H_t and an element h_{t+1} is added to $\boldsymbol{h}_t$. Thus,

$$H_{t+1} = \begin{bmatrix} H_t \\ \eta_t^T \end{bmatrix}, \quad \boldsymbol{h}_{t+1} = \begin{bmatrix} \boldsymbol{h}_t \\ h_{t+1} \end{bmatrix}. \tag{8.2.1}$$

This means that at each new time point we add an equation to the least-squares problem (3.7.3). The unknown parameter θ is assumed to be constant for the time being.

We will be interested in the special weight matrix

$$W_t = \begin{bmatrix} \ddots & \vdots & \vdots & \vdots \\ \cdots & \lambda^2 & 0 & 0 \\ \cdots & 0 & \lambda & 0 \\ \cdots & 0 & 0 & 1 \end{bmatrix}, \tag{8.2.2}$$

where $\lambda \leq 1$. The case $\lambda = 1$ gives the identity weight matrix. When $\lambda < 1$ we assign the largest weight to the most recent equation and exponentially decaying weights to past equations. In other words, $\lambda < 1$ causes the past to be gradually forgotten. Naturally, λ is called the *forgetting factor*.

Let us denote the weighted least-squares estimate of θ at time t by $\hat{\theta}_t$. Then we get, from (3.7.5),

$$\hat{\theta}_t = P_t H_t^T W_t \boldsymbol{h}_t, \tag{8.2.3}$$

where

$$P_t = [H_t^T W_t H_t]^{-1}. \tag{8.2.4}$$

The estimate at time $t + 1$ can be derived as follows.

$$\hat{\theta}_{t+1} = P_{t+1} H_{t+1}^T W_{t+1} \boldsymbol{h}_{t+1} = P_{t+1} \begin{bmatrix} H_t^T & \eta_t \end{bmatrix} \begin{bmatrix} \lambda W_t & 0 \\ 0 & 1 \end{bmatrix} \begin{bmatrix} \boldsymbol{h}_t \\ h_{t+1} \end{bmatrix}$$

$$= P_{t+1} [\lambda H_t^T W_t \boldsymbol{h}_t + \eta_t h_{t+1}]. \tag{8.2.5}$$

The matrix P_{t+1} can be expressed in terms of P_t using the matrix identity [A19],

$$P_{t+1} = [\lambda H_t^T W_t H_t + \eta_t \eta_t^T]^{-1} = [\lambda P_t^{-1} + \eta_t \eta_t^T]^{-1}$$

$$= \lambda^{-1} P_t - \frac{\lambda^{-2} P_t \eta_t \eta_t^T P_t}{1 + \lambda^{-1} \eta_t^T P_t \eta_t} = \lambda^{-1} [\mathbf{1} - K_t \eta_t^T] P_t, \tag{8.2.6}$$

where

$$K_t = \frac{P_t \eta_t}{\lambda + \eta_t^T P_t \eta_t}. \tag{8.2.7}$$

Substitution of (8.2.6) in (8.2.5) gives

$$\hat{\theta}_{t+1} = \lambda^{-1} [\mathbf{1} - K_t \eta_t^T] P_t [\lambda H_t^T W_t \mathbf{h}_t + \eta_t h_{t+1}]. \tag{8.2.8}$$

We have (Prob. 8.1)

$$[\mathbf{1} - K_t \eta_t^T] P_t \eta_t = \lambda K_t. \tag{8.2.9}$$

Therefore,

$$\hat{\theta}_{t+1} = [\mathbf{1} - K_t \eta_t^T] \hat{\theta}_t + K_t h_{t+1} = \hat{\theta}_t + K_t \epsilon_{t+1}, \tag{8.2.10}$$

where

$$\epsilon_{t+1} = h_{t+1} - \eta_t^T \hat{\theta}_t. \tag{8.2.11}$$

Equations (8.2.11), (8.2.10), (8.2.7,) and (8.2.6) constitute the *recursive least-squares* (RLS) algorithm.

To initialize the RLS algorithm, it must be given an initial matrix P_{t_0} that is the inverse of a nonsingular matrix. This can be achieved by collecting a number of equations equal to the dimension of θ and setting

$$P_{t_0} = [H_{t_0}^T W_{t_0} H_{t_0}]^{-1}, \tag{8.2.12}$$

where the inversion is carried out explicitly. Then we set

$$\hat{\theta}_{t_0} = P_{t_0} H_{t_0}^T W_{t_0} \mathbf{h}_{t_0} \tag{8.2.13}$$

and run the algorithm from t_0 on. We will refer to this scheme as *exact initialization*. An alternative, much simpler initialization, is to set $\hat{\theta}_{-1} = 0$ and $P_{-1} = \varepsilon \mathbf{1}$, where ε is a large number. We will refer to this scheme as *approximate initialization*. Table 8.1 summarizes the RLS algorithm with both initialization options.

When the algorithm is used with $\lambda = 1$, it gives the exact least-squares solution at each time point, provided exact initialization is employed. This means that its convergence properties are identical to those of the corresponding nonrecursive algorithm. When the algorithm is used with $\lambda < 1$, it does not converge to a constant value. Instead, it keeps fluctuating at random, adjusting its output to the fresh data. This enables the algorithm to track slowly varying models; hence the name "adaptive." The parameter λ is used to control the algorithm fluctuations and tracking ability. It can be shown that the effective number of equations used in the weighted least-squares cost function is about $2/(1 - \lambda)$. The closer is λ to 1, the better the data averaging, but at the expense of less tracking ability. Practical values of λ are 0.98 and above. When $\lambda < 1$, exact initialization is not important, since the algorithm will forget the initial conditions anyway. In this case it is therefore common to use approximate initialization.

$$
\begin{aligned}
&\text{Exact initialization:} \\
&P_{t_0} = [H_{t_0}^T W_{t_0} H_{t_0}]^{-1} \\
&\hat{\theta}_{t_0} = P_{t_0} H_{t_0}^T W_{t_0} \boldsymbol{h}_{t_0} \\
&\qquad \text{or} \\
&\text{Approximate initialization:} \\
&t_0 = -1,\ P_{-1} = \varepsilon \boldsymbol{1},\ \hat{\theta}_{-1} = 0 \\
&\text{For } t = t_0, t_0 + 1, \ldots \text{ do} \\
&\qquad \epsilon_{t+1} = h_{t+1} - \eta_t^T \hat{\theta}_t \\
&\qquad K_t = P_t \eta_t (\lambda + \eta_t^T P_t \eta_t)^{-1} \\
&\qquad P_{t+1} = \lambda^{-1}[\boldsymbol{1} - K_t \eta_t^T] P_t \\
&\qquad \hat{\theta}_{t+1} = \hat{\theta}_t + K_t \epsilon_{t+1}
\end{aligned}
$$

Table 8.1. A summary of the RLS algorithm.

A major drawback of the RLS algorithm is its numerical instability. To understand this problem, observe that the update of P_{t+1} in (8.2.6) involves subtraction. While P_t is theoretically positive definite (being the inverse of a positive definite matrix), accumulation of numerical errors in (8.2.6) may eventually lead to a loss of positive definiteness. If P_t loses its positive definiteness for some t, signs of some components of the gain vector K_t are usually reversed with respect to their theoretical values. This causes some components of $\hat{\theta}_{t+1}$ in (8.2.10) to build up in the wrong direction. When this happens, the algorithm usually diverges very fast.

The loss of positive definiteness of P_t can be avoided using a *square-root* version of the algorithm. The square-root RLS algorithm does not keep track of P_t but of its lower triangular square root $P_t^{1/2}$. Since $P_t = P_t^{1/2} P_t^{T/2}$, it remains positive semidefinite regardless of any numerical errors in the square root. To derive the algorithm, assume we already have $P_t^{1/2}$ and we wish to compute $P_{t+1}^{1/2}$. Let

$$
M = \begin{bmatrix} \lambda^{1/2} & \eta_t^T P_t^{1/2} \\ 0 & \lambda^{-1/2} P_t^{1/2} \end{bmatrix}. \tag{8.2.14}
$$

Let Q be an orthonormal matrix and R an upper triangular matrix such that $M^T = Q^T R$. The existence of such a decomposition is a well-known result in linear algebra, and good computational algorithms for it can be found (e.g., in [Golub and Van Loan, 1983]). Partition R compatibly with M^T and denote

$$
R = \begin{bmatrix} R_{11}^T & R_{12}^T \\ 0 & R_{22}^T \end{bmatrix}.
$$

From the orthogonality of Q, we get that $MM^T = R^T Q Q^T R = R^T R$. But

$$
MM^T = \begin{bmatrix} \lambda + \eta_t^T P_t \eta_t & \lambda^{-1/2} \eta_t^T P_t \\ \lambda^{-1/2} P_t \eta_t & \lambda^{-1} P_t \end{bmatrix} \tag{8.2.15}
$$

and

$$R^T R = \begin{bmatrix} R_{11} R_{11}^T & R_{11} R_{12}^T \\ R_{12} R_{11}^T & R_{12} R_{12}^T + R_{22} R_{22}^T \end{bmatrix}. \tag{8.2.16}$$

Comparison of (8.2.15) and (8.2.16) shows that

$$R_{11} = (\lambda + \eta_t^T P_t \eta_t)^{1/2}, \quad R_{12} = \frac{\lambda^{-1/2} P_t \eta_t}{(\lambda + \eta_t^T P_t \eta_t)^{1/2}}, \tag{8.2.17a}$$

$$R_{22} R_{22}^T = \lambda^{-1} P_t - \frac{\lambda^{-1} P_t \eta_t \eta_t^T P_t}{\lambda + \eta_t^T P_t \eta_t}. \tag{8.2.17b}$$

Comparing (8.2.17b) to (8.2.6) and recalling that R_{22} is lower triangular, we conclude that $R_{22} = P_{t+1}^{1/2}$. The QR decomposition of M therefore yields the updated square root in the lower-right block of the matrix R. Moreover, the gain vector K_t is easily obtained from (8.2.17a) and (8.2.7) as $K_t = \lambda^{1/2} R_{12}/R_{11}$. The square-root RLS algorithm is summarized in Table 8.2 (approximate initialization only). Note that P_t does not appear explicitly, only its square root. At the heart of the algorithm lies the QR decomposition that has to be carried out by a special routine. Problem 8.2 discusses an alternative square-root algorithm that does not require QR decomposition.

$$
\begin{array}{l}
\text{Set } P_{-1}^{1/2} = \varepsilon^{1/2} \mathbf{1}, \hat{\theta}_{-1} = 0 \\[2pt]
\text{For } t = -1, 0, \ldots \text{ do} \\[2pt]
\quad \epsilon_{t+1} = h_{t+1} - \eta_t^T \hat{\theta}_t \\[2pt]
\quad \text{Compute the QR decomposition} \\[2pt]
\quad \begin{bmatrix} \lambda^{1/2} & \eta_t^T P_t^{1/2} \\ 0 & \lambda^{-1/2} P_t^{1/2} \end{bmatrix}^T = Q^T \begin{bmatrix} R_{11}^T & R_{12}^T \\ 0 & R_{22}^T \end{bmatrix} \\[2pt]
\quad K_t = \lambda^{1/2} R_{12} R_{11}^{-1} \\[2pt]
\quad P_{t+1}^{1/2} = R_{22} \\[2pt]
\quad \hat{\theta}_{t+1} = \hat{\theta}_t + K_t \epsilon_{t+1}
\end{array}
$$

Table 8.2. A summary of the square-root RLS algorithm.

The RLS algorithm can be applied to the forward covariance method for AR estimation in a straightforward manner. The estimated parameter $\hat{\theta}$ is $-\mathbf{a}_f$, and $(H, \mathbf{h})$ correspond to $(\mathbf{Y}_f, \mathbf{y}_f)$, respectively. Therefore, $\eta_t = [y_t, y_{t-1}, \ldots, y_{t-p+1}]^T$ and $h_{t+1} = y_{t+1}$. The residual error ϵ_{t+1} is in this case given by

$$\epsilon_{t+1} = \sum_{k=0}^{p} \hat{a}_{k,t} y_{t+1-k}, \tag{8.2.18}$$

so ϵ_{t+1} is the estimated innovation at time $t+1$. The asymptotic properties of the algorithm are identical to those of the forward covariance method if $\lambda = 1$, so the estimate is asymptotically efficient in this case. As explained before, for $\lambda < 1$ the algorithm is not even consistent, but, on the other hand, it can track slow time variations of the AR model.

Example 8.1. 2000 data points of the AR(2) process with parameters $\{1.0, -0.8, 0.8\}$ were generated and fed into the AR/RLS algorithm (using approximate initialization). Figure 8.1 shows the estimated parameters as a function of t for four values of λ: 1.0, 0.996, 0.98, and 0.95. As expected, the estimates become more "noisy" as λ decreases.   ∎

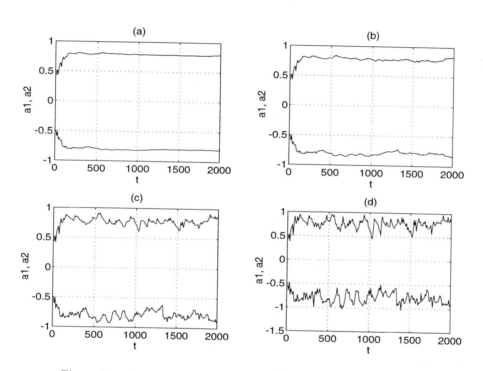

Figure 8.1. The estimates $\hat{a}_1$ and $\hat{a}_2$ as a function of time for Example 8.1: (a) $\lambda = 1.0$; (b) $\lambda = 0.996$; (c) $\lambda = 0.98$; (d) $\lambda = 0.95$.

8.3. THE EXTENDED LEAST-SQUARES ALGORITHM

The RLS algorithm cannot handle ARMA processes directly, because there is no linear least-squares batch algorithm to start with in the ARMA case (such as the forward covariance method for the AR problem). However, some modification of the basic idea enables the extension of the RLS algorithm to ARMA processes, as we will now show.

Recall the defining equation of an ARMA process, and rewrite it in the form

$$u_t = \sum_{k=0}^{p} a_k y_{t-k} - \sum_{k=1}^{q} b_k u_{t-k}. \tag{8.3.1}$$

Defining $n = \max\{p, q\}$, noting that $a_0 = 1$, and collecting these equations up to time t gives

$$\begin{bmatrix} u_n \\ u_{n+1} \\ \vdots \\ u_t \end{bmatrix} = \begin{bmatrix} y_n \\ y_{n+1} \\ \vdots \\ y_t \end{bmatrix} - \begin{bmatrix} y_{n-1} & \cdots & y_{n-p} & u_{n-1} & \cdots & u_{n-q} \\ y_n & \cdots & y_{n-p+1} & u_n & \cdots & u_{n-q+1} \\ \vdots & \ddots & \vdots & \vdots & \ddots & \vdots \\ y_{t-1} & \cdots & y_{t-p} & u_{t-1} & \cdots & u_{t-q} \end{bmatrix} \begin{bmatrix} -a_1 \\ \vdots \\ -a_p \\ b_1 \\ \vdots \\ b_q \end{bmatrix}.$$

$$\tag{8.3.2}$$

Recall also that the approximate time domain maximum likelihood algorithm described in Sec. 7.6 is based on the minimization of the sum of squares of the $\{u_t\}$. Superficially, (8.3.2) looks like a linear least-squares problem whose solution for $\theta = [-a_1, \ldots, -a_p, b_1, \ldots, b_q]^T$ would minimize $\sum_{i=n}^{t} u_i^2$. The difficulty, of course, is that the matrix in the right side of (8.3.2) contains the unknown values of $\{u_i\}$, so this is not really a linear least-squares problem.

The idea in extending the RLS algorithm to ARMA processes is simply to "cheat" on Eq. (8.3.2) by using *estimated* values of $\{u_i\}$ in the matrix in the right side. Such estimates can be recursively obtained by taking

$$\hat{u}_{t+1} = \sum_{k=0}^{p} \hat{a}_{k,t} y_{t+1-k} - \sum_{k=1}^{q} \hat{b}_{k,t} \hat{u}_{t+1-k} = y_{t+1} - \eta_t^T \hat{\theta}_t, \tag{8.3.3}$$

where

$$\eta_t = [y_t, \ldots, y_{t-p+1}, \hat{u}_t, \ldots, \hat{u}_{t-q+1}]^T. \tag{8.3.4}$$

The algorithm thus derived is called the *extended least-squares* (ELS) algorithm. In the statistical literature it is also known as *pseudolinear regression*. It is implemented exactly like the RLS algorithm in Table 8.1, with $h_{t+1} = y_{t+1}$ and η_t as in (8.3.4). Note that $\hat{u}_{t+1}$ in (8.3.3) is identical to ϵ_{t+1} in Table 8.1, so its computation is already built into the algorithm. Since the algorithm is based on a "cheat," exact initialization is neither possible nor necessary, so the ELS method is commonly used with approximate initialization.

In contrast to the AR/RLS algorithm, the ARMA/ELS algorithm is not guaranteed to be consistent when $\lambda = 1$. A necessary and sufficient condition for consistency is

$$\mathrm{Re}\{b(\mu^{-1})\} > 0 \text{ for all } \mu \text{ such that } a(\mu) = 0, \tag{8.3.5}$$

where $b(\mu^{-1}) = \sum_{k=0}^{q} b_k \mu^k$ and $a(\mu) = \sum_{k=0}^{p} a_k \mu^{-k}$. The method of proof of this condition is beyond the scope of this book; see Ljung and Söderström [1983, Ch. 4].

Example 8.2. 2000 data points of the MA(2) process with parameters $\{1.0, -0.8, 0.8\}$ were generated and inputted to the ARMA/ELS algorithm. Figure 8.2 shows the estimated parameters as a function of t for four values of λ: 1.0, 0.996, 0.98, and 0.95. For $\lambda = 1$, the algorithm is slow to converge and the estimates remain biased even after 2000 data points. Convergence is faster for $\lambda < 1$, but the estimates are more noisy. ∎

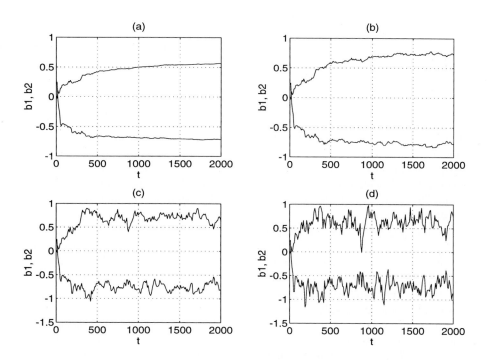

Figure 8.2. The estimates $\hat{b}_1$ and $\hat{b}_2$ as a function of time for Example 8.2: (a) $\lambda = 1.0$; (b) $\lambda = 0.996$; (c) $\lambda = 0.98$; (d) $\lambda = 0.95$.

8.4. THE RECURSIVE MAXIMUM LIKELIHOOD ALGORITHM

The recursive maximum likelihood algorithm is best motivated by considering first an iterative (but nonrecursive) approach to minimization of nonlinear least-squares problems. Consider the system of equations

$$e = \boldsymbol{h}(Y, \theta), \tag{8.4.1}$$

where Y is the data vector, θ the unknown parameter, $\boldsymbol{e}$ the error vector, and $\boldsymbol{h}(Y, \theta)$ a nonlinear vector function describing the dependence of the error on Y and θ. Suppose we wish to find $\hat{\theta}$ so as to minimize the square of the weighted norm of the error $\boldsymbol{e}^T W \boldsymbol{e}$. This is an example of a *nonlinear weighted least-squares problem*.

Suppose we have already obtained an estimate $\hat{\theta}_m$ of θ (the index m stands for the iteration number, not for time). If the estimate is reasonably close to the true θ, we can approximate (8.4.1) by

$$\boldsymbol{e} \approx \boldsymbol{h}(Y, \hat{\theta}_m) - H(Y, \hat{\theta}_m)(\theta - \hat{\theta}_m), \tag{8.4.2}$$

where $H(Y, \hat{\theta}_m)$ is the negative of the gradient of $\boldsymbol{h}$ evaluated at $\hat{\theta}_m$. We can then take the next estimate to be the minimizer of the weighted square error in the approximation (8.4.2); that is,

$$\hat{\theta}_{m+1} = \hat{\theta}_m + [H^T(Y, \hat{\theta}_m) W H(Y, \hat{\theta}_m)]^{-1} H^T(Y, \hat{\theta}_m) W \boldsymbol{h}(Y, \hat{\theta}_m). \tag{8.4.3}$$

Iterating (8.4.3) from some initial estimate $\hat{\theta}_0$ till convergence should yield the minimum of $\boldsymbol{e}^T W \boldsymbol{e}$. This algorithm is, in fact, just the Gauss-Newton algorithm for the nonlinear least-squares problem (8.4.1) [Fletcher, 1987]. The Gauss-Newton algorithm is not guaranteed to converge in general, so it is seldom used in this raw form. However, we are not interested in this algorithm per se, only to motivate its recursive version.

The recursive maximum likelihood algorithm (RML) is similar to the RLS algorithm, except that it uses the partial derivatives of the error $\boldsymbol{e}$ with respect to the parameters in building the matrix H, as follows from (8.4.3). Since (8.4.3) implies recalculation of the entire $H(Y, \hat{\theta})$ each time $\hat{\theta}$ is updated, it cannot be used as is in the recursive version. Instead, we simply retain all previously computed rows of this matrix (despite their being based on old estimates). Only the new row η_t is built from the most recent estimate at each time point. A similar remark holds for $\boldsymbol{h}(Y, \theta)$: only its new entry at each time point is calculated from the most recent estimate. The name "recursive maximum likelihood" is somewhat misleading—a more natural name would be *recursive nonlinear least-squares* algorithm. However, the name RML has become standard, so we shall continue to use it.

Let us now specialize the RML algorithm to the recursive estimation of ARMA models. The parameter vector is $\theta = [-a_1, \ldots, -a_p, b_1, \ldots, b_q]^T$. The cost function to be minimized is the weighted sum of squares of the innovations. The innovation u_t is related to y_t through

$$u_t = \sum_{\ell=0}^{p} a_\ell y_{t-\ell} - \sum_{\ell=1}^{q} b_\ell u_{t-\ell}. \tag{8.4.4}$$

Differentiation with respect to a_k and b_k yields

$$\frac{\partial u_t}{\partial a_k} = y_{t-k} - \sum_{\ell=1}^{q} b_\ell \frac{\partial u_{t-\ell}}{\partial a_k}, \quad \frac{\partial u_t}{\partial b_k} = -u_{t-k} - \sum_{\ell=1}^{q} b_\ell \frac{\partial u_{t-\ell}}{\partial b_k}. \tag{8.4.5}$$

Let $\{v_t\}$ and $\{w_t\}$ denote, respectively, the responses of $\{y_t\}$ and $\{u_t\}$ to the linear time-invariant filter $1/b(e^{j\omega})$. Then we get, from (8.4.5),

$$\frac{\partial u_t}{\partial a_k} = v_{t-k}, \quad \frac{\partial u_t}{\partial b_k} = -w_{t-k}. \tag{8.4.6}$$

The ARMA/RML algorithm estimates u_t, v_t, w_t at each new time point using the recursions

$$\hat{u}_{t+1} = y_{t+1} + \sum_{k=1}^{p} \hat{a}_{k,t} y_{t+1-k} - \sum_{k=1}^{q} \hat{b}_{k,t} \hat{u}_{t+1-k}, \tag{8.4.7a}$$

$$\hat{v}_{t+1} = y_{t+1} - \sum_{k=1}^{q} \hat{b}_{k,t} \hat{v}_{t+1-k}, \tag{8.4.7b}$$

$$\hat{w}_{t+1} = \hat{u}_{t+1} - \sum_{k=1}^{q} \hat{b}_{k,t} \hat{w}_{t+1-k}. \tag{8.4.7c}$$

Set $P_{-1} = \varepsilon \mathbf{1}$, $\hat{\theta}_{-1} = 0$

Set $\hat{u}_t = \hat{v}_t = \hat{w}_t = 0$, $-q \leq t \leq -1$

For $t = -1, 0, \ldots$ do

$\hat{u}_{t+1} = y_{t+1} - [y_t, \ldots, y_{t-p+1}, \hat{u}_t, \ldots, \hat{u}_{t-q+1}]^T \hat{\theta}_t$

$\begin{bmatrix} \hat{v}_{t+1} \\ \hat{w}_{t+1} \end{bmatrix} = \begin{bmatrix} y_{t+1} \\ \hat{u}_{t+1} \end{bmatrix} - \begin{bmatrix} \hat{v}_t & \cdots & \hat{v}_{t-q+1} \\ \hat{w}_t & \cdots & \hat{w}_{t-q+1} \end{bmatrix} \begin{bmatrix} \hat{b}_{1,t} \\ \vdots \\ \hat{b}_{q,t} \end{bmatrix}$

$\eta_t = [\hat{v}_t, \ldots, \hat{v}_{t-p+1}, \hat{w}_t, \ldots, \hat{w}_{t-q+1}]^T$

$K_t = P_t \eta_t (\lambda + \eta_t^T P_t \eta_t)^{-1}$

$P_{t+1} = \lambda^{-1} [\mathbf{1} - K_t \eta_t^T] P_t$

$\hat{\theta}_{t+1} = \hat{\theta}_t + K_t \hat{u}_{t+1}$

Table 8.3. A summary of the ARMA/RML algorithm.

Some caution is needed in applying the recursions (8.4.7), since $\hat{b}(z)$ may become unstable at some point, causing the recursions to diverge. A conservative approach is to stabilize $\hat{b}(z)$ at each time point by reflecting its unstable zeros into the unit circle. This is computationally expensive, however, so it is not always implemented. The ARMA/RML algorithm is summarized in Table 8.3. A square-root version, similar to the one in Table 8.2, can be easily obtained.

Surprisingly, the ARMA/RML algorithm is consistent for any ARMA model whose zeros are inside the unit circle, despite all the approximations made in its derivation (for $\lambda = 1$, of course). Moreover, the algorithm is asymptotically efficient as well, although the convergence of the normalized variance of the estimates to its asymptotic value may be very slow. A proof of these properties is beyond the scope of this book (see Ljung and Söderström [1983, Ch. 4]).

Example 8.3. The experiment described in Example 8.2 was repeated with the ARMA/RML algorithm. Figure 8.3 shows the estimated parameters. It is seen that in the case of $\lambda = 1$ the estimates converge faster than with the ELS algorithm. For $\lambda < 1$, the two algorithms behave similarly.   ∎

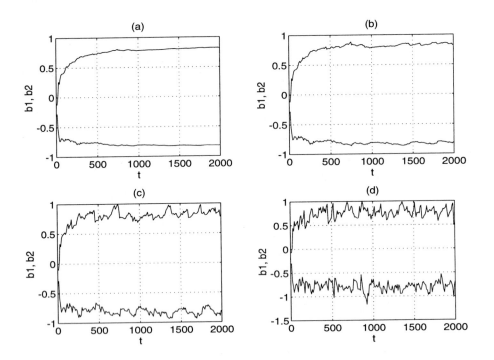

Figure 8.3. The estimates $\hat{b}_1$ and $\hat{b}_2$ as a function of time for Example 8.3: (a) $\lambda = 1.0$; (b) $\lambda = 0.996$; (c) $\lambda = 0.98$; (d) $\lambda = 0.95$.

8.5. STOCHASTIC GRADIENT ALGORITHMS

The algorithms discussed so far are essentially recursive versions of the Gauss-Newton method. As such, they rely on the matrix P_t, which is the approximate inverse Hessian of the cost function. The update of P_t takes a number of operations proportional to the square of the dimension of θ per time point. In real-time applications, the number of operations per time point may be a critical factor. In such applications we only have a fixed time to carry out the computations, the time between two successive measurements. It is therefore highly desirable to develop algorithms that require a number of operations proportional to the dimension of θ (rather than its square) at each time point.

The answer to this problem is provided by forfeiting the Gauss-Newton approach in favor of a simpler gradient descent method. The gradient of the cost function $e^T W e$ corresponding to the nonlinear error model (8.4.1) is given by $-2H^T(Y, \theta) W h(Y, \theta)$. Suppose we have an estimate $\hat{\theta}_t$ at time t and we compute the estimate at time $t+1$ by adding to $\hat{\theta}_t$ a negative multiple of the gradient. Such an updating scheme is called *gradient descent* or *steepest descent*. The explicit update formula is

$$\hat{\theta}_{t+1} = \hat{\theta}_t + 2\mu_t \begin{bmatrix} H_t^T & \eta_t \end{bmatrix} \begin{bmatrix} \lambda W_t & 0 \\ 0 & 1 \end{bmatrix} \begin{bmatrix} h_t \\ h_{t+1} \end{bmatrix} = \hat{\theta}_t + 2\mu_t \lambda H_t^T W_t h_t + 2\mu_t \eta_t h_{t+1},$$
$$(8.5.1)$$

where μ_t is the *step size* at time t. We now approximate (8.5.1) by assuming that *the algorithm has reached the minimum point at time t*. This approximation is likely to be rather crude, but it enables us to substitute zero for the term $H_t^T W_t h_t$ in (8.5.1). We then get the simplified update formula

$$\hat{\theta}_{t+1} = \hat{\theta}_t + 2\mu_t \eta_t h_{t+1}. \qquad (8.5.2)$$

This is known as the *stochastic gradient* method for the nonlinear least-squares problem (8.4.1). The word *stochastic* in the name has to do with the fact that the term $2\eta_t h_{t+1}$ in the formula is the gradient of the random quantity h_{t+1}^2, the instantaneous square error (as contrasted with its expectation).

Comparison of the stochastic gradient method to the recursive maximum likelihood method shows the following.

1. The update formula (8.5.2) does not involve the inverse Hessian matrix P_t, only the gradient vector η_t. The number of operations is thus proportional to the dimension of θ, not to its square.
2. The parameter λ does not appear. Instead, we have a sequence of scalar parameters $\{\mu_t\}$.

It can be shown that the sequence μ_t controls the algorithm's convergence similarly to λ in the recursive Gauss-Newton method. Specifically, $\mu_t = \mu_0/t$, where μ_0 is constant, roughly corresponds to $\lambda = 1$, while $\mu_t = \mu_0$ (a constant) corresponds to $\lambda < 1$. The choice $\mu_t = \mu_0/t$ can guarantee, under additional model-dependent conditions, convergence of the algorithm to some final value (which is not necessarily the true value). The first choice is due to Robbins and Monro [1951] and is known by the name *stochastic approximation*. The second is due to Widrow [Widrow and Stearns, 1985, and the references within] and is known as the *least mean-square* (LMS) algorithm.

Table 8.4 summarizes the stochastic gradient algorithm for ARMA processes. When $\mu_t = \mu_0$, we get the LMS algorithm. In the case of AR processes, the algorithm is simpler, since $\hat{v}_t = y_t$ and $\hat{w}_t$ is not needed.

$$
\begin{aligned}
&\text{Set } \hat{\theta}_{-1} = 0 \\
&\text{Set } \hat{u}_t = \hat{v}_t = \hat{w}_t = 0, -q \le t \le -1 \\
&\text{For } t = -1, 0, \ldots \text{ do} \\
&\quad \hat{u}_{t+1} = y_{t+1} - [y_t, \ldots, y_{t-p+1}, \hat{u}_t, \ldots, \hat{u}_{t-q+1}]\hat{\theta}_t \\
&\quad \begin{bmatrix} \hat{v}_{t+1} \\ \hat{w}_{t+1} \end{bmatrix} = \begin{bmatrix} y_{t+1} \\ \hat{u}_{t+1} \end{bmatrix} - \begin{bmatrix} \hat{v}_t & \cdots & \hat{v}_{t-q+1} \\ \hat{w}_t & \cdots & \hat{w}_{t-q+1} \end{bmatrix} \begin{bmatrix} \hat{b}_{1,t} \\ \vdots \\ \hat{b}_{q,t} \end{bmatrix} \\
&\quad \eta_t = [\hat{v}_t, \ldots, \hat{v}_{t-p+1}, \hat{w}_t, \ldots, \hat{w}_{t-q+1}]^T \\
&\quad \hat{\theta}_{t+1} = \hat{\theta}_t + 2\mu_t \eta_t \hat{u}_{t+1}
\end{aligned}
$$

Table 8.4. A summary of the ARMA stochastic gradient algorithm.

In passing we should mention a variant of the above method, called the *normalized stochastic gradient* method. The update formula for this method is

$$\hat{\theta}_{t+1} = \hat{\theta}_t + \frac{2\mu_t}{\sqrt{\eta_t^T \eta_t}} \eta_t h_{t+1}. \tag{8.5.3}$$

The normalized stochastic gradient algorithm enables better control of its convergence, since it bounds the norm of the change at each step by $2\mu_t|h_{t+1}|$, independently of the steepness of the gradient. On the other hand, its insensitivity to the steepness is likely to slow the convergence.

Example 8.4. The experiment described in Example 8.1 was repeated with the stochastic gradient algorithms. Figure 8.4(a) shows the estimated parameters as a function of t for the Robbins-Monro algorithm, with $\mu_t = 1/t$. Figures 8.4(b), (c), and (d) are for the LMS algorithm, with $\mu = 0.02, 0.002$, and 0.0002, respectively. The value 0.02 is the largest for which the algorithm is still stable in this example. It is seen that the Robbins-Monro algorithm converges reasonably fast, but is inferior to the RLS algorithm with $\lambda = 1$. In the LMS algorithm we see the trade-off between speed of convergence and smoothness of the estimates, controlled by μ. ∎

The study of the stability and convergence of stochastic gradient algorithms has been a subject of extensive research, but is outside the scope of this book (see, however, Probs. 8.5 and 8.6 for a simple special case). These algorithms are generally inferior to the ones based on the Gauss-Newton method, but their computational simplicity has made them popular in many applications [Haykin, 1991].

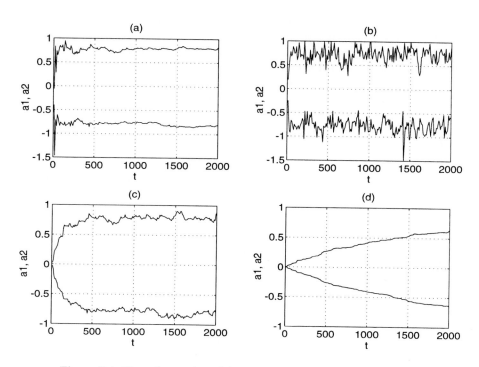

Figure 8.4. The estimates $\hat{a}_1$ and $\hat{a}_2$ as a function of time for Example 8.4: (a) $\mu_t = 1.0/t$; (b) $\mu = 0.02$; (c) $\mu = 0.002$; (d) $\mu = 0.0002$.

8.6. PROJECTION OPERATORS IN EUCLIDEAN SPACES

The rest of this chapter is devoted to a special class of adaptive algorithms known as *lattice algorithms*. In this section we present some background material on projection operators in Euclidean spaces, which will be needed for the derivation of the lattice algorithms.

Let H be a full rank $(t \times n)$-dimensional matrix, where $t \geq n$. Let $\boldsymbol{R}^t$ be the t-dimensional Euclidean space and W a positive definite square matrix of dimensions $t \times t$. For any two vectors $\boldsymbol{x}, \boldsymbol{y} \in \boldsymbol{R}^t$, define their inner product as

$$\langle \boldsymbol{x}, \boldsymbol{y} \rangle = \boldsymbol{x}^T W \boldsymbol{y}.$$

It is easy to check that $\boldsymbol{R}^t$ equipped with this inner product is a (finite dimensional) Hilbert space.

Let $\boldsymbol{V}_H$ be the subspace of $\boldsymbol{R}^t$ spanned by the columns of H. Since H is full rank, the dimension of this subspace is n. By Theorem B.1, every vector $\boldsymbol{x} \in \boldsymbol{R}^t$ has a unique orthogonal projection on $\boldsymbol{V}_H$. Thus, there exists a unique operator mapping any $\boldsymbol{x}$ to its projection on $\boldsymbol{V}_H$. This operator is easily verified to be linear (i.e., additive and homogeneous), so it can be expressed as a matrix P_H of dimensions $t \times t$. By this we mean that the projection of any $\boldsymbol{x}$ on $\boldsymbol{V}_H$ is given by $P_H \boldsymbol{x}$. The rank of this matrix is n, as the dimension of $\boldsymbol{V}_H$. We now wish to characterize the projection matrix.

Theorem 8.1. Under the above assumptions on H and W, the projection matrix P_H is given by

$$P_H = H(H^T W H)^{-1} H^T W. \qquad (8.6.1)$$

Proof. The first requirement on P_H is that $P_H \boldsymbol{x}$ be in the column space of H for all $\boldsymbol{x} \in \boldsymbol{R}^t$. This implies that P_H must have the form $P_H = HM$ for some matrix M of dimensions $n \times t$. Furthermore, the rank of M must be n if P_H is to have rank n [A21]. The second requirement is that P_H satisfy the projection theorem B.1. This is equivalent to requiring that

$$\langle P_H \boldsymbol{y}, (\boldsymbol{x} - P_H \boldsymbol{x}) \rangle = \boldsymbol{y}^T P_H^T W (\boldsymbol{1}_t - P_H) \boldsymbol{x} = 0$$

[cf. (b.2)]. Since this must hold for all $\boldsymbol{x}$ and $\boldsymbol{y}$, the second requirement translates to

$$P_H^T W = P_H^T W P_H,$$

or

$$M^T H^T W = M^T H^T W H M. \qquad (8.6.2)$$

Let us show that (8.6.2) is equivalent to

$$H^T W = H^T W H M. \qquad (8.6.3)$$

Indeed, the latter equality certainly implies the former. Since M has full rank, we can multiply (8.6.2) by $(MM^T)^{-1}M$ to get (8.6.3), so the former implies the latter as well. We finally get from (8.6.3)

$$M = (H^T W H)^{-1} H^T W,$$

from which (8.6.1) follows. ∎

Formula (8.6.1) characterizes the projection operator onto the subspace spanned by the columns of H fully and uniquely. It also characterizes the projection operator onto the orthogonal complement of $\boldsymbol{V}_H$. Since the result of the projection of $\boldsymbol{x}$ on the orthogonal complement must give $\boldsymbol{x} - P_H\boldsymbol{x}$ for all $\boldsymbol{x}$, we conclude that this projection is given by

$$P_H^\perp = \mathbf{1}_t - H(H^TWH)^{-1}H^TW. \tag{8.6.4}$$

We will call $P_H^\perp$ the *complementary projection* of P_H.

Next we consider the following question. Suppose we augment the matrix H by another matrix Z to form $[H\ Z]$. Z must have t rows, and we assume that the number of columns of $[H\ Z]$ is still not greater than t and that it has full rank. What then is the projection operator on the space spanned by the columns of $[H\ Z]$, and what is its complementary projection? The answer is provided by the following theorem.

Theorem 8.2. Under the above assumptions,

$$P_{[H\ Z]} = P_H + P_H^\perp Z(Z^TWP_H^\perp Z)^{-1}Z^T(P_H^\perp)^TW, \tag{8.6.5a}$$

$$P_{[H\ Z]}^\perp = P_H^\perp - P_H^\perp Z(Z^TWP_H^\perp Z)^{-1}Z^T(P_H^\perp)^TW. \tag{8.6.5b}$$

Proof. We have

$$\begin{bmatrix} H^T \\ Z^T \end{bmatrix} W \begin{bmatrix} H & Z \end{bmatrix} = \begin{bmatrix} H^TWH & H^TWZ \\ Z^TWH & Z^TWZ \end{bmatrix}.$$

By the matrix identity [A20],

$$\begin{bmatrix} H^TWH & H^TWZ \\ Z^TWH & Z^TWZ \end{bmatrix}^{-1} = \begin{bmatrix} (H^TWH)^{-1} & 0 \\ 0 & 0 \end{bmatrix}$$
$$+ \begin{bmatrix} -(H^TWH)^{-1}H^TWZ \\ 1 \end{bmatrix} \Delta^{-1} \begin{bmatrix} -Z^TWH(H^TWH)^{-1} & 1 \end{bmatrix},$$

where

$$\Delta = Z^TWZ - Z^TWH(H^TWH)^{-1}H^TWZ = Z^TWP_H^\perp Z.$$

Therefore,

$$P_{[H\ Z]} = \begin{bmatrix} H & Z \end{bmatrix} \begin{bmatrix} H^TWH & H^TWZ \\ Z^TWH & Z^TWZ \end{bmatrix}^{-1} \begin{bmatrix} H^T \\ Z^T \end{bmatrix} W$$
$$= H(H^TWH)^{-1}H^TW + P_H^\perp Z(Z^TWP_H^\perp Z)^{-1}Z^T(P_H^\perp)^TW.$$

This proves formula (8.6.5a). Formula (8.6.5b) follows upon subtracting both sides of (8.6.5a) from $\mathbf{1}_t$.

Corollary. Let X and Y be arbitrary matrices, each with t rows. Then

$$X^TWP_{[H\ Z]}^\perp Y = X^TWP_H^\perp Y - (X^TWP_H^\perp Z)(Z^TWP_H^\perp Z)^{-1}(Y^TWP_H^\perp Z)^T. \tag{8.6.6}$$

∎

Formula (8.6.6) is the fundamental identity upon which lattice algorithms are based. We name it the *projection update formula*.

The projection update formula has a normalized version, which will be used to derive normalized lattice algorithms. Define

$$\Gamma_H(X, Y) = (X^T W P_H^\perp X)^{-1/2} (X^T W P_H^\perp Y)(Y^T W P_H^\perp Y)^{-T/2}, \qquad (8.6.7)$$

where the matrix square roots are chosen as lower triangular (Prob. 8.9). Then:

Theorem 8.3.

$$\Gamma_{[H\ Z]}(X, Y) = [\mathbf{1} - \Gamma_H(X, Z)\Gamma_H^T(X, Z)]^{-1/2}$$
$$[\Gamma_H(X, Y) - \Gamma_H(X, Z)\Gamma_H^T(Y, Z)][\mathbf{1} - \Gamma_H(Y, Z)\Gamma_H^T(Y, Z)]^{-T/2}. \qquad (8.6.8)$$

Proof. We have, from (8.6.6),

$$X^T W P_{[H\ Z]}^\perp X = X^T W P_H^\perp X - (X^T W P_H^\perp Z)(Z^T W P_H^\perp Z)^{-1}(X^T W P_H^\perp Z)^T$$
$$= (X^T W P_H^\perp X)^{1/2}(X^T W P_H^\perp X)^{T/2}$$
$$- (X^T W P_H^\perp X)^{1/2}\Gamma_H(X, Z)\Gamma_H^T(X, Z)(X^T W P_H^\perp X)^{T/2}$$
$$= (X^T W P_H^\perp X)^{1/2}[\mathbf{1} - \Gamma_H(X, Z)\Gamma_H^T(X, Z)](X^T W P_H^\perp X)^{T/2}.$$

Therefore,

$$(X^T W P_{[H\ Z]}^\perp X)^{1/2} = (X^T W P_H^\perp X)^{1/2}[\mathbf{1} - \Gamma_H(X, Z)\Gamma_H^T(X, Z)]^{1/2}.$$

Similarly,

$$(Y^T W P_{[H\ Z]}^\perp Y)^{1/2} = (Y^T W P_H^\perp Y)^{1/2}[\mathbf{1} - \Gamma_H(Y, Z)\Gamma_H^T(Y, Z)]^{1/2}.$$

Using (8.6.6) again, we get

$$\Gamma_{[H\ Z]}(X, Y) = (X^T W P_{[H\ Z]}^\perp X)^{-1/2}(X^T W P_{[H\ Z]}^\perp Y)(Y^T W P_{[H\ Z]}^\perp Y)^{-T/2}$$
$$= [\mathbf{1} - \Gamma_H(X, Z)\Gamma_H^T(X, Z)]^{-1/2}(X^T W P_H^\perp X)^{-1/2}$$
$$[X^T W P_H^\perp Y - (X^T W P_H^\perp Z)(Z^T W P_H^\perp Z)^{-1}(Y^T W P_H^\perp Z)^T]$$
$$(Y^T W P_H^\perp Y)^{-T/2}[\mathbf{1} - \Gamma_H(Y, Z)\Gamma_H^T(Y, Z)]^{-T/2}$$
$$= [\mathbf{1} - \Gamma_H(X, Z)\Gamma_H^T(X, Z)]^{-1/2}[\Gamma_H(X, Y) - \Gamma_H(X, Z)\Gamma_H^T(Y, Z)]$$
$$[\mathbf{1} - \Gamma_H(Y, Z)\Gamma_H^T(Y, Z)]^{-T/2}. \qquad \blacksquare$$

Formula (8.6.7) will be called the *normalized projection update formula*. This formula (in a scalar form) dates back to Yule [1907].

So far we have considered augmentation of H in the "horizontal direction." As our final topic in this section, we examine augmentation in the "vertical direction," but only in a special case. Let us change the notation H to H_t to emphasize its row dimension, and let

$$H_{t+1} = \begin{bmatrix} H_t \\ \eta_t^T \end{bmatrix},$$

where η_t^T is a row vector. Let W_t be the exponential weight matrix given in (8.2.2). Let $\boldsymbol{p}$ be a vector of dimension $t+1$ with 1 in the last position and zeros elsewhere, and

$$G = [\, H_{t+1} \quad \boldsymbol{p}\,] = \begin{bmatrix} H_t & 0 \\ \eta_t^T & 1 \end{bmatrix} = \begin{bmatrix} H_t & 0 \\ 0 & 1 \end{bmatrix}\begin{bmatrix} \boldsymbol{1}_t & 0 \\ \eta_t^T & 1 \end{bmatrix}. \qquad (8.6.9)$$

Then we have:

Theorem 8.4.

$$X^T W_{t+1} \begin{bmatrix} P_{H_t}^\perp & 0 \\ 0 & 0 \end{bmatrix} Y = X^T W_{t+1} P_{H_{t+1}}^\perp Y$$
$$- (X^T W_{t+1} P_{H_{t+1}}^\perp \boldsymbol{p})(\boldsymbol{p}^T W_{t+1} P_{H_{t+1}}^\perp \boldsymbol{p})^{-1}(Y^T W_{t+1} P_{H_{t+1}}^\perp \boldsymbol{p})^T. \qquad (8.6.10)$$

Proof. The second matrix on the right side of (8.6.9) is nonsingular. Therefore, by Prob. 8.10, P_G is equal to the projection operator of the first matrix on the right side. Therefore, we get, using the special structure of the weight matrix,

$$P_G = \begin{bmatrix} H_t & 0 \\ 0 & 1 \end{bmatrix}\begin{bmatrix} \lambda H_t^T W_t H_t & 0 \\ 0 & 1 \end{bmatrix}^{-1}\begin{bmatrix} \lambda H_t^T W_t & 0 \\ 0 & 1 \end{bmatrix} = \begin{bmatrix} P_{H_t} & 0 \\ 0 & 1 \end{bmatrix},$$

so

$$P_G^\perp = \begin{bmatrix} P_{H_t}^\perp & 0 \\ 0 & 0 \end{bmatrix}.$$

On the other hand, we have, by (8.6.5b),

$$P_G^\perp = P_{H_{t+1}}^\perp - P_{H_{t+1}}^\perp \boldsymbol{p}(\boldsymbol{p}^T W_{t+1} P_{H_{t+1}}^\perp \boldsymbol{p})^{-1}\boldsymbol{p}^T (P_{H_{t+1}}^\perp)^T W_{t+1}.$$

From these two expressions for $P_G^\perp$, we get (8.6.10). ∎

Formula (8.6.10) is called the *time update formula*, and the vector $\boldsymbol{p}$ is sometimes called the *pinning vector*. As we see, augmenting H_{t+1} by $\boldsymbol{p}$ effectively wipes out the last row of this matrix, "sending it back" to time t. As the reader may have guessed, this formula will be used in the reverse direction, to proceed from t to $t+1$. The vector $\boldsymbol{p}$ plays a major role in the derivation of lattice algorithms, as we shall see in the next section.

8.7. LATTICE ALGORITHMS FOR AUTOREGRESSIVE ESTIMATION

In this section we apply the theory of projection operators presented above to recursive AR estimation. The algorithms to be derived essentially implement the recursive least-squares method of Sec. 8.2, but they do so in a smaller number of operations. The lattice approach is to compute the partial correlation coefficients directly, as in the Burg and Itakura-Saito methods (see Sec. 6.7). However, instead of computing them once for the entire data set, they are computed recursively and updated at each time point. Moreover, the updating can be made adaptive by using an exponential forgetting factor λ as in the RLS method.

The role of the matrix H in Sec. 8.6 will be played by the Toeplitz data matrix

$$
\boldsymbol{Y}_{n,t-1} = \begin{bmatrix} 0 & 0 & \cdots & 0 \\ y_0 & 0 & \cdots & 0 \\ \vdots & \vdots & & \vdots \\ y_{t-1} & y_{t-2} & \cdots & y_{t-n} \end{bmatrix}. \tag{8.7.1}
$$

Note that this is the same as the matrix in the prewindowed method for AR parameter estimation discussed in Sec. 6.6, see Eq. (6.6.6). The indexes n and $t-1$ in the definition designate the column dimension of the matrix and the most recent time point it contains, respectively. Note that the row dimension of this matrix is $t+1$, due to the row of zeros at the top. Both n and t are varying, so together they yield a two-parameter family of data matrices.

The crucial property of the matrix $\boldsymbol{Y}_{n,t-1}$ is its *shift invariance* along the diagonals. In particular, deleting the first row and column of $\boldsymbol{Y}_{n,t-1}$ yields the same matrix as deletion of the last row and column, the result being $\boldsymbol{Y}_{n-1,t-2}$ in both cases.

The roles of X, Y, and Z in Sec. 8.6 will be taken by the following three vectors in different permutations:

$$
\boldsymbol{y}_t = [y_0, y_1, \ldots, y_t]^T, \ \boldsymbol{y}_{t-n-1} = [0, \ldots, 0, y_0, \ldots, y_{t-n-1}]^T, \ \boldsymbol{p} = [0, \ldots, 0, 1]^T. \tag{8.7.2}
$$

All three vectors have dimension $t+1$ (viz. the same as the data matrix); $\boldsymbol{y}_{t-n-1}$ is padded by $n+1$ zeros.

Before we explore the details of the algorithms, let us make the following observations.

- The data matrix and the vector $\boldsymbol{y}_t$ give the nonadaptive AR estimate of the prewindowed method through

$$
\hat{\boldsymbol{a}}_{n,t} = -(\boldsymbol{Y}_{n,t-1}^T \boldsymbol{Y}_{n,t-1})^{-1} \boldsymbol{Y}_{n,t-1}^T \boldsymbol{y}_t.
$$

- The vector of residual errors corresponding to the estimated AR model is given by

$$
\boldsymbol{e}_{n,t} = \boldsymbol{y}_t - \boldsymbol{Y}_{n,t-1} \hat{\boldsymbol{a}}_{n,t} = [\boldsymbol{1}_{t+1} - \boldsymbol{Y}_{n,t-1} (\boldsymbol{Y}_{n,t-1}^T \boldsymbol{Y}_{n,t-1})^{-1} \boldsymbol{Y}_{n,t-1}^T] \boldsymbol{y}_t
$$
$$
= P_{n,t-1}^{\perp} \boldsymbol{y}_t,
$$

where $P_{n,t-1}$ is the projection operator of $\boldsymbol{Y}_{n,t-1}$ (corresponding to the identity weight matrix).

- The vector
$$\tilde{\boldsymbol{a}}_{n,t} = -(\boldsymbol{Y}_{n,t-1}^T \boldsymbol{Y}_{n,t-1})^{-1} \boldsymbol{Y}_{n,t-1}^T \boldsymbol{y}_{t-n-1}$$
  is the backward prewindowed estimate, in analogy with the backward covariance estimation defined in Sec. 6.6. Correspondingly, the vector of backward residual errors is
$$\tilde{\boldsymbol{e}}_{n,t-n-1} = P_{n,t-1}^{\perp} \boldsymbol{y}_{t-n-1}.$$

- Either $\boldsymbol{y}_t$ or $\boldsymbol{y}_{t-n-1}$ can be used as the vector Z in the framework of Sec. 8.6 to augment the rank of the projection. Augmenting with $\boldsymbol{y}_t$ gives $\boldsymbol{Y}_{n+1,t}$ (except for the row of zeros at the top), while augmenting with $\boldsymbol{y}_{t-n-1}$ gives $\boldsymbol{Y}_{n+1,t-1}$.

- The vector $\boldsymbol{p}$, when used as Z, acts to perform time update, as we have seen in Sec. 8.6. On the other hand, when we use this vector as X or Y, it acts to pick up the last component of the vector it multiplies.

Let us now define the actual variables appearing in the algorithm.

$$e_{n,t} = \boldsymbol{p}^T W P_{n,t-1}^{\perp} \boldsymbol{y}_t, \quad \tilde{e}_{n,t-n-1} = \boldsymbol{p}^T W P_{n,t-1}^{\perp} \boldsymbol{y}_{t-n-1}, \qquad (8.7.3a)$$

$$r_{n,t} = \boldsymbol{y}_t^T W P_{n,t-1}^{\perp} \boldsymbol{y}_t, \quad \tilde{r}_{n,t-n-1} = \boldsymbol{y}_{t-n-1}^T W P_{n,t-1}^{\perp} \boldsymbol{y}_{t-n-1}, \qquad (8.7.3b)$$

$$g_{n,t} = \boldsymbol{p}^T W P_{n,t-1}^{\perp} \boldsymbol{p}, \qquad \Delta_{n+1,t} = \boldsymbol{y}_t^T W P_{n,t-1}^{\perp} \boldsymbol{y}_{t-n-1}. \qquad (8.7.3c)$$

These definitions can be interpreted as follows.

- $e_{n,t}$ is the last component of $\boldsymbol{e}_{n,t}$, the forward residual error of order n at time t.
- $\tilde{e}_{n,t-n-1}$ is the last component of $\tilde{\boldsymbol{e}}_{n,t-n-1}$, the backward residual error.
- $r_{n,t}$ is the square norm of $\boldsymbol{e}_{t,n}$, which is (up to a factor $t+1$) an estimate of d_n, the variance of the partial innovation of order n.
- $\tilde{r}_{n,t-n-1}$ is the square norm of $\tilde{\boldsymbol{e}}_{n,t-n-1}$, which is (up to a factor $t+1$) an estimate of $\tilde{d}_n$, the variance of the backward partial innovation of order n.
- $\Delta_{n+1,t}$ is the inner product of $\boldsymbol{e}_{n,t}$ and $\tilde{\boldsymbol{e}}_{n,t-n-1}$, which is (up to a factor $t+1$) an estimate of Δ_{n+1}, the partial correlation of order $n+1$; see (6.2.3).
- We will not be concerned with the interpretation of $g_{n,t}$.

The lattice algorithm computes the above six entities for all $0 \le n \le p-1$, where p is some prechosen order, and for $t = 0, 1, \ldots$, as long as new data keep coming. The order update is the inner loop, and the time update is the outer one. The order update loop consists of six equations, five of which are straightforward substitutions in the projection update formula (8.6.6), and the sixth uses the time update formula (8.6.10) (Prob. 8.11). Here are the equations and the substitutions leading to them. Note that we are using transpositions where appropriate, although all quantities are scalar, in preparation for Sec. 8.8.

X	Y	Z		Eq. (8.7.4)	
y_t	y_{t-n-1}	p	$\Longrightarrow$	$\Delta_{n+1,t} = \lambda\Delta_{n+1,t-1} + e_{n,t}^T g_{n,t-1}^{-1}\tilde{e}_{n,t-n-1}$	(a)
p	p	y_t	$\Longrightarrow$	$g_{n+1,t} = g_{n,t-1} - e_{n,t}r_{n,t}^{-1}e_{n,t}^T$	(b)
p	y_{t-n-1}	y_t	$\Longrightarrow$	$\tilde{e}_{n+1,t-n-1} = \tilde{e}_{n,t-n-1} - e_{n,t}r_{n,t}^{-1}\Delta_{n+1,t}$	(c)
y_{t-n-1}	y_{t-n-1}	y_t	$\Longrightarrow$	$\tilde{r}_{n+1,t-n-1} = \tilde{r}_{n,t-n-1} - \Delta_{n+1,t}^T r_{n,t}^{-1}\Delta_{n+1,t}$	(d)
p	y_t	y_{t-n-1}	$\Longrightarrow$	$e_{n+1,t} = e_{n,t} - \tilde{e}_{n,t-n-1}\tilde{r}_{n,t-n-1}^{-1}\Delta_{n+1,t}^T$	(e)
y_t	y_t	y_{t-n-1}	$\Longrightarrow$	$r_{n+1,t} = r_{n,t} - \Delta_{n+1,t}\tilde{r}_{n,t-n-1}^{-1}\Delta_{n+1,t}^T$	(f)

The equations are written in the correct order of computation for each t. It should be emphasized that $\tilde{e}_{n,t-n-1}$ actually "belongs" to time $t-1$, and $\tilde{e}_{n+1,t-n-1}$ "belongs" to time t. If we assign a horizontal axis to the order n and a vertical axis to the time t, then Δ is updated in the vertical direction, e and r in the horizontal direction, and g, $\tilde{e}$, and $\tilde{r}$ in a diagonal direction [from $(t-1,n)$ to $(t,n+1)$]; see Figure 8.5. As long as $t < p$, the order update loop is done from 0 to $t-1$ only. The initial conditions can be derived from the definitions of the six entities (Prob. 8.12), giving

$$e_{0,t} = \tilde{e}_{0,t} = y_t, \quad \forall t, \tag{8.7.5a}$$

$$r_{0,0} = \tilde{r}_{0,0} = y_0^2, \quad r_{0,t} = \tilde{r}_{0,t} = \lambda r_{0,t-1} + y_t^2, \quad \forall t, \tag{8.7.5b}$$

$$g_{0,t} = 1, \quad \forall t, \tag{8.7.5c}$$

$$\Delta_{n+1,n} = 0, \quad \forall n. \tag{8.7.5d}$$

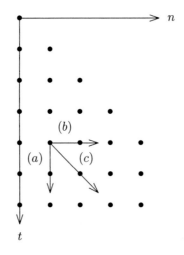

Figure 8.5. The directions of update of the lattice algorithm: (a) update of Δ; (b) update of e and r; (c) update of g, $\tilde{e}$, and $\tilde{r}$.

The partial correlation coefficients are computed as (Itakura-Saito style)

$$K_{n+1,t} = \frac{\Delta_{n+1,t}}{\sqrt{r_{n,t}\tilde{r}_{n,t-n-1}}}, \qquad (8.7.6a)$$

or (Burg's style)

$$K_{n+1,t} = \frac{2\Delta_{n+1,t}}{r_{n,t} + \tilde{r}_{n,t-n-1}}. \qquad (8.7.6b)$$

The first choice is the more common one.

The algorithm derived above is due to Lee [1980]; see also Lee, Morf, and Friedlander [1981]. It is known as the *prewindowed AR least-squares lattice*. The algorithm is summarized in Table 8.5.

Set $r_{0,0} = \tilde{r}_{0,0} = y_0^2$

Set $e_{0,0} = \tilde{e}_{0,0} = y_0$

Set $\Delta_{n+1,n} = 0, 0 \le n \le p-1$

Set $g_{0,0} = 1$

For $t = 1, 2, \ldots$ do

$\quad e_{0,t} = \tilde{e}_{0,t} = y_t$

$\quad r_{0,t} = \tilde{r}_{0,t} = \lambda r_{0,t-1} + y_t^2$

$\quad g_{0,t} = 1$

$\quad$ For $0 \le n \le \min\{p,t\} - 1$ do

$\qquad \Delta_{n+1,t} = \lambda\Delta_{n+1,t-1} + e_{n,t}\tilde{g}_{n,t-1}^{-1}e_{n,t-n-1}$

$\qquad g_{n+1,t} = g_{n,t-1} - e_{n,t}^2 r_{n,t}^{-1}$

$\qquad \tilde{e}_{n+1,t-n-1} = \tilde{e}_{n,t-n-1} - \Delta_{n+1,t}r_{n,t}^{-1}e_{n,t}$

$\qquad \tilde{r}_{n+1,t-n-1} = \tilde{r}_{n,t-n-1} - \Delta_{n+1,t}^2 r_{n,t}^{-1}$

$\qquad e_{n+1,t} = e_{n,t} - \Delta_{n+1,t}\tilde{r}_{n,t-n-1}^{-1}\tilde{e}_{n,t-n-1}$

$\qquad r_{n+1,t} = r_{n,t} - \Delta_{n+1,t}^2\tilde{r}_{n,t-n-1}^{-1}$

Table 8.5. A summary of the prewindowed AR lattice algorithm.

The storage requirements for the algorithm are as follows. The arrays g, $\tilde{e}$, and $\tilde{r}$ require $2(p+1)$ cells each, because both old and new values need to be stored. The array Δ needs p cells, and $r_{0,t}$ needs one cell. Finally, e and r need only one cell each, because neither their old values nor their new values need to be stored. The number of operations per time point is proportional to p. The algorithm gives the residual errors and the partial correlation coefficients of all orders, but not the estimated AR vector $\hat{a}_{p,t}$, at least not directly. In some applications, $\hat{a}_{p,t}$ is not needed anyway, only the parameters $e_{n,t}$ and $K_{n+1,t}$ (e.g., in speech applications; see Ex. 8.5 below). Both the whitening filter $\hat{a}(z)$ and the synthesis filter $1/\hat{a}(z)$ can be implemented in lattice form, as was shown in Sec. 6.4, avoiding the need for the explicit AR coefficients. In case the AR estimate is needed explicitly, there are two options as follows.

1. The Levinson recursion can be used to compute $\hat{a}_{p,t}$ from $K_{n+1,t}$. This, however, takes p^2 operations, so it spoils the computational efficiency of the algorithm if done at each time point. In fact, it never makes sense to compute $\hat{a}_{p,t}$ at each time point, because the rate of variation of the estimate is relatively slow: the effective time constant is on the order $2/(1-\lambda)$. A conservative approach is to compute $\hat{a}_{p,t}$ once every p time points. This way, the additional computational load increases by p operations per time point on the average.

2. A set of pure time update formulas has been developed for $\hat{a}_{p,t}$, in terms of the lattice variables. These formulas require a number of operations proportional to p at each time point and are discussed in Sec. 8.9.

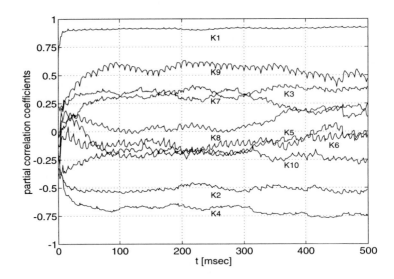

Figure 8.6. The estimated partial correlation coefficients as a function of time for the speech signal of the phoneme "ae."

Example 8.5. The speech signal of the phoneme "ae," described in Example 6.4, was inputted to the lattice algorithm using $p = 10$ and $\lambda = 0.995$. Figure 8.6 shows the 10 partial correlation coefficients as a function of time. As we see, most of the coefficients are approximately constant during the time interval of the record (the initial settling time of the algorithm notwithstanding). This means that the signal is approximately stationary in this interval.   ∎

The prewindowed AR lattice algorithm has a normalized version, which can be derived from the normalized projection update formula (8.6.7). Let us define

$$\varepsilon_{n,t} = \Gamma_{n,t-1}(\boldsymbol{p}, \boldsymbol{y}_t) = \frac{e_{n,t}}{\sqrt{g_{n,t} r_{n,t}}}, \tag{8.7.7a}$$

$$\tilde{\varepsilon}_{n,t-n-1} = \Gamma_{n,t-1}(\boldsymbol{p}, \boldsymbol{y}_{t-n-1}) = \frac{\tilde{e}_{n,t-n-1}}{\sqrt{g_{n,t} \tilde{r}_{n,t-n-1}}}, \tag{8.7.7b}$$

$$K_{n+1,t} = \Gamma_{n,t-1}(\boldsymbol{y}_{t-n-1}, \boldsymbol{y}_t) = \frac{\Delta_{n+1,t}}{\sqrt{r_{n,t} \tilde{r}_{n,t-n-1}}}. \tag{8.7.7c}$$

The following three equations (one time update and two order update formulas) are obtained from (8.6.7) analogously to (8.7.4a, c, d).

$$K_{n+1,t} = \lambda K_{n+1,t-1} \sqrt{1 - \varepsilon_{n,t}^2} \sqrt{1 - \tilde{\varepsilon}_{n,t-n-1}^2} + \varepsilon_{n,t} \tilde{\varepsilon}_{n,t-n-1}, \tag{8.7.8a}$$

$$\varepsilon_{n+1,t} = \frac{\varepsilon_{n,t} - K_{n+1,t} \tilde{\varepsilon}_{n,t-n-1}}{\sqrt{1 - \tilde{\varepsilon}_{n,t-n-1}^2} \sqrt{1 - \tilde{K}_{n+1,t}^2}}, \tag{8.7.8b}$$

$$\tilde{\varepsilon}_{n+1,t-n-1} = \frac{\tilde{\varepsilon}_{n,t-n-1} - K_{n+1,t} \varepsilon_{n,t}}{\sqrt{1 - \varepsilon_{n,t}^2} \sqrt{1 - \tilde{K}_{n+1,t}^2}}. \tag{8.7.8c}$$

As we see, the normalized prewindowed AR algorithm has half as many variables and half as many equations as the unnormalized one. It turns out that $r_{0,t}$ is also needed at each time point, but not the higher-order variance estimates. Also, the variable $g_{n,t}$ is absent from the normalized algorithm. The gain in number of computations is not clear, however, since the square root operations (three for each t and n) are usually more time consuming than rational operations. The organization and the ranges of the time and order loops are the same as for the unnormalized algorithm. The initial conditions are

$$r_{0,0} = y_0^2, \quad r_{0,t} = \lambda r_{0,t-1} + y_t^2, \quad \forall t, \tag{8.7.9a}$$

$$\varepsilon_{0,t} = \tilde{\varepsilon}_{0,t} = \frac{y_t}{\sqrt{r_{0,t}}}, \quad \forall t, \tag{8.7.9b}$$

$$K_{n+1,n} = 0, \quad \forall n. \tag{8.7.9c}$$

The normalized algorithm may be more attractive than the unnormalized one in cases where only the partial correlation coefficients $K_{n+1,t}$ are needed. The normalized algorithm is not very useful if the residuals $e_{n,t}$ are needed as well, since they cannot be easily obtained from the normalized residuals $\varepsilon_{n,t}$.

8.8. LATTICE ALGORITHMS FOR ARMA ESTIMATION

The lattice machinery is essentially limited to solving linear least-squares problems. The ARMA problem is, as we have seen, inherently nonlinear. Adaptive ARMA estimation by lattice is possible though by adopting the pseudolinear regression approach of Sec. 8.3. Of the several methods proposed in the literature, we limit ourselves to one, the *two-channel AR embedding* approach of Lee, Friedlander, and Morf [1982]. This approach is neither the most general nor the most efficient, but it is the easiest to derive and the one that can be expressed most concisely. The most significant limitation of two-channel AR embedding is that the MA order q is required to be equal to p.

As implied by its name, the approach to be presented replaces the ARMA problem with an equivalent two-channel AR problem. To see how this works, let us first assume that the innovation sequence $\{u_t\}$ is known and rewrite the pseudolinear regression equation (8.3.2) as (with $q = p$)

$$
\begin{bmatrix} u_p & u_p \\ u_{p+1} & u_{p+1} \\ \vdots & \vdots \\ u_t & u_t \end{bmatrix} = \begin{bmatrix} y_p & u_p \\ y_{p+1} & u_{p+1} \\ \vdots & \vdots \\ y_t & u_t \end{bmatrix} + \begin{bmatrix} y_{p-1} & u_{p-1} & \cdots & y_0 & u_0 \\ y_p & u_p & \cdots & y_1 & u_1 \\ \vdots & \vdots & \ddots & \vdots & \vdots \\ y_{t-1} & u_{t-1} & \cdots & y_{t-p} & u_{t-p} \end{bmatrix} \begin{bmatrix} -a_1 & 0 \\ b_1 & 0 \\ \vdots & \vdots \\ -a_p & 0 \\ b_p & 0 \end{bmatrix}.
$$

$$\tag{8.8.1}$$

We neglect the prewindowing for the time being, since it is not essential to the understanding of the basic idea. We now observe that the data matrix in (8.8.1) has the shift-invariance property, provided we shift by *two columns and one row* at a time. We can therefore use the projection update formula on the data matrix, letting the matrix Z consist of two columns, one of y's and one of u's. As $t \to \infty$ (and with $\lambda = 1$), we can expect the solution of the two-channel least-squares problem (8.8.1) to converge to the true parameters. Denoting the residual errors of the two channels by $e_t^{(1)}$ and $e_t^{(2)}$ and expressing the input-output relationships using formal z transforms, we can write the limiting relation as

$$
\begin{bmatrix} e^{(1)}(z) \\ e^{(2)}(z) \end{bmatrix} = \begin{bmatrix} a(z) & -\bar{b}(z) \\ 0 & 1 \end{bmatrix} \begin{bmatrix} y(z) \\ u(z) \end{bmatrix},
$$

$$\tag{8.8.2}$$

where $a(z) = 1 + a_1 z^{-1} + \cdots + a_p z^{-p}$, and $\bar{b}(z) = b_1 z^{-1} + \cdots + b_p z^{-p}$.

We now use the same device as in the extended least-squares method: we identify $u(z)$ with $e^{(1)}(z)$ by feeding the first output back to the second input. We then get from (8.8.2)

$$
u(z) = a(z)y(z) - \bar{b}(z)u(z) \quad \Longrightarrow \quad u(z) = \frac{a(z)y(z)}{1 + \bar{b}(z)} = \frac{a(z)y(z)}{b(z)},
$$

$$\tag{8.8.3}$$

which is precisely the desired relationship. The above calculation proves nothing, of course, as it is merely based on a plausibility argument. In fact, the ARMA lattice algorithm to be presented below has the same convergence properties as the extended least-squares method. In particular, its convergence to the true parameter values can only be guaranteed under restrictive conditions, such as (8.3.5).

Before we can translate the above idea to a working algorithm, we need to solve the problem of feeding back the error $e_t^{(1)}$ to the input. Since this error is computed at time t, it cannot be used as an input at time t because of the implied algebraic (delay-free) loop. Feeding back $e_{t-1}^{(1)}$ instead is wrong, since the additional delay will spoil the relation (8.8.3). A little reflection reveals that the feedback entity we really need at time t is $\hat{u}_t = y_t - \hat{y}_t$, where $\hat{y}_t$ is the pth-order predictor of y_t based on the measurements up to time $t-1$. $\hat{y}_t$ should be computed at time $t-1$, so $\hat{u}_t$ can be computed immediately when y_t becomes available, before starting the tth time update of the algorithm.

The following lemma tells us how to compute $\hat{y}_t$ as a function of the lattice variables. For simplicity of notation, we state and prove it for the one-channel lattice, but we will take care to use matrix transposes and inverses as necessary, so the result will be directly applicable to the two-channel case.

Lemma 8.1. The pth-order predictor of y_t is given by

$$\hat{y}_t = \sum_{n=0}^{p-1} \tilde{e}_{n,t-n-1}\tilde{r}_{n,t-n-1}^{-1}\Delta_{n+1,t}^T. \tag{8.8.4}$$

Proof. The crucial observation is that the set of vectors $\{\boldsymbol{y}_{t-1}, P_{1,t-1}^\perp \boldsymbol{y}_{t-2}, \cdots,$ $P_{p-1,t-1}^\perp \boldsymbol{y}_{t-p}\}$ forms an orthogonal basis to the column space of $\boldsymbol{Y}_{p,t-1}$ (Prob. 8.15). The vector $\hat{\boldsymbol{y}}_t$, the projection of $\boldsymbol{y}_t$ on this space, can therefore be expressed as a linear combination of the basis vectors; that is,

$$\hat{\boldsymbol{y}}_t = \sum_{n=0}^{p-1} P_{n,t-1}^\perp \boldsymbol{y}_{t-n-1}\alpha_n. \tag{8.8.5}$$

Moreover, due to the orthogonality of the basis, the coefficient α_n is given by the inner product of $\boldsymbol{y}_t$ and the corresponding basis vector, divided by the square norm of the basis vector; that is,

$$\alpha_n = \langle \boldsymbol{y}_{t-n-1}, P_{n,t-1}^\perp \boldsymbol{y}_{t-n-1}\rangle^{-1}\langle P_{n,t-1}^\perp \boldsymbol{y}_{t-n-1}, \boldsymbol{y}_t\rangle = \tilde{r}_{n,t-n-1}^{-1}\Delta_{n+1,t}^T.$$

Substitution in (8.8.5) gives

$$\hat{\boldsymbol{y}}_t = \sum_{n=0}^{p-1} P_{n,t-1}^\perp \boldsymbol{y}_{t-n-1}\tilde{r}_{n,t-n-1}^{-1}\Delta_{n+1,t}^T. \tag{8.8.6}$$

The last component of $\hat{\boldsymbol{y}}_t$ is obtained by forming the inner product of (8.8.6) with the vector $\boldsymbol{p}$, yielding (8.8.4). ∎

Formula (8.8.4) is still not computable at time $t-1$ because of the term $\Delta_{n+1,t}$. Instead, we can use $\Delta_{n+1,t-1}$ to a good approximation, since this quantity changes very little over one time point. In conclusion, $\hat{u}_t$ is taken as

$$\hat{u}_t = y_t - \left(\sum_{n=0}^{p-1} \tilde{e}_{n,t-n-1} \tilde{r}_{n,t-n-1}^{-1} \Delta_{n+1,t-1}^T \right)_1. \tag{8.8.7}$$

$$\text{Set } r_{0,0} = \tilde{r}_{0,0} = \begin{bmatrix} 1+y_0^2 & y_0^2 \\ y_0^2 & 1+y_0^2 \end{bmatrix}$$

$$\text{Set } e_{0,0} = \tilde{e}_{0,0} = \begin{bmatrix} y_0 & y_0 \end{bmatrix}$$

$$\text{Set } \Delta_{n+1,n} = \begin{bmatrix} 0 & 0 \\ 0 & 0 \end{bmatrix}, \quad 0 \le n \le p-1$$

$$\text{Set } g_{0,0} = 1$$

$$\text{Set } \hat{y}_1 = 0$$

$$\text{For } t = 1, 2, \ldots \text{ do}$$

$$\quad \hat{u}_t = y_t - \hat{y}_t$$

$$\quad e_{0,t} = \tilde{e}_{0,t} = \begin{bmatrix} y_t & \hat{u}_t \end{bmatrix}$$

$$\quad r_{0,t} = \tilde{r}_{0,t} = \lambda r_{0,t-1} + \begin{bmatrix} y_t^2 & y_t \hat{u}_t \\ y_t \hat{u}_t & \hat{u}_t^2 \end{bmatrix}$$

$$\quad g_{0,t} = 1$$

$$\quad \bar{n} = \min\{p, t\}$$

$$\quad \text{For } 0 \le n \le \bar{n}-1 \text{ do}$$

$$\quad\quad \Delta_{n+1,t} = \lambda \Delta_{n+1,t-1} + e_{n,t}^T g_{n,t-1}^{-1} \tilde{e}_{n,t-n-1}$$

$$\quad\quad g_{n+1,t} = g_{n,t-1} - e_{n,t} r_{n,t}^{-1} e_{n,t}^T$$

$$\quad\quad \tilde{e}_{n+1,t-n-1} = \tilde{e}_{n,t-n-1} - e_{n,t} r_{n,t}^{-1} \Delta_{n+1,t}$$

$$\quad\quad \tilde{r}_{n+1,t-n-1} = \tilde{r}_{n,t-n-1} - \Delta_{n+1,t}^T r_{n,t}^{-1} \Delta_{n+1,t}$$

$$\quad\quad e_{n+1,t} = e_{n,t} - \tilde{e}_{n,t-n-1} \tilde{r}_{n,t-n-1}^{-1} \Delta_{n+1,t}^T$$

$$\quad\quad r_{n+1,t} = r_{n,t} - \Delta_{n+1,t} \tilde{r}_{n,t-n-1}^{-1} \Delta_{n+1,t}^T$$

$$\quad \hat{y}_{t+1} = \left(\sum_{n=0}^{\bar{n}-1} \tilde{e}_{n,t-n} \tilde{r}_{n,t-n}^{-1} \Delta_{n+1,t}^T \right)_1$$

Table 8.6. A summary of the prewindowed ARMA lattice algorithm.

The two-channel prewindowed lattice is very similar to the basic lattice algorithm. We have only to replace the divisions by matrix inversions (the matrices in question are 2×2) and take care to put the matrix transposes in the proper places. The algorithm is summarized in Table 8.6, including the feedback of $\hat{u}_t$ to the input. Data prewindowing is implicit in the algorithm, and exponential weighting can be employed identically to the one-channel case. We note that the addition of $\mathbf{1}_2$ to the initial value of $\tilde{r}_{0,0}$ is arbitrary and was done to avoid singularity at $t = 0$.

A closer look at the two-channel embedding approach raises an intriguing question. The second channel is degenerate, in the sense of converging to the identity filter, whose input and output are both $\hat{u}_t$. From this it follows (Prob. 8.16) that the partial correlations $\Delta_{n,t}$ in the ARMA lattice algorithm converge to a form $\begin{bmatrix} * & * \\ 0 & 0 \end{bmatrix}$, where the * denote nonzero entries. We therefore have two options: (a) to force the entries in the second row of $\Delta_{n,t}$ to zero at each time step, thus saving computations, while being in agreement with the theory, or (b) to let these entries be computed as dictated by the algorithm until they converge to zero naturally. It is not clear which of the two options leads to better behavior, since the algorithm is highly nonlinear. In fact, this question has not been given a satisfactory answer. The Mathematica procedure described in Sec. 8.10 does not try to force zeros, but the reader may wish to modify it so as to implement the zero forcing and experiment with both options.

The purpose of the ARMA lattice algorithm is to recursively estimate the ARMA parameters, but up to this point we have only shown how to compute the lattice variables, and these do not include the ARMA parameters. In order to obtain the ARMA parameters, let us define the auxiliary matrices

$$K_{n+1,t} = \Delta_{n+1,t}^T r_{n,t}^{-1}, \quad \tilde{K}_{n+1,t} = \Delta_{n+1,t} \tilde{r}_{n,t}^{-1}. \tag{8.8.8}$$

As in the one-channel case, the two-channel AR predictor coefficients can be computed from $\{K_{n+1,t}, \tilde{K}_{n+1,t}\}$ by a multichannel version of the Levinson-Durbin algorithm. The multichannel Levinson-Durbin algorithm was developed in Prob. 6.4, and its application to the adaptive two-channel AR lattice yields

$$\begin{bmatrix} \mathbf{1}_2 & A_{n+1,1,t} & \cdots & A_{n+1,n,t} & A_{n+1,n+1,t} \\ \tilde{A}_{n+1,n+1,t} & \tilde{A}_{n+1,n,t} & \cdots & \tilde{A}_{n+1,1,t} & \mathbf{1}_2 \end{bmatrix}$$
$$= \begin{bmatrix} \mathbf{1}_2 & -\tilde{K}_{n+1,t} \\ -K_{n+1,t} & \mathbf{1}_2 \end{bmatrix} \begin{bmatrix} \mathbf{1}_2 & A_{n,1,t} & \cdots & A_{n,n,t} & 0 \\ \tilde{0} & \tilde{A}_{n,n,t} & \cdots & \tilde{A}_{n,1,t} & \mathbf{1}_2 \end{bmatrix}. \tag{8.8.9}$$

Recalling (8.8.2), we see that the estimated ARMA parameters are given by

$$\hat{a}_{k,t} = (A_{p,k,t})_{1,1}, \quad 0 \le k \le p, \tag{8.8.10a}$$

$$\hat{b}_{k,t} = -(A_{p,k,t})_{1,2}, \quad 1 \le k \le p, \quad \hat{b}_{0,t} = 1. \tag{8.8.10b}$$

Equation (8.8.9) takes a number of operations proportional to p^2, but there is no need to compute it at each time point, since the rate of variation of the estimated parameters is rather slow. As in the AR case of Sec. 8.7, a conservative approach is to compute (8.8.9) and (8.8.10) once every p time points.

Example 8.6. 10,000 data points were generated from the ARMA$(2,2)$ process

$$y_t = 1.4y_{t-1} - 0.9y_{t-2} + u_t + 0.8u_{t-1} + 0.5u_{t-2}$$

and inputted to the ARMA lattice algorithm described above, using $\lambda = 1$. Figure 8.7 shows the estimated parameters as a function of time. The estimates are seen to converge to their true values, but the convergence is rather slow.   ∎

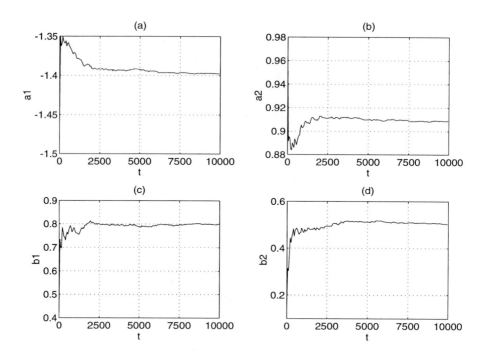

Figure 8.7. The estimates in Example 8.6 as a function of time: (a) $\hat{a}_1$; (b) $\hat{a}_2$; (c) $\hat{b}_1$; (d) $\hat{b}_2$.

We finally remark that Karlsson and Hayes [1987] have developed an alternative ARMA lattice algorithm, which does not rely on two-channel embedding. Their algorithm breaks each order update step to AR update followed by MA update. It is more general than the one presented here, since is does not require that $q = p$, but its derivation is rather lengthy, so it is not included here. Friedlander, Ljung, and Morf [1981] derived an ARMA/RML lattice algorithm.

8.9. EXTENSIONS OF LATTICE ALGORITHMS

This section includes several extensions to adaptive lattice algorithms. These extensions represent only a small fraction of the literature in this area and are not necessarily the most important ones. The truth is that, despite the huge literature on this subject, very little practical use has been made of lattice algorithms. As a result, it is difficult to judge, at the time this book is written, which lattice algorithms are better than others. Therefore, the contents of this section merely reflects the author's own prejudices.

THE FORWARD COVARIANCE LATTICE ALGORITHM

The forward covariance method for AR estimation was introduced in Sec. 6.6, see Eq. (6.6.1). A lattice algorithm for this method was derived by Porat, Friedlander, and Morf [1982]. We will not repeat the derivation given there, since it is somewhat tedious. Instead, we will capitalize on our investment in the multichannel prewindowed case and derive the covariance lattice as a special case of a two-channel prewindowed lattice. The covariance lattice is important only when $\lambda = 1$, since if $\lambda < 1$ the difference between the prewindowed and the covariance methods will vanish after about $2/(1 - \lambda)$ time points. We will therefore assume, for simplicity, that $\lambda = 1$, hence $W = \mathbf{1}$.

Let S be the $(t + 1) \times n$ matrix

$$S = \begin{bmatrix} 1 & 0 & \cdots & 0 \\ 0 & 1 & \cdots & 0 \\ \vdots & \vdots & \ddots & \vdots \\ 0 & 0 & \cdots & 1 \\ \vdots & \vdots & \vdots & \vdots \\ 0 & 0 & \cdots & 0 \end{bmatrix}.$$

The key observation is that augmenting the prewindowed matrix $\boldsymbol{Y}_{n,t-1}$ by S effectively "wipes out" its lower triangular part at the top, leaving us with the forward covariance data matrix depicted in (6.6.1). For the purpose of this section, let us denote the latter matrix by $\overline{\boldsymbol{Y}}_{n,t-1}$. Then we can prove, using the same method as in the proof of (8.6.10) (Prob. 8.17), that

$$P^{\perp}_{[\boldsymbol{Y}_{n,t-1}\ S]} = \begin{bmatrix} 0 & 0 \\ 0 & P^{\perp}_{\overline{\boldsymbol{Y}}_{n,t-1}} \end{bmatrix}. \tag{8.9.1}$$

Next observe that the column space of $[\boldsymbol{Y}_{n,t-1}\ S]$ is invariant to column permutation. Therefore, we can replace the projection operator of $[\boldsymbol{Y}_{n,t-1}\ S]$ by the projection operator of the matrix

$$\begin{bmatrix} 0 & 1 & 0 & 0 & \cdots & 0 & 0 \\ y_0 & 0 & 0 & 1 & \cdots & 0 & 0 \\ \vdots & \vdots & \vdots & \vdots & \ddots & \vdots & \vdots \\ y_{n-2} & 0 & y_{n-3} & 0 & \cdots & y_1 & 1 \\ y_{n-1} & 0 & y_{n-2} & 0 & \cdots & y_0 & 0 \\ \vdots & \vdots & \vdots & \vdots & \ddots & \vdots & \vdots \\ y_{t-1} & 0 & y_{t-2} & 0 & \cdots & y_{t-n} & 0 \end{bmatrix}.$$

This is precisely a two-channel prewindowed data matrix of the type encountered in the ARMA lattice of Sec. 8.8. The work is therefore done; all we need is to use the sequence $\{[y_t, 0]\}$ as the new data, prepending it with the "data point" $[0, 1]$ (corresponding to $t = -1$). Feeding this data sequence to the two-channel prewindowed lattice will yield the covariance method estimates. We should finally note that this two-channel algorithm is degenerate: the second component of the forward residual $e_{n,t}$ is identically zero (unlike the second component in the ARMA lattice, which only decays to zero asymptotically). Therefore, only two of the four entries of $\Delta_{n,t}$ are nonzero, a fact that can be used to eliminate some computations (Prob. 8.18).

WHITENING FILTER UPDATE FORMULAS

We already mentioned several times in the previous sections that the basic lattice algorithm does not yield the estimated predictor coefficients. The method proposed in Secs. 8.7 and 8.8 was to compute the predictor coefficients by carrying out the Levinson-Durbin recursions with the estimated partial correlations, see (8.8.9). It may be surprising to learn that this procedure *does not* yield the exact solution to the original least-squares problem from which the lattice algorithm was derived. This observation was made by Porat, Friedlander, and Morf [1982]. The alert reader may wish to reflect a little on why this is the case. We will not attempt here to give heuristic explanations. Instead, we simply derive the formulas for the *exact* solution of the least-squares problem in terms of the lattice variables. Once this derivation is complete, we will be able to compare the results with the procedure suggested above and interpret the difference. The derivation is done for the multichannel case, for the sake of generality. We limit ourselves to the growing-memory case in which $\lambda = 1$ and $W = \mathbf{1}$.

As preliminary material, we will derive order and time update formulas for certain matrices that appear in the least-squares solution. This material is very similar to (in fact relies on) the projection update formulas presented in Sec. 8.6, which the reader may want to review at this point.

Let H, X, Y, and Z be as in Sec. 8.6, and define

$$H^{\#} = H(H^T H)^{-1}. \tag{8.9.2}$$

This matrix satisfies $H^T H^{\#} = \mathbf{1}$, so it is in fact a right inverse of H^T. Let us consider the augmented matrix $[H\ Z]$, as in Sec. 8.6, and also the augmented matrix $[Z\ H]$. The following theorem provides expressions for the corresponding right inverses.

Theorem 8.5.

$$[H\ Z]^{\#} = [H^{\#},\ 0] + P_H^{\perp} Z (Z^T P_H^{\perp} Z)^{-1} [-Z^T H^{\#},\ \mathbf{1}], \tag{8.9.3a}$$

$$[Z\ H]^{\#} = [0,\ H^{\#}] + P_H^{\perp} Z (Z^T P_H^{\perp} Z)^{-1} [\mathbf{1},\ -Z^T H^{\#}]. \tag{8.9.3b}$$

Proof. We have, by the right inverse property,

$$\begin{bmatrix} H^T \\ Z^T \end{bmatrix} [H\ Z]^\# = \begin{bmatrix} \mathbf{1} & 0 \\ 0 & \mathbf{1} \end{bmatrix}. \tag{8.9.4}$$

Recall (8.6.5a) and multiply it on the right by $[H\ Z]^\#$ to get

$$P_{[H\ Z]}[H\ Z]^\# = P_H[H\ Z]^\# + P_H^\perp Z(Z^T P_H^\perp Z)^{-1} Z^T(\mathbf{1} - P_H^T)[H\ Z]^\#. \tag{8.9.5}$$

By (8.9.4), we have (Prob. 8.19)

$$P_{[H\ Z]}[H\ Z]^\# = [H\ Z]^\#, \qquad P_H[H\ Z]^\# = [H^\#, 0], \tag{8.9.6a}$$

$$Z^T[H\ Z]^\# = [0, \mathbf{1}], \quad Z^T P_H^T[H\ Z]^\# = [Z^T H^\#, 0]. \tag{8.9.6b}$$

Substitution of these four identities in (8.9.5) yields (8.9.3a). Formula (8.9.3b) is obtained in a dual manner.

Corollary.

$$X^T[H\ Z]^\# = [X^T H^\#, 0] + X^T P_H^\perp Z(Z^T P_H^\perp Z)^{-1}[-Z^T H^\#, \mathbf{1}], \tag{8.9.7a}$$

$$X^T[Z\ H]^\# = [0, X^T H^\#] + X^T P_H^\perp Z(Z^T P_H^\perp Z)^{-1}[\mathbf{1}, -Z^T H^\#]. \tag{8.9.7b}$$

∎

The time update formula for the right inverse is similarly obtained, as in the proof of (8.6.10) (Prob. 8.20).

Theorem 8.6.

$$X_{t+1}^T \begin{bmatrix} H_t^\# \\ 0 \end{bmatrix} = X_{t+1}^T H_{t+1}^\# - (X_{t+1}^T P_{H_{t+1}}^\perp \boldsymbol{p})(\boldsymbol{p}^T P_{H_{t+1}}^\perp \boldsymbol{p})^{-1} \boldsymbol{p}^T H_{t+1}^\#. \tag{8.9.8}$$

∎

We now define the following entities, which we term *whitening filters.*

$$A_{n,t} = [\mathbf{1}, -\boldsymbol{y}_t^T \boldsymbol{Y}_{n,t-1}^\#], \tag{8.9.9a}$$

$$B_{n,t-1} = [-\boldsymbol{y}_{t-n-1}^T \boldsymbol{Y}_{n,t-1}^\#, \mathbf{1}], \tag{8.9.9b}$$

$$C_{n,t} = -\boldsymbol{p}^T \boldsymbol{Y}_{n,t}^\#. \tag{8.9.9c}$$

The second part of $A_{n,t}$ is recognized as the solution of our original least-squares problem (actually the transpose thereof), so the entries of $A_{n,t}$ are the coefficients of the optimal predictor polynomial. The corresponding transfer function $A_{n,t}(z)$ represents the filter converting the input data to the partial innovation sequence, this being the reason for the name "whitening filter." Similarly, the entries of $B_{n,t-1}$ are the coefficients of the optimal backward predictor polynomial. The vector $C_{n,t}$ will be needed for the derivation, but it does not have a special interpretation.

We now use the update formulas of the right inverses to derive order and time updates for the whitening filters. By formulas (8.9.7a, b), we have

$$-\boldsymbol{y}_t^T \boldsymbol{Y}_{n+1,t-1}^{\#} = -[\boldsymbol{y}_t^T \boldsymbol{Y}_{n,t-1}^{\#}, \ 0] - \Delta_{n+1,t} \tilde{r}_{n,t-n-1}^{-1}[-\boldsymbol{y}_{t-n-1}^T \boldsymbol{Y}_{n,t-1}^{\#}, \ 1],$$

$$-\boldsymbol{y}_{t-n-1}^T \boldsymbol{Y}_{n+1,t}^{\#} = -[0, \ \boldsymbol{y}_{t-n-1}^T \boldsymbol{Y}_{n,t-1}^{\#}] - \Delta_{n+1,t}^T r_{n,t-n-1}^{-1}[1, \ -\boldsymbol{y}_t^T \boldsymbol{Y}_{n,t-1}^{\#}],$$

$$-\boldsymbol{p}^T \boldsymbol{Y}_{n+1,t}^{\#} = -[\boldsymbol{p}^T \boldsymbol{Y}_{n,t}^{\#}, \ 0] - \tilde{e}_{n,t-n} \tilde{r}_{n,t-n}^{-1}[-\boldsymbol{y}_{t-n}^T \boldsymbol{Y}_{n,t}^{\#}, \ 1].$$

Substitution in (8.9.9) then gives

$$A_{n+1,t} = [A_{n,t}, \ 0] - \Delta_{n+1,t} \tilde{r}_{n,t-n-1}^{-1}[0, \ B_{n,t-1}], \qquad (8.9.10)$$

$$B_{n+1,t} = [0, \ B_{n,t-1}] - \Delta_{n+1,t}^T r_{n,t-n-1}^{-1}[A_{n,t}, \ 0], \qquad (8.9.11)$$

$$C_{n+1,t} = [C_{n,t}, \ 0] - \tilde{e}_{n,t-n} \tilde{r}_{n,t-n}^{-1} B_{n,t}. \qquad (8.9.12)$$

We need one more relationship in order to step back from $B_{n,t}$ to $B_{n,t-1}$. We get, upon substituting $X = \boldsymbol{y}_{t-n}$ in (8.9.8),

$$\boldsymbol{y}_{t-n-1} \boldsymbol{Y}_{n,t-1}^{\#} = \boldsymbol{y}_{t-n} \boldsymbol{Y}_{n,t}^{\#} - \tilde{e}_{n,t-n} g_{n,t}^{-1} \boldsymbol{p}^T \boldsymbol{Y}_{n+1,t}^{\#},$$

which translates to

$$B_{n,t-1} = B_{n,t} - \tilde{e}_{n,t-n}^T g_{n,t}^{-1}[C_{n,t}, \ 0]. \qquad (8.9.13)$$

At this point it is instructive to compare the new update formulas with (8.8.9). As we see, (8.8.9) becomes identical to (8.9.10) and (8.9.11) upon identifying $B_{n,t-1}$ with $B_{n,t}$. In other words, (8.8.9) is an *approximation* of the exact relationships. These relationships, in turn, involve the additional equations (8.9.12) and (8.9.13), the auxiliary variable $C_{n,t}$, and the intermediate variable $B_{n,t-1}$. The whitening filter update formulas are local to time t. We start with

$$A_{0,t} = \mathbf{1}, \quad B_{0,t} = \mathbf{1}, \quad C_{0,t} = 0 \qquad (8.9.14)$$

and proceed to the whitening filters of maximum order $A_{p,t}, B_{p,t}$, and $C_{p,t}$. The update formulas depend on the lattice variables available after the tth time step. The entire computation takes a number of operations proportional to p^2. However, as we have said before, there is no need to carry it out at every time point.

We finally remark that the variables $\{C_{n,t}\}$ decay to zero as t increases (explain why). Therefore, the exact and the approximate filter formulas approach each other in the limit. From a practical point of view, there is probably no reason to prefer (8.9.10)–(8.9.13) to (8.8.9). Nevertheless, the theoretical difference between the two is worth being understood.

PURE TIME UPDATES OF THE WHITENING FILTERS

The whitening filter formulas derived above are suitable for cases where the whitening filters are needed relatively infrequently (say, once every p time points). In rare cases, we may wish to compute the pth-order filters at every data point. This is exactly what the classical RLS algorithm does, see Sec. 8.2. The update formulas derived above can be used to obtain pure time updates for the pth-order filters, as follows (see Porat [1983] for further discussion).

Suppose we have computed $A_{p,t-1}, B_{p,t-1}$, and $C_{p,t-1}$ at time $t-1$. From the time update formula (8.9.8), we get, upon substituting $X = \boldsymbol{y}_t$,

$$A_{p,t} = A_{p,t-1} - e_{p,t}^T g_{p,t-1}^{-1}[0, \, C_{p,t-1}]. \tag{8.9.15}$$

From the order update formula (8.9.7b), we get, upon substituting $X = \boldsymbol{p}$, $Z = \boldsymbol{y}_t$,

$$C_{p+1,t} = [0, \, C_{p,t-1}] - e_{p,t} r_{p,t}^{-1} A_{p,t}. \tag{8.9.16}$$

Finally, we get, upon rearrangement of (8.9.12) and (8.9.13),

$$\begin{bmatrix} B_{p,t} \\ [C_{p,t}, \, 0] \end{bmatrix} = \begin{bmatrix} 1 & -\tilde{e}_{p,t-p}^T g_{p,t}^{-1} \\ \tilde{e}_{p,t-p} \tilde{r}_{p,t-p}^{-1} & 1 \end{bmatrix}^{-1} \begin{bmatrix} B_{p,t-1} \\ [0, \, C_{p+1,t}] \end{bmatrix}. \tag{8.9.17}$$

This computation takes a number of operations proportional to p, so it is faster (for large enough p) than the RLS algorithm, while accomplishing the same task (the computation of the pth-order whitening filter at each time point). The algorithm can be initialized by carrying out the order update formulas (8.9.10)–(8.9.13) once, say at time $t = p$.

JOINT PROCESS ESTIMATION

Consider the model

$$x_t = \sum_{k=0}^{p-1} h_k y_{t-k} + u_t, \tag{8.9.18}$$

where $\{x_t, y_t\}$ are measured sequences, $\{u_t\}$ an unknown white noise sequence, and $\{h_k\}$ a set of unknown coefficients. The estimation of the $\{h_k\}$ is known as the *joint process* problem and is a classical problem in adaptive estimation (see, e.g., Widrow and Stearns [1985] for its numerous applications). The RLS and the stochastic gradient algorithms can handle this problem with trivial modifications; however, the lattice approach needs some work. We now describe the extension of the basic prewindowed algorithm to the case of joint process estimation.

The data matrix for the joint process problem is the same as for the AR problem, but the vector that needs to be projected on this matrix is now $\boldsymbol{x}_t = [x_0, \ldots, x_t]^T$. We already saw in Sec. 8.8 that the backward residual vectors form an orthogonal basis to the column space of the data matrix, so this basis can be used to express the required projection. We will need three more lattice variables and one more whitening filter in addition to the variables of the basic prewindowed lattice. The definitions of these variables are

$$\bar{e}_{n,t} = \boldsymbol{p}^T W P_{n,t}^\perp \boldsymbol{x}_t, \tag{8.9.19a}$$

$$\bar{r}_{n,t} = \boldsymbol{x}_t^T W P_{n,t}^\perp \boldsymbol{x}_t, \tag{8.9.19b}$$

$$\overline{\Delta}_{n+1,t} = \boldsymbol{x}_t^T W P_{n,t}^\perp \boldsymbol{y}_{t-n}, \tag{8.9.19c}$$

$$H_{n,t} = \boldsymbol{x}_t^T \boldsymbol{Y}_{n,t}^\# \quad \text{(only when } \lambda = 1\text{)}. \tag{8.9.19d}$$

Note that $H_{p,t}$ is the desired estimate at time t.

The time update formula for $\overline{\Delta}_{n+1,t}$ is

$$\overline{\Delta}_{n+1,t} = \lambda \overline{\Delta}_{n+1,t-1} + \bar{e}_{n,t}^T g_{n,t}^{-1} \tilde{e}_{n,t}. \tag{8.9.20}$$

The order update formulas for $\bar{e}_{n,t}$ and $\bar{r}_{n,t}$ are

$$\bar{e}_{n+1,t} = \bar{e}_{n,t} - \tilde{e}_{n,t}\tilde{r}_{n,t}^{-1}\overline{\Delta}_{n+1,t}^T, \tag{8.9.21}$$

$$\bar{r}_{n+1,t} = \bar{r}_{n,t} - \overline{\Delta}_{n+1,t}\tilde{r}_{n,t}^{-1}\overline{\Delta}_{n+1,t}^T. \tag{8.9.22}$$

These three equations should be added to the algorithm of Table 8.5 in the proper places.

The order update formula for $H_{n,t}$ is

$$H_{n+1,t} = [H_{n,t},\ 0] + \overline{\Delta}_{n+1,t}\tilde{r}_{n,t}^{-1}B_{n,t}. \tag{8.9.23}$$

The time update formula for $H_{p,t}$ is

$$H_{p,t} = H_{p,t-1} + \bar{e}_{n,t}^T g_{n,t}^{-1} C_{p,t}. \tag{8.9.24}$$

The reader is asked to provide the details of these formulas (Prob. 8.22).

SUMMARY

Although our discussion of lattice algorithms has been quite extensive, many topics were omitted. Judging by the amount of literature in this area, the subject of lattice filters (and fast algorithms in general) is apparently inexhaustible. Some notable omissions are:

- Normalized forms of the various extensions, as in (8.7.8)
- Extensions to forward-covariance forms (not via two-channel embedding), in growing-memory and sliding-memory variants
- Fast Transversal Filter forms (see, e.g., Carayannis, Manolakis, Kalouptsidis [1983]; Cioffi and Kailath [1984])
- Direct ARMA forms (see, e.g., Karlsson and Hayes [1987])
- Lattice forms based on QR decompositions of the data matrix (see, e.g., Cioffi [1990])
- Gradient lattice forms (see, e.g., Griffiths [1977])
- Close-to-Toeplitz matrices and their inversion formulas [Friedlander et al., 1979].

8.10 MATHEMATICA PACKAGES

The package `Rls.m` contains the recursive least-squares and the stochastic gradient algorithms, as follows.

Rls[data_, p_, q_, lambda_, rate_, opts___] implements the following three algorithms: the recursive least-squares for AR estimation, the extended least-squares for ARMA estimation, and the recursive maximum likelihood for ARMA estimation. The parameter data is the data sequence, p and q are the AR and MA orders, respectively. The special case $q = 0$ gives the AR/RLS algorithm. The parameter lambda is the forgetting factor. The parameter rate determines the rate at which the estimates are collected for output (e.g., every 10 data points when rate = 10). Options and defaults to the procedure are as follows. Method chooses the algorithms for the ARMA case; the default value is RML, and the alternative is ELS. ForceStab chooses whether to force a stable $b(z)$ at each iteration; the default value is False. The returned values are lists of ahat and bhat, according to the number of data points and the parameter rate.

RlsSqrt[data_, p_, q_, lambda_, rate_, opts___] is identical to Rls, except that it implements the P_t matrix update in square-root form.

StGrad[data_, p_, q_, mu_, rate_, opts___] implements the stochastic gradient algorithm for AR or ARMA estimation. The parameter data is the data sequence, p and q are the AR and MA orders, respectively. The special case $q = 0$ gives the AR algorithm. The parameter mu determines the step size: μ itself in the LMS version or μ/t in the Robbins-Monro version. The parameter rate determines the rate at which the estimates are collected for output. Options and defaults to the procedure are as follows. Method chooses the algorithm; the default value is LMS and the alternative is RM. ForceStab chooses whether to force a stable $b(z)$ at each iteration; the default value is False. The returned values are lists of ahat and bhat, according to the number of data points and the parameter rate.

The package Lattice.m contains the lattice algorithms, as follows.

LatticeAr[data_, p_, lambda_, rate_] implements the prewindowed AR unnormalized lattice algorithm. The parameter data is the data sequence, p is the order, lambda is the forgetting factor, and rate is as above. The returned values are the partial correlation coefficients K.

NorLatticeAr[data_, p_, lambda_, rate_] is similar to LatticeAr, except that it implements the prewindowed AR normalized lattice algorithm.

LatticeArma[data_, p_, lambda_, rate_] implements the prewindowed ARMA unnormalized lattice algorithm. The input parameters are as above. The returned values are the partial correlation coefficient matrices K and Kt, the latter denoting $\tilde{K}$.

KtoArma[{K_, Kt_}] computes the ARMA parameters from the partial correlation coefficient matrices K and Kt provided by the lattice ARMA algorithm. The returned values are ahat and bhat.

REFERENCES

Carayannis, G., Manolakis, D. G., and Kalouptsidis, N., "A Fast Sequential Algorithm for Least-squares Filtering and Prediction," *IEEE Trans. Acoustics, Speech, Signal Processing,* ASSP-31, pp. 1394–1402, 1983.

Cioffi, J. M., "The Fast Adaptive Rotor's RLS Algorithm", *IEEE Trans. Acoustics, Speech and Signal Processing,* Vol. ASSP-38, pp. 631–653, 1990.

Cioffi, J. M., and Kailath, T., "Fast, Recursive Least-squares Transversal Filters for Adaptive Filtering," *IEEE Trans. Acoustics, Speech, Signal Processing,* ASSP-32, pp. 304–337, 1984.

Feuer, A., and Weinstein, E., "Convergence Analysis of LMS Filters with Uncorrelated Gaussian Data," *IEEE Trans. Acoustics, Speech, Signal Processing,* ASSP-33, pp. 222–230, February 1985.

Fletcher, R., *Practical Methods of Optimization,* 2nd ed., Wiley, New York, 1987.

Friedlander, B., Ljung, L., and Morf, M., "Lattice Implementation of the Recursive Maximum Likelihood Algorithm," *Proc. 20th IEEE Conf. Decision and Control,* San Diego, CA, 1981.

Friedlander, B., Morf, M., Kailath, T., and Ljung, L., "New Inversion Formulas for Matrices Classified in Terms of Their Distance from Toeplitz Matrices", *Linear Algebra and Its Applications,* 27, pp. 31–60, 1979.

Golub, G. H., and Van Loan, C. F., *Matrix Computations,* Johns Hopkins University Press, Baltimore, MD, 1983.

Griffiths, L. J., "A Continuous Adaptive Filter Implemented as a Lattice Structure," *Proc. IEEE Int. Conf. Acoustics, Speech, Signal Processing,* Hartford, CT, pp. 683–686, 1977.

Haykin, S., *Adaptive Filter Theory,* Prentice Hall, Englewood Cliffs, NJ, 1991.

Karlsson, E., and Hayes, M. H., "Least-squares ARMA Modeling of Linear Time-varying Systems: Lattice Filter Structures and Fast RLS Algorithms," *IEEE Trans. Acoustics, Speech, Signal Processing,* ASSP-35, pp. 994–1014, 1987.

Lee, D. T. L., "Canonical Ladder Form Realization and Fast Estimation Algorithms,", Ph.D. dissertation, Stanford University, CA, 1980.

Lee, D. T. L., Friedlander, B., and Morf, M., "Recursive Ladder Algorithms for ARMA Modeling," *IEEE Trans. Automatic Control,* AC-27, pp. 753–764, 1982.

Lee, D. T. L., Morf, M., and Friedlander, B., "Recursive Least Squares Ladder Estimation Algorithms," *IEEE Trans. Acoustics, Speech, Signal Processing,* ASSP-29, pp. 627–641, 1981.

Ljung, L., and Söderström, T., *Theory and Practice of Recursive Identification,* MIT Press, Boston, 1983.

Porat, B., "Pure Time Updates for Least Squares Lattice Algorithms," *IEEE Trans. Automatic Control,* AC-28, pp. 865–866, 1983.

Porat, B., Friedlander, B., and Morf, M., "Square-root Covariance Ladder Algorithms," *IEEE Trans. Automatic Control,* AC-27, pp. 813–829, 1982.

Porat, B., Morf, M., and Morgan, D. R., "On the Relationship among Several Square-root Normalized Ladder Algorithms," *Proc. Conf. Information Systems and Sciences,* pp. 496–501, Johns Hopkins University, Baltimore, MD, 1981.

Robbins, H., and Monro, S., "A Stochastic Approximation Method," *Annals of Mathematical Statistics,* 22, pp. 400-407, 1951.

Söderström, T., and Stoica, P., *System Identification,* Prentice Hall, Englewood Cliffs, NJ, 1989.

Widrow, B., and Stearns, S. D., *Adaptive Signal Processing,* Prentice Hall, Englewood Cliffs, NJ, 1985.

Yule, G. U., "On the Theory of Correlation for Any Number of Variables, Treated by a New System of Notation," (1907), in *Statistical Papers of George Udny Yule,* pp. 85–96, Griffin, London, 1971.

PROBLEMS

8.1. Prove Eq. (8.2.9).

8.2. The following algorithm provides an alternative to the square-root RLS algorithm presented in Sec. 8.2. It updates the square root of P_t without using QR decomposition [Söderström and Stoica, 1989, p. 350].

$$f_t = \lambda^{-1/2} P_t^{T/2} \eta_t,$$
$$L_t = \lambda^{-1/2} P_t^{1/2} f_t,$$
$$\beta_t = 1 + f_t^T f_t$$
$$\alpha_t = \frac{1}{\beta_t + \sqrt{\beta_t}},$$

$$P_{t+1}^{1/2} = \lambda^{-1/2} P_t^{1/2} - \alpha_t L_t f_t^T,$$

$$K_t = \frac{L_t}{\beta_t}.$$

(a) Prove this algorithm by verifying that $P_{t+1}^{1/2} P_{t+1}^{T/2}$ satisfies (8.2.6).

(b) Compare the number of operations per time point in this algorithm with that of the algorithm given in Sec. 8.2.

(c) Implement the algorithm in a Mathematica procedure.

8.3. Time varying autoregressive processes.

(a) Implement a *time-varying* AR(2) model in a Mathematica procedure. The inputs are $\{a_{1,i}, a_{2,i}, a_{1,f}, a_{2,f}\}$, where the i-subscripted parameters are the initial values, and the f-subscripted ones are the final value. The AR parameters should vary linearly in time between the initial and the final values. Other inputs are the innovation variance and the number of data points. The process satisfies the time-varying difference equation

$$y_t = -a_1(t) y_{t-1} - a_2 y_{t-2} + u_t.$$

(b) Generate data sequences from the time-varying AR model, and feed them to the AR/RLS algorithm. Observe the tracking ability of the algorithm as a function of λ and the rate of change of the AR parameters. Summarize your results and conclusions.

8.4. Find $a(z)$ and $b(z)$ such that condition (8.3.5) is violated. Generate an ARMA data sequence with these $a(z)$ and $b(z)$, feed them to the ARMA/ELS algorithm, and verify that it does not converge to the right parameters. Then feed the same sequence to the ARMA/RML algorithm and see if it converges to the right values.

8.5. The purpose of the next two problems is to analyze the convergence of the LMS algorithm in a simple special case. This problem analyzes the convergence of the mean, and the next one the convergence of the mean-square error. The model for the two problems is

$$y_{t+1} = \eta_t^T \theta + u_{t+1},$$

where $\{\eta_t\}$ is a sequence of zero mean Gaussian m-dimensional vectors, satisfying

$$E\eta_t \eta_t^T = R, \quad E\eta_t \eta_s^T = 0, \quad t \neq s,$$

and $\{u_t\}$ is a zero mean i.i.d. sequence, with variance σ_u^2, independent of $\{\eta_t\}$. The measurements consist of the sequences $\{y_t\}$ and $\{\eta_t\}$. The LMS algorithm estimates θ recursively by

$$\hat{\theta}_{t+1} = \hat{\theta}_t + 2\mu\eta_t(y_{t+1} - \eta_t^T \hat{\theta}_t).$$

Let $R = Q\Lambda Q^T$ be the eigenvalue/eigenvector decomposition of R. Since Q is orthonormal, we can rewrite the model as

$$y_{t+1} = \eta_t^T Q Q^T \theta + u_{t+1}.$$

Thus we can replace η_t by $Q^T\eta_t$ and θ by $Q^T\theta$ and analyze the convergence properties of the algorithm in terms of the new variables. Let us therefore assume, without loss of generality, that $R = \Lambda$, a diagonal matrix with entries $\{\lambda_i, 1 \le i \le m\}$. Also, let $\tilde{\theta}_t = \hat{\theta}_t - \theta$ be the estimation error at time t.

(a) Derive the following difference equation for $\tilde{\theta}_t$:

$$\tilde{\theta}_{t+1} = [\mathbf{1}_m - 2\mu\eta_t\eta_t^T]\tilde{\theta}_t + 2\mu\eta_t u_{t+1}.$$

(b) Derive the following difference equation for the mean of the estimation error:

$$E\tilde{\theta}_{t+1} = (\mathbf{1}_m - 2\mu\Lambda)E\tilde{\theta}_t.$$

Hence prove that a necessary and sufficient condition for convergence of $E\tilde{\theta}_t$ is

$$\mu < \frac{1}{\lambda_{\max}}.$$

If this condition is satisfied, what will the mean converge to?

8.6. Let $V_t = E\tilde{\theta}_t\tilde{\theta}_t^T$, and let $\boldsymbol{v}_t$ be the vector of elements along the main diagonal of V_t. Clearly, V_t converges if and only if $\boldsymbol{v}_t$ converges.

(a) Prove that V_t satisfies the difference equation

$$V_{t+1} = E[(\mathbf{1}_m - 2\mu\eta_t\eta_t^T)V_t(\mathbf{1}_m - 2\mu\eta_t\eta_t^T)] + 4\mu\sigma_u^2\Lambda.$$

(b) The expectation on the right side of the above involves the fourth-order moments of η_t. Using (2.4.6), they can be expressed in terms of the second-order moments (since η_t is Gaussian). Carry out this computation and show that the diagonal elements of the above equation are given by

$$(V_{t+1})_{i,i} = (1 - 4\mu\lambda_i + 8\mu^2\lambda_i^2)(V_t)_{i,i} + 4\mu^2\sum_{k=1}^{m}\lambda_i\lambda_k(V_t)_{k,k} + 4\mu\sigma_u^2\lambda_i.$$

Hence conclude that

$$\boldsymbol{v}_{t+1} = B\boldsymbol{v}_t + 4\mu\sigma_u^2\boldsymbol{\lambda},$$

where $\boldsymbol{\lambda}$ is the vector of λ_i and B is the symmetric matrix

$$B = \text{diag}\{1 - 4\mu\lambda_i + 8\mu^2\lambda_i^2\} + 4\mu^2\boldsymbol{\lambda}\boldsymbol{\lambda}^T.$$

(c) Conclude that a necessary and sufficient condition for convergence of the mean-square error is that all eigenvalues of B be in the open interval $(-1, 1)$. If this condition is satisfied, what will $\boldsymbol{v}_t$ converge to?

Remark: It can be shown that the condition on the eigenvalues of B is equivalent to the two simultaneous conditions

$$\mu < \frac{1}{2\lambda_{\max}}, \quad \sum_{i=1}^{m}\frac{\mu\lambda_i}{1 - 2\mu\lambda_i} < 1.$$

The interested reader may wish to prove it, or see Feuer and Weinstein [1985]. From these conditions we can easily infer the *sufficient* condition

$$\mu < \frac{1}{(m+2)\lambda_{\max}}.$$

8.7. The purpose of this problem is to explore the stability and convergence rate of the LMS algorithm as a function of the eigenvalue spread of the covariance matrix of the input process.

(a) Find an expression for the ratio of the two eigenvalues of the 2×2 Toeplitz covariance matrix of an AR(2) process as a function of the parameters a_1, a_2. Hint: This is very easy to do using the symbolic algebraic capabilities of Mathematica (and the procedure `Rarma`), but can also be done by hand.

(b) Experiment with a_1, a_2 and find a way to control the ratio of the eigenvalues to a desired number. Hint: The closer the poles to each other, the greater the ratio.

(c) For each of the ratios 10, 100, and 1000, simulate the LMS algorithm and experiment with μ, as was done in Example 8.4. Find the maximum μ for which the algorithm is stable in each case. Give the approximate number of data points needed for convergence for each case. Write down your conclusions from this experiment.

8.8. The normalized stochastic gradient algorithm.

(a) Modify the procedure `StGrad` to implement the normalized version of the stochastic gradient algorithm.

(b) Repeat Prob. 8.7 with the normalized algorithm.

8.9. Show that the matrices whose square roots appear in (8.6.7) are indeed symmetric and positive definite.

8.10. Let U be a nonsingular matrix of dimensions $n \times n$. Prove that the projection on the column space of AU is identical to P_A.

8.11. Explain in detail how the time update formula (8.6.10) leads to (8.7.4a). Pay particular attention to the appearance of λ on the right side of this equation.

8.12. Show how the initial conditions in (8.7.5) are obtained from the definitions of the corresponding entities.

8.13. Rewrite the algorithm of Table 8.5 explicitly in terms of the necessary storage variables. Pay attention to the order of the computations and the substitutions to avoid premature overwriting of old values. Keep in mind the direction of update of each variable (order and/or time).

8.14. Show that the prewindowed AR lattice algorithm never exhibits singularity (i.e., division by zero does not occur) as long as $y_0 \neq 0$. Suggest a solution to the potential singularity problem when $y_0 = 0$.

8.15. In the proof of Lemma 8.1, explain why the vectors $\{y_{t-1}, P_{1,t-1}^{\perp} y_{t-2}, \dots, P_{p-1,t-1}^{\perp} y_{t-p}\}$ form an orthogonal basis to the column space of $Y_{p,t-1}$.

8.16. Explain why the partial correlations $\Delta_{n,t}$ in the ARMA lattice algorithm converge to a form $\begin{bmatrix} * & * \\ 0 & 0 \end{bmatrix}$.

8.17. Prove formula (8.9.1) using the method of proof of (8.6.10).

8.18. "Scalarize" the covariance lattice algorithm presented in Sec. 8.9 by working explicitly with the two scalar components of $\Delta_{n,t}$ and the two scalar components of $\tilde{e}_{n,t-n-1}$. Note that $e_{n,t}$ has only one nonzero component. If you need help, see Porat, Morf, and Morgan [1981].

8.19. Prove the four identities in (8.9.6) in detail, using (8.9.2), (8.6.1) and (8.9.4).

8.20. Prove formula (8.9.8).

8.21. Equations (8.9.10)–(8.9.13) can be used to construct a lattice-form realization of the whitening filter, similar to the Levinson lattice depicted in Figures 6.2 and 6.3. One way of obtaining this realization is to multiply the above equations by the column vector $[\mathbf{1}, z^{-1}\mathbf{1}, \ldots, z^{-n}\mathbf{1}, z^{-(n+1)}\mathbf{1}]^T$ and interpret the resulting equations as relations among transfer functions. Show that this leads to a three-input, three-output lattice filter, with three lines, one delay element and two butterfly interconnections per section. If you need help, see Porat, Friedlander, and Morf [1982].

8.22. Explain the substitutions leading to the update formulas of the joint process lattice algorithm, and specify the initial conditions for the three additional lattice variables.

8.23. Augment the Mathematica procedure `LatticeAr` to handle the joint process problem and test it.

CHAPTER 9

Estimation of Deterministic Processes

9.1. INTRODUCTION

In Sec. 2.12 we introduced the model (2.12.16) for stationary deterministic processes. Processes obeying this model are not very interesting from an estimation point of view, because their parameters can be computed *exactly* from a finite number of measurements. In fact, as we shall see in Sec. 9.4 below, $4M$ consecutive measurements are sufficient for exact computation of the parameters.

To make the problem of parameter estimation for deterministic processes both more interesting and more realistic, we will modify the model (2.12.16) by assuming that *the measurements contain additive noise*. The noise is assumed purely indeterministic and statistically independent of the deterministic process. Thus, we will be concerned with processes of the type

$$y_t = \sum_{m=1}^{M} A_m \cos(\omega_m t - \phi_m) + u_t. \tag{9.1.1}$$

Throughout this chapter, $\{u_t\}$ is assumed to be white Gaussian noise with zero mean and variance σ_u^2.

The model (9.1.1) can be interpreted in three different ways, leading to different statistical descriptions. The frequencies $\{\omega_m\}$ are assumed to be nonrandom unknown parameters in all three interpretations.

1. We can assume that the both the phases and the amplitudes are random. If the phases are uniform in $[0, 2\pi]$ and the amplitudes are Rayleigh, the process is Gaussian (see Prob. 2.11). In this case the process is not ergodic. When a single realization is given, only $\{A_m\}$ of that particular realization can be

estimated, but not $\{E(A_m^2)\}$. Since this interpretation leads nowhere from an estimation viewpoint, we shall not pursue it.

2. We can assume that the phases are random, while the amplitudes are non-random, but unknown parameters. If the phases are uniform in $[0, 2\pi]$, the process is wide-sense stationary but non-Gaussian. The probability distribution of the process in (9.1.1) is then intractable, so the Cramér-Rao lower bound is unknown, and maximum likelihood estimation is not possible. However, the method of moments can still be used for parameter estimation. This approach is pursued in Secs. 9.6 and 9.7 and in Chapter 10.

3. We can assume that both the amplitudes and the phases are nonrandom unknown parameters. The only random part in $\{y_t\}$ is then the noise $\{u_t\}$. Equation (9.1.1) then represents a *nonstationary* Gaussian process, where the nonstationarity is only in the mean. In this case the parameters to be estimated include the amplitudes, phases, and frequencies. The Cramér-Rao bound and maximum likelihood estimation are tractable and will be developed in Secs. 9.2 and 9.3. Other estimation methods based on this interpretation will be presented in Secs. 9.4 and 9.5.

The problem of estimating the parameters of sinusoids in noise is a classical one, dating back over 200 years. The statistical approach to the problem dates back to Fisher [1929]. Given this time span, we might assume that the problem has been satisfactorily solved long ago. Surprisingly, this is not so: estimation of sinusoidal waveforms in noise remains an active research area to date, and new estimation techniques continue to emerge. This chapter cannot possibly do justice to the rich literature on the problem. Instead, it tries to emphasize modern contributions and, in particular, the ones that appear most successful. The most prominent classical method—periodogram analysis with optional data windowing—was treated in Chapter 4 and will appear again in Sec. 9.3.

We should mention that many works on this subject deal with complex exponential waveforms instead of real sinusoids; that is, they use the model

$$y_t = \sum_{m=1}^{M} A_m e^{j(\omega_m t - \phi_m)} + u_t, \tag{9.1.2}$$

where $\{u_t\}$ is complex white Gaussian noise. This model appears to have been originated (or at least popularized) in the classical works of Rife and Boorstyn [1974, 1976]. The main reason for the use of a complex model is probably that it simplifies the algebraic derivations (e.g., summation of products of complex exponentials is easier than summation of products of sinusoids). Also, complex signals appear naturally in certain applications, such as radar or communication. Since real signals are probably more common, we will use them throughout this chapter. The reader should be able to extend the results to complex signals without difficulty (see the problems at the end of the chapter).

9.2. THE CRAMÉR-RAO BOUND FOR SINUSOIDS IN WHITE NOISE

In this section we derive the Cramér-Rao lower bound for unbiased estimates of the parameters of multiple sinusoids in Gaussian white noise under the non-random interpretation mentioned above. The parameter vector has dimension $3M + 1$ and consists of $\{\sigma_u^2, \boldsymbol{a}, \boldsymbol{\phi}, \boldsymbol{\omega}\}$, the noise variance and the vectors of amplitudes, phases, and radian frequencies. We will derive the exact information matrix first, then use it to derive an approximation to the CRB, and finally obtain the asymptotic form of the CRB.

The exact form of the information matrix is rather complicated, although there is no inherent difficulty in its derivation. It turns out that the expressions can be considerably simplified if we assume that the number of measurements N is odd, and they stretch from $-(N-1)/2$ to $(N-1)/2$. The results will apply to even N as well, but this needs to be derived separately and we wish to avoid the double work. The estimates of $\boldsymbol{a}$ and $\boldsymbol{\omega}$ are invariant to the range and depend only on the total number of measurements. The estimate of $\boldsymbol{\phi}$ is affected by this choice, since the initial phase is now relative to the middle of the interval, rather than to the beginning. However, it is easy to transform the result to get the bound on the initial phases using the general formula (3.3.3) (Prob. 9.1).

A general CRB formula for a signal with unknown parameters in Gaussian white noise was derived in Prob. 3.6. Using the result of that problem, we can write

$$J_N(\theta) = \frac{1}{\sigma_u^2} \begin{bmatrix} \frac{N}{2\sigma_u^2} & 0 & 0 & 0 \\ 0 & J_{aa} & J_{a\phi} & J_{a\omega} \\ 0 & J_{a\phi}^T & J_{\phi\phi} & J_{\phi\omega} \\ 0 & J_{a\omega}^T & J_{\phi\omega}^T & J_{\omega\omega} \end{bmatrix}. \tag{9.2.1}$$

The entries of the individual submatrices of (9.2.1) can be derived using standard trigonometric identities and summation formulas. We get, for $k \neq \ell$,

$$(J_{aa})_{k,\ell} = \sum_{t=-(N-1)/2}^{(N-1)/2} \cos(\omega_k t - \phi_k) \cos(\omega_\ell t - \phi_\ell)$$

$$= \frac{1}{2} \left[\frac{\cos(\phi_k + \phi_\ell) \sin[0.5N(\omega_k + \omega_\ell)]}{\sin[0.5(\omega_k + \omega_\ell)]} + \frac{\cos(\phi_k - \phi_\ell) \sin[0.5N(\omega_k - \omega_\ell)]}{\sin[0.5(\omega_k - \omega_\ell)]} \right], \tag{9.2.2}$$

$$(J_{a\phi})_{k,\ell} = A_\ell \sum_{t=-(N-1)/2}^{(N-1)/2} \cos(\omega_k t - \phi_k) \sin(\omega_\ell t - \phi_\ell)$$

$$= \frac{A_\ell}{2} \left[\frac{\sin(\phi_k - \phi_\ell) \sin[0.5N(\omega_k - \omega_\ell)]}{\sin[0.5(\omega_k - \omega_\ell)]} - \frac{\sin(\phi_k + \phi_\ell) \sin[0.5N(\omega_k + \omega_\ell)]}{\sin[0.5(\omega_k + \omega_\ell)]} \right], \tag{9.2.3}$$

$$(J_{a\omega})_{k,\ell} = -A_\ell \sum_{t=-(N-1)/2}^{(N-1)/2} t \cos(\omega_k t - \phi_k) \sin(\omega_\ell t - \phi_\ell)$$

$$= \frac{A_\ell}{2} \left[\frac{N \cos(\phi_k + \phi_\ell) \cos[0.5N(\omega_k + \omega_\ell)]}{2 \sin[0.5(\omega_k + \omega_\ell)]} - \frac{N \cos(\phi_k - \phi_\ell) \cos[0.5N(\omega_k - \omega_\ell)]}{2 \sin[0.5(\omega_k - \omega_\ell)]} \right.$$

$$- \frac{\cos(\phi_k + \phi_\ell) \sin[0.5N(\omega_k + \omega_\ell)] \cos[0.5(\omega_k + \omega_\ell)]}{2 \sin^2[0.5(\omega_k + \omega_\ell)]}$$

$$\left. + \frac{\cos(\phi_k - \phi_\ell) \sin[0.5N(\omega_k - \omega_\ell)] \cos[0.5(\omega_k - \omega_\ell)]}{2 \sin^2[0.5(\omega_k - \omega_\ell)]} \right], \qquad (9.2.4)$$

$$(J_{\phi\phi})_{k,\ell} = A_k A_\ell \sum_{t=-(N-1)/2}^{(N-1)/2} \sin(\omega_k t - \phi_k) \sin(\omega_\ell t - \phi_\ell)$$

$$= \frac{A_k A_\ell}{2} \left[\frac{\cos(\phi_k - \phi_\ell) \sin[0.5N(\omega_k - \omega_\ell)]}{\sin[0.5(\omega_k - \omega_\ell)]} - \frac{\cos(\phi_k + \phi_\ell) \sin[0.5N(\omega_k + \omega_\ell)]}{\sin[0.5(\omega_k + \omega_\ell)]} \right], \qquad (9.2.5)$$

$$(J_{\phi\omega})_{k,\ell} = -A_k A_\ell \sum_{t=-(N-1)/2}^{(N-1)/2} t \sin(\omega_k t - \phi_k) \sin(\omega_\ell t - \phi_\ell)$$

$$= \frac{A_k A_\ell}{2} \left[\frac{N \sin(\phi_k - \phi_\ell) \cos[0.5N(\omega_k - \omega_\ell)]}{2 \sin[0.5(\omega_k - \omega_\ell)]} - \frac{N \sin(\phi_k + \phi_\ell) \cos[0.5N(\omega_k + \omega_\ell)]}{2 \sin[0.5(\omega_k + \omega_\ell)]} \right.$$

$$- \frac{\sin(\phi_k - \phi_\ell) \sin[0.5N(\omega_k - \omega_\ell)] \cos[0.5(\omega_k - \omega_\ell)]}{2 \sin^2[0.5(\omega_k - \omega_\ell)]}$$

$$\left. + \frac{\sin(\phi_k + \phi_\ell) \sin[0.5N(\omega_k + \omega_\ell)] \cos[0.5(\omega_k + \omega_\ell)]}{2 \sin^2[0.5(\omega_k + \omega_\ell)]} \right], \qquad (9.2.6)$$

$$(J_{\omega\omega})_{k,\ell} = A_k A_\ell \sum_{t=-(N-1)/2}^{(N-1)/2} t^2 \sin(\omega_k t - \phi_k) \sin(\omega_\ell t - \phi_\ell)$$

$$= \frac{A_k A_\ell}{2} \left[\frac{N^2 \cos(\phi_k - \phi_\ell) \sin[0.5N(\omega_k - \omega_\ell)]}{4 \sin[0.5(\omega_k - \omega_\ell)]} - \frac{N^2 \cos(\phi_k + \phi_\ell) \sin[0.5N(\omega_k + \omega_\ell)]}{4 \sin[0.5(\omega_k + \omega_\ell)]} \right.$$

$$+ \frac{N \cos(\phi_k - \phi_\ell) \cos[0.5N(\omega_k - \omega_\ell)] \cos[0.5(\omega_k - \omega_\ell)]}{2 \sin^2[0.5(\omega_k - \omega_\ell)]}$$

$$- \frac{N \cos(\phi_k + \phi_\ell) \cos[0.5N(\omega_k + \omega_\ell)] \cos[0.5(\omega_k + \omega_\ell)]}{2 \sin^2[0.5(\omega_k + \omega_\ell)]}$$

$$- \frac{\cos(\phi_k - \phi_\ell) \sin[0.5N(\omega_k - \omega_\ell)]\{1 + \cos^2[0.5(\omega_k - \omega_\ell)]\}}{4 \sin^3[0.5(\omega_k - \omega_\ell)]}$$

$$\left. + \frac{\cos(\phi_k + \phi_\ell) \sin[0.5N(\omega_k + \omega_\ell)]\{1 + \cos^2[0.5(\omega_k + \omega_\ell)]\}}{4 \sin^3[0.5(\omega_k + \omega_\ell)]} \right]. \qquad (9.2.7)$$

For $k = \ell$, we get

$$(J_{aa})_{k,k} = \frac{1}{2}\left[N + \frac{\cos(2\phi_k)\sin(N\omega_k)}{\sin(\omega_k)}\right], \tag{9.2.8}$$

$$(J_{a\phi})_{k,k} = -\frac{A_k\sin(2\phi_k)\sin(N\omega_k)}{2\sin(\omega_k)}, \tag{9.2.9}$$

$$(J_{a\omega})_{k,k} = \frac{A_k}{2}\left[\frac{N\cos(2\phi_k)\cos(N\omega_k)}{2\sin(\omega_k)} - \frac{\cos(2\phi_k)\sin(N\omega_k)\cos(\omega_k)}{2\sin^2(\omega_k)}\right], \tag{9.2.10}$$

$$(J_{\phi\phi})_{k,k} = \frac{A_k^2}{2}\left[N - \frac{\cos(2\phi_k)\sin(N\omega_k)}{\sin(\omega_k)}\right], \tag{9.2.11}$$

$$(J_{\phi\omega})_{k,k} = \frac{A_k^2}{2}\left[-\frac{N\sin(2\phi_k)\cos(N\omega_k)}{2\sin(\omega_k)} + \frac{\sin(2\phi_k)\sin(N\omega_k)\cos(\omega_k)}{2\sin^2(\omega_k)}\right], \tag{9.2.12}$$

$$(J_{\omega\omega})_{k,k} = \frac{A_k^2}{2}\left[\frac{N(N^2-1)}{12} - \frac{N^2\cos(2\phi_k)\sin(N\omega_k)}{4\sin(\omega_k)}\right.$$
$$\left. - \frac{N\cos(2\phi_k)\cos(N\omega_k)\cos(\omega_k)}{2\sin^2(\omega_k)} + \frac{\cos(2\phi_k)\sin(N\omega_k)[1+\cos^2(\omega_k)]}{4\sin^3(\omega_k)}\right]. \tag{9.2.13}$$

Formulas (9.2.1) through (9.2.13) constitute a complete procedure for evaluation of the exact Fisher information matrix for any number of measurements. These formulas, however, are difficult to interpret, except by substituting numerical values and examining the results. Next we wish to assume that N is large and derive an approximation to the CRB under this assumption. The question is then: how large must N be in order to be able to approximate formulas (9.2.1) through (9.2.13)? The terms that need be examined are the various factors of the form $\sin(N\omega)/\sin(\omega)$. This function, known as the *Dirichlet kernel*, is equal to N when ω is an integer multiple of π, and its main lobe has width $\pm\pi/N$. A reasonable requirement is therefore that all ω appearing in the Dirichlet kernels be outside the main lobes centered at 0 and π. This happens when all the conditions

$$\frac{\pi}{N} < \omega_k, \quad 0.5|\omega_k - \omega_\ell|, \quad 0.5(\omega_k + \omega_\ell) < \pi\left(1 - \frac{1}{N}\right)$$

are satisfied simultaneously. Therefore, the approximate analysis is applicable when N satisfies

$$N > \max_{1 \le k < \ell \le M}\left\{\max\left\{\frac{\pi}{\omega_k}, \frac{\pi}{\pi - \omega_k}, \frac{2\pi}{|\omega_k - \omega_\ell|}, \frac{2\pi}{2\pi - \omega_k - \omega_\ell}\right\}\right\}. \tag{9.2.14}$$

When N satisfies this condition, we can proceed with the approximation as follows. Let D_N be the diagonal matrix

$$D_N = \operatorname{diag}\{N^{1/2}\mathbf{1}_{2M+1}, N^{3/2}\mathbf{1}_M\}.$$

Then we see from formulas (9.2.1) through (9.2.13) that

$$D_N^{-1} J_N(\theta) D_N^{-1} = 0.5\sigma_u^{-2} \begin{bmatrix} \sigma_u^{-2} & 0 & 0 & 0 \\ 0 & \mathbf{1}_M & 0 & 0 \\ 0 & 0 & \mathbf{A} & 0 \\ 0 & 0 & 0 & \frac{1}{12}\mathbf{A} \end{bmatrix} + 0.5\sigma_u^{-2}\mathbf{B},$$

where $\mathbf{A} = \text{diag}\{A_1^2, \ldots, A_M^2\}$ and $\mathbf{B}$ is $O(N^{-1})$. The entries of B can be read from (9.2.2)–(9.2.13). They include everything on the right side of these formulas except for the dominant terms along the main diagonal. Therefore,

$$D_N J_N^{-1}(\theta) D_N = 2\sigma_u^2 \begin{bmatrix} \sigma_u^2 & 0 & 0 & 0 \\ 0 & \mathbf{1}_M & 0 & 0 \\ 0 & 0 & \mathbf{A}^{-1} & 0 \\ 0 & 0 & 0 & 12\mathbf{A}^{-1} \end{bmatrix}$$

$$\cdot \left\{ \mathbf{1}_{3M+1} - \mathbf{B} \begin{bmatrix} \sigma_u^2 & 0 & 0 & 0 \\ 0 & \mathbf{1}_M & 0 & 0 \\ 0 & 0 & \mathbf{A}^{-1} & 0 \\ 0 & 0 & 0 & 12\mathbf{A}^{-1} \end{bmatrix} \right\} + o(N^{-1}). \qquad (9.2.15)$$

In particular, we get for the diagonal terms

$$\text{CRB}\{\sigma_u^2\} = \frac{2\sigma_u^4}{N}, \qquad (9.2.16a)$$

$$\text{CRB}\{A_k\} = \frac{2\sigma_u^2}{N} \left[1 - \frac{\cos(2\phi_k)\sin(N\omega_k)}{N\sin(\omega_k)} + o(N^{-1}) \right], \qquad (9.2.16b)$$

$$\text{CRB}\{\phi_k\} = \frac{2\sigma_u^2}{N A_k^2} \left[1 + \frac{\cos(2\phi_k)\sin(N\omega_k)}{N\sin(\omega_k)} + o(N^{-1}) \right], \qquad (9.2.16c)$$

$$\text{CRB}\{\omega_k\} = \frac{24\sigma_u^2}{N^3 A_k^2} \left[1 + \frac{3\cos(2\phi_k)\sin(N\omega_k)}{N\sin(\omega_k)} + o(N^{-1}) \right]. \qquad (9.2.16d)$$

As we see, the bounds on the parameters of the individual sinusoidal components are asymptotically decoupled from each other (see Prob. 9.1 in this regard). The bounds on the phase and frequency of a particular component are inversely proportional to the signal-to-noise ratio of that component. Note, in particular, the dependence of the asymptotic bound of the frequency on N^{-3}. This behavior is quite uncommon in time-series analysis: so far we have encountered only N^{-1} dependence. We should emphasize again that this behavior is asymptotic and applies only when N is large enough to satisfy (9.2.14).

Example 9.1. The exact CRB was computed for a process with three sinusoidal components, having amplitudes $\{1.0, 0.5, 0.25\}$, frequencies $\{0.1, 0.11, 0.12\}$, and phases $\{0, 0, 0\}$ and with noise variance $\sigma_u^2 = 1$. Figure 9.1 shows the bounds (standard deviations) for $100 \le N \le 10{,}000$ on a log-log scale. The threshold value of N, given by (9.2.14), is 630 in this case. It is seen that for values of N greater than the threshold the curves approach straight lines, as predicted by formulas (9.2.16). Below the threshold value of N, the bounds are much larger than the asymptotic approximations (9.2.16).   ∎

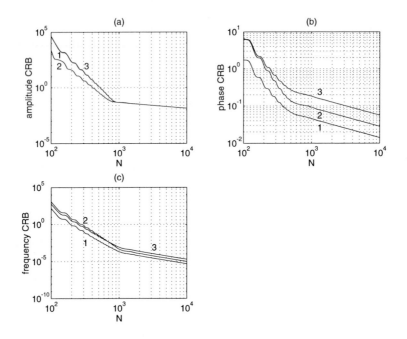

Figure 9.1. The Cramér-Rao bounds (standard deviations) as a function of N in Example 9.1: (a) the amplitudes; (b) the phases; (c) the frequencies.

9.3. MAXIMUM LIKELIHOOD ESTIMATION

In order to derive the maximum likelihood estimates of the parameters of the model (9.1.1), it is useful to make a preliminary variable transformation. Let

$$a_m = A_m \cos\phi_m, \quad b_m = A_m \sin\phi_m, \quad 1 \le m \le M. \tag{9.3.1}$$

Then (9.1.1) is equivalent to

$$y_t = \sum_{m=1}^{M} (a_m \cos\omega_m t + b_m \sin\omega_m t) + u_t. \tag{9.3.2}$$

The invariance property of the maximum likelihood estimate (Theorem 3.9) permits us to first find the maximum likelihood estimates of $\{\sigma_u^2, a_m, b_m, \omega_m\}$ and then obtain the estimates of the original parameters as

$$\hat{A}_m = \sqrt{\hat{a}_m^2 + \hat{b}_m^2} \quad ; \quad \hat{\phi}_m = \tan^{-1}\left(\frac{\hat{b}_m}{\hat{a}_m}\right). \tag{9.3.3}$$

Let us redefine $\boldsymbol{a}$ as the vector of $\{a_m, b_m\}$ and let

$$H(\boldsymbol{\omega}) = [H_c, H_s], \quad (H_c)_{t,k} = \cos\omega_k t, \quad (H_s)_{t,k} = \sin\omega_k t.$$

Then we can express (9.3.2) as

$$Y_N = H(\boldsymbol{\omega})\boldsymbol{a} + U_N. \tag{9.3.4}$$

Therefore, the negative of the log likelihood function is given by

$$-\frac{2}{N}\log f_\theta(Y_N) = \log(2\pi) + \log\sigma_u^2 + \frac{1}{N\sigma_u^2}[Y_N - H(\boldsymbol{\omega})\boldsymbol{a}]^T[Y_N - H(\boldsymbol{\omega})\boldsymbol{a}].$$

Let $\{\hat{\sigma}_u^2, \hat{\boldsymbol{a}}, \hat{\boldsymbol{\omega}}\}$ denote the maximum likelihood estimates of the parameters. Then the partial derivatives of the log likelihood with respect to σ_u^2 and $\boldsymbol{a}$ must be zero at the maximum point. Differentiating and equating the derivatives to zero gives

$$\hat{\boldsymbol{a}} = [H(\hat{\boldsymbol{\omega}})^T H(\hat{\boldsymbol{\omega}})]^{-1} H^T(\hat{\boldsymbol{\omega}}) Y_N, \tag{9.3.5}$$

$$\hat{\sigma}_u^2 = \frac{1}{N}[Y_N - H(\hat{\boldsymbol{\omega}})\hat{\boldsymbol{a}}]^T[Y_N - H(\hat{\boldsymbol{\omega}})\hat{\boldsymbol{a}}]. \tag{9.3.6}$$

Substitution of these values in the negative of the log likelihood gives

$$-\frac{2}{N}\log f_\theta(Y_N) = \log(2\pi) + \log\left\{\frac{1}{N}Y_N^T[\mathbf{1}_N - H(\hat{\boldsymbol{\omega}})(H(\hat{\boldsymbol{\omega}})^T H(\hat{\boldsymbol{\omega}}))^{-1}H^T(\hat{\boldsymbol{\omega}})]^T\right.$$

$$\left.[\mathbf{1}_N - H(\hat{\boldsymbol{\omega}})(H(\hat{\boldsymbol{\omega}})^T H(\hat{\boldsymbol{\omega}}))^{-1}H^T(\hat{\boldsymbol{\omega}})]Y_N\right\} + 1.$$

The matrix $\mathbf{1}_N - H(H^T H)^{-1}H^T$ is the projection operator on the orthogonal complement of the column space of H (cf. Sec. 8.6); therefore, it is symmetric and idempotent (i.e., is equal to its square). Hence,

$$-\frac{2}{N}\log f_\theta(Y_N) = \log(2\pi) + \log\left\{\frac{1}{N}Y_N^T[\mathbf{1}_N - H(\hat{\boldsymbol{\omega}})(H(\hat{\boldsymbol{\omega}})^T H(\hat{\boldsymbol{\omega}}))^{-1}H^T(\hat{\boldsymbol{\omega}})]Y_N\right\} + 1.$$

Minimization of this function is equivalent to maximization of the cost function

$$C(\boldsymbol{\omega}) = Y_N^T H(\boldsymbol{\omega})[H(\boldsymbol{\omega})^T H(\boldsymbol{\omega})]^{-1}H^T(\boldsymbol{\omega})Y_N. \tag{9.3.7}$$

In summary, the maximum likelihood estimate of $\boldsymbol{\omega}$ is obtained through nonlinear maximization of $C(\boldsymbol{\omega})$. Then the estimate $\hat{\boldsymbol{\omega}}$ is substituted in (9.3.5) to yield $\hat{\boldsymbol{a}}$ and finally in (9.3.6) to yield $\hat{\sigma}_u^2$.

As we see, the crucial step of the procedure is the nonlinear maximization, so we will discuss it in detail. The cost function can be written as

$$C(\boldsymbol{\omega}) = Y_N^T \begin{bmatrix} H_c & H_s \end{bmatrix} \begin{bmatrix} H_c^T H_c & H_c^T H_s \\ H_s^T H_c & H_s^T H_s \end{bmatrix}^{-1} \begin{bmatrix} H_c^T \\ H_s^T \end{bmatrix} Y_N, \tag{9.3.8}$$

where

$$(H_c^T H_c)_{k,\ell} = \begin{cases} \dfrac{\sin[0.5N(\omega_k - \omega_\ell)]}{2\sin[0.5(\omega_k - \omega_\ell)]} + \dfrac{\sin[0.5N(\omega_k + \omega_\ell)]}{2\sin[0.5(\omega_k + \omega_\ell)]}, & k \neq \ell \\[3mm] \dfrac{N}{2} + \dfrac{\sin(N\omega_k)}{2\sin(\omega_k)}, & k = \ell \end{cases} \qquad (9.3.9a)$$

$$(H_s^T H_s)_{k,\ell} = \begin{cases} \dfrac{\sin[0.5N(\omega_k - \omega_\ell)]}{2\sin[0.5(\omega_k - \omega_\ell)]} - \dfrac{\sin[0.5N(\omega_k + \omega_\ell)]}{2\sin[0.5(\omega_k + \omega_\ell)]}, & k \neq \ell \\[3mm] \dfrac{N}{2} - \dfrac{\sin(N\omega_k)}{2\sin(\omega_k)}, & k = \ell \end{cases} \qquad (9.3.9b)$$

$$(H_c^T H_s)_{k,\ell} = (H_s^T H_c)_{k,\ell} = 0. \qquad (9.3.9c)$$

Since $H_c^T H_s = 0$, formula (9.3.8) simplifies to

$$C(\boldsymbol{\omega}) = Y_N^T[H_c(H_c^T H_c)^{-1}H_c^T + H_s(H_s^T H_s)^{-1}H_s^T]Y_N. \qquad (9.3.10)$$

The exact maximum likelihood estimate of the frequencies is obtained by maximizing the nonlinear cost function (9.3.10). This maximization can be carried out numerically, as in the case of the ARMA algorithms of Chapter 7. However, the cost function typically displays multiple maxima at intervals of approximately $2\pi/N$ (Prob. 9.3). It is therefore imperative to supply the nonlinear minimization routine with good initial conditions differing from the frequencies of the global maximum by π/N or less. The algorithms described in the subsequent sections can be used for this purpose.

When N is large enough to satisfy (9.2.14), we have the approximation

$$H_c^T H_c \approx H_s^T H_s \approx \frac{N}{2}\mathbf{1}.$$

Then we can replace the cost function $C(\boldsymbol{\omega})$ by the approximation

$$\begin{aligned} \tilde{C}(\boldsymbol{\omega}) &= \frac{2}{N}Y_N^T[H_c H_c^T + H_s H_s^T]Y_N \\ &= \frac{2}{N}\sum_{m=1}^{M}\left[\left(\sum_{t=-(N-1)/2}^{(N-1)/2} y_t \cos\omega_m t\right)^2 + \left(\sum_{t=-(N-1)/2}^{(N-1)/2} y_t \sin\omega_m t\right)^2\right] \\ &= \frac{2}{N}\sum_{m=1}^{M}\left|\sum_{t=-(N-1)/2}^{(N-1)/2} y_t e^{-j\omega_m t}\right|^2. \end{aligned}$$

We now observe that the square magnitude of the sum on the right side does not change if the sum is multiplied by $\exp\{-j\omega_m(N-1)/2\}$, since this term has unit magnitude. Doing so yields

$$\tilde{C}(\boldsymbol{\omega}) = \frac{2}{N}\sum_{m=1}^{M}\left|\sum_{t=0}^{N-1} y_{t-(N-1)/2}e^{-j\omega_m t}\right|^2 = 2\sum_{m=1}^{M} I(\omega_m), \qquad (9.3.11)$$

where $I(\omega_m)$ is the periodogram of the measurements at the frequency ω_m. The interpretation of this result is this: when N is large enough, the maximum likelihood estimate of $\boldsymbol{\omega}$ is approximately the vector of M frequencies for which the corresponding sum of the periodogram values is maximized.

The approximate maximum likelihood is rarely used in the form (9.3.11) since, like (9.3.10), it requires multidimensional search. Instead, it is common to make further approximation, searching for the M *largest local maxima* of the periodogram $I(\omega)$. This method is likely to succeed only if three conditions are met:

- N satisfies (9.2.14).
- The square amplitude of the weakest sinusoid is sufficiently large relative to the average noise density.
- The weaker sinusoids are not masked by the side lobes of the Dirichlet kernels of the stronger sinusoids.

The latter condition can be alleviated by using a windowed periodogram. The window suppresses the side lobes, at the expense of widening the main lobe of each sinusoid. The art of using windows in periodogram analysis is well covered in the literature [Harris, 1978] and will not be pursued here.

The normalized log likelihood function converges almost surely to a function whose global maximum is at the true value of the parameter vector. This is the counterpart of Theorem 5.4 for ARMA processes, and its proof is left to the reader (Prob. 9.12). The asymptotic properties of the maximum likelihood estimates are the same as for Gaussian ARMA processes: it is strongly consistent, asymptotically efficient, and asymptotically normal. The proofs of these results are similar in principle to the ones used for the i.i.d. case (Chapter 3) and the ARMA case (Chapter 5). However, they are much more tedious, because the asymptotic dependence of the frequency estimation errors on N is different from that of the amplitude and the phase errors. For the former it is $O(N^{-3/2})$, while for the latter it is $O(N^{-1/2})$. This necessitates caution in expanding the log likelihood in Taylor series around θ, using different orders of magnitudes for the components of $\psi - \theta$. We will not give the details of the proofs, but see Rao and Zhao [1993].

Example 9.2. The exact maximum likelihood algorithm was tested with the two-sinusoid signal

$$y_t = \cos 2t + 0.5 \cos 2.1t + u_t, \quad \sigma_u = 0.1.$$

The number of data points was 64. This number is equal to the threshold value given by (9.2.14). Table 9.1 shows the results of 100 Monte-Carlo simulations. It is seen that the estimates are practically unbiased. The empirical standard deviations are slightly larger than the corresponding CRBs. ∎

Parameter	True value	Empirical bias	Empirical std. dev.	CRB
a_1	1.0	-0.00040	0.03945	0.02681
a_2	0.5	0.00252	0.04070	0.02763
ϕ_1	0.0	-0.00045	0.02534	0.01779
ϕ_2	0.0	0.00376	0.05061	0.03556
ω_1	2.0	-0.00009	0.00327	0.00224
ω_2	2.1	-0.00004	0.00632	0.00432

Table 9.1. Results of Monte-Carlo simulation of Example 9.2.

9.4. THE PRONY METHOD

The method of Prony [1795] is based on the observation that deterministic processes of the type (2.12.15) are perfectly predictable not only from an infinite number of past values (this follows from Wold's theorem), but from *a finite number* of past values. The method is facilitated by the following theorem.

Theorem 9.1. Let

$$y_t = \sum_{m=1}^{M} A_m \cos(\omega_m t - \phi_m)$$

and define the polynomial $c(z)$ by

$$c(z) = \prod_{m=1}^{M} (1 - 2\cos\omega_m z^{-1} + z^{-2}) = \sum_{k=0}^{2M} c_k z^{-k}, \quad c_0 = 1. \tag{9.4.1}$$

Then

$$y_t = -\sum_{k=1}^{2M} c_k y_{t-k}, \quad \forall t. \tag{9.4.2}$$

Proof. Let us first restrict ourselves to time points $t \geq 0$. Then we can express y_t by the one-sided z transform

$$Y(z) = \sum_{t=0}^{\infty} y_t z^{-t} = \sum_{m=1}^{M} \frac{A_m [\cos\phi_m - \cos(\omega_m + \phi_m) z^{-1}]}{1 - 2\cos\omega_m z^{-1} + z^{-2}},$$

where the sum converges absolutely for all $|z| > 1$. Multiplying both sides by $c(z)$ gives

$$c(z)Y(z) = d(z),$$

where $d(z)$ contains powers of z^{-1} in the range $[0, 2M - 1]$ only, the rest being identically zero. This means that the same is true for $c(z)Y(z)$, so

$$\sum_{k=0}^{2M} c_k y_{t-k} = 0, \quad t \geq 2M.$$

This proves (9.4.2) for $t \geq 2M$. However, the one-sided z-transform can be started at any initial point t_0 (positive or negative), not just at $t_0 = 0$. Therefore, (9.4.2) holds for all $t \geq 2M + t_0$ and for all t_0, hence in fact for all t. ∎

As we have said, y_t is deterministic; therefore, it is perfectly predictable from its past samples. Equations (9.4.1) and (9.4.2) reaffirm this property and show that the predictor has a finite order $(2M)$ and provide an explicit expression for it.

Prony's method uses (9.4.2) by estimating $c(z)$ as the solution of the set of equations

$$\begin{bmatrix} y_{2M} \\ y_{2M+1} \\ \vdots \\ y_{N-1} \end{bmatrix} + \begin{bmatrix} y_{2M-1} & y_{2M-2} & \cdots & y_0 \\ y_{2M} & y_{2M-1} & \cdots & y_1 \\ \vdots & \vdots & & \vdots \\ y_{N-2} & y_{N-3} & \cdots & y_{N-2M-1} \end{bmatrix} \begin{bmatrix} c_1 \\ c_2 \\ \vdots \\ c_{2M} \end{bmatrix} = 0. \quad (9.4.3)$$

Examination of this equation shows that Prony's method is identical to the forward covariance method for AR estimation (6.6.1). Prony's suggestion was to use (9.4.3) with $N = 4M$, for which the equation is exactly determined. In the noise-free case, the solution yields the exact polynomial $c(z)$. Taking $N > 4M$ causes this set to be overdetermined, and then the solution is obtained in the least-squares sense, as in (6.6.4). Furthermore, due to the symmetry of the polynomial $c(z)$ (Prob. 9.5), we can replace (9.4.3) by a backward Prony method or by a forward-backward method, as in (6.6.5).

Having obtained the estimate $\hat{c}(z)$, the frequencies $\{\hat{\omega}_m\}$ are computed by finding the roots of the polynomial. By (9.4.2), the roots should all be on the unit circle at conjugate pairs $\{e^{\pm j\omega_m}\}$. Therefore, the frequencies are simply the phase angles of the roots. The amplitudes and the phase angles are computed as in the maximum likelihood method; see Eqs. (9.3.5) and (9.3.6).

Prony's method and its variants are based on the assumption that y_t is free of additive noise, in which case the parameter estimates are exact. When noise is present, the accuracy of these methods deteriorates rapidly. Prony's original method (with $N = 4M$) is extremely sensitive to noise and is almost never used in practice. The overdetermined variants are better behaved in this sense, but are known to be biased, and the bias typically increases with N. The forward-backward method is best in this sense. When noise is present, the roots of $\hat{c}(z)$ are no longer guaranteed to be on the unit circle. However, unless the noise is very high, there will still be M pairs of complex conjugate roots. As before, the preferred method of estimating the frequencies is by taking the phase angles of the roots, while ignoring their magnitudes.

9.5. THE TRUNCATED SINGULAR VALUE DECOMPOSITION METHOD

Our aim in this section is to improve the basic Prony method as much as possible, while staying within the linear prediction framework. To this end, it will be helpful to interpret the method in the linear prediction terminology. The filter $c(z)$ can be viewed as the whitening (or innovation) filter for the input signal. As such, it minimizes the variance of the output subject to the constraint $c_0 = 1$. When there is no noise, the filter $c(z)$ given in (9.4.1) accomplishes this task in an ideal manner: it completely eliminates the input signal and yields zero output. Indeed, the innovation of a deterministic process is zero, so this is precisely what we expect.

When noise is added to the signal, the filter $c(z)$ in (9.4.1) is no longer optimal in the sense of minimizing the prediction error variance. The reason is that this filter has a *notch* characteristic: it sharply attenuates the energy in the vicinity of the frequencies $\{\omega_m\}$, but does very little to attenuate the energy at other frequencies. Since the noise has uniform energy density over the entire bandwidth, it will not be much affected by the filter $c(z)$. As we know from Chapter 6, the solution to the least-squares problem (9.4.3) minimizes the total output error energy. Therefore, when the signal-to-noise ratio is low, it gives more attention to the noise and less to the sinusoidal signal. The resulting estimate $\hat{c}(z)$ is therefore biased, and the bias increases as the SNR decreases.

The above interpretation also hints at a possible solution: increase the order of $c(z)$ by a sufficient amount so that some of its zeros will attenuate the sinusoidal components, while the rest will attenuate the noise. If our intuition is correct, then $2M$ zeros of the increased-order polynomial will have phase angles in the vicinity of the frequencies $\{\omega_m\}$, while the other will be approximately uniformly distributed over the range $[-\pi, \pi]$. As long as the SNR is not too low, we may hope to find the $2M$ zeros of interest among the roots of the increased-order polynomial. Indeed, the idea of increasing the order of the polynomial $c(z)$ in the overdetermined Prony method has been proposed by several authors and proved to be quite successful. A method for selecting the desired zeros among the roots of the estimated polynomial will be discussed later in this section.

The truncated singular value decomposition method, which is the topic of this section, exploits the above idea in a sophisticated manner. The method was proposed by Tufts and Kumaresan [1982] and is commonly referred to as the *Tufts-Kumaresan method*. It is generally considered to be the best among the methods based on the Prony (or linear prediction) approach.

In order to describe the Tufts-Kumaresan method, we will need the following notation. Let $\boldsymbol{Y}$ be a Toeplitz matrix as depicted in (9.4.3), but with L columns, where $2M \leq L \leq N - 2M$. Let $\boldsymbol{y}$ be the vector $[y_L, \ldots, y_{N-1}]^T$, and let $\overline{\boldsymbol{Y}} = [\boldsymbol{y}, \boldsymbol{Y}]$. We will continue to use the letter c for the polynomial of degree $2M$ and its coefficients, and introduce the letter p for any polynomial of degree L and its coefficients. In particular, $\boldsymbol{c} = [c_1, \ldots, c_{2M}]^T$, $\overline{\boldsymbol{c}} = [1, \boldsymbol{c}^T]^T$, $\boldsymbol{p} = [p_1, \ldots, p_L]^T$, and

$\bar{\boldsymbol{p}} = [1, \boldsymbol{p}^T]^T$. The notation $\boldsymbol{C}$ will be used for the $L \times (L - 2M)$ Toeplitz matrix

$$C_{k,\ell} = \begin{cases} c_{k-\ell}, & 0 \le k - \ell \le 2M \\ 0, & \text{otherwise.} \end{cases}$$

The notation $\overline{\boldsymbol{C}}$ will be used for a matrix similar to $\boldsymbol{C}$, but of dimensions $(L+1) \times (L+1-2M)$.

We first examine the noise-free case. When $L = 2M$, the matrix $\boldsymbol{Y}$ is full rank. In this case the solution to the least-squares problem $\boldsymbol{y} + \boldsymbol{Y}\boldsymbol{c} = \overline{\boldsymbol{Y}}\,\bar{\boldsymbol{c}} = 0$ is unique, and it yields the polynomial $c(z)$. When $L > 2M$, the rank of $\boldsymbol{Y}$ is still $2M$. In fact, it follows from (9.4.2) that any column of this matrix beyond the first $2M$ ones is a linear combination of the $2M$ preceding columns. Therefore, all the columns of $\boldsymbol{Y}$ are linear combinations of the first $2M$ ones. This means that, for $L > 2M$, the equation $\boldsymbol{y} + \boldsymbol{Y}\boldsymbol{p} = \overline{\boldsymbol{Y}}\,\bar{\boldsymbol{p}} = 0$ will have an infinite number of solutions. The following theorem characterizes these solutions.

Theorem 9.2. Any solution of $\overline{\boldsymbol{Y}}\,\bar{\boldsymbol{p}} = 0$ for $L > 2M$ is of the form

$$\bar{\boldsymbol{p}} = \overline{\boldsymbol{C}}\,\bar{\boldsymbol{d}}, \tag{9.5.1}$$

where $\bar{\boldsymbol{d}}$ is an $(L+1-2M)$-dimensional vector with 1 in the first position.

Proof. It is clear from (9.4.2) and from the definition of $\overline{\boldsymbol{C}}$ that

$$\overline{\boldsymbol{Y}}\,\overline{\boldsymbol{C}} = 0. \tag{9.5.2}$$

The rank of $\overline{\boldsymbol{Y}}$ is $2M$ and its rows are $(L+1)$-dimensional vectors. Therefore, the null space of the rows of $\overline{\boldsymbol{Y}}$ has dimension $L+1-2M$. The matrix $\overline{\boldsymbol{C}}$ has $L+1-2M$ columns and is clearly full rank. It follows from (9.5.2) that the columns of $\overline{\boldsymbol{C}}$ form a basis to the null space of the rows of $\overline{\boldsymbol{Y}}$. Any solution to $\overline{\boldsymbol{Y}}\,\bar{\boldsymbol{p}} = 0$ must lie in this null space, so it can be expressed as a linear combination of the columns of $\overline{\boldsymbol{C}}$. Therefore, there exists a vector $\bar{\boldsymbol{d}}$ such that (9.5.1) holds. The fact that the element in the first position of $\bar{\boldsymbol{d}}$ is 1 follows immediately upon equating the elements in the first position of the two sides of (9.5.1).

Corollary. The polynomial $p(z)$ corresponding to any solution of the least-squares problem with $L > 2M$ is divisible by $c(z)$.

Proof. This follows directly from (9.5.1) upon writing it in the polynomial form $c(z)d(z) = p(z)$. ∎

Theorem 9.2 guarantees that, in the noise-free case, any increased-order solution of the linear prediction equations will include the $2M$ desired roots among its L roots. The additional roots, those of $d(z)$, can be located anywhere in the complex plane. However, there is a special case in which the roots of $d(z)$ are guaranteed to be inside the unit circle. This happens when $\boldsymbol{p}$ is taken to be the *minimum-norm* or the *pseudoinverse* solution of $\boldsymbol{y} + \boldsymbol{Y}\boldsymbol{p} = 0$. Since the concept of pseudoinverse is crucial to the understanding of the Tufts-Kumaresan method, we will now give a brief summary of it.

Any real matrix $\boldsymbol{Y}$ admits a *singular value decomposition* (SVD), given by

$$\boldsymbol{Y} = \boldsymbol{U}\overline{\Sigma}\boldsymbol{V}^T = [\boldsymbol{U}_1 \quad \boldsymbol{U}_2] \begin{bmatrix} \Sigma & 0 \\ 0 & 0 \end{bmatrix} [\boldsymbol{V}_1^T \quad \boldsymbol{V}_2^T]$$

(any complex matrix admits a similar decomposition, with Hermitian transposes instead of regular transposes everywhere). The matrices $\boldsymbol{U}$ and $\boldsymbol{V}$ are orthonormal. The columns of $\boldsymbol{U}$ are called the *left singular vectors*, and the columns of $\boldsymbol{V}$ the *right singular vectors*. The matrix Σ is square and diagonal, and its dimension is equal to the rank of $\boldsymbol{Y}$. The elements on the diagonal of Σ are called the *singular values* of $\boldsymbol{Y}$. They are strictly positive and arranged in a descending order of magnitude. Note that $\boldsymbol{Y}$ can be expressed as $\boldsymbol{Y} = \boldsymbol{U}_1\Sigma\boldsymbol{V}_1^T$, so the number of left and right singular vectors needed to express $\boldsymbol{Y}$ is determined by the rank of this matrix, not by its dimensions.

Consider now the equation $\boldsymbol{Y}\boldsymbol{p} + \boldsymbol{y} = 0$, and assume that it has at least one solution, so there exists $\boldsymbol{p}$ such that $\boldsymbol{U}_1\Sigma\boldsymbol{V}_1^T\boldsymbol{p} + \boldsymbol{y} = 0$. Multiplying by $\boldsymbol{U}_1^T$ and using the orthonormality of $\boldsymbol{U}$ gives $\Sigma\boldsymbol{V}_1^T\boldsymbol{p} + \boldsymbol{U}_1^T\boldsymbol{y} = 0$. Since Σ is square and nonsingular, this gives $\boldsymbol{V}_1^T\boldsymbol{p} = -\Sigma^{-1}\boldsymbol{U}_1^T\boldsymbol{y}$. This imposes a constraint on $\boldsymbol{V}_1^T\boldsymbol{p}$, while allowing $\boldsymbol{V}_2^T\boldsymbol{p}$ to be arbitrary, say $\boldsymbol{V}_2^T\boldsymbol{p} = -\boldsymbol{x}$. Then we get (Prob. 9.7)

$$\boldsymbol{p} = -\boldsymbol{V}_1\Sigma^{-1}\boldsymbol{U}_1^T\boldsymbol{y} - \boldsymbol{V}_2\boldsymbol{x}, \tag{9.5.3}$$

as the family of all solutions to $\boldsymbol{Y}\boldsymbol{p} + \boldsymbol{y} = 0$. The square norm of $\boldsymbol{p}$ is obtained as (since $\boldsymbol{V}_1^T\boldsymbol{V}_2 = 0$)

$$\boldsymbol{p}^T\boldsymbol{p} = \boldsymbol{y}^T\boldsymbol{U}_1\Sigma^{-2}\boldsymbol{U}_1^T\boldsymbol{y} + \boldsymbol{x}^T\boldsymbol{x}.$$

In particular, the special choice $\boldsymbol{x} = 0$ gives a solution $\boldsymbol{p}$ of the least possible norm. This solution, given by

$$\boldsymbol{p} = -\boldsymbol{V}_1\Sigma^{-1}\boldsymbol{U}_1^T\boldsymbol{y},$$

is called the *minimum-norm* solution of the given equation. The matrix $\boldsymbol{V}_1\Sigma^{-1}\boldsymbol{U}_1^T$ is called the *pseudoinverse* of $\boldsymbol{Y}$. The interested reader can find detailed treatments of least-squares problems and pseudoinverses in many books; (e.g., Lawson and Hanson [1974]). The above discussion is sufficient for our purposes.

Returning to the least-squares problem for sinusoidal signals, we now prove the following theorem, due to Kumaresan [1982].

Theorem 9.3. If $\boldsymbol{p}$ is taken as the minimum-norm solution of $\boldsymbol{y} + \boldsymbol{Y}\boldsymbol{p} = 0$, the corresponding $d(z)$ has all its roots inside the unit circle.

Proof. Let us first show that the minimum-norm solution satisfies $\boldsymbol{C}^T\boldsymbol{p} = 0$. We have, similarly to (9.5.2), $\boldsymbol{Y}\boldsymbol{C} = 0$ for all $L > 2M$. Expressing $\boldsymbol{Y}$ in terms of its SVD gives $\boldsymbol{U}_1\Sigma\boldsymbol{V}_1^T\boldsymbol{C} = 0$. Premultiplying by $\boldsymbol{U}_1^T$ and then by Σ^{-1} gives $\boldsymbol{V}_1^T\boldsymbol{C} = 0$. Therefore,

$$\boldsymbol{C}^T\boldsymbol{p} = -\boldsymbol{C}^T\boldsymbol{V}_1\Sigma^{-1}\boldsymbol{U}_1^T\boldsymbol{y} = 0.$$

Next we have by (9.5.1)

$$C^T \overline{C} \overline{d} = C^T \overline{p} = 0.$$

So

$$\overline{C}^T \overline{C} \overline{d} = \begin{bmatrix} s \\ 0 \end{bmatrix}, \tag{9.5.4}$$

where s is nonzero scalar. The matrix $\overline{C}^T \overline{C}$ is positive definite Toeplitz (cf. Theorem 6.1), so (9.5.4) has the form of the Yule-Walker equation. Furthermore, the scalar s is just the square norm of p, so it is strictly positive. By Theorem 2.15, the polynomial $d(z)$ corresponding to the solution d of (9.5.3) is stable; that is, its roots are strictly inside the unit circle.   ∎

When the signal contains noise, the low-rank property of Y is lost. Instead, this matrix will have full rank with probability 1 for any L. If the SNR is relatively high, the matrix Σ will contain $2M$ singular values of relatively high magnitudes, while the rest, whose presence is due to the noise, will have relatively low magnitudes. Since p is affected by the inverses of the singular values [cf. (9.5.3)], the noise singular values will tend to destabilize the solution. The idea of Tufts and Kumaresan was to replace the noise singular values by zeros and then use the minimum-norm solution for the modified matrix. The Tufts-Kumaresan estimate is thus:

1. Compute the SVD of the data matrix Y.
2. Retain the first $2M$ singular values and force the rest to zero. Let Σ be the matrix of first $2M$ singular values and U_1, V_1 the corresponding matrices of left and right singular vectors. Then compute p as

$$p = -V_1 \Sigma^{-1} U_1^T y. \tag{9.5.5}$$

The corresponding polynomial $p(z)$ is still of degree L, but its coefficients are much "cleaner" than the coefficients of the polynomial obtained from the standard covariance method, which does not employ SVD truncation. The question is then how to extract the $2M$ "true" roots (the ones corresponding to the frequencies of the sinusoids) from the L roots of $p(z)$. By Theorem 9.3, the additional roots are inside the unit circle if there is no noise. Tufts and Kumaresan argued that, if the noise is not too high, it is very likely that the true roots will be close to the unit circle, while the noise roots will be closer to the origin. The next steps of the method are therefore:

3. Compute the roots of $p(z)$ and select the subset of $2M$ roots whose magnitudes are closest to 1. The phase angles of these roots are the estimated frequencies.
4. Compute the amplitudes and the phases as in the maximum likelihood and Prony methods.

The truncated SVD method, despite being based on heuristic arguments, has proved very successful. It does not achieve the accuracy of maximum likelihood, but it is often close to it. Even more important, it does not require initial conditions or nonlinear optimization and was experimentally found to be very robust even at relatively low signal-to-noise ratios. Of course, if we are interested in achieving the ultimate performance, it is advisable to use the truncated SVD estimates as initial conditions to the exact maximum likelihood. Since these two estimates are likely to be close to each other in the first place, we may hope to achieve fast convergence of the nonlinear maximum likelihood algorithm.

The root selection procedure suggested in step 3 above may not be the best if the SNR is relatively low. Two alternative approaches have been proposed, the first of which is as follows. Do not discard any of the roots of $p(z)$, but use all of them to estimate $L/2$ frequencies (assume that L is even and all the roots are complex; otherwise, discard only the real roots). After estimating the amplitudes, retain only the M sinusoids whose amplitudes are the largest and discard the rest. This approach is based on the premise that the noise energy is approximately evenly distributed among the "noise frequencies," so none of the individual noise components will be larger than the smallest signal component. Of course, when the SNR becomes sufficiently low, this assumption will be violated eventually.

A second alternative to step 3 involves the use of both the forward and the backward covariance methods with truncated SVD. Although we have not shown it, the backward covariance method satisfies Theorems 9.2 and 9.3. The proposed procedure is then as follows [Porat and Friedlander, 1986].

1. Carry out the Tufts-Kumaresan estimation procedure twice, once using the forward method and once the backward method. Let $\{\mu_\ell\}$ be the set of roots obtained from the first part and $\{\nu_\ell\}$ the set of roots obtained from the second part.

2. Replace the roots $\{\nu_\ell\}$ by their conjugate reciprocals $\{(\nu_\ell^*)^{-1}\}$. By Theorem 9.3, the noise roots will tend to migrate outside the unit circle away from their counterparts in the first set. On the other hand, the signal roots will tend to be invariant to this operation.

3. Find all L^2 distances between the roots in the first set and those in the second. Then choose the $2M$ pairs corresponding to the shortest distances. Take the arithmetic average of the phase angles of each pair as the final estimate of the corresponding frequency. This has the additional advantage of decreasing the bias of the estimates.

As the reader may recall, the problem of root selection was left open in the discussion of the increased-order covariance method in Sec. 9.4 (without SVD truncation). Of the above three methods of root selection, the method based on relative amplitudes can be used for that problem, but not the two methods based on root locations.

The Tufts-Kumaresan method involves a user-chosen parameter, the order

L. For the method to perform well, L should be a fairly large multiple of M, say $10M$ or more. It has been claimed, based on empirical evidence, that $L = 3N/4$ is optimal or nearly so. This is a reasonable choice only if N is relatively small, say on the order of 100 or less. Also, the optimal L is necessarily dependent on the number of sinusoids, their frequencies, and the SNR, so an excessive reliance on this thumb rule is not recommended.

Example 9.3. The Tufts-Kumaresan method was tested with the same signal as in Example 9.2. The degree of the polynomial was taken as 20. Table 9.2 shows the results of 500 Monte-Carlo simulations with 64 data points each. The estimates are practically unbiased, and their empirical standard deviations are somewhat larger than the corresponding CRBs. The standard deviations are also slightly larger than those of the maximum likelihood estimates for the same test case (cf. Table 9.1). ∎

Parameter	True value	Empirical bias	Empirical std. dev.	CRB
a_1	1.0	0.00634	0.05916	0.02681
a_2	0.5	−0.00370	0.05669	0.02763
ϕ_1	0.0	−0.00108	0.02585	0.01779
ϕ_2	0.0	−0.00186	0.05151	0.03556
ω_1	2.0	0.00045	0.00516	0.00224
ω_2	2.1	0.00285	0.01170	0.00432

Table 9.2. Results of Monte-Carlo simulation of Example 9.3.

9.6. ESTIMATION FROM THE SAMPLE COVARIANCES

The estimation methods presented in the previous sections are useful when the number of data points is relatively small. On the one hand, we are then usually interested in getting the most information out of the available data; on the other hand, the computational complexity of those methods (especially the maximum likelihood and the truncated SVD) is manageable only if the number of data points is not too large.

When the number of data is large (in the hundreds or more), it is reasonable to use the method of moments, and the sample covariances are the obvious candidates. As we have said in the introduction, the reliance on the sample covariances behooves us to regard the phases as random, but the amplitudes as nonrandom parameters. The true covariances of the process y_t are then given by

$$r_k = \frac{1}{2} \sum_{m=1}^{M} A_m^2 \cos \omega_m k + \sigma_u^2 \delta(k). \tag{9.6.1}$$

The sample covariances are defined, as usual, by

$$\hat{r}_k = \frac{1}{N} \sum_{t=k}^{N-1} y_t y_{t-k}, \quad k \geq 0; \quad \hat{r}_{-k} = \hat{r}_k. \tag{9.6.2}$$

The asymptotic properties of the sample covariances cannot be inferred from the results in Sec. 4.2, since the covariances in (9.6.1) are not absolutely summable. Therefore, we will derive them directly, as follows.

The asymptotic mean of $\hat{r}_k$ is given by

$$E\hat{r}_k = \frac{1}{N} \sum_{t=k}^{N-1} E\left[\sum_{m=1}^{M} A_m \cos(\omega_m t - \phi_m) + u_t \right]\left[\sum_{n=1}^{M} A_n \cos(\omega_n(t-k) - \phi_n) + u_{t-k} \right]$$

$$= \frac{1}{2} \sum_{m=1}^{M} A_m^2 \cos \omega_m k + \sigma_u^2 \delta(k) + O(N^{-1}). \tag{9.6.3}$$

Therefore, the sample covariances are asymptotically unbiased.

The asymptotic covariances of the sample covariances are given by the following theorem [Stoica, Söderström, and Ti, 1989a].

Theorem 9.4. Let $y_t = x_t + u_t$, where x_t is the noise-free signal, and let $\{r_{x,k}\}$ be the covariances of x_t. Then

$$\text{cov}\{\hat{r}_k, \hat{r}_\ell\} = \frac{\sigma_u^2}{N}[2r_{x,k-\ell} + 2r_{x,k+\ell} + \sigma_u^2 \delta(k-\ell) + \sigma_u^2 \delta(k+\ell)] + O(N^{-2}) =$$

$$\frac{\sigma_u^2}{N}\left[2\sum_{m=1}^{M} A_m^2 \cos \omega_m k \cos \omega_m \ell + \sigma_u^2 \delta(k-\ell) + \sigma_u^2 \delta(k+\ell) \right] + O(N^{-2}). \tag{9.6.4}$$

The proof is straightforward in principle, but rather tedious. Problem 9.8 provides the necessary guidelines for the proof and leaves the details to the reader. ∎

An interesting conclusion from (9.6.4) is that only the additive noise makes the asymptotic covariances behave according to the law $O(N^{-1})$. In the noise-free case, the behavior is as $O(N^{-2})$, but the quantities $N^2\text{cov}\{\hat{r}_k, \hat{r}_\ell\}$ do not approach a limit as N tends to infinity (they are only bounded). This has some intriguing effects on the accuracy of estimates in the noise-free case, but this issue will not be pursued here. In other words, we will assume that $\sigma_u^2 > 0$ throughout.

Examination of (9.6.1) reveals that the covariances of the process, with the exception of r_0, obey a model very similar to the process itself. The phases $\{\phi_m\}$ are absent, while the amplitudes are $\{0.5A_m^2\}$. We can thus regard the sample covariances $\{\hat{r}_k, 1 \leq k \leq K\}$ as a *new data set*, whose mean obeys (asymptotically) the sinusoidal model (9.6.3) and whose covariances are given by (9.6.4). Note that in this interpretation the "additive noise" in $\{\hat{r}_k\}$ is neither white nor stationary. The crucial point is that, while the number of "data points" is relatively small (since typically $K \ll N$), the "additive noise" is also small, since by (9.6.4) it is inversely proportional to N for large N. Therefore, the method of moments can be used to yield consistent estimates of $\{A_m\}$ and $\{\omega_m\}$ (but not of the phases). In the remainder of this section we explore various ways to exploit this idea.

Recall Prony's method (say, in its overdetermined version) and apply it to the covariances instead of the data. We will not use r_0, due to the additive term σ_u^2. This gives

$$
\begin{bmatrix}
r_{2M} & r_{2M-1} & \cdots & r_1 \\
r_{2M+1} & r_{2N} & \cdots & r_2 \\
\vdots & \vdots & \ddots & \vdots \\
r_{K-1} & r_{K-2} & \cdots & r_{K-2M}
\end{bmatrix}
\begin{bmatrix}
c_1 \\
c_2 \\
\vdots \\
c_{2M}
\end{bmatrix}
= -
\begin{bmatrix}
r_{2M+1} \\
r_{2M+2} \\
\vdots \\
r_K
\end{bmatrix}.
\qquad (9.6.5)
$$

This is identical to the overdetermined version of the modified Yule-Walker method for ARMA estimation [cf. (7.4.4)]. In particular, the choice $K = 4M$ leads to the usual modified Yule-Walker method [cf. (7.4.3)]. The conclusion is that the modified Yule-Walker method applies to sinusoidal signals in noise, as well as to ARMA processes. In practice the sample covariances $\{\hat{r}_k\}$ are used in (9.6.5) in place of the true covariances.

Cadzow [1980] was the first to apply the overdetermined Yule-Walker method to sinusoids in noise and to show its good performance in this application. Cadzow also observed the improvement offered by increasing the order of the polynomial $c(z)$ beyond $2M$. Friedlander and Porat [1984] extended Cadzow's work by suggesting a method for root selection and amplitude estimation. Since their method is aimed at general ARMA processes, not necessarily sinusoidal signals, it will not be pursued here. For the sinusoidal signal problem, the method of root selection and amplitude estimation described in Sec. 9.5 (in the context of the increased-order covariance method) is sufficient. Statistical analysis of the overdetermined Yule-Walker method for sinusoids in noise is given in Stoica, Söderström, and Ti [1989b].

Truncated SVD can be applied to the overdetermined Yule-Walker method. Equation (9.6.5) has to be replaced by an Lth-order equation (where $2M < L \leq K - 2M$), and then the estimate is obtained exactly as in the Tufts-Kumaresan method. The improvement obtained by using truncated SVD is not necessarily dramatic, since the basic statistical model of the sample covariances is different from that of the original data. In particular, the fact that the equivalent additive noise is nonstationary and correlated has an adverse effect on the performance of this method.

Parameter	True value	Empirical bias	Empirical std. dev.	CRB
a_1	1.0	-0.1213	0.1871	0.004457
a_2	0.5	-0.2323	0.1671	0.004453
ϕ_1	0.0	0.00206	0.02777	0.004430
ϕ_2	0.0	0.00190	0.1448	0.008866
ω_1	2.0	-0.00419	0.08212	0.000015
ω_2	2.1	-0.00056	0.00659	0.000031

Table 9.3. Results of Monte-Carlo simulation of Example 9.4.

Example 9.4. The modified Yule-Walker algorithm was tested with the two-sinusoid signal

$$y_t = \cos 2t + 0.5 \cos 2.1t + u_t, \quad \sigma_u = 0.1.$$

The number of data points was 1024. The number of sample covariances was 64 and the degree of the linear prediction polynomial was 16. No SVD truncation was employed (see Prob. 9.16 in this regard). Table 9.3 shows the results of 500 Monte-Carlo simulations. The estimates of the amplitudes are considerably biased, but not the estimates of the phases and frequencies. The standard deviations of the estimates are much larger than the corresponding CRBs, a phenomenon typical to this method. ∎

The asymptotically minimum variance method described in Sec. 7.5 for ARMA processes can be used for the present problem [Stoica, Händel, and Söderström, 1992]. Let θ denote the true vector of $\{\sigma_u^2, A_1^2, \ldots, A_M^2, \omega_1, \ldots, \omega_M\}$ and ψ any other vector in the parameter space. Let $r(\psi)$ be the vector of $\{r_k, 0 \le k \le K\}$ and $\hat{r}$ the corresponding vector of $\{\hat{r}_k\}$. As we will soon show, it is necessary to include $\hat{r}_0$ in this case. Let $\Sigma(\psi)$ be the normalized asymptotic covariance of $\hat{r}$. The asymptotically minimum variance estimate of θ based on $\hat{r}$ is the global minimizer of the cost function

$$C(\psi) = 0.5[r(\psi) - \hat{r}]^T \Sigma^{-1}(\psi)[r(\psi) - \hat{r}]. \tag{9.6.6}$$

Let H_c be the $(K+1) \times M$ matrix whose (k,m)th element is $[H_c]_{k,m} = \cos \omega_m k$. Let a be the vector of $\{0.5A_1^2, \ldots, 0.5A_M^2\}$ and A the diagonal matrix with $\{2A_1^2, \ldots, 2A_M^2\}$ along the main diagonal. Let e_1 be the first standard unit vector. Then we get, from (9.6.1) and (9.6.4),

$$r(\psi) = 0.5 H_c a + \sigma_u^2 e_1, \quad \Sigma(\psi) = \sigma_u^2 [\sigma_u^2(\mathbf{1}_{K+1} + e_1 e_1^T) + H_c A H_c^T]. \tag{9.6.7}$$

It is now clear why $\hat{r}_0$ is necessary. Without it, the cost function (9.6.6) would approach zero as $\sigma_u^2 \to \infty$, so it would not have a global minimum. The presence of σ_u^2 in r_0 prevents this from happening.

An algorithm based on the above expressions is developed in Prob. 9.9, and its normalized asymptotic covariance is derived in Prob. 9.10. The following remarks should be made about this method:

1. If $M \ll K + 1$, the matrix $\Sigma(\psi)$ can be efficiently inverted using an identity similar to [A19]. The reader is advised to use this identity in the solution of Problems 9.9 and 9.10.
2. Initial conditions for the nonlinear minimization can be obtained from the modified Yule-Walker method or one of its variants presented above.
3. When the SNR is sufficiently low, the matrix $\Sigma(\psi)$ can be approximated by $\sigma_u^4(\mathbf{1}_{K+1} + \boldsymbol{e}_1 \boldsymbol{e}_1^T)$. Then the cost function (9.6.6) can be approximated by

$$C(\psi) \approx 0.5[\boldsymbol{r}(\psi) - \hat{\boldsymbol{r}}]^T[\mathbf{1}_{K+1} - 0.5\boldsymbol{e}_1\boldsymbol{e}_1^T][\boldsymbol{r}(\psi) - \hat{\boldsymbol{r}}], \qquad (9.6.8)$$

the minimization of which is much simpler than the minimization of (9.6.6).

9.7. ESTIMATION FROM THE SAMPLE COVARIANCE MATRIX

The methods described in this section belong to the class of estimates based on the sample covariances. They are the counterparts of methods that have become popular in a different field: that of direction finding of point sources by sensor arrays, the *array processing problem*. The array processing problem is outside the scope of this book, but we will devote some space to these methods in the context of estimation of sinusoids in white noise.

Let R_{K+1} be the Toeplitz covariance matrix of $\{r_k, 0 \le k \le K\}$. We have

$$[R_{K+1}]_{k,\ell} = r_{k-\ell} = \sigma_u^2 \delta(k - \ell) + 0.5 \sum_{m=1}^{M} A_m^2 \cos \omega_m(k - \ell)$$

$$= \sigma_u^2 \delta(k - \ell) + 0.5 \sum_{m=1}^{M} A_m^2 (\cos \omega_m k \cos \omega_m \ell + \sin \omega_m k \sin \omega_m \ell). \quad (9.7.1)$$

Therefore, R_{K+1} can be expressed as

$$R_{K+1} = \sigma_u^2 \mathbf{1}_{K+1} + H_c A H_c^T + H_s A H_s^T, \qquad (9.7.2)$$

where H_c and A is as defined in the previous section, and H_s is similar to H_c but with sines instead of cosines.

The sum of the second and the third terms in (9.7.2) is a matrix of rank $2M$. If $K + 1 > 2M$, this matrix is singular. Let its eigenvalue/eigenvector decomposition be

$$H_c A H_c^T + H_s A H_s^T = V \overline{\Lambda} V^T = [V_1 \quad V_2] \begin{bmatrix} \Lambda & 0 \\ 0 & 0 \end{bmatrix} \begin{bmatrix} V_1^T \\ V_2^T \end{bmatrix}, \qquad (9.7.3)$$

where V_1 is a $(K+1) \times (2M)$ orthonormal matrix and $\Lambda = \text{diag}\{\lambda_1, \dots, \lambda_{2M}\}$. Then the eigenvalue/eigenvector decomposition of R_{K+1} is

$$R_{K+1} = V[\bar{\Lambda} + \sigma_u^2 \mathbf{1}_{K+1}]V^T = [V_1 \quad V_2] \begin{bmatrix} \Lambda + \sigma_u^2 \mathbf{1}_{2M} & 0 \\ 0 & \sigma_u^2 \mathbf{1}_{K+1-2M} \end{bmatrix} \begin{bmatrix} V_1^T \\ V_2^T \end{bmatrix}.$$
(9.7.4)

The columns of V_1 are called the *signal eigenvectors*, and they form an orthonormal basis for the *signal subspace*. The columns of V_2 are called the *noise eigenvectors*, and they form an orthonormal basis for the *noise subspace*. The two sets of vectors are easy to distinguish: as is clear from (9.7.3), the signal eigenvectors are those associated with the $2M$ largest eigenvalues of R_{K+1}, while the noise eigenvectors are those associated with the $K+1-2M$ smallest ones. It follows from (9.7.3) and (9.7.4) that

$$[H_c \ H_s] A \begin{bmatrix} H_c^T \\ H_s^T \end{bmatrix} V_1 \Lambda^{-1} = V_1, \quad [H_c \ H_s] V_2 = 0.$$
(9.7.5)

This means that the columns of V_1 span the same space as the columns of $[H_c \ H_s]$, and the columns of V_2 are orthogonal to this space.

We now describe several methods for estimating the frequencies $\{\omega_m\}$ based on the properties of the covariance matrix R_{K+1}. The methods are developed in terms of the true covariance matrix, but in practice they are used with the sample covariance matrix $\hat{R}_{K+1}$.

PISARENKO HARMONIC DECOMPOSITION

The method of Pisarenko [1972, 1973] is of historical interest, since it was the first to exploit the eigenstructure of the covariance matrix. We give a brief description of the method, but the reader should be aware that its performance is often relatively poor.

Let $K = 2M$, so the dimension of R_{K+1} is greater by 1 than its rank. The matrix V_2 then consists of a single column, and we denote the elements of this column by $\{v_k, 0 \le k \le 2M\}$. By the second equality in (9.7.5) we have

$$\sum_{k=0}^{2M} v_k \cos \omega_m k = \sum_{k=0}^{2M} v_k \sin \omega_m k = 0, \quad 1 \le m \le M.$$

Therefore,

$$\sum_{k=0}^{2M} v_k e^{j\omega_m k} = \sum_{k=0}^{2M} v_k e^{-j\omega_m k} = 0, \quad 1 \le m \le M.$$
(9.7.6)

This means that the $2M$ roots of the polynomial $v(z) = \sum_{k=0}^{2M} v_k z^k$ are $\{e^{\pm j\omega_m}, 1 \le m \le M\}$. The Pisarenko method is therefore as follows. Find the eigendecomposition of $\hat{R}_{2M+1}$, construct a polynomial from the elements of the eigenvector corresponding to the smallest eigenvalue, find the roots of this polynomial, and take the estimated frequencies to be the phase angles of the roots.

THE MUSIC ALGORITHM

The MUSIC algorithm (an acronym for *multiple signal classification*) was developed by Schmidt [1979, 1981] and independently by Bienvenu and Kopp [1979] for array processing problems. Their idea was to exploit the entire noise subspace by using its orthogonality to the columns of $[H_c, H_s]$. Let the *null spectrum* be defined as

$$S_n(\omega) = \frac{e^H(\omega)V_2 V_2^T e(\omega)}{(K+1)}, \quad e(\omega) = [1, e^{j\omega}, \dots, e^{j\omega K}]^H. \qquad (9.7.7)$$

Since $e(\omega_m)$ lies in the signal subspace, it is orthogonal to the noise subspace, so $\{1/S_n(\omega_m), 1 \le m \le M\}$ is infinite, while $1/S_n(\omega)$ is finite for all other frequencies. Therefore, plotting $1/S_n(\omega)$ as a function of ω in the range $[0, \pi]$ will exhibit M theoretically infinite peaks. When using $\hat{R}_{K+1}$ instead of R_{K+1}, these peaks will be finite, and their locations will be close to the true frequencies. The MUSIC algorithm is therefore as follows. Choose $K + 1 > 2M$ and compute the eigen-decomposition of $\hat{R}_{K+1}$. Let V_2 be an orthonormal basis for the noise subspace. Compute the null spectrum as a function of ω and find the M largest local maxima of its inverse. The locations of the maxima are taken as the frequency estimates. A mathematically equivalent algorithm is given as follows. Let

$$S_s(\omega) = \frac{e^H(\omega)V_1 V_1^T e(\omega)}{(K+1)}.$$

Then, since $S_s(\omega) + S_n(\omega) = 1$, the local maxima of $1/S_n(\omega)$ will coincide with those of $S_s(\omega)$. From a computational point of view, it is advised to work with $S_s(\omega)$ whenever the dimension of the signal subspace is smaller than that of the noise subspace.

Statistical analysis of the MUSIC algorithm for sinusoidal parameter estimation (with Pisarenko harmonic decomposition as a special case) was given in Stoica and Söderström [1991].

Another approach, which avoids the need for nonlinear maximization, is given as follows. Extend the definition of the null spectrum to the entire complex plane by taking

$$S_n(z) = [1, z^{-1}, \dots, z^{-K}]V_2 V_2^T [1, z, \dots, z^K]^T. \qquad (9.7.8)$$

$z^K S_n(z)$ is a symmetric polynomial of degree $2K$. When the true noise subspace is used in (9.7.8), it will have $4M$ roots on the unit circle, corresponding to $\{e^{\pm j\omega_m}\}$, each with multiplicity 2. When V_2 is computed from $\hat{R}_{K+1}$, these roots will move away from the unit circle. However, due to the symmetry of $z^K S_n(z)$, the roots will split into two sets, one inside the unit circle and one outside it. The algorithm thus computes the roots of $z^K S_n(z)$ and then selects among the K roots inside the unit circle the $2M$ roots of largest magnitudes. The phase angles of these roots are taken as the estimated frequencies. This algorithm is known as *Root-MUSIC* and can be regarded as a natural extension of the Pisarenko method.

THE ESPRIT METHOD

   The ESPRIT method (an acronym for *estimation of signal parameters via rotational invariance techniques*) was developed by Paulraj, Roy, and Kailath [1985] (also Roy and Kailath [1989]). As in the case of MUSIC, its primary use is for array processing, but it can be used for sinusoidal frequency estimation as well [Roy, Paulraj, and Kailath, 1986].
   Let H_u be the submatrix of $[H_c\ H_s]$ consisting of the first K rows and H_ℓ the submatrix of $[H_c\ H_s]$ consisting of the last K rows. Let Q_1 be the diagonal matrix with $\{\cos\omega_1,\ldots,\cos\omega_M\}$ along the main diagonal, and Q_1 the diagonal matrix with $\{\sin\omega_1,\ldots,\sin\omega_M\}$ along the main diagonal. Then it is easy to verify, using trigonometric identities for $\cos(\omega_m k+\omega_m)$ and $\sin(\omega_m k+\omega_m)$, that

$$H_\ell = H_u Q, \text{ where } Q = \begin{bmatrix} Q_1 & Q_2 \\ -Q_2 & Q_1 \end{bmatrix}. \tag{9.7.9}$$

The matrix Q is clearly orthonormal, so H_u and H_ℓ are related by an orthonormal transformation (this is the reason for the term "rotational invariance" in the name). The eigenvalues of Q are $\{e^{\pm j\omega_m}, 1 \le m \le M\}$. Now recall that

$$V_1 = [\,H_c\quad H_s\,]\,C, \tag{9.7.10}$$

where C is the matrix appearing on the left side of the first equality in (9.7.5). Denoting the upper and lower submatrices of V_1 by V_u and V_ℓ (compatibly with H_u and H_ℓ), we get from (9.7.10)

$$V_u = H_u C, \quad V_\ell = H_\ell C.$$

Therefore we get, using (9.7.9),

$$V_\ell = V_u C^{-1} Q C. \tag{9.7.11}$$

Equation (9.7.11) is the key to the ESPRIT method. The matrix of signal eigenvectors V_1 is obtained from the covariance matrix R_{K+1} as in the MUSIC algorithm. The matrix $C^{-1}QC$ can be obtained from V_1 as

$$C^{-1}QC = (V_u^T V_u)^{-1} V_u^T V_\ell. \tag{9.7.12}$$

This matrix is similar to Q; hence they both have the same eigenvalues, $\{e^{\pm j\omega_m}, 1 \le m \le M\}$. Therefore, the frequencies can be obtained from the phase angles of the eigenvalues of $C^{-1}QC$.
   In practice, V_1 is obtained from the sample covariance matrix $\hat{R}_{K+1}$. Correspondingly, (9.7.12) is the least-squares solution of (9.7.11), so it is known as *least-squares ESPRIT*. A slightly better estimate can be obtained if (9.7.11) is solved for $C^{-1}QC$ using total least-squares. The theory of total least-squares is outside the scope of the book, so we refer the reader to Golub and Van Loan [1983, Sec. 12.3] for the details. We just state the actual numerical procedure, as follows.

Let S be the matrix of *right singular vectors* in the SVD of $[V_u \ V_\ell]$. Note that this matrix has dimensions $4M \times 4M$. Partition it into four blocks of $2M \times 2M$ each:

$$S = \begin{bmatrix} S_{11} & S_{12} \\ S_{21} & S_{22} \end{bmatrix}.$$

Then the total least-squares solution of (9.7.11) is

$$C^{-1}QC = -S_{12}S_{22}^{-1}. \tag{9.7.13}$$

This is known as the *total least-squares ESPRIT* (see Roy and Kailath [1989]). We finally remark that the asymptotic statistical properties of ESPRIT were analyzed in Stoica and Söderström [1991].

OTHER METHODS

The aforementioned methods use the eigenvectors of the covariance matrix, but not its eigenvalues. For completeness, we mention two methods that use the eigenvalues as well. The two methods take the point of view of spectral density estimation and estimate the frequencies as the points of local maxima of the estimated spectral density. Of course, a deterministic process has no spectral density, so the "spectral density" constructed by these methods cannot be regarded as a valid estimate of anything. Nevertheless, the points of its local maxima are consistent estimates of the sinusoid frequencies.

Let us define

$$S_1(\omega) = \frac{1}{K+1} e^H V_1 \Lambda V_1^T e, \quad S_2(\omega) = \frac{K+1}{e^H V_1 \Lambda^{-1} V_1^T e}, \tag{9.7.14}$$

where e is defined in (9.7.7). $S_1(\omega)$ can be regarded as a "clean" version of the windowed periodogram corresponding to the triangular (Bartlett) window. By "clean" we mean that the noise component $\sigma_u^2 \mathbf{1}$ has been removed from $S_1(\omega)$. Similarly, $S_2(\omega)$ can be regarded as a clean version of the minimum variance spectral density estimator described in Sec. 6.9. As we have said, the estimated frequencies are taken as the points of local maxima of $S_1(\omega)$ or $S_2(\omega)$. Neither of the two estimates appears to offer an advantage over the MUSIC algorithm, but they are of historical interest, as well as of interest to the array processing problem.

9.8. CONCLUDING REMARKS

The methods described in this chapter were developed for the case of additive white noise. Most of them work adequately even when the noise is not Gaussian, but in general they are not suitable to the case of colored noise. The maximum likelihood method is an exception, at least for large data samples. Assume that the noise obeys an ARMA model and that the standard maximum likelihood algorithm (which *ignores* the noise's lack of whiteness) is applied to the noise-corrupted signal.

It turns out that the asymptotic covariance of the estimates is identical to the CRB for the *true model*, the one that includes the noise parameters. The proof of this result is given in Stoica and Nehorai [1989]. For small data samples, however, no such conclusions can be drawn.

In general, the problem of estimating sinusoids in colored noise is rather difficult. The reader may wish to ponder the following approaches:

1. Assigning a parametric model to the noise (e.g., autoregressive or ARMA) and using maximum likelihood for the full model. In this case the parameters will appear in both the mean and covariance matrix.
2. Estimating the innovation sequence by a preliminary long AR model (estimated by the Yule-Walker method), using it to separate the deterministic and purely indeterministic components of Wold decomposition, then estimating the signal parameters and the noise spectrum from the respective components.
3. Obtaining preliminary estimates of the frequencies from the periodogram (or from an increased-order overdetermined linear prediction), filtering out the sinusoidal components by a notch filter, estimating the noise spectrum from the filter's output, and using it to improve the sinusoidal components estimates.

In Chapter 10 we will present an estimation method for sinusoids in Gaussian colored noise.

9.9. MATHEMATICA PACKAGES

The package `SinData.m` contains a data generator for the sinusoids-in-white-noise model.

`SinData[a_, phi_, omega_, sigma_, Nsam_]` generates `Nsam` data points corresponding to the vectors of amplitudes `a`, phases `phi` and frequencies `omega`, and to the noise standard deviation `sigma`. The phase `phi` is referred to the middle of the interval.

`CovSeq[data_, nr_]` is identical to the corresponding function in the package `ArmaData.m`, computing the sample covariances of `data` of orders 0 through `nr`.

`NAsymptotic[omega_]` computes the minimum N for which the approximate asymptotic analysis in Secs. 9.2 and 9.3 applies, according to formula (9.2.14). `omega` is the list of radian frequencies.

The package `SinAlg.m` implements some of the estimation algorithms described in this chapter and the Cramér-Rao lower bound for the sinusoids-in-white-noise model.

`SinCrb[a_, phi_, omega_, n_]` computes the Cramér-Rao lower bound. `a` is the vector of amplitudes, `phi` the vector of phases, `omega` the vector of frequencies and `n` the number of data points (or a list of numbers of data points). Due to the block diagonal structure of $J_N(\theta)$ [cf. (9.2.1)], the part corresponding to the parameter σ_u^2 is trivial and was omitted. The procedure assumes that $\sigma_u^2 = 1$. If this parameter has a different value, the user should use the ratios $\{A_m/\sigma_u\}$ as the entries of **a**. The returned values are the square roots of the diagonal elements of the inverse Fisher information matrix. If **n** is a list, a list of lists is returned.

LocalPeaks[p_, n_] finds the **n** largest local peaks in the periodogram **p**. It returns the list of frequencies and the corresponding list of periodogram ordinates. LocalPeaks[data_, zpfactor_, window_, n_] is similar, but accepts the following inputs: **data**, the data sequence; **zpfactor**, an integer factor for zero-padding the data in order to increase the frequency resolution; and **window**, a function name for optional windowing of the data before computing the periodogram (must be one of the window functions in **Period.m**). This procedure can be used for coarse frequency estimation if the number of data points is sufficiently large so that the peaks are separated.

SinMl[data_, init_, opts___] computes the maximum likelihood estimates for multiple sinusoids in white noise. **data** is the data sequence, and **init** is the initial condition for the frequencies. The nonlinear minimization is by conjugate directions (without gradient). **Method** is an optional rule for choosing the type of estimate. **Method->Exact** (the default) chooses the exact maximum likelihood, using Eqs. (9.3.9) and (9.3.10). **Method->Approx** chooses the approximate method (9.3.11). The returned values are the estimated amplitudes, phases, frequencies, and the noise standard deviation.

TruncatedSvd[data_, l_, m_, opts___] estimates the parameters by the truncated SVD (Tufts-Kumaresan) method. **data** is the data sequence, **l** the order of the linear prediction polynomial, and **m** the number of sinusoids. **Method** determines the method of constructing the data matrix. **FCov**, the default, is for forward covariance, **BCov** for backward covariance, and **FBCov** for forward-backward covariance. The returned values are the estimated amplitudes, phases, frequencies, and the noise standard deviation.

ModYuleWalker[data_, n_, l_, m_] estimates the parameters of multiple sinusoids in white noise by the modified Yule-Walker method. The degree of the linear prediction polynomial is allowed to be larger than $2M$, and then root selection is employed, as in **TruncatedSvd**. However, SVD truncation is not performed (see Prob. 9.16.). **data** is the data sequence, **n** the number of sample covariances (must be twice the value of **l** at least), **l** the degree of the linear prediction polynomial, and **m** the number of sinusoids. The returned values are the estimated amplitudes, phases, frequencies, and the noise standard deviation.

Music[data_, l_, m_] estimates the parameters by the Root-MUSIC method. **data** is the data sequence, **l**+1 the dimension of the sample covariance matrix, and **m** the number of sinusoids. The special case **l** = 2**m** gives the Pisarenko estimate. The returned values are the estimated amplitudes, frequencies, and the noise standard deviation.

The package includes three auxiliary functions: **TheSum** for computing the sums in (9.2.2)–(9.2.13), **Dirichlet** for computing the Dirichlet kernel $(\sin nx)/(\sin x)$, and **EstRest** for estimating the amplitudes, phases, and noise variance when the frequencies are known, using (9.3.5) and (9.3.6).

The linear prediction method described in Sec. 9.4 (Prony's method and its variants) was not included since equivalent procedures were implemented in `ArAlg.m` and are described in Chapter 6.

REFERENCES

Bienvenu, G., and Kopp, L., "Principè de la Goniomètre Passive Adaptive," *Proc. 7'eme Colloque GRESIT,* pp. 106/1–106/10, Nice, France, 1979.

Cadzow, J., "High Performance Spectral Estimation—A New ARMA Method," *IEEE Trans. Acoustics, Speech, Signal Processing,* ASSP-28, pp. 524–529, 1980.

Fisher, R. A., "Tests of Significance in Harmonic Analysis," *Proc. Royal Society,* 125, pp. 54–59, 1929.

Friedlander, B., and Porat, B., "The Modified Yule-Walker Method of ARMA Spectral Estimation," *IEEE Trans. Aerospace Electronic Systems,* AES-20, pp. 158–173, 1984.

Golub, G. H., and Van Loan, C. F., *Matrix Computations,* Johns Hopkins University Press, Baltimore, MD, 1983.

Harris, F. J., "On the Use of Windows for Harmonic Analysis with the Discrete Fourier Transform," *Proc. IEEE,* 66, pp. 51–83, 1978.

Kumaresan, R., "Estimating the Parameters of Exponentially Damped or Undamped Sinusoidal Signals in Noise," Ph.D. dissertation, University of Rhode Island, RI, 1982.

Lawson, C., and Hanson, R., *Solving Least Squares Problems,* Prentice Hall, Englewood Cliffs, NJ, 1974.

Paulraj, A., Roy, R. H., and Kailath, T., "Estimation of Signal Parameters via Rotational Invariance Techniques—ESPRIT," *Proc. 19th Asilomar Conf. on Circuits, Systems, Computers,* Pacific Groves, CA, pp. 83–89, 1985.

Pisarenko, V. F., "On the Estimation of Spectra by Means of Nonlinear Functions of the Covariance Matrix," *Geophys. J. Roy. Astron. Soc.,* 28, pp. 511–531, 1972.

Pisarenko, V. F., "The Retrieval of Harmonics from a Covariance Function," *Geophys. J. Roy. Astron. Soc.,* 33, pp. 3471–366, 1973.

Porat, B., and Friedlander, B., "A Modification of the Kumaresan-Tufts Method for Estimating Rational Impulse Responses," *IEEE Trans. Acoustics, Speech, Signal Processing,* ASSP-34, pp. 1336–1338, 1986.

Prony, R., "Essai Experimental et Analytique, etc.," *L'ecole Polytechnique, Paris,* 1, pp. 24–76, 1795.

Rao, C. R., and Zhao, L. C., "Asymptotic Behavior of the Maximum Likelihood Estimates of Superimposed Exponential Signals," *IEEE Trans. Signal Processing,* 41, pp. 1461–1464, 1993.

Rife, D. C., and Boorstyn, R. R., "Single Tone Parameter Estimation from Discrete-Time Observations," *IEEE Trans. Information Theory,* IT-20, pp. 591–598, 1974.

Rife, D. C., and Boorstyn, R. R., "Multiple Tone Parameter Estimation from Discrete-Time Observations," *Bell System Tech. J.,* 55, pp. 1389–1410, 1976.

Roy, R. H., and Kailath, T., "ESPRIT—Estimation of Signal Parameters via Rotational Invariance Techniques," *IEEE Trans. Acoustics, Speech, Signal Processing,* 37, pp. 984–995, 1989.

Roy, R. H., Paulraj, A., and Kailath, T., "ESPRIT—A Subspace Rotational Approach to Estimation of Parameters of Cisoids in Noise," *IEEE Trans. Acoustics, Speech, Signal Processing,* 34, pp. 1340–1342, 1986.

Schmidt, R. O., "Multiple Emitter Location and Signal Parameter Estimation," *Proc. RADC Spectral Estimation Workshop,* pp. 243–258, Rome, NY, 1979.

Schmidt, R. O., "A Signal Subspace Approach to Multiple Emitter Location and Spectral Estimation," Ph.D. dissertation, Stanford University, CA, 1981.

Stoica, P., Händel, P., and Söderström, T., "Approximate Maximum Likelihood Frequency Estimation," Tech. Rep. UPTEC 92142R, Teknikum, Uppsala University, Uppsala, Sweden, 1992.

Stoica, P., and Nehorai, A., "Statistical Analysis of Two Nonlinear Least-squares Estimators of Sine-wave Parameters in the Colored-noise Case," *Circuits, Systems, Signal Processing,* 8, pp. 3–15, 1989.

Stoica, P., and Söderström, T., "Statistical Analysis of MUSIC and Subspace Rotation Estimates of Sinusoidal Frequencies," *IEEE Trans. Signal Processing,* 39, pp. 1836–1847, 1991.

Stoica, P., Söderström, T., and Ti, F. N., "Asymptotic Properties of the High-order Yule-Walker Estimates of Sinusoidal Frequencies," *IEEE Trans. Acoustics, Speech, Signal Processing,* 37, pp. 1721–1734, 1989a.

Stoica, P., Söderström, T., and Ti, F. N., "Overdetermined Yule-Walker Estimation of the Frequencies of Multiple Sinusoids: Accuracy Aspects," *Signal Process-ing,* 16, pp. 155–174, 1989b.

Tufts, D. W., and Kumaresan, R., "Estimation of Frequencies of Multiple Sinusoids: Making Linear Prediction Perform Like Maximum Likelihood," *Proc. IEEE,* 70, pp. 975–989, 1982.

PROBLEMS

9.1. The CRB formula (9.2.16c) applies to the phase at the middle of the ob-servation interval. Using (3.3.3), prove that the asymptotic CRB for the phase at the beginning of the interval is four times larger than the value in (9.2.16c). Also show that the asymptotic cross-term of ϕ_k and ω_k does not vanish in this case.

9.2. Derive the expressions for the gradient of the cost function $C(\omega)$ in (9.3.10).

9.3. Write the cost function of the exact maximum likelihood for the case of a single sinusoid. Explain why this function exhibits local maxima at intervals of approximately $2\pi/N$.

9.4. Consider the signal model

$$y_t = \sum_{m=1}^{M} A_m \cos(m\omega_0 t - \phi_m) + u_t,$$

where ω_0 and $\{A_m, \phi_m\}$ are the unknown parameters. Assume that $\omega_0 < \pi/M$, so none of the sinusoidal components is aliased. This model is a special case of a periodic nonsinusoidal signal (the general model involves an infinite number of harmonics). $\{u_t\}$ is white Gaussian noise with zero mean and variance σ_u^2.

   (a) Derive the asymptotic CRB for the parameters of this model.
   (b) Derive the maximum likelihood procedure for this model.
   (c) Using the function `SinM1` as an example, implement the maximum likelihood procedure in Mathematica.

9.5. Prove that $c(z)$ in Prony's method is symmetric; that is, $c_k = c_{2M-k}$.

9.6. Use the symmetry of $c(z)$ to devise a Prony-like estimator with only M unknowns.

9.7. Show how (9.5.3) follows from the preceding argument.

9.8. Prove the asymptotic formula (9.6.4) by the following steps.

   (a) Let $x_t = \sum_{m=1}^{M} A_m \cos(\omega_m t - \phi_m)$ and

$$\hat{r}_k = \frac{1}{N}\sum_t (x_t+u_t)(x_{t-k}+u_{t-k}) \; ; \; \hat{r}_\ell = \frac{1}{N}\sum_s (x_s+u_s)(x_{s-\ell}+u_{s-\ell}).$$

   In $E\hat{r}_k\hat{r}_\ell$ there are sixteen terms, eight of which are not identically zero. Show that the nonzero terms give

$$E\hat{r}_k\hat{r}_\ell = \frac{1}{N^2}\sum_t\sum_s E[x_t x_{t-k}x_s x_{s-\ell} + x_t x_{t-k}u_s u_{s-\ell}$$

$$+ x_t u_{t-k}x_s u_{s-\ell} + x_t u_{t-k}u_s x_{s-\ell} + u_t x_{t-k}x_s u_{s-\ell}$$

$$+ u_t x_{t-k}u_s x_{s-\ell} + u_t u_{t-k}x_s x_{s-\ell} + u_t u_{t-k}u_s u_{s-\ell}],$$

where

$$x_t x_{t-k}x_s x_{s-\ell} = \sum_m\sum_n\sum_p\sum_q A_m A_n A_p A_q \cos(\omega_m t - \phi_m)$$

$$\cos(\omega_n t - \omega_n k - \phi_n)\cos(\omega_p s - \phi_p)\cos(\omega_q s - \omega_q \ell - \phi_q)$$

and similarly for the other terms.

(b) The first term in the above is the most difficult to handle. Use the trigonometric identity

$$\cos\alpha\cos\beta\cos\gamma\cos\delta = \frac{1}{8}[\cos(\alpha+\beta+\gamma+\delta) + \cos(\alpha+\beta+\gamma-\delta)$$

$$+ \cos(\alpha+\beta-\gamma+\delta) + \cos(\alpha-\beta+\gamma+\delta) + \cos(\alpha-\beta-\gamma-\delta)$$

$$+ \cos(\alpha+\beta-\gamma-\delta) + \cos(\alpha-\beta+\gamma-\delta) + \cos(\alpha-\beta-\gamma+\delta)]$$

to break it into eight terms. Using the uniform distribution of the phases, show that only the last three terms of the trigonometric identity are not identically zero.

(c) Use the uniform distribution of the phases again to find when the three terms are nonzero. Show that these terms add up to $r_{x,k}r_{x,\ell} + O(N^{-2})$. This will be easier if you break factors such as $\cos(\alpha+\beta)$ to products of sines and cosines, thus facilitating the summations over t and s.

(d) The other seven terms in the formula for $E\hat{r}_k\hat{r}_\ell$ are much easier to handle. In the last term use the formula for the fourth-order moments of Gaussian random variables. The other terms involve only second-order moments, so they should pose no problem.

9.9. Develop the asymptotically minimum variance algorithm based on the sample covariances using (9.6.6) and (9.6.7). Find expressions for the partial derivatives, write a Mathematica procedure for this algorithm, and test it.

9.10. Write a Mathematica procedure for computing the normalized asymptotic covariance of the algorithm developed in the previous problem.

9.11. The following extension of the Pisarenko method is due to Kumaresan [1982]. Let the dimension of the noise subspace be greater than 1. Then every column of V_2 satisfies (9.7.6). Let $\bar{p} = V_2\alpha$ be any *nondegenerate* linear combination of the columns of V_2. By this we mean that $p(z)$, the polynomial constructed from the elements of $\bar{p}$, has a degree $2M$ at least. Then this polynomial satisfies (9.7.6) as well. This means that the frequencies can be found by picking the $2M$ roots of $p(z)$ on the unit circle (or the one closest to the unit circle when $\hat{R}_{K+1}$ is used). Now suppose we wish to pick $\boldsymbol{\alpha}$ such that (1) $p_0 = 1$ and (2) $\sum_{k=1}^{K} p_k^2$ is minimum.

(a) Let $\boldsymbol{v}^T$ be the first row of V_2 and $\boldsymbol{V}$ the matrix of all rows of V_2 except the first. Let $\boldsymbol{p}$ be the vector of all elements of $\bar{\boldsymbol{p}}$ except the first. Show that $\boldsymbol{\alpha}$ is obtained as the minimum of $\boldsymbol{\alpha}^T \boldsymbol{V}^T \boldsymbol{V} \boldsymbol{\alpha}$ subject to the constraint $\boldsymbol{v}^T \boldsymbol{\alpha} = 1$.

(b) Using the method of Lagrange multipliers, show that the solution of the constrained minimization problem is

$$ \boldsymbol{\alpha} = \frac{(\boldsymbol{V}^T \boldsymbol{V})^{-1} \boldsymbol{v}}{\boldsymbol{v}^T (\boldsymbol{V}^T \boldsymbol{V})^{-1} \boldsymbol{v}}, \quad \boldsymbol{p} = \frac{\boldsymbol{V} (\boldsymbol{V}^T \boldsymbol{V})^{-1} \boldsymbol{v}}{\boldsymbol{v}^T (\boldsymbol{V}^T \boldsymbol{V})^{-1} \boldsymbol{v}}. $$

(c) Using [A19], show that the solution can be simplified to

$$ \boldsymbol{p} = \frac{\boldsymbol{V} \boldsymbol{v}}{\boldsymbol{v}^T \boldsymbol{v}}. $$

9.12. Prove that the normalized log likelihood of the sinusoids-in-white-noise model (9.1.1) converges almost surely to a function whose global maximum is at the true values of the parameters. Assume for simplicity that the noise variance σ_u^2 is known, so it is not included in θ. Proceed according to the following steps.

(a) Let $x_t(\psi)$ be the noise-free signal corresponding to any value ψ of the parameter vector and $x_t(\theta)$ the one corresponding to the true parameter vector. Prove that the negative of the normalized log likelihood can be expressed as

$$ -\frac{2}{N} \log f_\psi(Y_N) = \log 2\pi + \log \sigma_u^2 $$
$$ + \frac{1}{N\sigma_u^2} \left\{ \sum_t [x_t(\theta) - x_t(\psi)]^2 + 2 \sum_t 2u_t[x_t(\theta) - x_t(\psi)] + \sum_t u_t^2 \right\}. $$

(b) Using the Borel-Cantelli lemma and Chebychev's inequality, as in the proof of Theorem C.5, prove that $N^{-1} \sum_t 2u_t[x_t(\theta) - x_t(\psi)]$ converges to zero almost surely. Also, $N^{-1} \sum_t u_t^2$ converges to σ_u^2 almost surely by the strong law of large numbers.

(c) The term $N^{-1} \sum_t [x_t(\theta) - x_t(\psi)]^2$ is obviously zero for $\psi = \theta$ and positive otherwise. However, to prove the asserted property, it is necessary to show that it is strictly positive in the limit as $N \to \infty$. Prove that $\lim_{N \to \infty} N^{-1} \sum_t [x_t(\theta) - x_t(\psi)]^2$ is strictly positive for all $\psi \neq \theta$.

9.13. Consider the second approach to the problem of estimating sinusoids in colored noise mentioned in Sec. 9.8. Assume that the innovation sequence of $\{y_t\}$ is known. Devise a method for separating the two components of the Wold decomposition of $\{y_t\}$.

9.14. Derive the Fisher information matrix and the CRB for the complex model (9.1.2), in analogy with the derivation in Sec. 9.2. Note that a complex white

Gaussian noise is defined as $u_t = u_t^{(r)} + ju_t^{(i)}$, where $\{u_t^{(r)}\}$ and $\{u_t^{(i)}\}$ are white Gaussian noise sequences, independent of each other and having zero means and equal variances. Also, the p.d.f. of a complex random variable is defined as the joint p.d.f. of the real part and the imaginary part.

9.15. Derive the maximum likelihood estimate and its approximation for the complex model (9.1.2).

9.16. Modify the procedure `ModYuleWalker` to include SVD truncation. Repeat Example 9.4 with SVD truncation and report the results. Compare with the results in Table 9.3 and draw conclusions.

9.17. Write a Mathematica procedure that implements the total least-squares ESPRIT algorithm. Use parts of the procedure `Music` as necessary.

CHAPTER 10

High-Order
Statistical Analysis

10.1. INTRODUCTION

All the problems considered in this book up to this point assumed Gaussian models. This does not mean that the methods presented cannot be used for non-Gaussian signals, only that the *theoretical analysis* is valid under the Gaussian assumption. In this chapter we depart from the Gaussian model and consider specific non-Gaussian processes and estimation methods for such processes.

Cumulants are the main tool in dealing with non-Gaussian processes. The cumulants are functions of the moments of the process and are nonzero only if the process is non-Gaussian. The role they play for non-Gaussian processes is similar to the role played by the sample covariances in the Gaussian case: they can be consistently estimated from the measurements (under certain assumptions) and contain information about the process parameters—information that can be used for parameter estimation. Cumulants of independent random variables are additive and those of Gaussian processes are zero. This implies that cumulants can be used to combat the effect of additive Gaussian noise on a desired non-Gaussian signal. The frequency domain counterparts of the cumulants are the *polyspectra*, which play a role similar to that of the power spectral density.

Maximum likelihood estimation, which is central to Gaussian parameter estimation, is seldom used for non-Gaussian random processes, because the likelihood function is almost always intractable in this case. Therefore, the estimation methods presented in this chapter use the method of moments exclusively. The next three sections serve as an introduction to the general theory of high-order statistics, while the rest of the chapter deals with estimation.

10.2. DEFINITION AND PROPERTIES OF CUMULANTS

Let X be an n-dimensional random vector with components $\{X_k\}$. Let $\boldsymbol{\lambda}$ be an n-dimensional real vector with components $\{\lambda_k\}$. The moment generating function of X is defined as

$$M_X(\boldsymbol{\lambda}) = E\{\exp(\boldsymbol{\lambda}^T X)\} = E\left\{\exp\left(\sum_{k=1}^n \lambda_k X_k\right)\right\} \qquad (10.2.1)$$

for all $\boldsymbol{\lambda}$ such that the right side exists. As is known from probability theory, the coefficients of the Taylor series of $M_X(\boldsymbol{\lambda})$ around zero yield all the moments of X. In particular, the coefficient of $\prod_{k=1}^n \lambda_k$ is the nth-order moment

$$\mu(X) = \mu(X_1, \ldots, X_n) = E\left\{\prod_{k=1}^n X_k\right\}. \qquad (10.2.2)$$

The *cumulant generating function* of X is defined as

$$C_X(\boldsymbol{\lambda}) = \log M_X(\boldsymbol{\lambda}) = \log E\{\exp(\boldsymbol{\lambda}^T X)\}. \qquad (10.2.3)$$

Clearly, the cumulant generating function is defined whenever the moment generating function is defined. The (nth-order) *cumulant* of X is defined as the coefficient of $\prod_{k=1}^n \lambda_k$ in the Taylor series of $C_X(\boldsymbol{\lambda})$ around zero. It will be denoted by $\kappa(X)$, or by $\kappa(X_1, \ldots, X_n)$ when we want to make the components of X explicit. Both moments and cumulants are invariant to permutations of the components of X. The above definitions clearly imply

$$\mu(X_1, \ldots, X_n) = \left.\frac{\partial^n M_X(\boldsymbol{\lambda})}{\partial\lambda_1 \ldots \partial\lambda_n}\right|_{\boldsymbol{\lambda}=0}, \quad \kappa(X_1, \ldots, X_n) = \left.\frac{\partial^n C_X(\boldsymbol{\lambda})}{\partial\lambda_1 \ldots \partial\lambda_n}\right|_{\boldsymbol{\lambda}=0}. \qquad (10.2.4)$$

Example 10.1. Let X be a Gaussian random vector with mean $\boldsymbol{v}$ and covariance matrix Γ. Then (Prob. 10.1)

$$M_X(\boldsymbol{\lambda}) = \exp(\boldsymbol{\lambda}^T\boldsymbol{v} + 0.5\boldsymbol{\lambda}^T\Gamma\boldsymbol{\lambda}).$$

Therefore,

$$C_X(\boldsymbol{\lambda}) = \boldsymbol{\lambda}^T\boldsymbol{v} + 0.5\boldsymbol{\lambda}^T\Gamma\boldsymbol{\lambda}.$$

Since $C_X(\boldsymbol{\lambda})$ is a quadratic function of $\boldsymbol{\lambda}$ in this case, it follows that the cumulant of an n-dimensional Gaussian vector X is zero if $n > 2$. ∎

The elements of the vector X need not be distinct. For example, if x is a scalar random variable, we can define the n-dimensional vector $X = [x, \ldots, x]^T$. The nth-order cumulant of x is then $\kappa(X)$ according the above definition. This is usually denoted by $\kappa_n(x)$.

Example 10.2. Let x have the shifted exponential distribution $f(x) = e^{-(x+1)}$, $x \geq -1$ (the shifting causes x to have zero mean). Let $X = [x, \ldots, x]^T$ (n times). Then

$$M_X(\boldsymbol{\lambda}) = E\left\{ \exp\left(\sum_{k=1}^{n} \lambda_k \right) x \right\} = \int_{-1}^{\infty} \exp\left\{ -1 - \left(1 - \sum_{k=1}^{n} \lambda_k \right) x \right\} dx$$

$$= \frac{\exp\{-\sum_{k=1}^{n} \lambda_k\}}{1 - \sum_{k=1}^{n} \lambda_k},$$

$$C_X(\boldsymbol{\lambda}) = -\sum_{k=1}^{n} \lambda_k - \log\left(1 - \sum_{k=1}^{n} \lambda_k \right).$$

For $n = 1$ we get

$$\frac{\partial C_X(\lambda)}{\partial \lambda} = -1 + \frac{1}{1 - \lambda},$$

while for $n > 1$,

$$\frac{\partial^n C_X(\boldsymbol{\lambda})}{\partial \lambda_1 \ldots \partial \lambda_n} = \frac{(n-1)!}{\left(1 - \sum_{k=1}^{n} \lambda_k \right)^n}.$$

Therefore, by (10.2.4),

$$\kappa(x) = 0, \quad \kappa(X) = \kappa_n(x) = (n-1)!.$$ ∎

Two important properties of cumulants are given by the following lemmas.

Lemma 10.1. Let X and Y be independent n-dimensional random variables. Then

$$\kappa(X + Y) = \kappa(X) + \kappa(Y). \tag{10.2.5}$$

Proof.

$$C_{X+Y}(\boldsymbol{\lambda}) = \log E\{\exp[\boldsymbol{\lambda}^T(X + Y)]\} = \log E\{\exp(\boldsymbol{\lambda}^T X)\exp(\boldsymbol{\lambda}^T Y)\}$$

$$= \log E\{\exp(\boldsymbol{\lambda}^T X)\} + \log\{\exp(\boldsymbol{\lambda}^T Y)\} = C_X(\boldsymbol{\lambda}) + C_Y(\boldsymbol{\lambda}).$$

This clearly implies (10.2.5). ∎

Lemma 10.2. Suppose X consists of two independent subvectors X_a and X_b. Then $\kappa(X) = 0$.

Proof. Independence of X_a and X_b implies

$$C_X(\boldsymbol{\lambda}) = \log E\{\exp(\boldsymbol{\lambda}_a^T X_a + \boldsymbol{\lambda}_b^T X_b)\} = \log E\{\exp(\boldsymbol{\lambda}_a^T X_a)\} + \log E\{\exp(\boldsymbol{\lambda}_b^T X_b)\}$$

$$= C_{X_a}(\boldsymbol{\lambda}_a) + C_{X_b}(\boldsymbol{\lambda}_b).$$

Since the first term is free of $\boldsymbol{\lambda}_b$ while the second is free of $\boldsymbol{\lambda}_a$, repeated differentiation with respect to all the components of $\boldsymbol{\lambda}$ gives zero identically. ∎

Next we wish to explore the relationship between the moments and the cumulants of a random vector. We will use boldface lowercase letters (e.g., $\boldsymbol{n}$) to denote sets of integers. Let $\boldsymbol{n}$ be the set of integers from 1 to n, and let $\boldsymbol{m}$ be a nonempty subset of $\boldsymbol{n}$. The notation $X_{\boldsymbol{m}}$ will stand for the vector consisting of the elements of X with indexes in $\boldsymbol{m}$. Thus $X_{\boldsymbol{n}}$ is identical to X. A *partition* of $\boldsymbol{n}$ is a collection of disjoint subsets of $\boldsymbol{n}$ whose union is $\boldsymbol{n}$. For example, $\{\{1,4\},\{2\},\{3,5\}\}$ is a partition of $\{1,2,3,4,5\}$. We will denote by $P(n)$ a generic partition of $\boldsymbol{n}$ and by $\mathcal{P}(n)$ the set of all partitions of $\boldsymbol{n}$.

Partitions can be generated inductively using the following procedure. The only partition of the set $\{1\}$ is $\{\{1\}\}$, so the corresponding set of partitions is $\{\{\{1\}\}\}$ (note the increasing number of braces as we proceed from a set to a partition of the set and then to the set of all partitions). Suppose we have generated the set of all partitions for n and we wish to generate the set corresponding to $n+1$. Clearly, we have only two choices: either $n+1$ will form a subset by itself, or it will be joined to an existing subset. Therefore, the procedure consists of the following steps.

1. Pick a partition $P(n) \in \mathcal{P}(n)$ and add to it the set $\{n+1\}$. Repeat for all members of $\mathcal{P}(n)$.
2. Pick a partition $P(n) \in \mathcal{P}(n)$ and a member of $P(n)$, and add to it the number $n+1$. Repeat for all members of $P(n)$ and for all $P(n) \in \mathcal{P}(n)$.
3. The set of all sets generated in steps 1 and 2 is the desired set of partitions for $n+1$.

Let us illustrate this procedure in going from $n=2$ to $n=3$. We have $\mathcal{P}(2) = \{\{\{1\},\{2\}\},\{\{1,2\}\}\}$. Step 1 yields the sets $\{\{1\},\{2\},\{3\}\}$ and $\{\{1,2\},\{3\}\}$. Step 2 yields the sets $\{\{1,3\},\{2\}\}$, $\{\{1\},\{2,3\}\}$, and $\{\{1,2,3\}\}$. Altogether we will get five partitions for $n=3$. The procedures in the Mathematica package `Partitio.m` generate the partitions of a given set by the inductive method described above.

The following lemma facilitates the expression of the moments in terms of the cumulants.

Lemma 10.3. Let $C(\boldsymbol{\lambda})$ be a function of the n-dimensional vector $\boldsymbol{\lambda}$ and $M(\boldsymbol{\lambda}) = \exp\{C(\boldsymbol{\lambda})\}$. Assume that $C(\boldsymbol{\lambda})$ is differentiable as needed. Then

$$\frac{\partial^n}{\partial \lambda_1 \ldots \partial \lambda_n} M(\boldsymbol{\lambda}) = M(\boldsymbol{\lambda}) F(\boldsymbol{\lambda}), \qquad (10.2.6a)$$

where

$$F(\boldsymbol{\lambda}) = \sum_{P(n) \in \mathcal{P}(n)} \prod_{\boldsymbol{m} \in P(n)} C_{\boldsymbol{m}}(\boldsymbol{\lambda}) \qquad (10.2.6b)$$

and $C_{\boldsymbol{m}}(\boldsymbol{\lambda})$ is the repeated partial derivative of $C(\boldsymbol{\lambda})$, differentiated once with respect to each member of $\boldsymbol{\lambda_m}$.

Proof. We will prove the above inductively on k, for $1 \leq k \leq n$. For $k=1$ we have

$$\frac{\partial M(\boldsymbol{\lambda})}{\partial \lambda_1} = M(\boldsymbol{\lambda}) \frac{\partial C(\boldsymbol{\lambda})}{\partial \lambda_1},$$

so (10.2.6a,b) hold. Assume they hold for k and differentiate with respect to λ_{k+1}:

$$\frac{\partial^{k+1}}{\partial \lambda_1 \dots \partial \lambda_{k+1}} M(\boldsymbol{\lambda}) = \frac{\partial M(\boldsymbol{\lambda})}{\partial \lambda_{k+1}} F(\boldsymbol{\lambda}) + M(\boldsymbol{\lambda}) \frac{\partial F(\boldsymbol{\lambda})}{\partial \lambda_{k+1}}$$

$$= M(\boldsymbol{\lambda}) \frac{\partial C(\boldsymbol{\lambda})}{\partial \lambda_{k+1}} F(\boldsymbol{\lambda}) + M(\boldsymbol{\lambda}) \frac{\partial F(\boldsymbol{\lambda})}{\partial \lambda_{k+1}} = M(\boldsymbol{\lambda}) \left[\frac{\partial C(\boldsymbol{\lambda})}{\partial \lambda_{k+1}} F(\boldsymbol{\lambda}) + \frac{\partial F(\boldsymbol{\lambda})}{\partial \lambda_{k+1}} \right].$$

We now observe that the two terms on the right side correspond exactly to steps 1 and 2 of the inductive procedure for partition generation. In the first term, $\partial C(\boldsymbol{\lambda}) / \partial \lambda_{k+1}$ corresponds to taking $\{k+1\}$ as a set by itself and adding it to each partition. In the second term, applying the rules of differentiation of sums and products to (10.2.6b) is equivalent to adding the number $k+1$ to all subsets of all the partitions in $\mathcal{P}(k)$. Therefore, (10.2.6a,b) hold for $k+1$ and the proof is complete.   ∎

Lemma 10.3 immediately implies the following result.

Theorem 10.1. The nth-order moment of X is related to the cumulants of the subvectors of X via

$$\mu(X) = \sum_{P(n) \in \mathcal{P}(n)} \prod_{\boldsymbol{m} \in P(n)} \kappa(X_{\boldsymbol{m}}). \qquad (10.2.7)$$

Proof. Substitute $\boldsymbol{\lambda} = 0$ in (10.2.6a,b) and recall that $M_X(0) = 1$ and that the moments/cumulants are the partial derivatives of the corresponding generating function evaluated at zero.   ∎

The inverse relationship, from moments to cumulants, is also of interest and is facilitated by the following lemma.

Lemma 10.4. Let $M(\boldsymbol{\lambda})$ be a function of the n-dimensional vector $\boldsymbol{\lambda}$ and $C(\boldsymbol{\lambda}) = \log\{M(\boldsymbol{\lambda})\}$. Assume that $M(\boldsymbol{\lambda})$ is differentiable as needed. Then

$$\frac{\partial^n}{\partial \lambda_1 \dots \partial \lambda_n} C(\boldsymbol{\lambda}) = \sum_{P(n) \in \mathcal{P}(n)} (-1)^{(r-1)} (r-1)! M^{-r}(\boldsymbol{\lambda}) G_{P(n)}(\boldsymbol{\lambda}), \qquad (10.2.8a)$$

where

$$G_{P(n)}(\boldsymbol{\lambda}) = \prod_{\boldsymbol{m} \in P(n)} M_{\boldsymbol{m}}(\boldsymbol{\lambda}) \qquad (10.2.8b)$$

and $M_{\boldsymbol{m}}(\boldsymbol{\lambda})$ is the repeated partial derivative of $M(\boldsymbol{\lambda})$, differentiated once with respect to each element of $\boldsymbol{\lambda}_{\boldsymbol{m}}$. The integer r is the number of sets in the partition $P(n)$.

Proof. We will prove the above inductively on k, for $1 \leq k \leq n$. For $k = 1$, we have

$$\frac{\partial C(\boldsymbol{\lambda})}{\partial \lambda_1} = M^{-1}(\boldsymbol{\lambda}) \frac{\partial M(\boldsymbol{\lambda})}{\partial \lambda_1},$$

so (10.2.8a,b) hold. Assume they hold for k and differentiate with respect to λ_{k+1}:

$$\frac{\partial^{k+1}}{\partial \lambda_1 \ldots \partial \lambda_{k+1}} C(\boldsymbol{\lambda}) = \sum_{P(n) \in \mathcal{P}(n)} (-1)^{(r-1)}(r-1)!(-r) M^{-(r+1)}(\boldsymbol{\lambda}) \frac{\partial M(\boldsymbol{\lambda})}{\partial \lambda_{k+1}} G_{P(n)}(\boldsymbol{\lambda})$$

$$+ \sum_{P(n) \in \mathcal{P}(n)} (-1)^{(r-1)}(r-1)! M^{-r}(\boldsymbol{\lambda}) \frac{\partial G_{P(n)}(\boldsymbol{\lambda})}{\partial \lambda_{k+1}}$$

$$= \sum_{P(n) \in \mathcal{P}(n)} (-1)^{r} r! M^{-(r+1)}(\boldsymbol{\lambda}) \frac{\partial M(\boldsymbol{\lambda})}{\partial \lambda_{k+1}} G_{P(n)}(\boldsymbol{\lambda})$$

$$+ \sum_{P(n) \in \mathcal{P}(n)} (-1)^{(r-1)}(r-1)! M^{-r}(\boldsymbol{\lambda}) \frac{\partial G_{P(n)}(\boldsymbol{\lambda})}{\partial \lambda_{k+1}}.$$

We now observe, similarly to the proof of Lemma 10.3, that the two terms on the right side correspond to steps 1 and 2 of the inductive procedure for partition generation. Therefore, (10.2.8a,b) hold for $k + 1$ and the proof is complete. ∎

Lemma 10.4 immediately implies the following result.

Theorem 10.2. The nth-order cumulant of X is related to the moments of the subvectors of X via

$$\kappa(X) = \sum_{P(n) \in \mathcal{P}(n)} (-1)^{(r-1)}(r-1)! \prod_{\boldsymbol{m} \in P(n)} \mu(X_{\boldsymbol{m}}). \qquad (10.2.9)$$

Proof. Substitute $\boldsymbol{\lambda} = 0$ in (10.2.8a,b) and repeat the argument in the proof of Theorem 10.1. ∎

Example 10.3. Theorems 10.1 and 10.2 give for $n = 2$,

$$\mu(X_1, X_2) = \kappa(X_1, X_2) + \kappa(X_1)\kappa(X_2), \quad \kappa(X_1, X_2) = \mu(X_1, X_2) - \mu(X_1)\mu(X_2).$$

Therefore, $\kappa(X_1, X_2)$ is the covariance of X_1 and X_2. In the special case $X_1 = X_2 = x$, $\kappa_2(x)$ is the variance of x.

For $n = 3$, we get

$$\mu(X_1, X_2, X_3) = \kappa(X_1, X_2, X_3) + \kappa(X_1)\kappa(X_2, X_3) + \kappa(X_2)\kappa(X_1, X_3)$$
$$+ \kappa(X_3)\kappa(X_1, X_2) + \kappa(X_1)\kappa(X_2)\kappa(X_3)$$
$$\kappa(X_1, X_2, X_3) = \mu(X_1, X_2, X_3) - \mu(X_1)\mu(X_2, X_3) - \mu(X_2)\mu(X_1, X_3)$$
$$- \mu(X_3)\mu(X_1, X_2) + 2\mu(X_1)\mu(X_2)\mu(X_3).$$ ∎

The expressions (10.2.7) and (10.2.9) yield considerably simpler results when X has zero mean. Recall that a set with exactly one member is called a *singleton*. All singleton subsets of a given partition contribute factors of the form $\mu(X_k)$ or $\kappa(X_k)$ to the term corresponding to that partition. If X has zero mean, the term vanishes. Thus, when X has zero mean, all partitions containing singletons can be removed from (10.2.7) and (10.2.9). For example, let $n = 4$ and assume that X has zero mean. Then the valid partitions are $\{\{1, 2, 3, 4\}\}$, $\{\{1, 2\}, \{3, 4\}\}$, $\{\{1, 3\}, \{2, 4\}\}$, and $\{\{1, 4\}, \{2, 3\}\}$. Substitution in (10.2.9) gives the fourth-order cumulant formula (2.4.6).

10.3. CUMULANTS AND POLYSPECTRA OF STATIONARY LINEAR PROCESSES

Let y_t be the linear process

$$y_t = \sum_k h_k w_{t-k} = \sum_k h_{t-k} w_k, \qquad (10.3.1)$$

where $\{h_k\}$ is an absolutely summable sequence and $\{w_t\}$ an i.i.d. sequence. Eventually, we will be interested in *causal* models, in which h_k is zero for $k < 0$. However, this restriction is not necessary for the time being, so the summation in (10.3.1) is from $-\infty$ to $+\infty$. The sequence $\{h_k\}$ can be regarded as the impulse response of a linear time-invariant system, and $\{y_t\}$ is the response of the system to the i.i.d. input process $\{w_t\}$. The absolute summability of the impulse response guarantees the BIBO (bounded input bounded output) stability of the linear system, hence the strict stationarity of the output process. We will assume that the moments of $|w_t|$ up to order n are finite for some $n > 2$.

Consider the cumulant $\kappa(y_t, y_{t+\tau_1}, \ldots, y_{t+\tau_{n-1}})$. Since the process $\{y_t\}$ is strictly stationary, this cumulant is independent of t, so it is a function of the $n-1$ integer variables $\{\tau_1, \ldots, \tau_{n-1}\}$. We will denote this function by $c_{\tau_1, \ldots, \tau_{n-1}}$ and call it the nth-order cumulant of the process. Note that the nth-order cumulant is a function of only $n-1$ variables. For example, the third- and fourth-order cumulants are denoted by c_{τ_1, τ_2} and $c_{\tau_1, \tau_2, \tau_3}$, respectively.

The basic relationship between the cumulants and the impulse response parameters is given by the following theorem.

Theorem 10.3.

$$c_{\tau_1,\ldots,\tau_{n-1}} = \kappa_n(w) \sum_k h_k h_{k+\tau_1} \ldots h_{k+\tau_{n-1}}, \tag{10.3.2}$$

where $\kappa_n(w)$ is the nth-order cumulant of w_t (which is independent of t by stationarity).

Partial Proof. Let us first assume that $\{h_k\}$ is finite, say $h_k \neq 0$ only for $-q \leq k \leq q$. By stationarity, there is no loss of generality in taking $t = 0$. Define $\tau_0 = 0$ for convenience. Then

$$y_{\tau_i} = \sum_k h_{k+\tau_i} w_{-k}.$$

Therefore,

$$\log E\left\{\exp\left(\sum_{i=0}^{n-1} \lambda_i y_{\tau_i}\right)\right\} = \log E\left\{\exp\left(\sum_k w_{-k} \sum_{i=0}^{n-1} \lambda_i h_{k+\tau_i}\right)\right\},$$

where the range of k is $[-q - \max\{\tau_i\}, q - \min\{\tau_i\}]$. Since this range is finite and since the w_k are i.i.d., we get

$$\log E\left\{\exp\left(\sum_{i=0}^{n-1} \lambda_i y_{\tau_i}\right)\right\} = \sum_k C_w\left(\sum_{i=0}^{n-1} \lambda_i h_{k+\tau_i}\right).$$

Denoting the sum over i by z and differentiating n times with respect to the $\{\lambda_i\}$ gives

$$\frac{\partial^n}{\partial \lambda_0 \ldots \partial \lambda_{n-1}} \log E\left\{\exp\left(\sum_{i=0}^{n-1} \lambda_i y_{\tau_i}\right)\right\} = \sum_k \left(\prod_{i=0}^{n-1} h_{k+\tau_i}\right) \frac{d^n C_w(z)}{dz^n}.$$

Substituting $z = 0$ yields (10.3.2) for the special case of finite impulse response.

The proof is completed by taking the limit $q \to \infty$ and using the absolute summability of the impulse response. It is not difficult to show that the moments of the partial sum $\sum_k h_{k+t} w_{-k}$ converge to the corresponding moments of y_t; hence the same is true for the cumulants. The details are omitted. ∎

The nth-order cumulant is a function of $n - 1$ variables, but this function possesses many symmetries and is therefore highly redundant. First, it is symmetric in its arguments, so we can rearrange them, for example, in an increasing order. Second, if one of the τ_i is negative, we can make the following transformation. Let $\tau_{\min} = \min\{\tau_i\}$, form the set $\{-\tau_{\min}, \tau_1 - \tau_{\min}, \ldots, \tau_{n-1} - \tau_{\min}\}$, and remove from it the element $\tau_{\min} - \tau_{\min}$ (which is zero). All the other elements in this set are nonnegative. Rearrange them in an increasing order and denote the resulting set by $\{\sigma_i\}$. It follows from (10.3.2) that $c_{\tau_1,\ldots,\tau_{n-1}} = c_{\sigma_1,\ldots,\sigma_{n-1}}$ (Prob. 10.6). Summing up, the cumulants corresponding to the indexes $0 \leq \tau_1 \leq \ldots \leq \tau_{n-1}$ form a minimal sufficient set.

Recall that the power spectral density of a stationary process is defined as the Fourier transform of the covariance sequence (whenever the transform exists). This notion can be extended to cumulants as follows. Let $\{\omega_i, 1 \leq i \leq n-1\}$ be real numbers in the range $[-\pi, \pi]$. Let

$$S(\omega_1, \ldots, \omega_{n-1}) = \sum_{\tau_1} \cdots \sum_{\tau_{n-1}} c_{\tau_1, \ldots, \tau_{n-1}} \exp\left\{ -j \sum_{i=1}^{n-1} \omega_i \tau_i \right\}. \tag{10.3.3}$$

$S(\omega_1, \ldots, \omega_{n-1})$ is called the nth-order *polyspectrum* of the process $\{y_t\}$. The polyspectrum is thus the $(n-1)$-dimensional Fourier transform of the nth-order cumulant sequence, provided the Fourier transform exists. The cumulants are then given by the inverse formula

$$c_{\tau_1, \ldots, \tau_{n-1}} = \frac{1}{(2\pi)^{(n-1)}} \int_{-\pi}^{\pi} \cdots \int_{-\pi}^{\pi} S(\omega_1, \ldots, \omega_{n-1}) \exp\left\{ j \sum_{i=1}^{n-1} \omega_i \tau_i \right\} d\omega_1 \ldots d\omega_{n-1}. \tag{10.3.4}$$

When $n = 3$, $S(\omega_1, \omega_2)$ is called the *bispectrum*. When $n = 4$, $S(\omega_1, \omega_2, \omega_3)$ is called the *trispectrum*. The reader is asked to explore the symmetries in $S(\omega_1, \omega_2)$ resulting from the symmetries in the cumulants (Prob. 10.7).

The next theorem establishes the existence of polyspectra for linear processes.

Theorem 10.4. The polyspectrum of a linear process exists and is continuous.

Proof.

$$\sum_{\tau_1} \cdots \sum_{\tau_{n-1}} \left| c_{\tau_1, \ldots, \tau_{n-1}} \exp\left\{ -j \sum_{i=1}^{n-1} \omega_i \tau_i \right\} \right| = \sum_{\tau_1} \cdots \sum_{\tau_{n-1}} \left| c_{\tau_1, \ldots, \tau_{n-1}} \right|$$

$$\leq |\kappa_n(w)| \sum_{\tau_1} \cdots \sum_{\tau_{n-1}} \sum_{k} |h_k h_{k+\tau_1} \ldots h_{k+\tau_{n-1}}| = |\kappa_n(w)| \left[\sum_{k} |h_k| \right]^n < \infty.$$

Therefore, the sum in the definition of the polyspectrum converges absolutely, so the existence and the continuity of $S(\omega_1, \ldots, \omega_{n-1})$ follow from Fourier transform theory. ∎

The polyspectrum can be expressed in terms of the Fourier transform of the impulse response as follows. Let

$$H(e^{j\omega}) = \sum_{k} h_k e^{-j\omega k}. \tag{10.3.5}$$

Then we have:

Theorem 10.5.

$$S(\omega_1, \ldots, \omega_{n-1}) = \kappa_n(w) H(e^{j\omega_1}) \ldots H(e^{j\omega_{n-1}}) H(e^{-j(\omega_1 + \cdots + \omega_{n-1})}). \tag{10.3.6}$$

The simple proof is left as an exercise to the reader (Prob. 10.8).   ∎

We now arrive at what is probably the most striking property of cumulants and polyspectra of linear processes: their relation to the phase of the transfer function $H(e^{j\omega})$. For better appreciation of this property, let us first recall the relationship of $H(e^{j\omega})$ to the spectral density:

$$S(\omega) = \sigma_w^2 H(e^{j\omega})H(e^{-j\omega}) = \sigma_w^2|H(e^{j\omega})|^2$$

[this is just a special case of (10.3.6) for $n = 2$]. It is clear that the knowledge of $S(\omega)$ (equivalently, of the covariance sequence) enables the determination of *the magnitude, but not the phase* of $H(e^{j\omega})$. Lii and Rosenblatt [1982] have discovered that, for non-Gaussian linear processes, it is possible (under certain conditions) to recover *both the magnitude and the phase* from the polyspectra (equivalently, the cumulants). This property fails to hold for Gaussian processes because their cumulants/polyspectra are identically zero for $n > 2$.

In preparation for the Lii-Rosenblatt theorem, we will prove the following lemma.

Lemma 10.5. Assume that $H(e^{j\omega}) \neq 0$ for all ω and that $\sum_k |kh_k| < \infty$. Then the phase of $H(e^{j\omega})$ is continuously differentiable with respect to ω.

Proof. $H(e^{j\omega})$ is continuous due to the absolute summability of the impulse response. Differentiation of (10.3.5) with respect to ω yields

$$\frac{dH(e^{j\omega})}{d\omega} = -j\sum_k kh_k e^{-j\omega k},$$

provided the right side converges absolutely. However, by our assumption, $\sum_k |kh_k e^{-j\omega k}| = \sum_k |kh_k| < \infty$. This proves that $H(e^{j\omega})$ is continuously differentiable. Denote $H(e^{j\omega}) = H_R(e^{j\omega}) + jH_I(e^{j\omega})$. Differentiation of $\tan^{-1}(H_I(e^{j\omega})/H_R(e^{j\omega}))$ with respect to ω yields

$$\frac{\frac{dH_I(e^{j\omega})}{d\omega}H_R(e^{j\omega}) - \frac{dH_R(e^{j\omega})}{d\omega}H_I(e^{j\omega})}{H_R^2(e^{j\omega}) + H_I^2(e^{j\omega})}.$$

Since $H(e^{j\omega}) \neq 0$ by assumption, it follows that this function exists and is continuous for all ω.   ∎

Now we prove the main theorem.

Theorem 10.6. Let $H(e^{j\omega})$ satisfy the assumptions of Lemma 10.5. Assume that $\kappa_n(w) \neq 0$ for some $n > 2$. Then $H(e^{j\omega})$ can be computed from $S(\omega_1, \ldots, \omega_{n-1})$ up to an unknown complex constant scale factor $Ae^{j\omega m}$, where A is real (either positive or negative) and m is an integer.

Proof. For any nonzero real number A and any integer m, the sequences $\{Aw_{t+m}\}$ and $\{(1/A)h_{k-m}\}$ yield the same output sequence $\{y_t\}$ as the sequences $\{w_t\}$ and $\{h_k\}$. Therefore, the sign, scale factor, and time origin of $\{h_k\}$ cannot be inferred from $S(\omega_1, \ldots, \omega_{n-1})$, and we might as well assume that $H(e^{j0}) = 1$. Having made this assumption, let us first prove the theorem for $n = 3$, assuming that $\kappa_3(w) \neq 0$.

Substituting $\omega_1 = \omega$ and $\omega_2 = 0$ in (10.3.6) yields $\kappa_3(w)|H(e^{j\omega})^2| = S(\omega, 0)$; hence

$$|H(e^{j\omega})| = \sqrt{\frac{S(\omega, 0)}{S(0, 0)}}.$$

Now let us denote the phase of $H(e^{j\omega})$ by $\phi(\omega)$ and the phase of $S(\omega_1, \omega_2)$ by $\psi(\omega_1, \omega_2)$. Also denote the derivative of $\phi(\omega)$ by $\dot{\phi}(\omega)$. By our assumption, $\phi(0) = 0$, and $\phi(\omega)$ is an odd function. Therefore, we get, from (10.3.6)

$$\phi(\omega_1) + \phi(\omega_2) - \phi(\omega_1 + \omega_2) = \psi(\omega_1, \omega_2).$$

Substitute $\omega_1 = \omega$ and $\omega_2 = d\omega$ to get

$$\phi(\omega) + \phi(d\omega) - \phi(\omega + d\omega) = \psi(\omega, d\omega),$$

so

$$\phi(\omega) + \phi(0) + \dot{\phi}(0)d\omega - \phi(\omega) - \dot{\phi}(\omega)d\omega + o(d\omega) = \left.\frac{\partial\psi(\omega, \omega_2)}{\partial\omega_2}\right|_{\omega_2=0} d\omega + o(d\omega).$$

Dividing by $d\omega$ and taking the limit as $d\omega \to 0$ yields

$$\dot{\phi}(\omega) = \dot{\phi}(0) - \left.\frac{\partial\psi(\omega, \omega_2)}{\partial\omega_2}\right|_{\omega_2=0}.$$

Therefore,

$$\phi(\omega) = \dot{\phi}(0)\omega - \int_0^\omega \left.\frac{\partial\psi(\nu, \omega_2)}{\partial\omega_2}\right|_{\omega_2=0} d\nu.$$

Denote

$$\gamma(\omega) = \int_0^\omega \left.\frac{\partial\psi(\nu, \omega_2)}{\partial\omega_2}\right|_{\omega_2=0} d\nu,$$

so

$$\phi(\omega) = \dot{\phi}(0)\omega - \gamma(\omega).$$

Since $H(e^{j\pi}) = \sum_k h_k(-1)^k$ is real, we must have $\phi(\pi) = m\pi$, where m is an integer. Therefore,

$$\phi(\pi) = m\pi = \dot{\phi}(0)\pi - \gamma(\pi).$$

This enables us to obtain $\dot{\phi}(0)$ up to an unknown integer, giving finally

$$\phi(\omega) = m\omega + \frac{\omega}{\pi}\gamma(\pi) - \gamma(\omega).$$

We have thus proved the theorem for $n = 3$. If $\kappa_3(w) = 0$ but $\kappa_n(w) \neq 0$ for some $n > 3$, we simply take $\omega_i = 0, i > 2$ and repeat the above argument. ∎

Theorem 10.6 can be interpreted as follows. Suppose we are given a realization of the output process $\{y_t, t \in \mathbf{Z}\}$, but not of the input process $\{w_t\}$. Since the process is ergodic, we can compute all the cumulants of all orders and use them to obtain the polyspectra (in principle, of course, not in practice). If $H(e^{j\omega})$ satisfies the conditions of Lemma 10.5, we can compute it up to a sign, scale factor, and a linear phase term. We can then (again, in principle) use $H(e^{j\omega})$ to "deconvolve" $\{y_t\}$ and obtain $\{w_t\}$ up to sign, scale, and time shift. The only restriction on this hypothetical procedure [except for the theorem's conditions on $H(e^{j\omega})$] is that the process be non-Gaussian. Thus, the class of Gaussian processes is really "underprivileged": it is the only class for which reconstruction of the input from the output is impossible under any circumstances.[†]

To use Theorem 10.6 in practice, approximations have to be made. The cumulants, hence the polyspectra, can be estimated from a finite sample and used to estimate the transfer function. Under additional hypotheses, the estimated transfer function can be used to perform approximate deconvolution of the output process. Procedures of this kind will be explored in later sections.

10.4. THE CUMULANTS OF ARMA PROCESSES

ARMA processes are a special case of linear processes satisfying the absolute summability condition of the impulse response. Therefore, the theory of the previous section applies to them. In this section we explore the relationships between the cumulants and the ARMA parameters—relationships that will prove useful in the subsequent sections.

Let us first consider moving average processes. Since MA processes have a finite impulse response, their cumulants are given as a special case of (10.3.2):

$$c_{\tau_1,\dots,\tau_{n-1}} = \kappa_n(w) \sum_{k=0}^{q-\tau_{\max}} b_k b_{k+\tau_1} \dots b_{k+\tau_{n-1}}, \qquad (10.4.1)$$

where we have assumed that all τ_k are nonnegative, and $\tau_{\max} = \max\{\tau_i\}$. In particular, since $b_0 = 1$,

$$c_{0,\dots,0,q} = \kappa_n(w)b_q, \qquad c_{0,\dots,k,q} = \kappa_n(w)b_k b_q.$$

[†] Strictly speaking, this is not true. If we further assume that $H(e^{j\omega})$ is *minimum phase*, then $\phi(\omega)$ is uniquely determined by $|H(e^{j\omega})|$ as shown, for example, in Oppenheim and Schafer [1975, Sec. 7.2]. The magnitude function $|H(e^{j\omega})|$, in turn, can be computed from $S(\omega)$. Hence deconvolution will be possible for Gaussian processes as well in this case. The minimum-phase condition is, however, a very restrictive one and is often violated in real-life problems.

Since $b_q \neq 0$ by definition, we get

$$b_k = \frac{c_{0,\ldots,k,q}}{c_{0,\ldots,0,q}}, \quad 1 \leq k \leq q, \tag{10.4.2}$$

provided $\kappa_n(w) \neq 0$. This result is due to Giannakis [1987]. Formulas (10.4.1) and (10.4.2) are complete and well-defined procedures for computing the cumulants from the MA parameters, and vice versa. Note, in particular, that there are no limitations on $b(z)$; its zeros can be anywhere in the complex plane, including the unit circle. The only requirements are that the *true* order q be known and that the input cumulant $\kappa_n(w)$ be nonzero.

In the case of ARMA processes, the sum in (10.3.2) involves an infinite number of terms. The following computational procedure for this sum, based on the use of Kronecker products and Liapunov equations, was proposed by Porat and Friedlander [1989]; see also Swami and Mendel [1989]. Let $\{A, B, C\}$ be any state-space realization of the ARMA process [e.g., (2.12.13)]. Then

$$h_t^{(i)} = h_{t+\tau_i} = CA^{t+\tau_i}B = CA^{\tau_i}A^t B;$$

thus $\{A, B, CA^{\tau_i}\}$ is a state-space realization of $h_t^{(i)}$. We have, by (10.3.2),

$$c_{\tau_1,\ldots,\tau_{n-1}} = \kappa_n(w) \sum_{t=0}^{\infty} \prod_{i=0}^{n-1} h_t^{(i)}. \tag{10.4.3}$$

Appendix D shows how to evaluate the sum (10.4.3) by solving a matrix Liapunov equation. As explained there, the state-space realization should be brought first to a complex Schur form by a similarity transformation.

The inverse problem, that of computing the ARMA parameters from the cumulants, can be solved in two steps, first computing the AR parameters and then the MA parameters. To compute the AR parameters we use the identity

$$h_\ell + \sum_{k=1}^{p} a_k h_{\ell-k} = 0, \quad \forall \ell \geq q+1. \tag{10.4.4}$$

Pick $\tau_1 \geq q+1$ and let the other τ's be arbitrary. Then substitute (10.4.4) in (10.3.2) to get

$$\sum_{k=1}^{p} a_k c_{\tau_1-k,\ldots,\tau_{n-1}} = -c_{\tau_1,\ldots,\tau_{n-1}}. \tag{10.4.5}$$

This identity can generate a set of equations as large as we wish for the unknowns $\{a_k\}$. Actually, we only need p independent equations. For example, when $n=2$, we get the modified Yule-Walker equations upon choosing $q+1 \leq \tau_1 \leq q+p$. Thus, (10.4.5) can be viewed as a generalization of the modified Yule-Walker equations to high-order cumulants. When $n > 2$, choosing a set of p independent equations from (10.4.5) is not a trivial matter. The following theorem, due to Swami [1988], gives a sufficient condition for independence of the generalized Yule-Walker equations.

Theorem 10.7. The set of equations (10.4.5) corresponding to $q+1 \leq \tau_1 \leq q+p$, $q - p \leq \tau_2 \leq q$, and $\tau_i = 0, i \geq 3$ has a unique solution. The solution is the vector of true AR parameters.

Proof. Assume that some set of numbers $\{\alpha_k, 0 \leq k \leq p, \alpha_0 = 1\}$ satisfies (10.4.5) with the τ's given in the theorem. Then

$$0 = \sum_{k=0}^{p} \alpha_k c_{\tau_1-k,\tau_2,0,\dots,0} = \kappa_n(w) \sum_{k=0}^{p} \alpha_k \sum_{t=0}^{\infty} h_t^{n-2} h_{t+\tau_1-k} h_{t+\tau_2}$$

$$= \kappa_n(w) \sum_{t=0}^{\infty} h_t^{n-2} \left[\sum_{k=0}^{p} \alpha_k h_{t+\tau_1-k} \right] h_{t+\tau_2} = \kappa_n(w) \sum_{t=0}^{\infty} g_t h_{t+\tau_2}, \quad (10.4.6)$$

where the definition of g_t is obvious from this equation. Since (10.4.6) holds for the $p + 1$ consecutive values $q - p \leq \tau_2 \leq q$, and due to the linear dependence (10.4.4), it holds in fact for all $\tau_2 \geq q - p$ [prove by induction on τ_2 using (10.4.4)]. The right side of (10.4.6) is the coefficient of $z^{-\tau_2}$ in $G(z^{-1})H(z)$, where $G(z)$ and $H(z)$ are the z transforms of $\{g_t\}$ and $\{h_t\}$, respectively. Therefore, we conclude that the terms in the expansion of $G(z^{-1})H(z)$ are identically zero for all $\tau_2 \geq q - p$. Recalling that $H(z) = b(z)/a(z)$ and multiplying by $a(z)$, we conclude that the terms in the expansion of $G(z^{-1})b(z)$ are identically zero for all $\tau_2 \geq q$. However, the coefficient of z^{-q} in this expansion is $g_0 b_q$, and $b_q \neq 0$. Therefore, necessarily

$$g_0 = h_0^{n-2} \left[\sum_{k=0}^{p} \alpha_k h_{\tau_1-k} \right] = \sum_{k=0}^{p} \alpha_k h_{\tau_1-k} = 0, \quad q+1 \leq \tau_1 \leq q+p. \quad (10.4.7)$$

Finally, it is a known fact from linear system theory that (10.4.7), together with the coprimeness of $a(z)$ and $b(z)$, implies that (10.4.5) has the unique solution $\alpha_k = a_k, 0 \leq k \leq p$ [Kailath, 1981, Sec. 5.1]. ∎

According to Theorem 10.7, we must collect $p(p + 1)$ equations to guarantee that p of them be independent. Note, however, that in the case $n = 2$, p equations are sufficient (the modified Yule-Walker equations). It is an open question whether the number $p(p + 1)$ can be reduced for $n > 2$.

Once the AR parameters have been computed, the impulse response sequence can be computed by the following formula, due to Swami [1988].

Lemma 10.6.

$$h_\ell = \frac{\sum_{k=0}^{p} a_k c_{0,\dots,\ell,q-k}}{\sum_{k=0}^{p} a_k c_{0,\dots,0,q-k}}. \quad (10.4.8)$$

Proof.

$$\sum_{k=0}^{p} a_k c_{0,\dots,\ell,q-k} = \kappa_n(w) \sum_{k=0}^{p} \sum_{t=0}^{\infty} a_k h_t^{n-2} h_{t+\ell} h_{t+q-k} = \kappa_n(w) \sum_{t=0}^{\infty} h_t^{n-2} h_{t+\ell} b_{t+q}$$

$$= \kappa_n(w) h_0^{n-2} h_\ell b_q.$$

Substitute $\ell = 0$ to get

$$\sum_{k=0}^{p} a_k c_{0,\dots,0,q-k} = \kappa_n(w) h_0^{n-1} b_q.$$

Dividing the first equality by the second and using the fact that $h_0 = 1$ yields (10.4.8).   ∎

The $\{b_\ell\}$ can be computed as

$$b_\ell = \sum_{k=0}^{\min\{p,\ell\}} a_k h_{\ell-k}, \quad 1 \le \ell \le q. \tag{10.4.9}$$

We should finally emphasize that the recipes given in this section for computing the ARMA (or the MA) parameters from the cumulants are important from a theoretical viewpoint, but are not necessarily useful as practical estimation algorithms. The estimation problem is discussed in the subsequent sections.

10.5. ESTIMATION OF THE CUMULANTS

This section explores the problem of estimating the cumulants of a stationary linear process from a realization of the process and develops the second-order moments of the estimated cumulants.

Suppose we are given a set of measurements $\{y_t, \; 0 \le t \le N - 1\}$. The standard way of estimating the cumulants is by two steps: first we estimate the moments and then we use the moments-to-cumulants formula (10.2.9) to estimate the cumulants. Let us denote, in analogy with the cumulant notation, the nth-order moments of a stationary process by

$$m_{\tau_1,\dots,\tau_{n-1}} = E\{y_t y_{t+\tau_1} \cdots y_{t+\tau_{n-1}}\}.$$

The *sample moments* are defined in analogy with the definition of sample covariances,

$$\hat{m}_{\tau_1,\dots,\tau_{n-1}} = \frac{1}{N} \sum_{t=0}^{N-\tau_{\max}-1} y_t y_{t+\tau_1} \cdots y_{t+\tau_{n-1}}, \tag{10.5.1}$$

where all the τ_i are nonnegative. The sample moments for negative time lags can be expressed by moments of positive time lags using the transformation described in Sec. 10.3.

The *sample cumulants* $\hat{c}_{\tau_1,\ldots,\tau_{n-1}}$ are defined by formula (10.2.9), where $X = [y_t, y_{t+\tau_1}, \ldots, y_{t+\tau_{n-1}}]$, and the moments in the right side are replaced by the corresponding sample moments.

Example 10.4. Let y_t be zero mean. Then the third-order sample cumulants are given by

$$\hat{c}_{\tau_1,\tau_2} = \hat{m}_{\tau_1,\tau_2} = \frac{1}{N} \sum_{t=0}^{N-\tau_{\max}-1} y_t y_{t+\tau_1} y_{t+\tau_2},$$

while the fourth-order sample cumulant is given by

$$\hat{c}_{\tau_1,\tau_2,\tau_3} = \hat{m}_{\tau_1,\tau_2,\tau_3} - \hat{m}_{\tau_1}\hat{m}_{\tau_3-\tau_2} - \hat{m}_{\tau_2}\hat{m}_{\tau_3-\tau_1} - \hat{m}_{\tau_3}\hat{m}_{\tau_2-\tau_1},$$

where

$$\hat{m}_{\tau_1,\tau_2,\tau_3} = \frac{1}{N} \sum_{t=0}^{N-\tau_{\max}-1} y_t y_{t+\tau_1} y_{t+\tau_2} y_{t+\tau_3}$$

and $\hat{m}_k = \hat{r}_k$, the sample covariance of lag k. ∎

In the case of linear processes, the sample moments, as well as the sample cumulants, have "good" asymptotic properties, provided all the moments of the input process $\{w_t\}$ are finite. Specifically, they are strongly consistent estimates of the true moments/cumulants and are asymptotically normal. The proofs of these results, however, are beyond the scope of this book: they rely on extensions of the strong law of large numbers and the central limit theorem that the reader of this book may not be familiar with.

In the preceding chapters, the covariances of the sample covariances played a central role, both for development of algorithms and for their analysis. In particular, we frequently used the asymptotic Bartlett formula (4.2.12). When dealing with algorithms based on the high-order cumulants, it will be equally important to know the covariances of the sample cumulants. The rest of this section is devoted to this task. We first derive the asymptotic covariances of the sample moments and then use them for the asymptotic covariances of the sample cumulants. Our derivation is an extension of the one in Porat and Friedlander [1989].

Let us introduce the notations $\boldsymbol{\tau} = \{0, \tau_1, \ldots, \tau_{n-1}\}$ and $\boldsymbol{\sigma} = \{0, \sigma_1, \ldots, \sigma_{k-1}\}$ and use them in entities such as $m_{\boldsymbol{\tau}}$. In particular, let

$$\hat{m}_{\boldsymbol{\tau}} = \frac{1}{N} \sum_{t=0}^{N-1} y_t y_{t+\tau_1} \cdots y_{t+\tau_{n-1}}, \quad \hat{m}_{\boldsymbol{\sigma}} = \frac{1}{N} \sum_{s=0}^{N-1} y_s y_{s+\sigma_1} \cdots y_{s+\sigma_{k-1}}.$$

We have assumed, in order to simplify the derivation, that the measurements are available in the interval $[0, N+r-1]$, where $r = \max\{\tau_{\max}, \sigma_{\max}\}$. This assumption does not affect the asymptotic results. Note that the sample moments thus defined are unbiased. Then

$$E\{(\hat{m}_{\boldsymbol{\tau}} - m_{\boldsymbol{\tau}})(\hat{m}_{\boldsymbol{\sigma}} - m_{\boldsymbol{\sigma}})\}$$

$$= \frac{1}{N^2} \sum_{t=0}^{N-1} \sum_{s=0}^{N-1} [E\{y_t y_{t+\tau_1} \cdots y_{t+\tau_{n-1}} y_s y_{s+\sigma_1} \cdots y_{s+\sigma_{k-1}}\} - m_{\boldsymbol{\tau}} m_{\boldsymbol{\sigma}}]$$

$$= \frac{1}{N^2} \sum_{t=0}^{N-1} \sum_{s=0}^{N-1} [E\{y_0 y_{\tau_1} \cdots y_{\tau_{n-1}} y_{s-t} y_{s-t+\sigma_1} \cdots y_{s-t+\sigma_{k-1}}\} - m_{\boldsymbol{\tau}} m_{\boldsymbol{\sigma}}]$$

$$= \frac{1}{N} \sum_{t=-(N-1)}^{N-1} \left(1 - \frac{|t|}{N}\right) [E\{y_0 y_{\tau_1} \cdots y_{\tau_{n-1}} y_t y_{t+\sigma_1} \cdots y_{t+\sigma_{k-1}}\} - m_{\boldsymbol{\tau}} m_{\boldsymbol{\sigma}}].$$

So, by [A10],

$$\lim_{N\to\infty} N\mathrm{cov}\{\hat{m}_{\boldsymbol{\tau}}, \hat{m}_{\boldsymbol{\sigma}}\} = \sum_{t=-\infty}^{\infty} [E\{y_0 y_{\tau_1} \cdots y_{\tau_{n-1}} y_t y_{t+\sigma_1} \cdots y_{t+\sigma_{k-1}}\} - m_{\boldsymbol{\tau}} m_{\boldsymbol{\sigma}}].$$

$$(10.5.2)$$

Formula (10.5.2) can be regarded as a generalization of Bartlett's asymptotic formula. The reader is asked to verify that it indeed yields Bartlett's formula for $n = k = 2$ (Prob. 10.11).

The general procedure for computing the asymptotic covariance of $\hat{m}_{\boldsymbol{\tau}}$ and $\hat{m}_{\boldsymbol{\sigma}}$ is as follows. The moments $E\{y_0 y_{\tau_1} \cdots y_{\tau_{n-1}} y_t y_{t+\sigma_1} \cdots y_{t+\sigma_{k-1}}\}$ have to be evaluated for all t. This is done by computing all the cumulants corresponding to subsets of $\boldsymbol{\tau} \cup (\boldsymbol{\sigma}+t)$ and using them in the cumulants-to-moments formula (10.2.7). The cumulants are computed as explained in Sec. 10.4 (depending on whether the given process is MA or ARMA). We then subtract $m_{\boldsymbol{\tau}} m_{\boldsymbol{\sigma}}$ and sum over all t. The cumulants-to-moments formula yields a number of terms equal to the number of partitions in $\mathcal{P}(n+k)$. For example, when $n = k = 3$ (case of third-order moments), the number of terms is 41 if the process is zero mean, and 203 otherwise. When $n = k = 4$ (case of fourth-order moments), the number of terms is 715 if the process is zero mean, and 4140 otherwise. A few of the terms add up to $m_{\boldsymbol{\tau}} m_{\boldsymbol{\sigma}}$, so they are canceled. The 41 terms corresponding to a zero mean process with $n = k = 3$ are enumerated in Porat and Friedlander [1989].

Example 10.5. For $n = 2$, $k = 3$, and zero mean processes, formula (10.5.2) expands to

$$\lim_{N\to\infty} N\mathrm{cov}\{\hat{m}_{\boldsymbol{\tau}}, \hat{m}_{\boldsymbol{\sigma}}\} = \sum_{t=-\infty}^{\infty} [c_{\tau_1,t,t+\sigma_1,t+\sigma_2} + c_t c_{t+\sigma_1-\tau_1,t+\sigma_2-\tau_1}$$

$$+ c_{t+\sigma_1} c_{t-\tau_1,t+\sigma_2-\tau_1} + c_{\sigma_1} c_{\tau_1,t+\sigma_2} + c_{t-\tau_1} c_{t+\sigma_1,t+\sigma_2} + c_{t+\sigma_1-\tau_1} c_{t,t+\sigma_2}$$

$$+ c_{t+\sigma_2-\tau_1} c_{t,t+\sigma_1} + c_{\sigma_2-\sigma_1} c_{\tau_1,t} + c_{\sigma_2} c_{\tau_1,t+\sigma_1} + c_{t+\sigma_2} c_{t-\tau_1,t+\sigma_1-\tau_1}]. \quad (10.5.3)$$

∎

In the case of an MA process, the sum in (10.5.2) includes a finite number of terms, corresponding to $-q-r \leq t \leq q+r$ (recall that r is the maximum of the τ's and the σ's). It can thus be computed directly, using the cumulant formula for MA processes given in Sec. 10.4. The situation is much more complicated in the case of ARMA processes, since we must derive a closed-form formula for the infinite sum. We now show how such a formula can be derived for any given choice of $\boldsymbol{\tau}$ and $\boldsymbol{\sigma}$.

1. The first step is to break the sum in (10.5.2) into three parts: from $-\infty$ to $-(r+1)$ (the lower sum), from $-r$ to r (the middle sum), and from $r+1$ to ∞ (the upper sum). We then make the variable substitution $t = -t' - r - 1$ in the lower sum and the substitution $t = t' + r + 1$ in the upper sum. This gives, after some calculations,

$$\lim_{N \to \infty} N\text{cov}\{\hat{m}_{\boldsymbol{\tau}}, \hat{m}_{\boldsymbol{\sigma}}\} = \sum_{t=-r}^{r} [E\{y_0 y_{\tau_1} \cdots y_{\tau_{n-1}} y_t y_{t+\sigma_1} \cdots y_{t+\sigma_{k-1}}\} - m_{\boldsymbol{\tau}} m_{\boldsymbol{\sigma}}]$$

$$+ \sum_{t=0}^{\infty} [E\{y_0 y_{\tau_1} \cdots y_{\tau_{n-1}} y_{t+r+1} y_{t+\sigma_1+r+1} \cdots y_{t+\sigma_{k-1}+r+1}\} - m_{\boldsymbol{\tau}} m_{\boldsymbol{\sigma}}]$$

$$+ \sum_{t=0}^{\infty} [E\{y_0 y_{\sigma_1} \cdots y_{\sigma_{k-1}} y_{t+r+1} y_{t+\tau_1+r+1} \cdots y_{t+\tau_{n-1}+r+1}\} - m_{\boldsymbol{\tau}} m_{\boldsymbol{\sigma}}]. \quad (10.5.4)$$

2. The first sum in (10.5.4) is computed term by term. The moment for the tth term is expressed as a sum of products of cumulants, and the cumulants are computed as described in Sec. 10.4.

3. The other two terms can be computed by one procedure, since they are identical except for the interchange of $\boldsymbol{\tau}$ and $\boldsymbol{\sigma}$. When expressing the tth moment as a sum of product of cumulants, a typical cumulant will include both t-dependent and t-independent indexes. The idea is to express each cumulant as the tth term of the impulse response of some state-space model. Then the infinite sum over t is brought to the form given in (d.26) (see App. D), and the summation is performed as described there. It only remains to show how to find the state-space matrices of the impulse responses.

Let the indexes of a typical cumulant in a typical product be denoted by $\{\beta_1, \ldots, \beta_\ell, \beta_{\ell+1} + t, \ldots, \beta_{\ell+m} + t\}$, where the β's are nonnegative. Subtract from each of these indexes the minimum of $\{\beta_1, \ldots, \beta_\ell\}$, and rename the resulting indexes as $\{\alpha_i, 1 \leq i \leq \ell + m\}$. Observing (10.5.4), we see that the presence of $r + 1$ in the t-dependent indexes guarantees that $\{\alpha_i\}$ be nonnegative. Also, at least one of $\{\alpha_i, 1 \leq i \leq \ell\}$ is zero. Now define

$$H(\boldsymbol{\alpha}) = \sum_{s=0}^{\infty} \prod_{i=1}^{\ell} h_{s+\alpha_i} \prod_{i=1}^{m} h_{s+\alpha_{\ell+i}+t}.$$

$H(\boldsymbol{\alpha})$ is, up to a κ-factor, the cumulant in question. Using the Kronecker product identities as in App. D, we get

$$H(\boldsymbol{\alpha}) = \sum_{s=0}^{\infty} \left[\bigotimes_{i=1}^{\ell} (CA^{s+\alpha_i} B) \right] \otimes \left[\bigotimes_{i=1}^{m} (CA^{s+\alpha_{\ell+i}+t} B) \right]$$

$$= \sum_{s=0}^{\infty} \boldsymbol{C} [(A^{\otimes \ell})^s \otimes (A^{\otimes m})^{s+t}] B^{\otimes(\ell+m)},$$

where

$$\boldsymbol{C} = \bigotimes_{i=1}^{\ell+m} (CA^{\alpha_i}).$$

Also,

$$(A^{\otimes \ell})^s \otimes (A^{\otimes m})^{s+t} = [\mathbf{1}^t (A^{\otimes \ell})^s] \otimes [(A^{\otimes m})^t (A^{\otimes m})^s] = (\mathbf{1} \otimes A^{\otimes m})^t (A^{\otimes(\ell+m)})^s,$$

where the dimension of the identity matrix is equal to that of $A^{\otimes \ell}$. Therefore,

$$H(\boldsymbol{\alpha}) = \sum_{s=0}^{\infty} \boldsymbol{C} (\mathbf{1} \otimes A^{\otimes m})^t (A^{\otimes(\ell+m)})^s B^{\otimes(\ell+m)}$$

$$= \boldsymbol{C} (\mathbf{1} \otimes A^{\otimes m})^t (\mathbf{1} - A^{\otimes(\ell+m)})^{-1} B^{\otimes(\ell+m)}.$$

In summary, $H(\boldsymbol{\alpha})$ can regarded as the tth term of the impulse response sequence of the state-space triplet $\{\mathbf{1} \otimes A^{\otimes m}, (\mathbf{1} - A^{\otimes(\ell+m)})^{-1} B^{\otimes(\ell+m)}, \boldsymbol{C}\}$. Those triplets are computed for all the cumulants in the product and then (d.26) is computed as described in App. D.

So far we have shown how to compute the asymptotic covariances of the sample moments. To compute the asymptotic covariances of the sample cumulants, we use the Jacobian transformation rule $\Sigma_c = G\Sigma_m G^T$, where Σ_m is the sample moment covariance, Σ_c the sample cumulant matrix, and G is the Jacobian matrix of the cumulants with respect to the moments. This matrix is obtained from the moments-to-cumulants formula (10.2.9).

Example 10.6. Suppose the process has nonzero mean μ. By Example 10.3,

$$c_{\tau_1,\tau_2} = m_{\tau_1,\tau_2} - (r_{\tau_1} + r_{\tau_2} + r_{\tau_1-\tau_2})\mu + 2\mu^3.$$

Hence c_{τ_1,τ_2} depends on five different moments (including μ), and the corresponding row of G is given by

$$G_{\tau_1,\tau_2} = [1, -\mu, -\mu, -\mu, 6\mu^2 - (r_{\tau_1} + r_{\tau_2} + r_{\tau_1-\tau_2})].$$

Thus, for example,

$$\lim_{N \to \infty} \mathrm{var}\{\hat{c}_{\tau_1,\tau_2}\} = G_{\tau_1,\tau_2} \Sigma_m G_{\tau_1,\tau_2},$$

where Σ_m is the 5×5 matrix of the normalized asymptotic covariances of $\{\hat{m}_{\tau_1,\tau_2}, \hat{r}_{\tau_1}, \hat{r}_{\tau_2}, \hat{r}_{\tau_1-\tau_2}, \hat{\mu}\}$.   ∎

Problem 10.13 discusses the implementation of this method in Mathematica.

10.6. MOVING AVERAGE PARAMETER ESTIMATION: LINEAR METHODS

As we saw in the chapters dealing with Gaussian processes, the parameters of moving average processes (or the MA parameters of ARMA processes) can be consistently estimated only if the zeros of $b(z)$ are inside the unit circle. On the other hand, we saw in Sec. 10.4 that the cumulants of non-Gaussian MA processes can be used to compute the MA parameters in a simple manner [cf. (10.4.2)]. It follows that consistent estimates of the cumulants can be used to get consistent estimates of the parameters. The method of (10.4.2) is not necessarily good for estimation purposes though, since its performance may be (and usually is) relatively poor. In this section we present several linear estimation methods for the parameters of non-Gaussian MA processes. Methods of this type were first proposed by Giannakis [1987], extended by Giannakis and Mendel [1989] and Tugnait [1990], and analyzed by Porat and Friedlander [1989].

The linear methods are based on the following idea. Let $\tau_1 = \tau$ and $\tau_{i+1} = \tau + \sigma_i, 1 \leq i \leq n - 2$. Let us denote $\boldsymbol{\sigma} = \{\sigma_1, \ldots, \sigma_{n-2}\}$. Define the sequence $g_t(\boldsymbol{\sigma})$ by

$$g_t(\boldsymbol{\sigma}) = h_t h_{t+\sigma_1} \ldots h_{t+\sigma_{n-2}}. \tag{10.6.1}$$

Then we get, from (10.3.2),

$$c_{\tau, \tau+\sigma_1, \ldots, \tau+\sigma_{n-2}} = \kappa_n(w) \sum_t h_t g_{t+\tau}(\boldsymbol{\sigma}). \tag{10.6.2}$$

The left side of (10.6.2) represents a noncausal sequence in the integer index τ. This sequence is given [up to the scale factor $\kappa_n(w)$] by the cross-correlation of the sequences $\{h_t\}$ and $\{g_t(\boldsymbol{\sigma})\}$. Denote the z transforms of the three sequences by $C(z; \boldsymbol{\sigma})$, $H(z)$, and $G(z; \boldsymbol{\sigma})$, respectively. Then (10.6.2) is equivalent to

$$C(z; \boldsymbol{\sigma}) = \kappa_n(w) H(z^{-1}) G(z; \boldsymbol{\sigma}). \tag{10.6.3}$$

Also, recall that the spectral density, regarded as a function of z, is given by

$$S(z) = \sigma_w^2 H(z) H(z^{-1}). \tag{10.6.4}$$

Equations (10.6.3) and (10.6.4) give, upon elimination of $H(z^{-1})$,

$$H(z) C(z; \boldsymbol{\sigma}) = \frac{\kappa_n(w)}{\sigma_w^2} S(z) G(z; \boldsymbol{\sigma}). \tag{10.6.5}$$

This equation is an extension of the one given originally in [Giannakis, 1987] and has come to be known as the Giannakis-Mendel equation. Its left side yields a set of linear combinations of the impulse response parameters whose coefficients are the cumulants. Its right side yields a set of linear combinations of nonlinear functions of the impulse response parameters (the elements of the sequence $\{g_t(\boldsymbol{\sigma})\}$) whose coefficients are the covariances.

The algorithm of Giannakis and Mendel exploits Eq. (10.6.5) by using it with $\sigma_i = 0, 1 \leq i \leq n-2$. In this case, $g_t(\boldsymbol{\sigma}) = h_t^{n-1}$. Let us also denote $c_{\tau,\tau,\ldots,\tau} = \gamma_\tau$ and $\varepsilon = -\kappa_n(w)/\sigma_w^2$. Since the process is assumed to be MA, (10.6.5) can be expressed by the equivalent time domain form

$$\sum_{\tau=0}^{q} b_\tau \gamma_{k-\tau} + \varepsilon \sum_{\tau=0}^{q} b_\tau^{n-1} r_{k-\tau} = 0. \qquad (10.6.6)$$

Collecting these equations for $-q \leq k \leq 2q$ gives

$$\begin{bmatrix} \gamma_{-q} \\ \gamma_{-q+1} \\ \vdots \\ \gamma_q \\ 0 \\ \vdots \\ 0 \end{bmatrix} + \begin{bmatrix} 0 & \cdots & 0 & r_{-q} & 0 & \cdots & 0 \\ \gamma_{-q} & \cdots & 0 & r_{-q+1} & r_{-q} & \cdots & 0 \\ \vdots & \ddots & \vdots & \vdots & \vdots & \ddots & \vdots \\ \gamma_{q-1} & \cdots & \gamma_0 & r_q & r_{q-1} & \cdots & r_0 \\ \gamma_q & \cdots & \gamma_1 & 0 & r_q & \cdots & r_1 \\ \vdots & \ddots & \vdots & \vdots & \vdots & \ddots & \vdots \\ 0 & \cdots & \gamma_q & 0 & 0 & \cdots & r_q \end{bmatrix} \begin{bmatrix} b_1 \\ \vdots \\ b_q \\ \varepsilon \\ \varepsilon b_1^{n-1} \\ \vdots \\ \varepsilon b_q^{n-1} \end{bmatrix} = 0. \qquad (10.6.7)$$

These are $3q+1$ equations in $2q+1$ unknowns. The estimate is obtained by substituting $\hat{\gamma}_\tau$ and $\hat{r}_\tau$ for the corresponding true values and solving (10.6.7) in the least-squares sense (or weighted least-squares). This yields estimates $\{\hat{b}_k\}$ and $\hat{\varepsilon}$, as well as estimates $\widehat{\varepsilon b_k^{n-1}}$. The latter can be discarded or used to improve the final estimates of the MA parameters. For example, we can divide by $\hat{\varepsilon}$ to obtain $\widehat{b_k^{n-1}}$ and then take the $(n-1)$th root. If n is even, this provides an additional estimate that can be averaged with $\hat{b}_k$. If n is odd, it provides an additional estimate of $|b_k|$.

A sufficient condition for consistency of the estimate obtained from (10.6.7) is that the matrix appearing in this equation be full rank. As was shown in Giannakis and Mendel [1989], this need not be the case: there exist values of $\{b_k\}$ for which this matrix is rank deficient. However, the set of such $\{b_k\}$ has measure zero, so this by itself is not a serious objection to using this method.

Example 10.7. Let $\{y_t\}$ be the non-Gaussian MA(2) process

$$y_t = w_t - 1.4142\rho w_{t-1} + \rho^2 w_{t-2},$$

where w_t has a shifted exponential distribution. The parameter ρ, the modulus of the zeros of $b(z)$, was given the values 0.8 and 1.25 (i.e., the zeros are inside the unit circle in the first case and outside it in the second case). The number of data points was 2000. The first row in Table 10.1 shows the results of 100 Monte-Carlo simulations of the Giannakis-Mendel algorithm for each of the two cases. As can be seen, the bias of the estimates is relatively small (except for b_2 in the second case). The standard deviations are comparable for the two cases. The conclusion is that the algorithm can handle zeros inside the unit circle and outside it about equally well. This example will be continued.   ∎

The Giannakis-Mendel estimate is a special case of the estimation problem discussed in Prob. 3.19, so its asymptotic variance is obtained by the expression derived in that problem. In this case, $s(Y_N)$ is the joint statistic of the cumulants and covariances appearing in (10.6.7), that is, $\{\hat{\gamma}_\tau, \hat{r}_\tau, -q \leq \tau \leq q\}$. The parameter vector θ consists of the MA parameters and ε, and the matrix $\Sigma(\theta)$ is the normalized asymptotic covariance of the statistics. This matrix can be computed by the method described in Sec. 10.5. This analysis was carried out by Porat and Friedlander [1989] for both identity-weighted least-squares and the optimal weighting derived in Prob. 3.2. Since we are dealing with non-Gaussian processes, the asymptotic variance of the estimates cannot be compared with the Cramér-Rao bound (whose form is unknown), but can be judged in absolute terms. In Porat and Friedlander [1989] it was shown, by numerical experiments, that the performance of the Giannakis-Mendel estimate is not always satisfactory, so other methods should be pursued. Nevertheless, this estimate is a good means of providing initial conditions for the nonlinear techniques described in the next section. In this sense it serves the same purpose as the modified Yule-Walker equation for ARMA estimation or Prony's method for sinusoids-in-noise estimation. We finally remark that an adaptive version of the Giannakis-Mendel estimate, in the spirit of the algorithms of Chapter 8, has been developed in Friedlander and Porat [1989].

Parameter	b_1	b_2	b_1	b_2
True value	-1.1314	0.64	-1.7677	1.5625
GM algorithm	-0.02091 0.1659	0.00109 0.06330	0.03546 0.3112	-0.1046 0.1852
Min. var. algorithm	0.00082 0.04279	-0.00755 0.04074	-0.00243 0.06973	-0.00729 0.08150

Table 10.1. Results of Monte-Carlo simulation of Example 10.7. Upper entry: empirical biases; lower entry: empirical standard deviations.

The modification of Tugnait [1990] to the Giannakis-Mendel estimate is given as follows.[†] Equation (10.6.3) gives, for two different set of σ's,

$$G(z; \boldsymbol{\sigma}_1)C(z; \boldsymbol{\sigma}_2) = G(z; \boldsymbol{\sigma}_2)C(z; \boldsymbol{\sigma}_1). \qquad (10.6.8)$$

Choose $\boldsymbol{\sigma}_1 = \{0, \ldots, 0, 0\}$ and $\boldsymbol{\sigma}_2 = \{0, \ldots, 0, q\}$. Accordingly, $g_k(\boldsymbol{\sigma}_1) = b_k^{n-1}$ and

---

[†] The algorithm given by Tugnait is somewhat different from the one described here, since he considers a problem in which there is additive noise, and his choice of equations avoids the noise-dependent cumulants. Our model does not include noise, so we have included all available equations.

$g_k(\boldsymbol{\sigma}_2) = b_k^{n-2} b_{k+q} = b_q \delta(0)$. Let γ_τ be as before and define $\eta_\tau = c_{\tau,\ldots,\tau,\tau+q}$. Note that η_τ is nonzero only for $-q \le \tau \le 0$. We get from (10.6.8), upon multiplication by ε and expressing it in the time domain,

$$\sum_{k=0}^{q} \varepsilon b_k^{n-1} \eta_{\tau-k} = \varepsilon b_q \gamma_\tau. \qquad (10.6.9)$$

Introducing the new unknown $(-\varepsilon b_q)$ and collecting these equations for $-q \le \tau \le q$ gives

$$\begin{bmatrix} \eta_{-q} & 0 & \cdots & 0 & \gamma_{-q} \\ \eta_{-q+1} & \eta_{-q} & \cdots & 0 & \gamma_{-q+1} \\ \vdots & \vdots & \ddots & \vdots & \vdots \\ \eta_0 & \eta_{-1} & \cdots & \eta_{-q} & \gamma_0 \\ 0 & \eta_0 & \cdots & \eta_{-q+1} & \gamma_1 \\ \vdots & \vdots & \ddots & \vdots & \vdots \\ 0 & 0 & \cdots & \eta_0 & \gamma_q \end{bmatrix} \begin{bmatrix} \varepsilon \\ \varepsilon b_1^{n-1} \\ \vdots \\ \varepsilon b_q^{n-1} \\ -\varepsilon b_q \end{bmatrix} = 0. \qquad (10.6.10)$$

When we join (10.6.10) to (10.6.7), we get $5q+2$ equations in $2q+2$ unknowns, so presumably the estimates will be improved. Indeed, the improvement was demonstrated in Tugnait [1990] for some numerical test cases. However, no comparative statistical analysis was provided by Tugnait, although such analysis would be easy to perform using the result of Prob. 3.19 (at least for the case of third-order cumulants). The aspiring reader is encouraged to carry out this analysis (Prob. 10.14).

Tugnait [1990] has also proposed the following algorithm. Start with (10.6.5) and choose $\boldsymbol{\sigma} = \{0, \ldots, 0, q\}$. Then we get, similarly to (10.6.9),

$$\sum_{k=0}^{q} b_k \eta_{\tau-k} = \varepsilon b_q r_\tau.$$

Introducing the new unknown $(-\varepsilon b_q)$ and collecting these equations for $-q \le \tau \le q$ gives

$$\begin{bmatrix} \eta_{-1} \\ \eta_{-q+1} \\ \vdots \\ \eta_0 \\ 0 \\ \vdots \\ 0 \end{bmatrix} + \begin{bmatrix} r_{-q} & 0 & \cdots & 0 \\ r_{-q+1} & \eta_{-q} & \cdots & 0 \\ \vdots & \vdots & \ddots & \vdots \\ r_0 & \eta_{-1} & \cdots & \eta_{-q} \\ r_1 & \eta_0 & \cdots & \eta_{-q+1} \\ \vdots & \vdots & \ddots & \vdots \\ r_q & 0 & \cdots & \eta_0 \end{bmatrix} \begin{bmatrix} -\varepsilon b_q \\ b_1 \\ \vdots \\ b_q \end{bmatrix} = 0. \qquad (10.6.11)$$

As before, we solve these equations in the least-squares sense to obtain the estimates. Note that the matrix in (10.6.11) is guaranteed to be full rank, since r_{-q} and η_{-q} are nonzero for an $MA(q)$ process. Therefore, the resulting estimates are always consistent. The performance of this algorithm has not been analyzed (only numerical simulations were given), but the reader is asked to do so (Prob. 10.14).

Presumably, many more algorithms can be derived using the basic idea of Giannakis and Mendel. The performance of these algorithms, however, should be carefully tested. Algorithms of this kind are extremely sensitive to the knowledge of the MA model order q. Moreover, they depend on b_q being not too close to zero. It often happens in reality that the impulse response coefficients decay to zero gradually, so if b_{q+1} is zero, b_q is likely to be rather small. The use of the above algorithms is questionable in such cases.

10.7. ARMA PARAMETER ESTIMATION: LINEAR METHODS

Linear methods for cumulant-based ARMA parameter estimation are necessarily multiple stage. Typically, the AR parameters are estimated first, followed by estimation of the MA parameters. There does not appear to exist a linear algorithm that estimates all ARMA parameters simultaneously. Many linear multistage algorithms have been reported in the literature. Mendel, Giannakis, and Swami have been particularly active in this area; see Mendel [1991] and the references within. Since none of the algorithms appears to offer a definite advantage over the others, we will settle for two straightforward methods and leave it to the interested reader to explore other methods given in the literature.

The first stage, common to the two methods, is an estimation of the AR parameters from the cumulants, covariances, or a combination thereof. This is accomplished by the extended Yule-Walker equations (10.4.5). As we already know, we can choose p covariance-based equations to get consistent estimates. If we wish to use only higher-order cumulants, more equations are needed, as given by Theorem 10.7.

Once the AR parameters have been estimated, the following two methods can be used to obtain the MA parameters.

1. Denote

$$S_{MA}(\omega_1, \ldots, \omega_{n-1}) = \kappa_n(w)b(e^{j\omega_1})\ldots b(e^{j\omega_{n-1}})b(e^{-j(\omega_1+\cdots+\omega_{n-1})}).$$

This is the polyspectrum of the moving average process $\sum_{k=0}^{q} b_k w_{t-k}$, related to the polyspectrum of y_t by

$$S_{MA}(\omega_1, \ldots, \omega_{n-1}) = a(e^{j\omega_1})\ldots a(e^{j\omega_{n-1}})a(e^{-j(\omega_1+\cdots+\omega_{n-1})})S(\omega_1, \ldots, \omega_{n-1}).$$
$$(10.7.1)$$

We can estimate as many cumulants of the ARMA process as we wish from the given measurements (assuming the number of measurements N is large enough) and use them on the right side of (10.7.1). The left side (when expressed as a multidimensional power series) consists of a finite number of terms. Each term can be computed as a finite sum of terms on the right side, since $a(z)$ has a finite number of terms. Therefore, we only need a finite number of terms of $S(\omega_1, \ldots, \omega_{n-1})$ to fully compute $S_{MA}(\omega_1, \ldots, \omega_{n-1})$. Once this has been done, we can use any of the methods of the previous section to estimate the MA parameters.

2. Use the estimated AR parameters to form the residual sequence

$$e_t = \sum_{k=0}^{p} \hat{a}_k y_{t-k}, \quad p \le t \le N - 1. \tag{10.7.2}$$

Then treat $\{e_t\}$ as an MA process and use any of the methods given in the previous section. Recall that this approach was also used in Sec. 7.4 to complete the ARMA estimate based on the modified Yule-Walker equations.

10.8. MA AND ARMA PARAMETER ESTIMATION: NONLINEAR METHODS

In dealing with Gaussian processes, we often emphasized the maximum likelihood estimate as the best from a performance point of view (computational difficulties notwithstanding). For the non-Gaussian estimation problems of interest to us here, maximum likelihood estimation is not feasible, since the likelihood function is either unknown or very difficult to compute. Therefore, the only approach at our disposal is the method of moments (of which the algorithms presented in the two previous sections are special cases). As we saw several times in this book, an asymptotically minimum variance estimate based on a vector of statistics $s_N(Y_N)$ can be obtained by global minimization of the cost function

$$C(\psi, s_N(Y_N)) = \frac{1}{2}(s_N(Y_N) - s(\psi))^T \Sigma^{-1}(\psi)(s_N(Y_N) - s(\psi)), \tag{10.8.1}$$

or the cost function

$$\hat{C}(\psi, s_N(Y_N)) = \frac{1}{2}(s_N(Y_N) - s(\psi))^T \hat{\Sigma}_N^{-1}(Y_N)(s_N(Y_N) - s(\psi)), \tag{10.8.2}$$

where $\Sigma(\psi)$ is the normalized asymptotic covariance of the statistics $s_N(Y_N)$ as a function of the parameter vector ψ, and $\hat{\Sigma}_N(Y_N)$ is a consistent estimate of $\Sigma(\theta)$ (θ being the true parameter vector). In this section we explore the application of (10.8.1) and (10.8.2) to the problem of estimating MA and ARMA parameters from high-order moments.

In order to use (10.8.1) or (10.8.2), we need:

1. A vector of statistics computed from the measurements. This vector includes either sample moments or sample cumulants of one or more orders. The most common choice is to include a set of sample covariances (usually $\hat{r}_0$ through $\hat{r}_K$ for some K) and either a set of third-order moments/cumulants or a set of fourth-order moments/cumulants. The choice of third-order statistics is limited to problems in which the distribution of w_t is skewed, so $\kappa_3(w)$ is nonzero. Many common distributions are symmetric (e.g., binary, uniform,

and Laplace), so they have $\kappa_3(w) = 0$. In such cases the fourth-order statistics is the most common choice.

2. An algorithm for computing the theoretical moments/cumulants corresponding to a given value ψ of the parameter vector. Algorithms for MA and ARMA processes were provided in Sec. 10.4 above.

3. For (10.8.1), an algorithm for computing the asymptotic covariance matrix of the statistics $s_N(Y_N)$ corresponding to a given value ψ of the parameter vector. Algorithms for MA and ARMA processes were provided in Sec. 10.5. For (10.8.2), a consistent estimate of the asymptotic covariance matrix of $s_N(Y_N)$ corresponding to the true θ.

4. Good initial conditions to increase as much as possible the probability of convergence to the local minimum closest to the true parameter values.

5. Some nonlinear minimization procedure, such as the ones used in previous chapters for Gaussian estimation problems.

A suboptimal nonlinear method is to replace $\Sigma(\psi)$ in the cost function (10.8.1) by the identity matrix, thus saving the need for step 3 above. This method was proposed by Tugnait [1987], while the asymptotically optimal method was developed by Porat and Friedlander [1989]; see also Friedlander and Porat [1990].

Consider first the estimation problem for moving average processes. All the ingredients for the algorithm based on the cost function (10.8.1) have already been given. The only question is: which sample moments should we choose for the statistic $s_N(Y_N)$? Porat and Friedlander [1989] have considered the choice $s_N(Y_N) = \{\hat{r}_k, 0 \leq k \leq K\} \cup \{\hat{c}_{k,\ldots,k}, -K \leq k \leq K\}$, where $K \geq q$ (actually, only the case $n = 3$ was treated there). This is the set of *diagonal cumulants*. Of course, the statistics can be made larger, which will improve the accuracy of the estimates at the expense of increased computational complexity. From a theoretical point of view, it may be advisable to add the set $\{\hat{c}_{q,k,\ldots,0}, 0 \leq k \leq q\}$, since by (10.4.2) this guarantees the identifiability of the MA parameters.

For the algorithm based on (10.8.2), we need to obtain $\hat{\Sigma}_N(Y_N)$. One way (as in the similar algorithm for Gaussian ARMA processes) is to use the initial estimate and compute $\Sigma(\hat{\theta})$ once, where $\hat{\theta}$ is the initial estimate. An alternative method was proposed by Friedlander and Porat [1990], as follows. Recall (10.5.2) and observe that, for MA processes, the only nonzero terms in the sum over t correspond to $-q - \max\{\sigma_i\} \leq t \leq q + \max\{\tau_i\}$. The number of these terms is finite; moreover, due to the stationarity of the process, we can use the approximation

$$E\{y_0 y_{\tau_1} \cdots y_{\tau_{n-1}} y_t y_{t+\sigma_1} \cdots y_{t+\sigma_{k-1}}\}$$

$$\approx \frac{1}{N} \sum_{\ell=0}^{N-1} y_\ell y_{\ell+\tau_1} \cdots y_{\ell+\tau_{n-1}} y_{\ell+t} y_{\ell+t+\sigma_1} \cdots y_{\ell+t+\sigma_{k-1}}, \quad (10.8.3)$$

where the values of $\{y_t\}$ outside the observation interval are taken as zero. The right side of (10.8.3) is a consistent estimate of the left side. We can therefore substitute the right side in (10.5.2) and sum over the active range of t to get a consistent estimate of $\lim_{N \to \infty} N \text{cov}\{\hat{m}_{\boldsymbol{\tau}}, \hat{m}_{\boldsymbol{\sigma}}\}$. Repeating for all range of $\boldsymbol{\tau}$ and $\boldsymbol{\sigma}$ of interest yields the desired matrix $\hat{\Sigma}_N(Y_N)$. This algorithm was implemented and tested in [Friedlander and Porat, 1990] for the case $n = 3$.

A potential problem with the above method is that the matrix $\hat{\Sigma}_N(Y_N)$ may fail to be positive definite due to the approximation involved in (10.8.3). Then the minimization of (10.8.2) may fail to converge. As a precaution, it is advised to compute the eigendecomposition of $\hat{\Sigma}_N(Y_N)$ and replace all negative eigenvalues (if any) by zeros. The modified matrix will be positive semidefinite, thus guaranteeing convergence of the algorithm.

Similar remarks apply to the case of ARMA estimation. The major difference (besides that of computational complexity, which is considerably higher for ARMA than for MA) is the computation of $\hat{\Sigma}_N(Y_N)$ for use in (10.8.2). For ARMA processes, the sum in (10.5.2) involves an infinite number of terms, so truncation is necessary if (10.8.3) is to be used. For this we need to know the "effective MA order" of the process, that is, the value of q_0 such that y_t and y_{t-q_0} are approximately independent. If this parameter is known and is not too large, we can use (10.8.3) for $-q_0 - \max\{\sigma_i\} \leq t \leq q_0 + \max\{\tau_i\}$; otherwise, this method cannot be used.

Example 10.7 (continued). The same simulated data used to test the Giannakis-Mendel algorithm were used with the algorithm based on (10.8.2). The covariance and cumulant statistics were as in the Giannakis-Mendel algorithm. The matrix $\hat{\Sigma}$ was computed from the initial (Giannakis-Mendel) estimate, as described in Sec. 10.5. The second row of Table 10.1 shows the results. The estimates are practically unbiased, and the empirical standard deviations are considerably smaller than the ones for the Giannakis-Mendel estimates. This improvement is obtained, however, at a great computational cost.   ∎

A simpler nonlinear method, due to Lii and Rosenblatt [1982, the MA case] and Tugnait [1986, the ARMA case] is also worth mentioning. The first step of this method consists of estimating the *innovation representation* of the process from the second-order statistics (or by Gaussian maximum likelihood). Let $\hat{\beta}(z)/\hat{a}(z)$ denote the estimated transfer function. By assumption, the true ARMA model is casual and stable, so $\hat{a}(z)$ can be taken as the final estimate of the AR part. This is not true for the MA part, since $\hat{\beta}(z)$ is minimum phase by construction (its zeros are inside the unit circle), while the same may not hold for the true $b(z)$. The second step of the method therefore consists of:

1. Computing a set of 2^q candidate estimates of $b(z)$, denoted by $\{\hat{b}_k(z), 1 \leq k \leq 2^q\}$. These estimates are obtained from $\hat{\beta}(z)$ by finding its q roots; each $\hat{b}_k(z)$ contains a subset of roots of $\hat{\beta}(z)$ and the reciprocals of the complement set.

2. Computing a set of theoretical cumulants for each of the processes $\hat{b}_k(z)/\hat{a}(z)$, using the methods of Sec. 10.4.

3. Forming the sum of squares of the differences between the theoretical cu-

mulants and the corresponding sample cumulants for each k, and choosing the model $\hat{b}_k(z)/\hat{a}(z)$ for which the sum of squares is minimal among the 2^k candidate models.

The Lii-Rosenblatt method is relatively simple to implement (provided q is not too large) and is a reasonable alternative to the linear methods for providing initial conditions to the nonlinear minimization algorithms. A minor drawback of this method is its inability to treat processes in which one (or more) zero is the reciprocal of one of the poles. This is because such a pair would cancel out in the innovation representation of the process.

10.9. DECONVOLUTION

In Sec. 10.3, while discussing Theorem 10.6, we alluded to the possibility of reconstructing the input sequence to a non-Gaussian linear process from output measurements. This reconstruction is important in various applications, one of which will be presented in Example 10.8 below. Reconstruction of the input sequence from output measurements is also known as *deconvolution*. In this section we discuss deconvolution for MA processes and leave it to the reader to extend the results to ARMA processes (Prob. 10.18).

Suppose we are given a sequence of measurements from the MA process $y_t = \sum_{k=0}^q b_k w_{t-k}$, where $\{w_t\}$ is i.i.d., and $b(z)$ is not necessarily minimum phase. However, assume that $b(z)$ does not contain zeros *on* the unit circle (recall that this condition is necessary for Theorem 10.6). In order to retrieve $\{w_t\}$ from $\{y_t\}$, it is necessary to perform the linear filtering operation described formally by $w(z) = y(z)/b(z)$. However, since $b(z)$ may contain zeros outside the unit circle, it may not possess a stable and causal inverse. Stability is imperative for deconvolution, so the inverse filter $1/b(z)$ must be made noncausal if $b(z)$ is not minimum phase. Recall that a stable impulse response can be obtained from $1/b(z)$ by integrating the inverse z transform integral (a.2) on the unit circle [Oppenheim and Schafer, 1975, Sec. 2.2]. The impulse response will be doubly infinite in general, so it cannot be convolved with $\{y_t\}$ in a causal manner. However, the convolution can be approximated by truncating the negative side of the impulse response and introducing enough delay to make the truncated sequence causal.

In practice, both sides of the impulse response of $1/b(z)$ are usually truncated, so the inverse filter is approximated by a finite impulse response filter, whose z transform will be denoted by $g(z)$. In other words, we synthesize an FIR filter of length L, such that

$$b(z)g(z) \approx z^{-L_1}, \tag{10.9.1}$$

and compute $\{w_t\}$ by passing $\{y_t\}$ through this FIR. The filter's length L and the delay L_1 can be determined as follows. We find the roots of $b(z)$ and let ρ_1 be the minimum modulus of the roots outside the unit circle and ρ_2 the maximum modulus of the roots inside it. We then make L_1 proportional to $1/\log\rho_1$ and L_2 proportional to $-1/\log\rho_2$. The proportionality constants are determined by the desired amount of relative decay of each side of the impulse response at which they can be truncated. If there are no roots outside the unit circle, we take $L_1 = 0$; if there are no roots inside it, we take $L_2 = 0$. Finally, we take $L = L_1 + L_2$. The reader is asked to supply the rationale for this procedure (Prob. 10.18).

In some applications, the above truncation procedure may be too cumbersome. In such cases it is common to choose L on an ad hoc basis and take $L_1 \approx L/2$ (which should be good "on the average"). Moreover, it is common to compute the coefficients of $g(z)$ so as to satisfy (10.9.1) in the least-squares sense. This leads to the minimization of the quadratic cost function

$$C(\boldsymbol{g}) = \sum_{k=0}^{L+q-1}\left[\sum_{\ell=\max\{0,k-q\}}^{\min\{k,L-1\}} g_\ell b_{k-\ell} - \delta(k-L_1)\right]^2. \qquad (10.9.2)$$

The FIR filter whose coefficients are obtained by minimizing (10.9.2) is then used to perform the deconvolution.

Example 10.8. The most common method of transmitting digital communication signals is probably by *linear modulation*. In linear modulation, each symbol of the alphabet is assigned certain amplitude and phase. The amplitude and phase together define a complex number, and the collection of the complex numbers corresponding to all the symbols in the alphabet is known as the *symbol constellation*. In the special case when the permitted phases are $0°$ and $180°$, the symbol constellations consist only of (positive and negative) real numbers. This modulation method is sometimes called *amplitude shift keying* (ASK). For example, in binary ASK the symbols are $\{-1,1\}$, in quaternary ASK they are $\{-3,-1,1,3\}$, and so on.

The communication channel, which contains as a minimum the transmitter, the receiver, and the propagation medium, is typically dispersive. As a result, the received signal exhibits *intersymbol interference* (ISI). By this we mean that the signal value at each time point contains contributions from several transmitted symbols. Mathematically, the signal can be represented as a convolution of the symbol stream (the input to the channel) with the impulse response of the channel. The channel's output is usually (but not always) sampled at a rate of one sample per symbol. In this case, the relationship between the input symbol sequence and the sampled channel output can be represented by a discrete-time convolution [Qureshi, 1985]. The impulse response of the equivalent discrete-time linear system is determined by the transmitter, the receiver, and the medium (whether it be wired communication or radio link). Ideally, the channel's impulse response should be a unit impulse. Any deviation from the ideal response introduces ISI, and the amount of ISI is usually specified by some measure of the deviation of the impulse response from a unit impulse.

In case of severe ISI, the receiver's output must be equalized before it is used to detect the symbol sequence. *Equalization* is a term equivalent to deconvolution; it means passing the receiver's output through a linear filter that ideally inverts the channel response and reconstructs the symbol sequence. *Blind equalization* refers to the case where the only information given to the equalizer is the channel's output: neither the corresponding input sequence nor the channel's impulse response is assumed to be known.

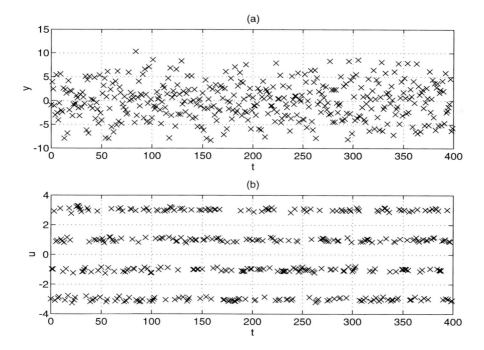

Figure 10.1. Simulation results of Example 10.9. (a) The channel's output; (b) the equalizer's output.

Blind equalization is a difficult problem, and considerable research has been devoted to its solution. Many blind equalization techniques take advantage of the non-Gaussian characteristic of the input sequence. Since the probability distribution of the symbols is discrete, it is necessarily non-Gaussian. In many cases it can be assumed uniform, and the sequence can be regarded as i.i.d. Here we illustrate an equalization method based on estimating the channel's response from the high-order cumulants of the output process and using the estimated parameters to construct an FIR equalizer, as described above. The channel is assumed to be FIR; that is, the output process is assumed to be MA of a known order. The MA parameters in the example (including b_0) are $\{1.0, -1.22, 0.37, 0.87\}$. The symbol sequence is quaternary ASK. This channel has zeros both inside and outside the unit circle and exhibits a severe ISI. A typical output sequence (of 400 measurements) is shown in Figure 10.1(a). Clearly, direct detection of the input symbols from the output sequence is impossible. We generated 10,000 samples of the output sequence and fed them to the Giannakis-Mendel algorithm. Since the distribution of the input symbols is symmetric, its third-order cumulants are zero, so we used fourth-order cumulants. The estimated channel parameters were $\{1.0, -1.2246, 0.3537, 0.8594\}$ in this example. We then computed an FIR equalizer of length 60 by solving the least-squares equation (10.9.2). Finally, we passed 400 output values (out of the 10,000) through the equalizer and plotted the equalizer's output. The result is shown in Figure 10.1(b). It is seen that the reconstructed sequence is distributed around the four true values, and can be detected safely.

Further discussion of this equalization technique and references to other blind equalization methods can be found in Porat and Friedlander [1991].   ∎

10.10. ESTIMATION OF SINUSOIDS IN GAUSSIAN NOISE

When discussing the problem of sinusoids in noise in Chapter 9, we left open the case of colored noise, except for some rudimentary suggestions at the end of that chapter. The approaches suggested there are difficult to implement and analyze and are not in common use. High-order statistics provides us with an alternative approach to this problem, in the case where the noise $\{u_t\}$ in the model (9.1.1) is Gaussian, but possibly colored. In this case, the true cumulants of $\{y_t\}$ will be free of $\{u_t\}$, since the cumulants of a Gaussian process of order greater than 2 are zero. Of course, this property does not hold for the sample cumulants. However, since the sample cumulants are consistent estimates of the true cumulants, we can expect to be able to consistently estimate the signal parameters from the measurements, regardless of the noise spectrum. We now elaborate on this idea.

Consider first a single sinusoid $x_t = A\cos(\omega t - \phi)$, and assume there is no noise. The parameters A and ω are deterministic and unknown, and ϕ is assumed to be uniformly distributed in $[0, 2\pi]$. We get, using the trigonometric identity in Prob. 9.8(b),

$$
\begin{aligned}
&x_t x_{t+\tau_1} x_{t+\tau_2} x_{t+\tau_3} \\
=&\frac{A^4}{8}\{\cos[4\omega t + \omega(\tau_1 + \tau_2 + \tau_3) - 4\phi] + \cos[2\omega t + \omega(\tau_1 + \tau_2 - \tau_3) - 2\phi] \\
&+ \cos[2\omega t + \omega(\tau_1 - \tau_2 + \tau_3) - 2\phi] + \cos[2\omega t + \omega(\tau_2 + \tau_3 - \tau_1) - 2\phi] \\
&+ \cos[2\omega t + \omega(\tau_1 + \tau_2 + \tau_3) - 2\phi] + \cos[\omega(\tau_1 - \tau_2 - \tau_3)] \\
&+ \cos[\omega(\tau_2 - \tau_1 - \tau_3)] + \cos[\omega(\tau_3 - \tau_1 - \tau_2)]\}.
\end{aligned}
$$

Taking expected value gives

$$m_{\tau_1,\tau_2,\tau_3} = \frac{A^4}{8}\{\cos[\omega(\tau_1 - \tau_2 - \tau_3)] + \cos[\omega(\tau_2 - \tau_1 - \tau_3)] + \cos[\omega(\tau_3 - \tau_1 - \tau_2)]\}.$$

(10.10.1)

Also recall that $m_\tau = (A^2/2)\cos\omega\tau$. Therefore, we get, upon substituting in (2.4.6) and simplifying,

$$c_{\tau_1,\tau_2,\tau_3} = -\frac{A^4}{8}\{\cos[\omega(\tau_1 - \tau_2 - \tau_3)] + \cos[\omega(\tau_2 - \tau_1 - \tau_3)] + \cos[\omega(\tau_3 - \tau_1 - \tau_2)]\}.$$

(10.10.2)

In particular, if we choose $\tau_1 = \tau_2 = \tau_3 = \tau$, we will get

$$c_{\tau,\tau,\tau} = -\frac{3A^4}{8}\cos\omega\tau.$$

(10.10.3)

Finally, since the cumulants of independent processes are additive (Lemma 10.1) and the cumulants of $\{u_t\}$ are zero, we get for the model (9.1.1)

$$c_{\tau,\tau,\tau} = -\frac{3}{8}\sum_{m=1}^{M} A_m^4 \cos\omega_m\tau.$$

(10.10.4)

As we see, the true cumulant $c_{\tau,\tau,\tau}$ has the same structure as the process $\{y_t\}$ itself, with different amplitudes, zero phases, and no additive noise. An analysis similar to the one in Theorem 9.4 would reveal that the sample cumulant will be $\hat{c}_{\tau,\tau,\tau} = c_{\tau,\tau,\tau} + O_p(N^{-1})$. The additive term $O_p(N^{-1})$ has a very complicated form, but it decreases to zero in probability as $N \to \infty$. Therefore, any of the methods presented in Chapter 9 can be applied to $\hat{c}_{\tau,\tau,\tau}$ to yield consistent estimates of $\{A_m, \omega_m\}$. The performance of the estimates (their asymptotic variances) is very difficult to analyze, but simulations have been carried out [Swami and Mendel, 1991]; [Shin and Mendel, 1992]. The analysis of cumulant-based estimation methods for sinusoids in colored noise is, at the time this book is written, an open research topic.

10.11. MATHEMATICA PACKAGES

The packages `Partitio.m`, `MomCum.m`, `NgData.m`, `Cumulant.m`, `CMarma.m` and `HoAlg.m` implement some of the algorithms described in this chapter. Here is a brief description of the procedures in these packages.

`PartitionNumberSet[n_]` and `PartitionAnySet[set_]` find all partitions of a set of n elements using the iterative procedure described in Sec. 10.2. The former finds the partitions of the set of integers from 1 to n, while the latter does the same for an arbitrary `set`. The two procedures can optionally exclude partitions containing singletons by including the optional parameter `NoOnes->True` (the default value is `False`). As explained in Sec. 10.2, this is useful when the random variables are known to have zero mean. Also included in the package `Partitio.m` are some internal auxiliary procedures.

`CumToMom[c_, set_, opts___]` and `MomToCum[m_, set_, opts___]` implement the cumulants-to-moments formula (10.2.7) and the moments-to-cumulants formula (10.2.9), respectively. `set` is an arbitrary set of indexes. `c` in the first procedure is an arbitrary function that computes the cumulant corresponding to a set of indices. Similarly, `m` in the second procedure is an arbitrary function that computes the moment corresponding to a set of indexes. The optional parameter `NoOnes->True` fulfills the same task as in the partition procedures.

`NgData[a_, b_, dist_, Nsam_]` generates `Nsam` data points of a non-Gaussian ARMA process with parameters `a` and `b`. The input is a random i.i.d. sequence taken from the distribution `dist`. This distribution can be taken from Mathematica's standard package `ContinuousDistributions` (which is automatically loaded by `NgData`) or supplied by the user. The initial conditions are identically zero, so it is advised to delete from the output sequence a certain number of values before using the rest. The number of deleted values depends on the time it takes the ARMA filter to reach stationarity, which in turn depends on the pole(s) of largest modulus. The implementation of this procedure is similar to that of `ArmaData`, described in Chapter 2.

`SampleMom[data_, stat_]` generates the sample moments of the sequence `data` specified by the list `stat` of lists of desired indexes. It does so by calling the internal procedure `OneSampleMom[data_, index_]` for each member of `stat`. The latter procedure computes the sample moments using the formula (10.5.1).

`SampleCum[data_, stat_]` generates the sample cumulants of the sequence `data` specified by the list `stat` of lists of desired indexes. It does so by calling the internal procedure `OneSampleCum[data_, index_]` for each member of `stat`. The latter procedure computes the sample cumulants using the moments-to-cumulants formula, as implemented in `MomToCum`. The optional parameter `NoOnes->True` should be used when `data` is known to have zero mean.

`Cum[{a_, b_, c_}, tau_, kappa_]` computes the cumulant of the process generated by the state-space model {a, b, c}, with input cumulant function `kappa`, corresponding to a list of indexes `tau`. `Cum[b_, tau_, kappa_]` is for a moving average process with coefficients `b`. `Cum[{a_, b_}, tau_, kappa_]` is for an ARMA process with coefficients {a, b}.

`kappaexp[n_]` returns the nth-order cumulant of the shifted exponential distribution.

`CMarma[{a_, b_}, stat_, kappa_, opts___]` computes the covariance matrix of the sample moments of an ARMA process, as explained in Sec. 10.5. {a, b} are the lists of ARMA parameters. `stat` is a list of lists of τ's (those lists can have different lengths). `kappa` is a function that computes the cumulants of the input process. The option `NoOnes->True` excludes partitions containing singletons in the case of a zero mean process. `CMarma[b_, stat_, kappa_, opts___]` is for an MA process with parameter list `b`.

The package `CMarma.m` also includes various auxiliary procedures. In particular, `SetGen` generates the sets of indexes for the individual terms in the cumulants-to-moments formula; `Set2SS` computes the state-space triplets for representing the t-dependent cumulants, as explained in Sec. 10.5; `UpLowSum` computes the two infinite sums in (10.5.4), and `MidSum` computes the finite sum; `CMone` computes one entry of the covariance matrix of the sample cumulants.

`PreGianMendelMa[data_, q_, order_, opts___]` prepares the lists of covariances and cumulants for the Giannakis-Mendel MA algorithm. `data` is the non-Gaussian data sequence. `q` is the order of the estimated MA model. `order` is the order of the cumulants used in the estimation algorithm. The option `NoOnes->True` should be used when the process is known to have zero mean. The procedure returns `{cov, cum}`, the lists of sample covariances and sample cumulants.

`GianMendelMa[{cov_, cum_}]` implements the Giannakis-Mendel MA algorithm. `{cov, cum}` are the lists of sample covariances and cumulants, prepared by `PreGianMendelMa`. The procedure returns the estimated MA polynomial b.

`MvcMa[stat_, samplestat_, binit_, kappa_, opts___]` implements the asymptotically minimum variance algorithm [based on (10.8.2)] for MA processes. `stat` is a list of lists of τ's, defining the moments to be used. `samplestat` is a vector of statistics compatible with `stat` (computed by `SampleMom`). `binit` is the initial value of the estimate to be provided by `GianMendelMa`. `kappa` is a function for computing the cumulants of the input process. The option `MaxIter` determines the maximum number of iterations; the default is 200. The option `RelAcc` determines the relative accuracy; the default is 10^{-6}. The option `NoOnes` is for taking advantage of the zero-mean property of the given process; the default is `True`. The matrix $\hat{\Sigma}$ is computed from `binit` as described in Sec. 10.5. The minimization is by conjugate directions (without gradient). The returned values are `{bhat, costmin, iter}`, the estimated MA parameters, the minimum value of the cost function, and the number of iterations, respectively.

`InverseFilter[b_, n_]` computes an approximate FIR inverse filter of length n for the MA filter b. The computation is by minimization of (10.9.2). The procedure returns the list of coefficients of the inverse filter.

REFERENCES

Friedlander, B., and Porat, B., "Adaptive IIR Algorithms Based on High-order Statistics," *IEEE Trans. Acoustics, Speech, Signal Processing*, ASSP-37, pp. 485–495, 1989.

Friedlander, B., and Porat, B., "Asymptotically Optimal Estimation of MA and ARMA Parameters of Non-Gaussian Processes from High-order Moments," *IEEE Trans. Automatic Control*, AC-35, pp. 27–35, 1990.

Giannakis, G. B., "Signal Processing Using Higher-order Statistics", Ph.D. dissertation, University of Southern California, CA, 1987.

Giannakis, G. B., and Mendel, J. M., "Identification of Non-minimum Phase Systems Using Higher-order Statistics," *IEEE Trans. Acoustics, Speech, Signal Processing,* 37, pp. 360–377, 1989.

Kailath, T., *Linear Systems,* Prentice Hall, Englewood Cliffs, NJ, 1981.

Lii, K. S., and Rosenblatt, M., "Deconvolution and Estimation of Transfer Function Phase and Coefficients for Non-Gaussian Linear Processes," *Ann. Statistics,* 10, pp. 1195–1208, 1982.

Mendel, J. M., "Tutorial on Higher-order Statistics (Spectra) in Signal Processing and System Theory: Theoretical Results and Some Applications," *Proc. IEEE,* 79, pp. 278–305, 1991.

Oppenheim, A. V., and Schafer, R. W., *Digital Signal Processing,* Prentice Hall, Englewood Cliffs, NJ, 1975.

Porat, B., and Friedlander, B., "Performance Analysis of Parameter Estimation Algorithms Based on High-order Moments," *Int. J. Adaptive Control Signal Processing,* 3, pp. 191–229, 1989.

Porat, B. and Friedlander, B., "Blind Adaptive Equalization of Digital Communication Channels Using High-order Moments," *IEEE Trans. Signal Processing,* 39, pp. 522–526, 1991.

Qureshi, S. U. H., "Adaptive Equalization," *Proc. IEEE,* 73, pp. 1349–1387, 1985.

Shin, D. C., and Mendel, J. M., "Assessment of Cumulant-based Approaches to Harmonic Retrieval," *Proc. Int. Conf. Acoustics, Speech, Signal Processing,* San Francisco, CA, pp. V-205–V-208, 1992.

Swami, A., "System Identification Using Cumulants," Ph.D. dissertation, University of Southern California, CA, 1988.

Swami, A., and Mendel, J. M., "Computation of Cumulants of ARMA Spectra," *Proc. Int. Conf. Acoustics, Speech, Signal Processing,* Glasgow, Scotland, pp. 2318–2321, 1989.

Swami, A., and Mendel, J. M., "Cumulant-based Approach to the Harmonic Retrieval and Related Problems," *IEEE Trans. Signal Processing,* 39, pp. 1099–1109, 1991.

Tugnait, J. K., "Identification of Nonminimum Phase Linear Stochastic Systems," *Automatica,* 22, pp. 454–464, 1986.

Tugnait, J. K., "Identification of Linear Stochastic Systems via Second- and Fourth-order Cumulant Matching," *IEEE Trans. Information Theory,* 33, pp. 393–407, 1987.

Tugnait, J. K., "Approaches to FIR System Identification with Noisy Data Using Higher Order Statistics," *IEEE Trans. Acoustics, Speech, Signal Processing*, 38, pp. 1307–1317, 1990.

PROBLEMS

10.1. Prove that the moment generating function of a Gaussian random vector with mean $\boldsymbol{v}$ and covariance matrix Γ is given by

$$M_X(\boldsymbol{\lambda}) = \exp(\boldsymbol{\lambda}^T \boldsymbol{v} + 0.5 \boldsymbol{\lambda}^T \Gamma \boldsymbol{\lambda}).$$

10.2. We have defined $\kappa_n(x)$, the nth-order cumulant of a scalar random variable, as the cumulant of the vector $[x, \ldots, x]$. Show that

$$\kappa_n(x) = \left. \frac{d^n C_x(\lambda)}{d\lambda^n} \right|_{\lambda=0},$$

where $C_x(\lambda)$ is the cumulant generating function of the scalar random variable x.

10.3. Let x have the distribution $f(x) = 0.5 e^{-|x|}$, $-\infty < x < \infty$. Find an expression for $\kappa_n(x)$ for $n \geq 2$. Hint: Compute the cumulant generating function and differentiate it once. Expand the derivative in Taylor series and then use Prob. 10.2.

10.4. Let X be the sum of squares of m i.i.d. Gaussian random variables with zero mean and unit variance. The distribution of X is called *chi-square* with m degrees of freedom and is denoted by χ_m^2. Compute $\kappa_n(X)$ for this distribution.

10.5. Partition generation.

(a) Modify the Mathematica procedure `PartitionNumberSet` so as to yield only partitions $P(n)$ such that every member of $P(n)$ has an even number of elements. This is useful in generating moments and cumulants of random variables whose densities are symmetric.

(b) Modify the procedure `PartitionNumberSet` so as to yield only partitions $P(n)$ such that every member of $P(n)$ has exactly two elements. This is useful in computing high-order moments of Gaussian zero-mean random variables, since then the only nonzero cumulants are of second order.

(c) Use the modified procedure in (b) for an efficient computation of the moments (of any order) of a Gaussian zero-mean random vector as a function of the covariance matrix of this vector. Write a Mathematica procedure that does this.

10.6. Prove that the transformation from the set $\{\tau_1, \ldots, \tau_{n-1}\}$ to the nonnegative set $\{\sigma_1, \ldots, \sigma_{n-1}\}$ (as described after Theorem 10.3) leaves the corresponding cumulant invariant.

10.7. The symmetries in c_{τ_1,τ_2} described in Sec. 10.3 imply certain symmetries in the bispectrum $S(\omega_1,\omega_2)$. There are also certain conjugate symmetries, since the cumulants are real. Explore those symmetries and specify the domain of (ω_1,ω_2) that is sufficient to uniquely determine the bispectrum. Remark: The symmetries in the trispectrum $S(\omega_1,\omega_2,\omega_3)$ are far more complicated.

10.8. Prove (10.3.6).

10.9. Let $\{x_t\}$ be a non-Gaussian stationary process, and suppose $\{x_t\}$ is passed through a linear time-invariant stable system with impulse response $\{h_t\}$ to yield the output process $\{y_t\}$.

    (a) Express the nth-order cumulants of $\{y_t\}$ in terms of those of $\{x_t\}$ and the impulse response parameters.

    (b) Express the polyspectrum of $\{y_t\}$ in terms of the polyspectrum of $\{x_t\}$ and the system's transfer function.

10.10. Let $\{x_t\}$ be a non-Gaussian MA(q) process, with $\kappa_3(w) \neq 0$. Find the values of τ_1,τ_2 for which $c_{\tau_1,\tau_2} \neq 0$. Draw the support region of c_{τ_1,τ_2} in the (τ_1,τ_2) plane.

10.11. Prove that (10.5.2) yields Bartlett's asymptotic formula (4.2.12) when $n = k = 2$.

10.12. The procedure based on (10.5.4) is "wasteful" in the sense that the middle sum contains too many terms. Show that two of the terms in the sum can be moved to the upper and lower sums if $r > 0$, and the middle sum can be eliminated if $r = 0$.

10.13. Let $\Sigma_m(\theta)$ denote the matrix computed by CRarma. Recall that this is the covariance matrix of a vector of sample *moments*. Now suppose we wish to compute the covariance matrix $\Sigma_c(\theta)$ of a vector of sample *cumulants*. Write a Mathematica procedure for this purpose. Hint: As was said in Sec. 10.5, the two matrices are related by $\Sigma_c(\theta) = G\Sigma_m(\theta)G^T$, where G is the Jacobian of the vector function describing the dependence of the vector of cumulants on a properly chosen vector of moments. Use the moments-to-cumulants formula (as implemented in MomCum.m) to identify the vector of moments needed for the given vector of cumulants. Then use the symbolic differentiation capability of Mathematica to compute the Jacobian G. Finally, substitute the given the numerical values of the process parameters (MA or ARMA, as the case may be) to compute $\Sigma_c(\theta)$.

10.14. Use the results of Prob. 3.19 to carry out the analysis of the algorithms given by (10.6.10) and (10.6.11). Use the Mathematica procedures supplied for this chapter as necessary, and complement them by procedures of your own. Your procedures should compute the normalized asymptotic covariances of the estimates given by these algorithms as a function of the MA parameters, the variance, and the corresponding cumulant of the input process. Compare the performance of these algorithms with that of the asymptotically minimum variance algorithm for some test cases of your choice and report the results.

10.15. Implement a linear ARMA algorithm as described in Sec. 10.7. Proceed according to the following steps: (1) write a procedure for solving a user-chosen set of Yule-Walker equations for the AR parameters; (2) obtain the residual sequence as in (10.7.2); (3) compute the MA parameters from the cumulants of the residual sequence, using the Giannakis-Mendel algorithm. Test your algorithm by performing some Monte-Carlo simulations and report the results.

10.16. Write a procedure MvcArma, in the style of MvcMa, that implements the asymptotically minimum variance algorithm for ARMA models. Use the algorithm developed in Prob. 10.15 for initial conditions. Test the procedure by performing some Monte-Carlo simulations and report the results.

10.17. Write a procedure that implements (10.8.3), and use it in a procedure SigmaHat that computes $\hat{\Sigma}$ as described in Sec. 10.8. Modify the algorithm MvcMa to use this matrix, repeat Example 10.7 with the modified algorithm, and compare with the results in Table 10.1.

10.18. Deconvolution.

    (a) Explain the rationale for the truncation procedure described in Sec. 10.9.

    (b) Show that the minimization of (10.9.2) leads to a linear system of equations whose coefficient matrix is Toeplitz.

    (c) Extend the deconvolution method described in Sec. 10.9 to ARMA processes.

CHAPTER 11

Time-Frequency
Signal Analysis:
Linear Transforms

11.1. INTRODUCTION

The processes discussed so far in this book were stationary, at least in the wide sense. This means, for example, that if we split a given sequence of measurements into two equal segments and plot a windowed periodogram of each segment the two plots are expected to be similar. If we use a parametric approach and estimate an ARMA(p, q) model for each segment, the two sets of estimated parameters are expected to be similar. Also, stationarity and ergodicity enable us to get consistent estimates by increasing the number of measurements as needed.

Many real-life processes are nonstationary. For example, if we record and digitize two speech segments and repeat the above experiment, the results for the two segments are likely to be rather different. Nonstationarity manifests itself by time-varying characteristics of the signal dynamics, whether it be the amplitude, the frequency range, the shape, or all of these.

Example 11.1. Suppose we want to write a computer program that will identify classical music works from short segments of digitized sound. Consider the signal

$$y(t) = \sin\left(2\pi \int_0^t f(\tau)d\tau\right),$$

where the function $f(t)$ is shown in Figure 11.1. [$f(t)$ is called the *instantaneous frequency* of $y(t)$]. This signal represents a grossly simplified version of the two opening bars of a famous symphony (which one?). The program should be able to identify the symphony from the above signal, sampled, say, at 5 kHz.

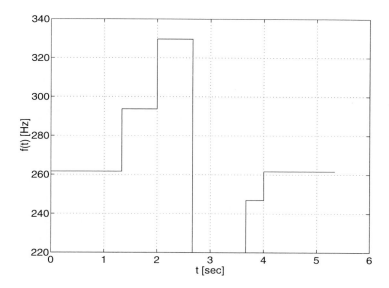

Figure 11.1. The instantaneous frequency of the signal in Example 11.1.

In order to write such a program, we may try to take the Fourier transform of the signal and deduce the information from its magnitude and/or phase. Figure 11.2 shows the magnitude and phase functions in this case. The frequencies of the signal (which represent the notes A, B, C, D, and E of the C major scale) are indeed visible in the spectrum. The information about the durations and relative timings is in the phase function. However, due to the 2π ambiguity of the phase, it is virtually impossible to use this information. What we really want is some mechanism that would reconstruct an approximation of the *frequency as a function of time*, as shown in Figure 11.1. The program should then consult its data base of classical works, identify the sequence of notes, their durations and timings, and report that the piece in question is the Great C Major symphony of Schubert. The reconstruction of frequency-versus-time characteristics of a signal is one of the major goals of time-frequency analysis. ∎

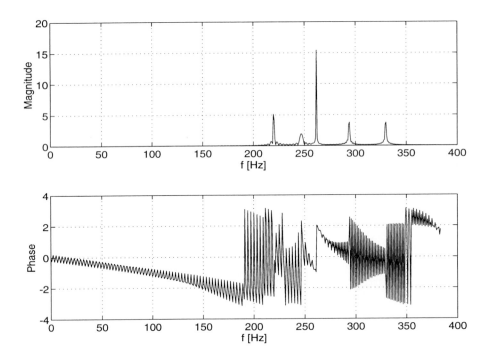

Figure 11.2. The Fourier transform of the signal in Example 11.1.

In this chapter and the next we will present the most important time-frequency methods for nonstationary signal analysis. This chapter is devoted to *linear* transforms and the next to *nonlinear* ones. In particular, we will discuss the short-time Fourier transform and its relative, the Gabor representation, and then the newly developed wavelet transform. In dealing with a linear transform, there are two problems that need to be explored: (a) the *analysis* problem (i.e., the operation performed on the signal), and (b) the *synthesis* or *reconstruction* problem (the operation that yields the signal from its transform).

The reader will encounter very little discussion on discrete-time signals and nothing on random signals in this chapter. Most of the exposition will use continuous-time signals whose randomness (if at all) is not explicitly characterized. If so, what does this subject have to do in a book on digital processing of random signals? The answer is that time-frequency analysis is, at the time the book is written, a developing field. While the general mathematical foundations are fairly well developed, specific applications, mainly to random signal processing, are still scarce. However, the *potential* importance of this field to random signal processing is too high to ignore. The inclusion of the material in this book represents the author's belief in the future of this area and is partly intended as encouragement to the reader to make his or her own original contributions in proper time.

11.2. THE SHORT-TIME FOURIER TRANSFORM

The origins of the short-time Fourier transform (STFT) are hard to trace, but at least in the electrical engineering context it was first introduced by Gabor [1946]. As was said in the introduction, the purpose of the STFT is to capture the time variation of the frequency contents of the signal. It does so by multiplying the signal $y(s)$ by a *sliding window* $\gamma^*(s-t)$, centered at time t, and taking the Fourier transform of the product. The result is a function of the variable t (the window's center) and the frequency variable ω. Mathematically,

$$Y(t,\omega) = \int_{-\infty}^{\infty} y(s)\gamma^*(s-t)e^{-j\omega s}ds. \qquad (11.2.1)$$

The conjugation of γ is mainly a "mathematical nicety," since in most cases it is a real-valued function anyway.

The function $\gamma(s)$ should be centered around time zero, and its Fourier transform $\Gamma(\omega)$ should be centered around frequency zero; that is, it is usually required that[†]

$$\int_{-\infty}^{\infty} s|\gamma(s)|^2 ds = 0, \quad \int_{-\infty}^{\infty} \omega|\Gamma(\omega)|^2 d\omega = 0. \qquad (11.2.2)$$

Intuitively, $\gamma(s)$ should be a "narrow" function so that the integral (11.2.1) will only be affected by the values of the function $y(s)$ in the vicinity of t. We also want the Fourier transform of $\gamma(s)$ to be "narrow." To explain the reason for this, recall that time domain multiplication is equivalent to frequency domain convolution. If the Fourier transform of $\gamma(s)$ is wide, it will smear the frequency contents of $y(s)$. Unfortunately, the time-bandwidth product of any function is bounded from below, as asserted by the famous *uncertainty principle* [Bracewell, 1986, p. 160]. More precisely, define the *effective duration and bandwidth* of a function $\gamma(s)$ satisfying (11.2.2) as

$$\tau_{\text{eff}} = \left[\frac{\int_{-\infty}^{\infty} s^2|\gamma(s)|^2 ds}{\int_{-\infty}^{\infty} |\gamma(s)|^2 ds}\right]^{1/2}, \quad \omega_{\text{eff}} = \left[\frac{\int_{-\infty}^{\infty} \omega^2|\Gamma(\omega)|^2 d\omega}{\int_{-\infty}^{\infty} |\Gamma(\omega)|^2 d\omega}\right]^{1/2}. \qquad (11.2.3)$$

Then

$$\tau_{\text{eff}}\omega_{\text{eff}} \geq 0.5, \qquad (11.2.4)$$

with equality for the Gaussian function $\gamma(s) = e^{-0.5s^2}$. Since the Gaussian function has the best (lowest) time-bandwidth product, it was proposed by Gabor as the most appropriate one for use in the STFT.

[†] Of course, the second requirement in (11.2.2) is trivially met when $\gamma(s)$ is real valued.

The inverse STFT is given by a formula similar to (11.2.1), but possibly with a different window function $g(u)$. Specifically,

$$y(t) = \frac{1}{2\pi} \int_{-\infty}^{\infty} \int_{-\infty}^{\infty} Y(u, \omega) g(t - u) e^{j\omega t} du\, d\omega. \tag{11.2.5}$$

Let us check under what conditions we will get back the original signal $y(t)$. Using the formula $\int_{-\infty}^{\infty} e^{j\omega \tau} d\omega = 2\pi \delta_D(\tau)$, we get[†]

$$\frac{1}{2\pi} \int_{-\infty}^{\infty} \int_{-\infty}^{\infty} \int_{-\infty}^{\infty} y(s) \gamma^*(s - u) e^{-j\omega s} g(t - u) e^{j\omega t} ds\, du\, d\omega$$

$$= \int_{-\infty}^{\infty} \int_{-\infty}^{\infty} y(s) \gamma^*(s - u) g(t - u) \delta_D(t - s) ds\, du = y(t) \int_{-\infty}^{\infty} \gamma^*(\tau) g(\tau) d\tau.$$

This is equal to $y(t)$ if and only if

$$\int_{-\infty}^{\infty} g(\tau) \gamma^*(\tau) d\tau = 1. \tag{11.2.6}$$

As (11.2.6) shows, the restriction on the relation between the two window functions is very mild. For example, we can take $g(\tau) = \gamma(\tau)$ if this function has unit norm. Another alternative is to take $\gamma(0) = 1$ and $g(\tau) = \delta_D(\tau)$. Then (11.2.5) will simplify to

$$y(t) = \frac{1}{2\pi} \int_{-\infty}^{\infty} Y(t, \omega) e^{j\omega t} d\omega. \tag{11.2.7}$$

The STFT has an interesting interpretation as a *filter bank*. To see this let $\Gamma(\omega)$ be the Fourier transform of $\gamma(t)$. Rewrite 11.2.1 for a fixed ω_0 as

$$Y(t, \omega_0) = e^{-j\omega_0 t} \int_{-\infty}^{\infty} y(s) \gamma^*(s - t) e^{-j\omega_0 (s-t)} ds. \tag{11.2.8}$$

Observe that $\gamma^*(-t) e^{j\omega_0 t}$ is the impulse response of a linear filter whose frequency response is $\Gamma^*(\omega - \omega_0)$. Thus, $Y(t, \omega_0)$ can be obtained from $y(t)$ by passing it through the filter $\Gamma^*(\omega - \omega_0)$, multiplying the filter's output by $e^{-j\omega_0 t}$, and observing the result at time t. This process is depicted in Figure 11.3. The complete STFT can obtained by using an infinite (and continuously varying) bank of filters, parameterized by ω_0. All these filters are obtained from a single filter $\Gamma^*(\omega)$ by frequency translations. Thus, they all have the same bandwidth. The STFT can therefore be interpreted as a time-varying spectral analysis of the given signal: each spectral component is monitored continuously in time. The postmultiplication by $e^{-j\omega_0 t}$ simply demodulates the filter's output (which is centered around ω_0) to the dc and its vicinity.

[†] Many of the proofs in this chapter and the next use the Dirac delta function in a nonrigorous manner. However, these proofs can be made rigorous by treating the Dirac delta as a *generalized function* [Kolmogorov and Fomin, 1975, Sec. 21].

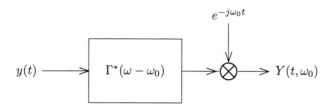

Figure 11.3. Filter interpretation of the STFT.

We will meet the STFT again in Sec. 11.7 when we discuss its implementation for discrete-time signals. For now we will turn our attention to its discrete-parameter version, the Gabor representation.

11.3. THE GABOR REPRESENTATION: ELEMENTARY DISCUSSION

The inverse STFT formula (11.2.5) expresses the signal $y(t)$ in terms of the function $Y(t, \omega)$ of the *continuous* time and frequency variables t and ω. This section is devoted to the question of representing the signal in terms of *discrete* time and frequency variables. Such a representation was originally proposed by Gabor [1946] and is given by

$$y(t) = \sum_{m=-\infty}^{\infty} \sum_{n=-\infty}^{\infty} c_{m,n} g(t - n\beta) e^{j2\pi m\alpha t}. \tag{11.3.1}$$

The function $g(t)$ is called the *Gabor window function*. It is assumed to be square integrable; that is, $\int_{-\infty}^{\infty} g^2(t)dt < \infty$. The integer variables n and m represent time and frequency, respectively, and α and β are positive constant scale factors. The complex numbers $\{c_{m,n}\}$ are the Gabor coefficients of $y(t)$.

An equivalent way of writing (11.3.1) is as

$$y(t) = \sum_{m=-\infty}^{\infty} \sum_{n=-\infty}^{\infty} c_{m,n} g_{m,n}(t), \tag{11.3.2}$$

where the set of functions $\{g_{m,n}(t)\}$ is given by

$$g_{m,n}(t) = g(t - n\beta) e^{j2\pi m\alpha t}. \tag{11.3.3}$$

The representation (11.3.2) formally looks like an expansion in terms of basis functions, but this interpretation may be misleading. As we will see later, the set $\{g_{m,n}(t)\}$ is usually not orthonormal and may not be a basis. An interesting property of this set is that all its members are obtained from a single function $g(t)$ by translations and modulations. The modulations, in turn, are translations of the Fourier transform of $g(t)$, so the $\{g_{m,n}(t)\}$ are translations of the fundamental function in the time-frequency grid. In physics they are called *coherent states*.

To illustrate the Gabor representation, consider the case $\alpha = 1$, $\beta = 1$, and let $g(t)$ be the unit-height rectangular window on $[-1/2, 1/2)$. Then the following holds trivially:

$$y(t) = \sum_{n=-\infty}^{\infty} y(t)g(t-n).\tag{11.3.4}$$

If $y(t)$ is sufficiently smooth, then $y(t)g(t-n)$ (which is supported on the interval $[n-0.5, n+0.5)$) can be represented by the Fourier series

$$y(t)g(t-n) = \sum_{m=-\infty}^{\infty} c_{m,n}e^{j2\pi mt}, \quad n-0.5 \le t \le n+0.5.\tag{11.3.5}$$

Substitution of (11.3.4) in (11.3.5) yields the representation (11.3.1) for the special case of the rectangular window function.

As in the case of the STFT, the window is assumed to be concentrated around zero in both time and frequency domains. The window function originally suggested by Gabor was the Gaussian function

$$g(t) = 2^{1/4}e^{-\pi t^2}.\tag{11.3.6}$$

As was explained in the previous section, this function has the minimum possible time-bandwidth product, so it yields the best possible time-frequency resolution. By comparison, the rectangular window mentioned above has good time resolution, but poor frequency resolution. Nevertheless, as we will see later, the Gaussian window has some severe drawbacks when used in the Gabor representation.

The main issues related to the Gabor representation are its existence for a given window function $g(t)$, ways of computing the coefficients $\{c_{m,n}\}$, and the stability of the representation. By "stability" we mean the extent to which finite approximations of the infinite sum (11.3.1) can approximate the given signal. We begin with an elementary treatment of these questions and then proceed to a more advanced treatment. We will assume in this section that $\alpha\beta = 1$. In this case there is no loss of generality in assuming that $\alpha = \beta = 1$ (otherwise, we scale the time variable t by β and work with $t' = t/\beta$).

A basic tool in the treatment of Gabor representation is the *Zak transform* [Zak, 1967]. This is defined as

$$H_z(t,f) = \sum_{k=-\infty}^{\infty} h(t-k)e^{-j2\pi kf}.\tag{11.3.7}$$

Both t and f are continuous variables. Thus, $H_z(t,f)$ is simply the Fourier transform of the sampled and time-reversed sequence $\{h(t-k)\}$.[†] The Zak transform is obviously periodic in f (with period 1), but is not periodic in t. However, it is "nearly periodic" in the sense of the following lemma.

[†] When dealing with the Gabor and Zak transforms, it is common to work with the cycle frequency f, rather than with the radian frequency ω.

Lemma 11.1. For any integer n,

$$H_z(t - n, f) = e^{j2\pi n f} H_z(t, f). \tag{11.3.8}$$

Proof.

$$H_z(t - n, f) = \sum_{k=-\infty}^{\infty} h(t - n - k)e^{-j2\pi k f} = \sum_{\ell=-\infty}^{\infty} h(t - \ell)e^{-j2\pi(\ell-n)f}$$

$$= e^{j2\pi n f} H_z(t, f). \qquad \blacksquare$$

Let $y(t)$ be a square-integrable signal. For the time being, we ignore the question whether the representation (11.3.1) exists and simply assume its existence. Then we get from (11.3.1) (since $e^{j2\pi k m} = 1$ for integer k and m)

$$y(t - k)e^{-j2\pi k f} = \sum_{m=-\infty}^{\infty} \sum_{n=-\infty}^{\infty} c_{m,n} g(t - k - n)e^{j2\pi m t}e^{-j2\pi k f}$$

$$= \sum_{m=-\infty}^{\infty} \sum_{n=-\infty}^{\infty} c_{m,n} g(t - k - n)e^{-j2\pi(k+n)f}e^{j2\pi(mt+nf)}.$$

Summation over k yields

$$Y_z(t, f) = \sum_{m=-\infty}^{\infty} \sum_{n=-\infty}^{\infty} c_{m,n} G_z(t, f)e^{j2\pi(mt+nf)},$$

so

$$\sum_{m=-\infty}^{\infty} \sum_{n=-\infty}^{\infty} c_{m,n}e^{j2\pi(mt+nf)} = \frac{Y_z(t, f)}{G_z(t, f)}.$$

The left side is the two-dimensional Fourier transform of $\{c_{m,n}\}$, evaluated at the frequencies $(-2\pi t, -2\pi f)$. Therefore, by the inverse Fourier transform formula,

$$c_{m,n} = \int_0^1 \int_0^1 \frac{Y_z(t, f)}{G_z(t, f)}e^{-j2\pi(mt+nf)} df\, dt. \tag{11.3.9}$$

This result is due to Janssen [1981a, b].

An alternative approach to the computation of the Gabor coefficients can be derived as follows. Assume there exists a function $\gamma(t)$ satisfying

$$\int_{-\infty}^{\infty} \gamma^*(t) g(t - \ell)e^{-j2\pi k t} dt = \delta(k)\delta(\ell). \tag{11.3.10}$$

Then $\gamma(t)$ is said to be *biorthogonal* to $g(t)$. We get, from (11.3.1),

$$y(t)\gamma^*(t - \ell)e^{-j2\pi k t} = \sum_{m=-\infty}^{\infty} \sum_{n=-\infty}^{\infty} c_{m,n}\gamma^*(t - \ell)g(t - n)e^{j2\pi(m-k)t}.$$

Integrating over t and using the biorthogonality relationship (11.3.10) gives

$$\int_{-\infty}^{\infty} y(t)\gamma^*(t-\ell)e^{-j2\pi kt}dt = \sum_{m=-\infty}^{\infty}\sum_{n=-\infty}^{\infty} c_{m,n}\delta(n-\ell)\delta(m-k).$$

Therefore,

$$c_{k,\ell} = \int_{-\infty}^{\infty} y(t)\gamma^*(t-\ell)e^{-j2\pi kt}dt = \int_{-\infty}^{\infty} y(t)\gamma_{k,\ell}^*(t)dt, \qquad (11.3.11)$$

where $\gamma_{k,\ell}(t) = \gamma(t-\ell)e^{j2\pi kt}$. This method of computing the Gabor coefficients was given by Bastiaans [1980]. Formulas (11.3.11) and (11.3.1) can be regarded as an analysis/synthesis pair, in analogy with the STFT analysis/synthesis pair (11.2.1) and (11.2.2). There is a minor difference in the *point of view*: The STFT is usually thought of as a transform, so the analysis formula is regarded as the direct one and the synthesis as the inverse one. In the Gabor case we usually regard the synthesis formula (11.3.1) as the direct one and the analysis formula (11.3.11) as the inverse one.

The formula (11.3.11) is much more convenient than (11.3.9) for computing the Gabor coefficients, provided the biorthogonal function can be obtained. To this end we multiply (11.3.10) by $e^{-j2\pi f\ell}$ and sum over ℓ to get

$$\int_{-\infty}^{\infty} \gamma^*(t)G_z(t,f)e^{-j2\pi kt}dt = \delta(k).$$

Multiply both sides by $e^{j2\pi k\tau}$ and sum over k to get

$$\int_{-\infty}^{\infty} \gamma^*(t)G_z(t,f)\left[\sum_{k=-\infty}^{\infty} e^{-j2\pi k(t-\tau)}\right]dt = 1.$$

Recall Poisson's formula,

$$\sum_{k=-\infty}^{\infty} e^{-j2\pi ka} = \sum_{k=-\infty}^{\infty} \delta_D(a+k), \qquad (11.3.12)$$

and use it to get

$$\int_{-\infty}^{\infty} \gamma^*(t)G_z(t,f)\left[\sum_{k=-\infty}^{\infty} \delta_D(t-\tau+k)\right]dt = \sum_{k=-\infty}^{\infty} \gamma^*(\tau-k)G_z(\tau-k,f) = 1.$$

By Lemma 11.1,

$$\sum_{k=-\infty}^{\infty} \gamma^*(\tau-k)G_z(\tau,f)e^{j2\pi fk} = 1.$$

So,

$$\sum_{k=-\infty}^{\infty} \gamma(\tau-k)e^{-j2\pi fk} = \frac{1}{G_z^*(\tau,f)}.$$

Therefore, $\gamma(\tau - k)$ is given by the inverse Fourier transform

$$\gamma(\tau - k) = \int_0^1 \frac{e^{j2\pi fk}}{G_z^*(\tau, f)} df.$$

In particular,

$$\gamma(\tau) = \int_0^1 \frac{1}{G_z^*(\tau, f)} df. \qquad (11.3.13)$$

In summary, if the biorthogonal function can be computed from (11.3.13), it can then be used to compute the Gabor coefficients via (11.3.11).

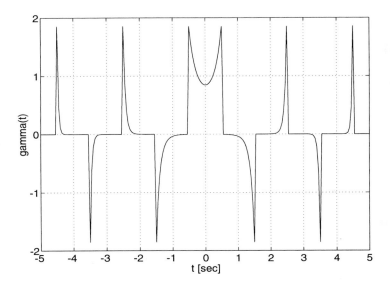

Figure 11.4. The biorthogonal function of the Gaussian window.

The biorthogonal function of the Gaussian window was computed by Bastiaans [1980] and found to be

$$\gamma(t) = 2^{-1/4} \left(\frac{K_0}{\pi} \right)^{-3/2} e^{\pi t^2} \sum_{n > |t| - 0.5} (-1)^n e^{-\pi(n+0.5)^2}, \qquad (11.3.14a)$$

where

$$K_0 = \int_0^{\pi/2} (1 - 0.5 \sin^2 \theta)^{-1/2} d\theta \approx 1.854075. \qquad (11.3.14b)$$

This is a rather irregular function; see Figure 11.4. In particular, it is not square integrable and its use for numerical calculation of the Gabor coefficients of an arbitrary signal $y(t)$ is dubious. The next section will provide an explanation for the irregularity of the biorthogonal function of the Gaussian window.

We finally emphasize again that the results of this section are valid only for $\alpha\beta = 1$.

11.4. THE GABOR REPRESENTATION: ADVANCED DISCUSSION

In this section we will deal with some advanced aspects of the Gabor representation. This material will be useful for the treatment of the Wavelet transform as well. Since the Gabor functions $\{g_{m,n}\}$ are not orthonormal in general, the usual orthonormal-basis tools of Hilbert space theory are not adequate for the treatment of the Gabor representation. It turns out that the proper tool in this case is the theory of frames. The theory of frames is due to Duffin and Schaeffer [1952] and was introduced to Gabor and Wavelet theory by Daubechies, Grossmann, and Meyer [1986] and Daubechies [1990]. We will first give a brief introduction to the theory of frames and then apply it to the Gabor representation. The application of the theory to the Wavelet transform will be given in Sec. 11.5.

Let L_2 be the space of complex-valued square-integrable (in the Lebesgue sense) functions on the real line, and K a countable set [which is, in the Gabor case, the set of integer pairs $\{(m, n), m, n \in \mathbf{Z}\}$]. We will denote the inner product in L_2 by

$$\langle h_1, h_2 \rangle = \int_{-\infty}^{\infty} h_1(t) h_2^*(t) dt \qquad (11.4.1)$$

(note that we omit the explicit dependence of the functions on t when regarding them as vectors in the Hilbert space L_2). Let $\{g_k, k \in K\}$ be a set of L_2 functions indexed on K. This set is said to be *a frame* if there exist numbers $0 < A \le B < \infty$ such that

$$A\|f\|^2 \le \sum_{k \in K} |\langle g_k, f \rangle|^2 \le B\|f\|^2, \quad \forall f \in L_2. \qquad (11.4.2)$$

The numbers A and B are called the *frame bounds* [Young, 1980, pp. 184–190].

A frame is necessarily a complete set in L_2 (Prob. 11.2). However, a frame is generally redundant, so it does not necessarily constitute a basis for L_2. Frames that are not redundant (i.e., for which the deletion of any member would leave a set that is not a frame) are said to be *exact*. A frame is exact if and only if it is a basis for L_2 (not necessarily an orthonormal one); see, for example, Young [1980, pp. 184–190] for a proof. A frame is said to be *tight* if the frame bounds A and B are equal. There is a special case in which a frame is in fact an orthonormal basis, as given by the following lemma.

Lemma 11.2. If the frame is tight, with $A = B = 1$, and if all the members of the frame have unit norm, the frame is an orthonormal basis.

Proof. Fix g_ℓ in the frame. Then, if $A = B = 1$, we get from (11.4.2),

$$\|g_\ell\|^2 = \sum_{k \in K} |\langle g_k, g_\ell \rangle|^2 = \|g_\ell\|^4 + \sum_{k \neq \ell} |\langle g_k, g_\ell \rangle|^2.$$

Since $\|g_\ell\|^2 = \|g_\ell\|^4 = 1$, necessarily $\langle g_k, g_\ell \rangle = 0$ for all $k \neq \ell$.   ∎

With any given frame $\{g_k, k \in K\}$, we can associate a *frame operator* $T : L_2 \to L_2$, defined by

$$Tf = \sum_{k \in K} \langle f, g_k \rangle g_k. \tag{11.4.3}$$

Lemma 11.3. The frame operator is (a) bounded, (b) self-adjoint, (c) positive, and (d) invertible.

Proof.

(a) Denote $h = Tf$. Then

$$\|h\|^4 = |\langle h, h \rangle|^2$$

$$= \left| \sum_{k \in K} \langle f, g_k \rangle \langle g_k, h \rangle \right|^2 \leq \sum_{k \in K} |\langle f, g_k \rangle|^2 \sum_{k \in K} |\langle g_k, h \rangle|^2 \leq B^2 \|f\|^2 \|h\|^2.$$

Therefore,

$$\|h\| = \|Tf\| \leq B\|f\|.$$

This proves the boundedness of T.

(b) To prove that T is self-adjoint, we have to show that $\langle f, Th \rangle = \langle Tf, h \rangle$ for all f and h. Now

$$\langle f, Th \rangle = \sum_{k \in K} \langle g_k, h \rangle \langle f, g_k \rangle, \quad \langle Tf, h \rangle = \sum_{k \in K} \langle f, g_k \rangle \langle g_k, h \rangle,$$

so the two are equal.

(c)

$$\langle f, Tf \rangle = \sum_{k \in K} \langle g_k, f \rangle \langle f, g_k \rangle = \sum_{k \in K} |\langle g_k, f \rangle|^2.$$

This proves the positiveness of T.

(d)

$$\|f\| \cdot \|Tf\| \geq \langle Tf, f \rangle \geq A\|f\|^2,$$

so $\|Tf\| \geq A\|f\|$, meaning that T is bounded from below. Now it is a known fact from Hilbert space theory [Taylor, 1958, p. 234] that a self-adjoint operator that is bounded from below is invertible.   ∎

Let $\{\gamma_k, k \in K\}$ be the set of functions obtained from the given frame by $\gamma_k = T^{-1}g_k$. As the next lemma shows, this set is also a frame, called the *dual frame* of $\{g_k\}$.

Lemma 11.4. The set $\{\gamma_k\}$ is a frame.

Proof. Since T is self-adjoint, so is T^{-1}. Since the norm of T is bounded between A and B, the norm of T^{-1} is bounded between B^{-1} and A^{-1}. Therefore,

$$\sum_{k \in K} |\langle f, \gamma_k \rangle|^2 = \sum_{k \in K} |\langle f, T^{-1}g_k \rangle|^2 = \sum_{k \in K} |\langle T^{-1}f, g_k \rangle|^2.$$

So

$$\frac{A}{B^2} \le A\|T^{-1}f\|^2 \le \sum_{k \in K} |\langle f, \gamma_k \rangle|^2 \le B\|T^{-1}f\|^2 \le \frac{B}{A^2}. \qquad \blacksquare$$

It should be noted that the sets $\{g_k\}$ and $\{\gamma_k\}$ are not biorthogonal in general; that is, $\langle g_k, \gamma_\ell \rangle = \delta(k - \ell)$ does not necessarily hold. However, when the frame $\{g_k\}$ is exact, biorthogonality holds. The proof of this result is not central to our discussion, so we will skip it (see, e.g., Young [1980, p. 188]).

The next theorem is our main reason for introducing frames.

Theorem 11.1. For any frame $\{g_k\}$ and its dual $\{\gamma_k\}$,

$$f = \sum_{k \in K} \langle f, \gamma_k \rangle g_k, \quad \text{and} \quad f = \sum_{k \in K} \langle f, g_k \rangle \gamma_k. \tag{11.4.4}$$

Proof. By the definition of the frame operator of $\{g_k\}$, we have

$$f = T T^{-1} f = \sum_{k \in K} \langle T^{-1}f, g_k \rangle g_k = \sum_{k \in K} \langle f, T^{-1}g_k \rangle g_k = \sum_{k \in K} \langle f, \gamma_k \rangle g_k.$$

The second equality is proved in a dual manner. $\qquad \blacksquare$

Formula (11.4.4) means that we can express any function in L_2 as an infinite linear combination of the frame elements, and the set of coefficients is square summable (this follows from the fact that the dual set is a frame). While (11.4.4) is formally similar to an expansion in terms of an orthogonal basis, it has different properties. In particular, the coefficients are not obtained as inner products with the frame elements, but as inner products with the elements of the dual frame. Also, neither of the expansions in (11.4.4) is claimed to be unique (in general, the expansion $f = \sum_{k \in K} a_k g_k$ is possible in an infinite number of ways).

In the special case of a tight frame (i.e., when $A = B$), the frame operator T is simply $A\mathbf{1}$, where $\mathbf{1}$ is the identity operator. In this case, $\gamma_k = A^{-1}g_k$, so the dual frame is obtained in a trivial manner. If the frame is not tight, the dual frame cannot be obtained by a simple procedure. However, there exists an algorithm that yields γ_k from g_k by a convergent iteration, as follows. Define $c = 2/(A+B)$ and write T as $T = c^{-1}[\mathbf{1} - (\mathbf{1} - cT)]$. Let us show that the norm of $(\mathbf{1} - cT)$ is strictly less than 1. Indeed,

$$A\mathbf{1} \leq T \leq B\mathbf{1} \Longrightarrow (1 - cB)\mathbf{1} \leq \mathbf{1} - cT \leq (1 - cA)\mathbf{1}$$

$$\Longrightarrow \frac{A-B}{A+B}\mathbf{1} \leq \mathbf{1} - cT \leq \frac{B-A}{A+B}\mathbf{1},$$

so

$$\|\mathbf{1} - cT\| \leq \frac{B-A}{A+B} < 1.$$

Therefore, we can expand the inverse operator T^{-1} as the infinite convergent series

$$T^{-1} = c\sum_{n=0}^{\infty}(\mathbf{1} - cT)^n. \tag{11.4.5}$$

In particular, we can apply this expansion to g_k to get

$$T^{-1}g_k = c\sum_{n=0}^{\infty}(\mathbf{1} - cT)^n g_k.$$

The explicit algorithm is therefore as follows: Set $f_0 = h_0 = g_k$; then let $f_{k+1} = f_k - cTf_k$ and $h_{\ell+1} = h_\ell + f_{\ell+1}$. The sequence $\{ch_\ell\}$ then converges to γ_k at an exponential rate. Clearly, the closer the ratio A/B to 1, the faster the convergence. This shows the importance of having as tight as possible frame bounds.

Let us now specialize the theory of frames to the Gabor representation. The set K is now the set of integer pairs $\{(m,n)\}$. For each such pair, let $W_{m,n} : L_2 \to L_2$ be the operator defined by

$$W_{m,n}f(t) = f(t - n\beta)e^{j2\pi m\alpha t}. \tag{11.4.6}$$

Following Daubechies [1990], we will call $W_{m,n}$ the *Weyl-Heisenberg* operator, since it was used by Weyl and Heisenberg in quantum mechanics before the work of Gabor. Let $g(t)$ be a fixed function in L_2, and *assume* that the set $\{g_{m,n}(t) = W_{m,n}g(t), m, n \in \mathbf{Z}\}$ is a frame. Then the expansion $f(t) = \sum_{m,n}\langle f, \gamma_{m,n}\rangle g_{m,n}(t)$ of Theorem 11.1 is the Gabor representation of $f(t)$, and $\{\langle f, \gamma_{m,n}\rangle\}$ are the Gabor coefficients.

Let us discuss the question of obtaining the dual frame, since it is crucial to the construction of the Gabor coefficients. It is easy to show (Prob. 11.4) that the Weyl-Heisenberg operator satisfies the composition rule

$$W_{m,n}W_{m',n'} = e^{-j2\pi\alpha\beta m'n}W_{m+m',n+n'}. \tag{11.4.7}$$

This can be used to derive the following relationship.

$$W_{m,n}Tf = \sum_{k,\ell}\langle f, g_{k,\ell}\rangle W_{m,n}g_{k,\ell} = \sum_{k,\ell}\langle f, W_{k,\ell}g\rangle W_{m,n}W_{k,\ell}g =$$

$$\sum_{k,\ell}\langle f, W_{k,\ell}g\rangle e^{-j2\pi\alpha\beta kn}W_{m+k,n+\ell}g = \sum_{k',\ell'}\langle f, W_{k'-m,\ell'-n}g\rangle e^{-j2\pi\alpha\beta(k'-m)n}W_{k',\ell'}g$$

$$= \sum_{k',\ell'}\langle f, e^{-j2\pi\alpha\beta k'n}W_{-m,-n}W_{k',\ell'}g\rangle e^{-j2\pi\alpha\beta(k'-m)n}W_{k',\ell'}g$$

$$= \sum_{k',\ell'}\langle W_{m,n}f, W_{k',\ell'}g\rangle W_{k',\ell'}g = TW_{m,n}f.$$

Hence,

$$T^{-1}W_{m,n} = W_{m,n}T^{-1}.$$

Applying this equality to g yields

$$T^{-1}g_{m,n} = \gamma_{m,n} = W_{m,n}\gamma. \qquad (11.4.8)$$

Therefore, the dual frame is completely defined by the single function $\gamma(t) = T^{-1}g(t)$. This function can be computed by the iterative algorithm described above. In summary, we have generalized the Gabor representation to α and β whose product may be different from 1, and for window functions $g(t)$ such that $\{g_{m,n}(t)\}$ constitute a frame. Janssen's formula for the Gabor coefficients and the computation via the biorthogonal function do not apply to the general case. Nevertheless, the coefficients can be computed from the translations and modulations of the dual function $\gamma(t)$. This function can be computed from $g(t)$ provided the frame bounds A and B are known.

All the above is true provided $g(t)$, α, and β are chosen so as to make $\{g_{m,n}(t)\}$ a frame. Let us discuss this issue separately for each of the cases $\alpha\beta < 1, = 1$, and > 1. The following theorem addresses the case $\alpha\beta = 1$.

Theorem 11.2. If the set $\{g_{m,n}\}$ corresponding to the L_2 function $g(t)$ is a frame for $\alpha\beta = 1$, then either the derivative of $g(t)$ is not in L_2 or $tg(t)$ is not in L_2 (or both).

This theorem is usually attributed to Balian, but as was pointed out by Daubechies [1990], its full proof was given by Coifman and Semmes. The proof relies on the property of the Zak transform treated in Prob. 11.6. It is rather technical and will not be given here, see Daubechies [1990]. ∎

Balian's theorem essentially says that, for $g(t)$ to be a frame generator, it is not allowed to be "too nice". For example, the one-sided exponential function discussed in Prob. 11.3 does give rise to a frame. On the other hand, the Gaussian window used by Gabor (and discussed in Sec. 11.3) *does not give rise to a frame*. This means that, although the construction of the coefficients $\{c_{m,n}\}$ presented in Sec. 11.3 is valid, the coefficients are not necessarily square summable, so any truncation of the Gabor representation to a finite sum may not be close to the given function. This instability of the Gaussian Gabor representation is rather disappointing. Fortunately, the Gaussian window can be salvaged by using $\alpha\beta < 1$, as will be discussed below.

To check whether a given L_2 function $g(t)$ generates a frame for $\alpha\beta = 1$, we must compute its Zak transform $G_z(t, f)$ and verify that (see Prob. 11.6)

$$0 < A \le |G_z(t, f)|^2 \le B < \infty \quad \text{almost everywhere on } [0, 1]^2. \tag{11.4.9}$$

The frame bounds are less important in this case, since $\gamma(t)$ can be computed directly from $G_z(t, f)$ through (11.3.13).

When $\alpha\beta > 1$, no $g(t)$ gives rise to a frame. This is shown in Daubechies [1990] for rational values of $\alpha\beta$ and was also proved for irrational values. The proofs are beyond the scope of this book.

It remains to treat the case $\alpha\beta < 1$, which is the best from a practical point of view, since it leads to stable reconstruction. Here we have two choices:

1. To rely on functions $g(t)$ that are known to yield tight frames. This has the advantage of obviating the need for computing the dual function $\gamma(t)$, since it is equal to $g(t)$ up to a proportionality factor.
2. To choose a function $g(t)$ that suits our needs and verify in some manner that it yields a frame.

Daubechies, Grossmann, and Meyer [1986] have constructed a family of tight frames for the Gabor representation. We give here their construction for the case $0.5 \le \alpha\beta < 1$. The first step is to choose a continuous function $\nu(t)$ such that $\nu(0) = 0$ and $\nu(1) = 0.5\pi$. It is reasonable, though not necessary, to let $\nu(t)$ be monotone increasing [$\nu(t) = 0.5\pi t$ being the simplest example]. Then we choose α and β such that $0.5 \le \alpha\beta < 1$ and define $\lambda = 1/\alpha - \beta = (1 - \alpha\beta)/\alpha$ (note that $\lambda \le 1/2\alpha$). The choice of α controls the time-frequency resolution trade-off. In general, large α (hence small β) causes $g(t)$ to be narrow in time and wide in frequency. The function $g(t)$ is defined to be zero outside the interval $[-1/2\alpha, 1/2\alpha]$. Inside this interval it is given by

$$g(t) = \beta^{-1/2} \begin{cases} \sin\left[\nu\left(\frac{1}{\lambda}\left(t + \frac{1}{2\alpha}\right)\right)\right], & -\frac{1}{2\alpha} \le t < -\frac{1}{2\alpha} + \lambda \\ 1, & -\frac{1}{2\alpha} + \lambda \le t \le \frac{1}{2\alpha} - \lambda \\ \cos\left[\nu\left(\frac{1}{\lambda}\left(t - \frac{1}{2\alpha} + \lambda\right)\right)\right], & \frac{1}{2\alpha} - \lambda < t \le \frac{1}{2\alpha}. \end{cases} \tag{11.4.10}$$

The function $g(t)$ is normalized to unit energy. Its crucial property, which makes it generate a frame, is (Prob. 11.8)

$$\beta \sum_{n=-\infty}^{\infty} |g(t - n\beta)|^2 = 1 \tag{11.4.11}$$

[this explains the careful choice of the sine and cosine functions at the two sides of $g(t)$].

Lemma 11.5. The functions $\{g_{m,n}(t)\}$ constitute a tight frame.

Proof. Using Poisson's formula (11.3.12) and (11.4.11), we get for any $f(t)$ in L_2

$$\sum_{m,n} |\langle f, g_{m,n}\rangle|^2$$

$$= \sum_{m,n} \int_{n\beta-1/2\alpha}^{n\beta+1/2\alpha} \int_{n\beta-1/2\alpha}^{n\beta+1/2\alpha} f(t)f^*(\tau)g(t-\beta n)g(\tau-\beta n)e^{j2\pi\alpha m(t-\tau)}dt\,d\tau$$

$$= \frac{1}{\alpha}\sum_{m,n} \int_{n\beta-1/2\alpha}^{n\beta+1/2\alpha} \int_{n\beta-1/2\alpha}^{n\beta+1/2\alpha} f(t)f^*(\tau)g(t-\beta n)g(\tau-\beta n)\delta_D(\alpha(t-\tau+m))dt\,d\tau$$

$$= \frac{1}{\alpha}\sum_{n} \int_{-\infty}^{\infty} |f(t)|^2|g(t-\beta n)|^2 dt = \frac{1}{\alpha\beta}\int_{-\infty}^{\infty} |f(t)|^2 dt.$$

This proves that $\{g_{m,n}(t)\}$ is indeed a tight frame, with $A = B = 1/\alpha\beta$. ∎

The above construction yields the rectangular window when $\lambda = 0$, that is, when $\alpha\beta = 1$. As we have said, this gives good time concentration, but very poor frequency concentration. Letting λ be strictly positive tapers the Gabor window, as is seen from (11.4.10), and enables the improvement of the frequency resolution (at the expense of decreased time resolution, of course).

When we wish to choose $g(t)$ according to some other criteria (convenience, similarity to the physical signals being analyzed, etc.), there is a need to check whether it yields a frame and estimate the corresponding frame bounds in order to compute the dual function $\gamma(t)$. If α and β are chosen so that $\alpha\beta = 1/r$, where r is an integer, then a simple extension of (11.4.9) can be used for both purposes. Indeed, we have in this case (Prob. 11.9)

$$0 < A \le \sum_{k=0}^{r-1} |G_z(t-k/r, f)|^2 \le B < \infty \quad \text{a.e. on } [0,1] \times [0,1] \qquad (11.4.12)$$

if and only if $g(t)$ generates a frame. Note the averaging operation in (11.4.12), which tends to tighten the bounds A and B with respect to the case $r = 1$.

For general values of $\alpha\beta$, the computation is much more tedious. It has been carried out by Daubechies [1990] for the Gaussian window and for the window $g(t) = e^{-|t|}$. For example, the "tightness parameter" $(B-A)/(A+B)$ of the Gaussian window was found to be about 0.2 for $\beta = 2, \alpha = 1/4$ and about 0.025 for $\beta = 1.5, \alpha = 1/6$. Recall that a smaller value of $(B-A)/(A+B)$ leads to a tighter frame, hence to a $\gamma(t)$ that is more similar to $g(t)$.

11.5. THE WAVELET TRANSFORM

Let $h(\tau)$ be an L_2 function satisfying

$$C_h = \int_{-\infty}^{\infty} |\omega|^{-1}|H(\omega)|^2 d\omega < \infty, \tag{11.5.1}$$

where $H(\omega)$ is the Fourier transform of $h(\tau)$. Such a function is called a *wavelet*, and (11.5.1) is called the *admissibility condition* for the wavelet. The need for this condition and its interpretation will be explained shortly.

Let us denote

$$h_{s,t}(\tau) = |s|^{-1/2}h(s^{-1}(\tau - t)), \quad -\infty < s, t < \infty, \quad s \neq 0. \tag{11.5.2}$$

The variables t and s are known as the *translation* and *scale* variables, respectively.[†] The *continuous wavelet transform* (CWT) of an L_2 signal $y(t)$ is defined as

$$Y_w(s,t) = \int_{-\infty}^{\infty} y(\tau)h_{s,t}^*(\tau)d\tau = |s|^{-1/2}\int_{-\infty}^{\infty} y(\tau)h^*(s^{-1}(\tau - t))d\tau. \tag{11.5.3}$$

Similarly to the STFT, the wavelet transform can be interpreted as a linear filtering operation. It clearly follows from (11.5.3) that, if we pass the signal $y(t)$ through a linear time-invariant filter whose transfer function is $s^{1/2}H^*(s\omega)$, the output of the filter at time t will be $Y_w(t, s)$. Thus, we can view the CWT as an infinite bank of filters, each yielding $Y_w(t, s)$ as a function of time for fixed s. The filters are obtained from a fixed filter $H^*(\omega)$ by scaling of the frequency axis, as well as amplitude scaling. The amplitude scaling causes all filters to have the same integral of square magnitude. Thus, while the filter bank in the STFT is characterized by constant bandwidth, here it is characterized by a *constant relative bandwidth* (i.e., relative to the center frequency). Such a filter bank is also called *constant Q*. This interpretation leads to the following conclusion: As we progress along the STFT filter bank (increasing f), both time and frequency resolutions remain constant. On the other hand, as we progress along the CWT filter bank (increasing s), the frequency resolution improves while the time resolution degrades in proportion.

Theorem 11.3. The inverse CWT is given by

$$y(t) = C_h^{-1} \int_{-\infty}^{\infty} \int_{-\infty}^{\infty} s^{-2}Y_w(s,\rho)h_{s,\rho}(t)d\rho\,ds$$

$$= C_h^{-1} \int_{-\infty}^{\infty} \int_{-\infty}^{\infty} |s|^{-5/2}Y_w(s,\rho)h(s^{-1}(t - \rho))d\rho\,ds. \tag{11.5.4}$$

[†] We have chosen to denote the translation and scale variables by t and s as a reminder of their role. The reader should be aware that most of the wavelet literature uses b and a, respectively, for these variables.

Proof. Denote the right side of (11.5.4) by $f(t)$. Then

$$f(t) = C_h^{-1} \int_{-\infty}^{\infty} \int_{-\infty}^{\infty} \int_{-\infty}^{\infty} |s|^{-3} y(\tau) h(s^{-1}(t-\rho)) h^*(s^{-1}(\tau-\rho)) d\rho \, ds \, d\tau.$$

By Parseval's theorem,

$$\int_{-\infty}^{\infty} h(s^{-1}(t-\rho)) h^*(s^{-1}(\tau-\rho)) d\rho = \frac{s^2}{2\pi} \int_{-\infty}^{\infty} |H(s\omega)|^2 e^{j\omega(t-\tau)} d\omega.$$

So

$$f(t) = \frac{C_h^{-1}}{2\pi} \int_{-\infty}^{\infty} \int_{-\infty}^{\infty} \int_{-\infty}^{\infty} |s|^{-1} y(\tau) |H(s\omega)|^2 e^{j\omega(t-\tau)} d\omega \, ds \, d\tau.$$

Also,

$$\int_{-\infty}^{\infty} |s|^{-1} |H(s\omega)|^2 ds = \int_{-\infty}^{\infty} |\nu|^{-1} |H(\nu)|^2 d\nu = C_h.$$

Therefore,

$$f(t) = \frac{1}{2\pi} \int_{-\infty}^{\infty} \int_{-\infty}^{\infty} y(\tau) e^{j\omega(t-\tau)} d\omega d\tau = y(t).$$

Therefore, $f(t) = y(t)$, as stated.   ∎

The need for the admissibility condition (11.5.1) is now clear: to guarantee the existence of the inverse transform. The condition

$$H(0) = \frac{1}{2\pi} \int_{-\infty}^{\infty} h(\tau) d\tau = 0 \qquad (11.5.5)$$

is necessary for admissibility (Prob. 11.11). If $H(\omega)$ behaves "sufficiently nice" near $\omega = 0$ [e.g., if it is continuous with $|H(\omega)| \approx K|\omega|^\varepsilon$ for small ω and some positive ϵ], this condition is also sufficient for admissibility. Thus, a wavelet is necessarily a "dc-free" function.

As in the case of the short-time Fourier transform, we do not have much to say about the continuous transform, so we will proceed straight to the discrete case. The idea is, as in the case of Gabor representation, to discretize the (t, s) plane. We will look for discrete (t, s) grids that guarantee *invertibility*, that is, the ability to represent the given signal in terms of the values of the wavelet transform at the grid points. To this end we choose a scale step a and a translation step b, which are the analog of the frequency and time steps α and β in the Gabor case. Both parameters are positive, and $a > 1$. The discrete scale steps are taken to be in the geometric progression $s = a^{-m}, m \in \mathbf{Z}$. The discrete translation steps are taken to be in arithmetic progression, with step size proportional to the scale, that is, $t = a^{-m}bn, n \in \mathbf{Z}$. Thus, the discrete wavelet transform (DWT) is defined as

$$c_{m,n} = \int_{-\infty}^{\infty} y(\tau) h_{m,n}^*(\tau) d\tau, \qquad (11.5.6)$$

where

$$h_{m,n}(\tau) = h_{m,n}^{(+)}(\tau) = a^{m/2}h(a^m\tau - bn).$$ (11.5.7)

Figure 11.5 illustrates the sampling of the (t, s) plane in the definition of the discrete wavelet transform.

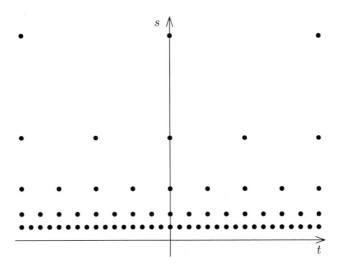

Figure 11.5. The DWT grid in the (t, s) plane (positive scales only).

Comparing (11.5.7) to (11.5.3) leads us to suspect that positive scales may not be sufficient for reconstruction, since the inverse CWT involves both positive and negative scales. Indeed, in some cases we will need the functions

$$h_{m,n}^{(-)}(\tau) = a^{m/2}h(-a^m\tau - bn)$$ (11.5.8)

as well, while in other cases the functions in (11.5.7) will be sufficient. More on this will be said later.

The choice of the translation step $a^{-m}b$ needs some explanation. Suppose the effective duration (by some measure) of the wavelet $h(\tau)$ is T_0. Then the effective duration of the wavelet scaled by a^{-m} is $a^{-m}T_0$. It intuitively makes sense to sample the time axis of the transform at an interval proportional to the wavelet duration, as in the Gabor case. So, since the duration is proportional to a^{-m}, so should be the translation step.

The questions we ask about the discrete transform (11.5.6) are basically the same as for the Gabor case: Does the inverse transform exist, and for what choices of a and b? Does it give the signal $y(t)$ in terms of the $h_{m,n}(\tau)$ or in terms of some dual functions? Is the inversion stable? How to choose "good" wavelets? We answer these questions in the remainder of this section and in the next.

The theory of frames discussed in the previous section applies directly to wavelets and saves much of the guesswork. We know that a reconstruction formula

$$y(t) = \sum_{m,n} c_{m,n} \eta_{m,n}(t) \tag{11.5.9}$$

exists for a square summable sequence $\{c_{m,n}\}$ if $\{h_{m,n}(t)\}$ is a frame and $\{\eta_{m,n}(t)\}$ is its dual frame. A similar statement holds for the extended family $\{h_{m,n}^{(+)}(t), h_{m,n}^{(-)}(t)\}$. It turns out that, unlike the Gabor case, there are no inherent limitations on a and b, but any $a > 1$ and $b > 0$ can give rise to a frame, and even to a tight frame. We will now give a construction of tight wavelet frames due to Daubechies, Grossmann, and Meyer [1986]. This construction makes use of both the $h^{(+)}$ and the $h^{(-)}$ functions. In the next section we will construct frames that use only the $h^{(+)}$ functions. Recall that for a tight frame $\eta_{m,n}(t)$ is equal to a multiple of $h_{m,n}(t)$, so the need for the dual frame is obviated.

Let $\nu(t)$ be a continuous function such that $\nu(0) = 0$ and $\nu(1) = 0.5\pi$ (same as in the construction for the Gabor case). Let $\lambda = 2\pi/b(a^2 - 1)$ and

$$H(\omega) = (\log a)^{-1/2} \begin{cases} \sin\left[\nu\left(\dfrac{\omega - \lambda}{\lambda(a-1)}\right)\right], & \lambda \le \omega \le a\lambda \\ \cos\left[\nu\left(\dfrac{\omega - a\lambda}{\lambda a(a-1)}\right)\right], & a\lambda < \omega \le a^2\lambda \\ 0, & \text{otherwise.} \end{cases} \tag{11.5.10}$$

The function $H(\omega)$ is normalized so that $C_h = 1$, where C_h is defined in (11.5.1). It is supported on an interval of length $2\pi/b$. Its crucial property, which makes it generate a frame, is (Prob. 11.12)

$$\log a \sum_{m=-\infty}^{\infty} |H(a^m \omega)|^2 = \begin{cases} 1, & \omega > 0 \\ 0, & \text{otherwise} \end{cases} \tag{11.5.11}$$

[this explains the careful choice of the sine and cosine functions at the two sides of $H(\omega)$, as in the construction of (11.4.10)]. Note that $H(\omega)$, hence all its scaled versions, are nonzero at positive frequencies only. Therefore, they cannot represent L_2 functions whose Fourier transforms are nonzero at negative frequencies. Nevertheless, by taking both the $h^{(+)}$ and the $h^{(-)}$ functions we do get a frame, as is proved in the following lemma.

Lemma 11.6. Let $h(t)$ be the inverse Fourier transform of $H(\omega)$ given in (11.5.10). Then the functions $\{h_{m,n}^{(+)}(t), h_{m,n}^{(-)}(t)\}$ constitute a tight frame.

Proof. By Parseval's theorem and Poisson's formula,

$$\sum_{m,n} |\langle f, h_{m,n}^{(+)} \rangle|^2$$

$$= \frac{1}{4\pi^2} \sum_{m,n} a_m \int_0^\infty \int_0^\infty F(\omega)F^*(\theta)H(a^m\omega)H^*(a^m\theta)e^{ja^mbn(\omega-\theta)} d\omega \, d\theta$$

$$= \frac{1}{2\pi b} \sum_{m,n} \int_0^\infty \int_0^\infty F(\omega)F^*(\theta)H(a^m\omega)H^*(a^m\theta)\delta_D(\omega - \theta + 2\pi b^{-1}n) d\omega \, d\theta$$

$$= \frac{1}{2\pi b} \sum_m \int_0^\infty |F(\omega)|^2|H(a^m\omega)|^2 d\omega = \frac{1}{2\pi b \log a} \int_0^\infty |F(\omega)|^2 d\omega.$$

In exactly the same way we get

$$\sum_{m,n} |\langle f, h_{m,n}^{(-)} \rangle|^2 = \frac{1}{2\pi b \log a} \int_{-\infty}^0 |F(\omega)|^2 d\omega.$$

So, finally,

$$\sum_{m,n} |\langle f, h_{m,n}^{(+)} \rangle|^2 + \sum_{m,n} |\langle f, h_{m,n}^{(-)} \rangle|^2 = \frac{1}{b \log a} \|f\|^2. \quad\blacksquare$$

11.6. ORTHONORMAL WAVELET BASES

In this section we develop wavelet functions that give rise to orthonormal bases for L_2. Furthermore, the elements of the basis will turn out to be *compactly supported* functions (a function is compactly supported if it is zero outside a finite closed interval). Orthonormal wavelets were developed by Meyer [1989], Mallat [1989a, b] and Daubechies [1988]. Our exposition is based on the multiresolution analysis approach of Mallat and its application to compactly supported orthonormal bases by Daubechies. The level of some of the mathematical details is beyond of the scope of this book, so some of the proofs will be omitted. Fortunately, the actual construction of Daubechies wavelets requires only elementary tools. The wavelets described here are for $a = 2$ and $b = 1$. We remark that, while b is arbitrary (it can always be normalized to 1 by scaling of the time axis), this is not so for a. As we will see, the value $a = 2$ is rather special.

We will begin with an important lemma, which will be used repeatedly in this section.

Lemma 11.7. Let $f(t)$ be an L_2 function and $F(\omega)$ its Fourier transform. Then $\{f(t-n), n \in \mathbf{Z}\}$ is an orthonormal set if and only if $\sum_{n=-\infty}^\infty |F(\omega + 2\pi n)|^2 = 1$.

Proof. For any two integers p, q we have by Parseval's theorem

$$\langle f(t-p), f(t-q) \rangle = \langle F(\omega)e^{-j\omega p}, F(\omega)e^{-j\omega q} \rangle = \frac{1}{2\pi} \int_{-\infty}^{\infty} |F(\omega)|^2 e^{-j\omega(p-q)} d\omega.$$

Let us break the integral to an infinite sum of integrals over the intervals $[(2n-1)\pi, (2n+1)\pi]$, $n \in \mathbf{Z}$. Then

$$\langle f(t-p), f(t-q) \rangle = \sum_{n=-\infty}^{\infty} \frac{1}{2\pi} \int_{-\pi}^{\pi} |F(\omega+2\pi n)|^2 e^{-j\omega(p-q)} d\omega$$

$$= \frac{1}{2\pi} \int_{-\pi}^{\pi} \left[\sum_{n=-\infty}^{\infty} |F(\omega+2\pi n)|^2 \right] e^{-j\omega(p-q)} d\omega.$$

By a basic property of Fourier transforms of sequences, the right side is equal to $\delta(p-q)$ if and only if the sum in the brackets is equal to 1. ∎

A *multiresolution analysis* of L_2 consists of a countable family $\{V_m, m \in \mathbf{Z}\}$ of subspaces of L_2 and an L_2 function $f(t)$ satisfying the following four properties:

(1) The family is *nested*, meaning that $V_m \subseteq V_{m+1}$ for all m.
(2) $\bigcap_{m=-\infty}^{\infty} V_m = \{0\}$ and $\overline{Sp}\left(\bigcup_{m=-\infty}^{\infty} V_m\right) = L_2$.
(3) For all m, $x(t) \in V_m$ if and only if $x(2t) \in V_{m+1}$.
(4) The set $\{f(t-n), n \in \mathbf{Z}\}$ is an orthonormal basis for V_0 (see Prob. 11.14 in this connection).

Let us define $f_{m,n}(t) = 2^{m/2} f(2^m t - n)$. By (4), $\{f_{0,n}, n \in \mathbf{Z}\}$ form an orthonormal basis for V_0, so by (3), $\{f_{m,n}, n \in \mathbf{Z}\}$ form an orthonormal basis for V_m for all m. The function $f(t)$ is called the *scaling function* of the multiresolution analysis.

Example 11.2. Let V_0 be the subspace of L_2 consisting of functions that are constant on the intervals $[n, n+1)$, and let $f(t) = 1$ if $0 \le t < 1$ and zero elsewhere. The translates of $f(t)$ are clearly orthonormal and they span V_0. Correspondingly, V_m is the subspace consisting of functions constant on the intervals $[2^{-m}n, 2^{-m}(n+1)]$. The projections of a given function $x(t)$ on the V_m represent approximations of $x(t)$ of different resolutions, the resolution of V_m being 2^{-m}. This explains the name "multiresolution analysis." ∎

Since $f(t) \in V_0 \subseteq V_1$, and since V_1 is spanned by the functions $\{f(2t-n)\}$, there exists a square summable sequence $\{c_n\}$ such that

$$f(t) = 2 \sum_{n=-\infty}^{\infty} c_n f(2t-n). \tag{11.6.1}$$

This is the fundamental equation of the multiresolution analysis. Its significance will be clear from the following derivations. Take the Fourier transform of (11.6.1) to get

$$F(\omega) = \sum_{n=-\infty}^{\infty} c_n F(0.5\omega) e^{-j0.5\omega n} = F(0.5\omega) C(e^{j0.5\omega}), \qquad (11.6.2)$$

where $C(e^{j\omega})$ is the Fourier transform of $\{c_n\}$. An immediate consequence of (11.6.2), obtained upon substituting $\omega = 0$, is

$$C(e^{j0}) = \sum_{n=-\infty}^{\infty} c_n = 1. \qquad (11.6.3)$$

By applying (11.6.2) inductively, we get the formal solution for $F(\omega)$:

$$F(\omega) = F(0) \prod_{m=1}^{\infty} C(e^{j2^{-m}\omega}). \qquad (11.6.4)$$

The question is, of course, does the infinite product in the right side converge? If it does, then (11.6.4) provides a solution for $F(\omega)$ [hence for $f(t)$] in terms of the function $C(e^{j\omega})$ [or of the sequence $\{c_n\}$]. The following lemma is a weaker version of the one given in Daubechies [1988], but is sufficient for our purposes.

Lemma 11.8. Assume that

$$\sum_{n=-\infty}^{\infty} |nc_n| = K < \infty.$$

Then the right side of (11.6.4) converges pointwise for all ω.

Proof. It is known (see Apostol [1974, p. 208]), that an infinite product $\prod_{m=1}^{\infty} a_m$ converges if and only if $\sum_{m=1}^{\infty} (1 - a_m)$ converges. In our case, $a_m = C(e^{j2^{-m}\omega})$. We have, due to (11.6.3),

$$1 - C(e^{j\omega}) = \sum_{n=-\infty}^{\infty} c_n (1 - e^{-j\omega n}) = \sum_{n=-\infty}^{\infty} 2j c_n e^{j0.5\omega n} \sin(0.5\omega n),$$

so, since $|\sin \alpha| \leq |\alpha|$,

$$|1 - C(e^{j\omega})| \leq \sum_{n=-\infty}^{\infty} |nc_n \omega| = K|\omega|.$$

Therefore,

$$\sum_{m=1}^{\infty} |1 - C(e^{j2^{-m}\omega})| \leq K|\omega| \sum_{m=1}^{\infty} 2^{-m} = K|\omega| < \infty. \qquad \blacksquare$$

When (11.6.4) converges, it implies the following recursive construction algorithm for $f(t)$ (obtained by taking the products as convolutions in the time domain):

1. Let ϕ_0 be the rectangular window on $[0, 1)$ (as in Example 11.2).
2. For $m \geq 0$, let

$$\phi_{m+1}(t) = \sum_{n=-\infty}^{\infty} c_n \phi_m(2t - n). \qquad (11.6.5)$$

Then $\lim_{m \to \infty} \phi_m(t) = f(t)$ up to a multiplicative constant. The convergence proof of this algorithm is given in Daubechies [1988]. An important consequence of the algorithm is the following lemma.

Lemma 11.9. If $\{c_n\}$ is nonzero only for $0 \leq n \leq N - 1$, then $f(t)$ is nonzero only for $0 \leq t < N - 1$.

Proof. Assume the support of $\phi_m(t)$ is $[0, T_m)$. This clearly holds for $m = 0$ with $T_0 = 1$. The support of $\phi_m(2t - n)$ is then $[0.5n, 0.5(T_m + n))$, so the support of the right side of (11.6.5) is $[0, 0.5(T_m + N - 1))$. We therefore get the difference equation $T_{m+1} = 0.5T_m + 0.5(N - 1)$. Therefore, T_m converges monotonically to $T_\infty = N - 1$. ∎

The following property of $C(e^{j\omega})$ will prove to be fundamental to the construction of multiresolution analysis and wavelet bases.

Lemma 11.10. The function $C(e^{j\omega})$ satisfies

$$|C(e^{j\omega})|^2 + |C(e^{j(\omega+\pi)})|^2 = 1, \quad \forall \omega. \qquad (11.6.6)$$

Proof. We have from (11.6.2),

$$F(2\omega) = F(\omega)C(e^{j\omega}).$$

By Lemma 11.7, the orthonormality of the set $\{f(t - n)\}$ implies $\sum_{n=-\infty}^{\infty} |F(\omega + 2\pi n)|^2 = 1$ for all ω. Since this holds for all ω, it holds for 2ω as well as for $\omega + \pi$. Therefore,

$$1 = \sum_{n=-\infty}^{\infty} |F(2\omega + 2\pi n)|^2 = \sum_{n=-\infty}^{\infty} |F(\omega + \pi n)|^2 |C(e^{j(\omega+\pi n)})|^2 =$$

$$\sum_{k=-\infty}^{\infty} |F(\omega + 2\pi k)|^2 |C(e^{j(\omega+2\pi k)})|^2 + \sum_{k=-\infty}^{\infty} |F(\omega + \pi + 2\pi k)|^2 |C(e^{j(\omega+\pi+2\pi k)})|^2$$

$$= |C(e^{j\omega})|^2 \sum_{k=-\infty}^{\infty} |F(\omega + 2\pi k)|^2 + |C(e^{j(\omega+\pi)})|^2 \sum_{k=-\infty}^{\infty} |F(\omega + \pi + 2\pi k)|^2$$

$$= |C(e^{j\omega})|^2 + |C(e^{j(\omega+\pi)})|^2. \qquad\qquad\blacksquare$$

We can now deduce that the seemingly arbitrary factor $F(0)$ in (11.6.4) must, in fact, be equal to 1.

Lemma 11.11. For any multiresolution analysis, $F(0) = 1$.

Proof. We have already seen in (11.6.3) that $C(e^{j0}) = 1$. It follows from (11.6.6) that $C(e^{j\pi}) = 0$. By the 2π periodicity of $C(e^{j\omega})$, $C(e^{j\ell\pi}) = 0$ for any odd ℓ. Substituting $\omega = 2\pi n$ in (11.6.4), we see that $F(2\pi n)$ must be zero for any $n \neq 0$. This is because repeated division of $2n$ by 2 must eventually yield an odd number, and then the corresponding factor in the right side of (11.6.4) will make the product zero. Finally, returning to Lemma 11.8, we conclude that $F(0)$ must be 1 in order for the sum of square magnitudes to equal 1. $\qquad\blacksquare$

The multiresolution analysis functions $\{2^{m/2} f(2^m t - n)\}$ are not wavelets, because by Lemma 11.11 they do not satisfy the admissibility condition (11.5.5). Nevertheless, multiresolution analysis is only one step away from wavelets, as we will now show.

By definition of multiresolution analysis, $V_m \subseteq V_{m+1}$ for all m. Let W_m denote the orthogonal complement of V_m in V_{m+1}, so $V_{m+1} = V_m \oplus W_m$. It easily follows that W_m and W_k are orthogonal for all $m \neq k$. Furthermore, from property (2) of the multiresolution analysis, it follows that L_2 can be decomposed as

$$L_2 = \bigoplus_{m=-\infty}^{\infty} W_m. \qquad\qquad (11.6.7)$$

Let us look for a function $h(t)$ that plays for $\{W_m\}$ a role similar to the one played by $f(t)$ for $\{V_m\}$: we want $\{h(t - n), n \in \mathbf{Z}\}$ to constitute an orthonormal basis for W_0. It will then follow, by the multiresolution property, that $\{h_{m,n}(t) = 2^{m/2} h(2^m t - n), n \in \mathbf{Z}\}$ constitute an orthonormal basis for W_m. But this implies, by (11.6.7), that $\{h_{m,n}(t), m, n \in \mathbf{Z}\}$ is an orthonormal basis for L_2. If, in addition, $h(t)$ satisfies the admissibility condition (11.5.5), it will give rise to an orthonormal wavelet basis.

Example 11.3. Let $f(t)$ be the rectangular window function of Example 11.2, and let

$$h(t) = f(t) - 2f(2t - 1) = \begin{cases} 1, & 0 \le t < 0.5 \\ -1, & 0.5 \le t < 1 \\ 0, & \text{otherwise.} \end{cases}$$

Then $h(t)$ is a dc-free function, and $\{h_{m,n}(t)\}$ is an orthonormal basis for L_2, known as the *Haar basis*. It is easy to check that $h(t)$ is orthogonal to $f(t)$ and obviously to all $\{f(t-n), n \neq 0\}$. ∎

Let us *assume* the existence of such a function $h(t)$ and see what consequences are implied by its existence. This will provide us with a recipe for constructing this function. Since $h(t) \in W_0 \subseteq V_1$, it can be represented as [cf. (11.6.1)]

$$h(t) = 2 \sum_{n=-\infty}^{\infty} d_n f(2t - n). \qquad (11.6.8)$$

for some square summable sequence $\{d_n\}$. Taking the Fourier transforms of both sides gives

$$H(\omega) = \sum_{n=-\infty}^{\infty} d_n F(0.5\omega) e^{-j0.5\omega n} = F(0.5\omega) D(e^{j0.5\omega}), \qquad (11.6.9)$$

where $D(e^{j\omega})$ is the Fourier transform of $\{d_n\}$. We can now prove, exactly as in Lemma 11.10, that

$$|D(e^{j\omega})|^2 + |D(e^{j(\omega+\pi)})|^2 = 1, \quad \forall \omega. \qquad (11.6.10)$$

Furthermore, the two functions $C(e^{j\omega})$ and $D(e^{j\omega})$ satisfy the following relationship.

Lemma 11.12.

$$C(e^{j\omega})D^*(e^{j\omega}) + C(e^{j(\omega+\pi)})D^*(e^{j(\omega+\pi)}) = 0. \qquad (11.6.11)$$

Proof. Since W_0 is orthogonal to V_0, necessarily $\langle f(t-n), h(t) \rangle = 0$ for all n, so $\langle F(\omega)e^{-j\omega n}, H(\omega) \rangle = 0$. It follows, exactly as in Lemma 11.7, that

$$\sum_{n=-\infty}^{\infty} F(\omega + 2\pi n) H^*(\omega + 2\pi n) = 0.$$

Equation (11.6.11) now follows from (11.6.9) and (11.6.2) exactly as in Lemma 11.9 (Prob. 11.15). ∎

To show that $\{h_{m,n}(t), m, n \in \mathbf{Z}\}$ is a wavelet basis, it only remains to verify the admissibility condition $H(0) = 0$. Indeed, we have seen that $C(e^{j0}) = 1$, so, by (11.6.6), $C(e^{j\pi}) = 0$. Therefore, by (11.6.11), $D(e^{j0}) = 0$, so, finally, $H(0) = 0$ by (11.6.9).

Equations (11.6.6), (11.6.10), and (11.6.11) can be expressed in the combined form:

$$\begin{bmatrix} C(e^{j\omega}) & C(e^{j(\omega+\pi)}) \\ D(e^{j\omega}) & D(e^{j(\omega+\pi)}) \end{bmatrix} \begin{bmatrix} C^*(e^{j\omega}) & D^*(e^{j\omega}) \\ C^*(e^{j(\omega+\pi)}) & D^*(e^{j(\omega+\pi)}) \end{bmatrix} = \begin{bmatrix} 1 & 0 \\ 0 & 1 \end{bmatrix}, \qquad (11.6.12)$$

meaning that the left matrix in (11.6.12) is unitary. These are the equations of a two-channel *quadrature mirror filter* (QMF) or *conjugate quadrature filter* (CQF). In conclusion, any two-channel CQF such that $\sum_{n=-\infty}^{\infty} |nc_n| < \infty$ will give rise to multiresolution analysis and an orthonormal wavelet basis. In particular, any *finite impulse response* CQF will serve this purpose. CQF synthesis is a well-studied subject in digital signal processing (see, e.g., Vaidyanathan [1993]). In fact, once $C(e^{j\omega})$ has been synthesized, $D(e^{j\omega})$ is no longer arbitrary. To see that, denote $C_0 = C(e^{j\omega})$, $C_1 = C(e^{j(\omega+\pi)})$ for convenience, and similarly for D_0, D_1. Then we get, from (11.6.11),

$$\frac{D_0}{D_1} = -\left(\frac{C_1}{C_0}\right)^*,$$

so

$$\frac{\dfrac{D_0}{D_1}}{\left[1 + \left|\dfrac{D_0}{D_1}\right|^2\right]^{1/2}} = \frac{-\left(\dfrac{C_1}{C_0}\right)^*}{\left[1 + \left|\dfrac{C_1}{C_0}\right|^2\right]^{1/2}}.$$

Multiply the numerator and the denominator of the left side by $|D_1|$ and those of the right side by $|C_0|$. Then use (11.6.6) and (11.6.10) to get

$$D_0 = -\frac{|C_0|}{|D_1|} \cdot \frac{D_1}{C_0^*} \cdot C_1^* = -A(e^{j\omega})C_1^*,$$

where $A(e^{j\omega})$ is an all-pass function (i.e., its magnitude is 1 for all ω). Substituting back in (11.6.11), we find that the condition

$$A(e^{j\omega}) + A(e^{j(\omega+\pi)}) = 0$$

is necessary and sufficient for (11.6.11) to hold. The choice $A(e^{j\omega}) = e^{-j\omega}$ is standard, leading to

$$D(e^{j\omega}) = -e^{-j\omega}C^*(e^{j(\omega+\pi)}), \quad \text{or} \quad d_n = (-1)^n c_{1-n}^*. \tag{11.6.13}$$

Let us now proceed to the construction of compactly supported orthonormal wavelet bases. By Lemma 11.9, any finite impulse response CQF yields a compactly supported scaling function $f(t)$. It immediately follows from (11.6.8) that the wavelet $h(t)$ is also compactly supported. Arbitrary choices of CQF may lead, however, to highly irregular wavelets. The construction of Daubechies [1988] imposes further conditions on the CQF so that $h(t)$ be made regular, with a "regularity parameter" N. Daubechies's construction is as follows.

1. Pick $N > 0$ (more on the choice of N later).
2. Pick an odd polynomial $R(x)$, subject to the restriction specified below.
3. Let

$$P(z) = \sum_{n=0}^{N-1} \binom{N-1+n}{n} \left(\frac{2-z-z^{-1}}{4}\right)^n + \left(\frac{2-z-z^{-1}}{4}\right)^N R\left(\frac{z+z^{-1}}{4}\right).$$

$$\tag{11.6.14}$$

$P(z)$ is a real symmetric polynomial in z and z^{-1}. The polynomial $R(x)$ has to be chosen so that $P(e^{j\omega})$ [which is real, due to the symmetry of $P(z)$] is nonnegative for all ω. For example, $R(x) = 0$ satisfies this requirement (Prob. 11.16).

4. Factor $P(z)$ as $P(z) = Q(z)Q(z^{-1})$, where $Q(z)$ is a polynomial in z. This is possible in 2^K ways, where K is the degree of $P(z)$. For example, we can take $Q(z)$ as the polynomial consisting of the roots of $P(z)$ inside the unit circle.

5. Take $C(z)$ as

$$C(z) = \left(\frac{1 + z^{-1}}{2}\right)^N Q(z) \tag{11.6.15}$$

and $D(z)$ as in (11.6.13).

6. Construct $f(t)$ to an arbitrary degree of accuracy by iterating through (11.6.5) and $h(t)$ from (11.6.8).

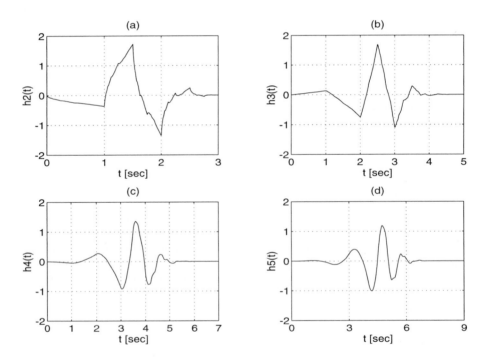

Figure 11.6. Daubechies's wavelets of orders 2 through 5.

As we have said, the parameter N controls the regularity of $f(t)$ and $h(t)$ in a sense made precise in Daubechies [1988]. We just remark that this Nth-order regularity is *weaker* than Nth-order differentiability. In any case, $f(t)$ is continuous for all $N \geq 1$. The length of the impulse response of $C(z)$ (hence the support of the scaling function and the wavelet) is at least $2N - 1$ [exactly $2N - 1$ for $R(x) = 0$]. Daubechies's family of wavelets correspond to $R(x) = 0$ and $Q(z)$ the minimum-phase factor of $P(z)$. The wavelets corresponding to $N = 2, 3, 4, 5$ are shown in Figure 11.6. As we see, the smoothness of the wavelet increases with the order.

11.7. IMPLEMENTATION OF LINEAR TRANSFORMS FOR DISCRETE-TIME SIGNALS

Up to this point we have dealt with continuous-time signals, specifically with L_2 (finite energy) signals. We know turn our attention to discrete-time signals and show how STFT/Gabor and wavelet analysis are implemented in this case.

Assume we are given a discrete-time signal $\{y_n\}$. The direct analog of the continuous-time STFT is given by

$$Y(n, \omega) = \sum_{k=-\infty}^{\infty} y_k \gamma_{k-n}^* e^{-j\omega k}. \qquad (11.7.1)$$

Here n is a discrete (integer) variable and ω a continuous (real) variable. This form of the STFT is not very useful, however. In order to have the full benefit of discrete signal processing techniques (FFT, digital filters, etc.), it is common to modify the above definition in two respects: (a) Let ω be a discrete variable; (b) Decimate the variable n by a certain factor, that is, restrict it to be an integer multiple of some number. With these two modifications, we get the definition

$$Y_{m,n} = \sum_{k=-\infty}^{\infty} y_k \gamma_{k-nN}^* e^{-j2\pi km/M}, \quad 0 \leq m \leq M - 1, \quad -\infty < n < \infty. \quad (11.7.2)$$

This is the analysis formula for the discrete STFT and is recognized to be the direct analog of Gabor's analysis formula (11.3.11). The numbers $1/M$ and N are the analogs of the parameters α and β in the Gabor representation. In particular, $N = M$ represents the critical case, as does the case $\alpha\beta = 1$ for the Gabor representation.

Motivated by the Gabor representation formula (11.3.1), we will guess the synthesis formula

$$y_n = \frac{1}{M} \sum_{m=0}^{M-1} \sum_{\ell=-\infty}^{\infty} Y_{m,\ell} g_{n-\ell N} e^{j2\pi mn/M}. \qquad (11.7.3)$$

Let us check under what conditions the right side of (11.7.3) indeed gives back the original signal. Substitute (11.7.2) in (11.7.3) to get

$$\frac{1}{M} \sum_{m=0}^{M-1} \sum_{\ell=-\infty}^{\infty} \sum_{k=-\infty}^{\infty} y_k g_{n-\ell N} \gamma_{k-\ell N}^* e^{j2\pi(n-k)m/M}$$

$$= \sum_{\ell=-\infty}^{\infty} \sum_{i=-\infty}^{\infty} y_{n-iM} g_{n-\ell N} \gamma_{n-iM-\ell N}^*.$$

Hence we get the following necessary and sufficient condition for (11.7.3):

$$\sum_{\ell=-\infty}^{\infty} g_{n-\ell N} \gamma_{n-\ell N-iM}^* = \delta(i), \quad \forall n. \tag{11.7.4}$$

This condition is due to Portnoff [1980]. Note that it only needs to be verified for $0 \leq n \leq N - 1$, since the left side of (11.7.4) is periodic in n with period N.

The design of impulse response pairs $\{\gamma_n, g_n\}$ that meet the constraint (11.7.4) is far from being straightforward. In addition to satisfying (11.7.4), we want $\Gamma(e^{j\omega})$, the Fourier transform of $\{\gamma_n\}$, to be a "good" low-pass filter with bandwidth of about $2\pi/M$. Filters of this kind are called M-channel quadrature mirror filters. If $N = M$ (this is the common case), they are called *critically decimated*. The two-channel QMF appearing in the discrete wavelet transform is just a special case. We will not discuss design methods for quadrature mirror filters here; see Vaidyanathan [1993].

The existence of finite impulse response (FIR) analysis/synthesis windows $\{\gamma_n, g_n\}$ was studied in Dembo and Malah [1988] and Farkash and Raz [1993]. Suppose both windows are nonzero only for $0 \leq n \leq Q - 1$. Then we get from (11.7.4),

$$\sum_{\ell} g_{n-\ell N} \gamma_{n-\ell N-iM}^* = \begin{cases} 1, & i=0, \ 0 \leq n \leq N-1 \\ 0, & 1 \leq i \leq I, \ 0 \leq n \leq N-1, \end{cases} \tag{11.7.5}$$

where $I = \lfloor (Q-1)/M \rfloor$. If the synthesis window $\{g_n\}$ is given, (11.7.5) yields $(2I - 1)N$ equations for the Q unknowns $\{\gamma_n\}$. Denote this set of equations by

$$U\boldsymbol{\gamma} = \boldsymbol{e}$$

[note that $\boldsymbol{e}$ consists of N ones and $2(I - 1)$ zeros]. Then a solution for $\boldsymbol{\gamma}$ exists if and only if $\boldsymbol{e}$ is in the column space of U. The solution is unique if and only if the null space of U is $\{0\}$. Problem 11.17 discusses some special cases of this general result.

Let us now turn our attention to wavelet analysis for discrete-time signals. Denote

$$s_{m,n} = \langle y, f_{m,n} \rangle, \quad w_{m,n} = \langle y, h_{m,n} \rangle. \tag{11.7.6}$$

The letters s and w are reminders of the scaling and wavelet functions in the respective inner products. The fundamental scaling and wavelet equations (11.6.1) and (11.6.8) give, upon replacing t by $2^m t - n$ and multiplying by $2^{m/2}$,

$$f_{m,n}(t) = \sqrt{2} \sum_{k=-\infty}^{\infty} c_k f_{m+1,2n+k}(t), \quad h_{m,n}(t) = \sqrt{2} \sum_{k=-\infty}^{\infty} d_k f_{m+1,2n+k}(t).$$

Taking inner products with the signal $y(t)$ gives

$$s_{m,n} = \sqrt{2} \sum_{k=-\infty}^{\infty} c_k s_{m+1,2n+k}, \quad w_{m,n} = \sqrt{2} \sum_{k=-\infty}^{\infty} d_k s_{m+1,2n+k}. \qquad (11.7.7)$$

These recursions enable us to compute the scaling and wavelet coefficients of scale m from those of scale $m + 1$, assuming that the latter are known for all values of the translation variable. Note that we need to carry the scaling coefficients even if we are eventually interested only in the wavelet coefficients, because they will be needed when we go from m to $m - 1$.

Equations (11.7.7) can be interpreted as convolutions followed by decimations. For example, the first equation is a discrete convolution of the sequence $\{s_{m+1,n}\}$ (where n is the sequence index) with the impulse response $\{\sqrt{2}c_{-k}\}$. The appearance of $2n$ (instead of n) on the right side implies decimation by a factor of 2 (i.e., retaining every other output value). A similar interpretation holds for the second equation. The two filter transfer functions are $\sqrt{2}C^*(e^{j\omega})$ and $\sqrt{2}D^*(e^{j\omega})$, respectively, the conjugates (or time-reversed) of the two channels of the CQF introduced in the previous section (up to a factor $\sqrt{2}$). Figure 11.7 illustrates the operation defined by (11.7.7).

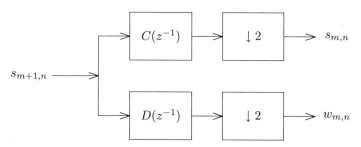

Figure 11.7. One stage of the discrete wavelet analysis by CQF.

In order to use the recursions (11.7.7), we need to assume some maximal scale M for which the scale coefficients $\{s_{M,n}\}$ are known. This is impossible in general for continuous-time signals, but is always possible for discrete-time signals. To see this, assume for convenience that the continuous-time signal was sampled 2^M times per second (we will soon see that this assumption is needed only for the explanation and has no effect on the actual algorithm). The resulting discrete-time signal $\{y_n\}$ represents time points 2^{-M} seconds apart. Recall that the scaling function $f(t)$ satisfies $\|f_{m,n}(t)\|^2 = 1$ for all m, and its support becomes narrower as m increases.

Thus, when m tends to infinity, it becomes similar to a "square-root delta function" (i.e., an energy-preserving sampler, in contrast to the usual delta function, which is area preserving). The support of $f_{M,n}(t)$ is on the order of 2^{-M}, so its inner product with $y(t)$ produces values close to the sampled values $\{y_n\}$. In other words, we can *identify* the sequence $\{y_n\}$ with $\{s_{M,n}\}$. This provides us with the desired initial sequence for the recursions (11.7.7). The implementation of the complete wavelet analyzer is depicted in Figure 11.8. Each section includes the two-channel CQF shown in Figure 11.7.

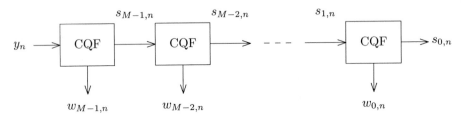

Figure 11.8. The discrete wavelet analysis scheme.

It is clear from the above description that the parameter M is arbitrary, and it never appears in the actual analyzer. On the other hand, the total number of sections is necessarily finite and has to be chosen by the user. Since m does not continue to $-\infty$, we have to retain the coarsest-scale scale coefficients, in addition to the wavelet coefficients of all scales. These scale coefficients contain the information about the signal at dc and near it and are necessary for reconstruction.

The synthesis scheme can be obtained by reversing the operations in (11.7.7). This is left to the reader (Prob. 11.18).

11.8. MATHEMATICA PACKAGES

Due to the theoretical nature of this chapter, only few procedures were implemented. The package `Wavelet.m` contains two procedures, as follows.

`DaubechiesCqf[order_]` computes the Daubechies conjugate quadratic filter, as described in Sec. 11.6. `order` is the parameter N. The returned value is the list of impulse response coefficients $\{c_n\}$.

`DaubechiesWavelet[order_, density_]` computes a discrete-time approximation to the Daubechies wavelet of the given `order`. The parameter `density` is the number of points per second. The returned values are {f, h}, the lists of samples of the scaling function $f(t)$, and the wavelet $h(t)$.

The implementation of the wavelet analysis and synthesis algorithms is left to the reader (Prob. 11.19).

The package `Gabor.m` contains the following procedures.

`Bastiaans[t_]` computes the biorthogonal function of the Gaussian window, according to Bastiaans's formula (11.3.14).

`GaborAnalysis[y_, n_, m_, gamma_]` and `GaborSynthesis[Y_, n_, m_, g_]` implement the discrete-time Gabor analysis and synthesis formulas (11.7.2) and (11.7.3). The computation is by the weighted overlap-add method, as described in Crochiere and Rabiner [1983, Sec. 7.2.5].

REFERENCES

Apostol, T. M., *Mathematical Analysis,* 2nd ed., Addison-Wesley, Reading, MA, 1974.

Bastiaans, M. J., "Gabor's Signal Expansion and Degrees of Freedom of a Signal," *Proc. IEEE,* 68, pp. 538–539, 1980.

Bracewell, R. N., *The Fourier Transform and Its Applications,* 2nd ed., McGraw-Hill, New York, 1986.

Crochiere, R. E., and Rabiner, L. R., *Multirate Digital Signal Processing,* Prentice Hall, Englewood Cliffs, NJ, 1983.

Daubechies, I., "Orthonormal Bases of Compactly Supported Wavelets," *Comm. Pure Appl. Math.,* XLI, pp. 909–996, 1988.

Daubechies, I., "The Wavelet Transform, Time-frequency Localization and Signal Analysis," *IEEE Trans. Information Theory,* 36, pp. 961–1005, 1990.

Daubechies, I., Grossmann, A., and Meyer, Y., "Painless Nonorthogonal Expansions," *J. Math. Phys.,* 27, pp. 1271–1283, 1986.

Dembo, A., and Malah, D., "Signal Synthesis from Modified Discrete Short-time Fourier Transform Analysis and Synthesis," *IEEE Trans. Acoustics, Speech, Signal Processing,* ASSP-36, pp. 168–181, 1988.

Duffin, R. J., and Schaeffer, A. C., "A Class of Nonharmonic Fourier Series," *Trans. Amer. Math. Soc.,* 72, pp. 341–366, 1952.

Farkash, S., and Raz, S., "The Discrete Gabor Expansion—Existence and Uniqueness," *Signal Processing,* to appear, 1993.

Friedlander, B., and Porat, B., "Detection of Transient Signals by Gabor Representation," *IEEE Trans. Acoustics, Speech, Signal Processing,* ASSP-37, pp. 169–180, 1989.

Gabor, D., "Theory of Communication", *J. Inst. Elec. Eng.,* 93, No. 3, pp. 429–457, 1946.

Janssen, A. J. E. M., "Gabor Representation and Wigner Distribution of Signals," *Proc. Int. Conf. Acoustics, Speech, Signal Processing,* Atlanta, GA, No. 41B.2, 1981a.

Janssen, A. J. E. M., "Gabor Representation of Generalized Functions," *J. Math. Anal. Appl.,* 83, pp. 377–394, 1981b.

Kolmogorov, A. N., and Fomin, S. V., *Introductory Real Analysis,* Dover Publications, New York, 1975.

Mallat, S., "A Theory of Multiresolution Signal Decomposition: The Wavelet Representation," *IEEE Trans. Pattern Anal. Machine Intell.,* 11, pp. 674–693, 1989a.

Mallat, S., "Multiresolution Approximation and Wavelet Orthonormal Bases of $L_2(R)$," *Trans. Amer. Math. Soc.,* 315, pp. 69–87, 1989b.

Meyer, Y., "Orthonormal Wavelets," in *Wavelets, Time-frequency Methods and Phase Space,* Proc. Int. Conf. Marseille, France, 1987. J. M. Combes et al. (eds.), pp. 21–37, 1989.

Portnoff, M. R., "Time-frequency Representation of Digital Signals and Systems Based on Short-time Fourier Analysis," *IEEE Trans. Acoustics, Speech, Signal Processing,* 28, pp. 55–69, 1980.

Taylor, A. E., *Introduction to Functional Analysis,* Wiley, New York, 1958.

Vaidyanathan, P. P., *Multirate Systems and Signal Processing,* Prentice Hall, Englewood Cliffs, NJ, 1993.

Young, R. M., *An Introduction to Nonharmonic Fourier Series,* Academic Press, New York, 1980.

Zak, J., "Finite Translations in Solid State Physics," *Phys. Rev. Lett.,* 19, pp. 1385–1397, 1967.

PROBLEMS

11.1. Find a dual formula to (12.2.1) that expresses $Y(t, \omega)$ in terms of the Fourier transforms of $y(t)$ and $\gamma(t)$. Use this formula to further justify the requirement that $\gamma(t)$ be narrow band.

11.2. Using the definition (11.4.2), show that any frame $\{g_k, k \in K\}$ is a complete set in L_2. Remark: Recall that a set is complete if $\langle g_k, f \rangle = 0$ for all $k \in K$ implies that $f = 0$.

11.3. Let $g(t) = \sqrt{2\lambda}\exp(-\lambda t), t \geq 0$.

    (a) Show that the Zak transform of $g(t)$ is given by

$$G_z(t, f) = \frac{\sqrt{2\lambda}\exp[-\lambda t + \lfloor t \rfloor(\lambda - j2\pi f)]}{1 - \exp(-\lambda + j2\pi f)}.$$

    (b) Show that the biorthogonal function of $g(t)$ is given by

$$\gamma(t) = \begin{cases} -\dfrac{\exp(\lambda t)}{\sqrt{2\lambda}}, & -1 \leq t < 0 \\ \dfrac{\exp(\lambda t)}{\sqrt{2\lambda}}, & 0 < t \leq 1 \\ 0, & \text{otherwise.} \end{cases}$$

    Plot this function (see Friedlander and Porat [1989] for more details).

11.4. Prove formula (11.4.7).

11.5. Prove the following two properties of the Zak transform [with $h(t), g(t)$ in L_2 and $g_{m,n}(t) = g(t - n)e^{j2\pi mt}$].

    (a) $[G_{m,n}]_z(t, f) = e^{j2\pi(mt - nf)}G_z(t, f).$

    (b) $\langle h, g_{m,n}\rangle = \int_0^1 \int_0^1 H_z(t, f)[G_{m,n}]_z^*(t, f)dt\, df.$

11.6. Use Prob. 11.5 and Poisson's summation formula (11.3.12) to derive the following identity:

$$\sum_{m,n} |\langle h, g_{m,n}\rangle|^2 = \int_0^1 \int_0^1 |H_z(t, f)|^2 |G_z(t, f)|^2 dt\, df.$$

Use this identity to show that $\{g_{m,n}(t)\}$ is a frame if and only if

$$A \leq |G_z(t, f)|^2 \leq B$$

almost everywhere on $[0, 1]^2$.

11.7. Use Prob. 11.6 to show that the function $g(t)$ in Prob. 11.3 generates a frame.

11.8. Show that $g(t)$ in (11.4.10) satisfies (11.4.11).

11.9. Prove (11.4.12) for the case $\alpha\beta = 1/r$, in a similar manner to Prob. 11.6. Hint: Write m as $m = m'r + k$, where $0 \leq k \leq r - 1$. Express $\sum_{m,n} |\langle h, g_{m,n}\rangle|^2$ as a triple sum over m', k and n and proceed as in 11.4.

11.10. Recall that in the STFT the synthesis function $\gamma(t)$ could be different from the analysis function $g(t)$, as long as they satisfied a certain common condition. Does the same hold for the wavelet transform? In other words, can we use a function $\eta_{\rho,s}(t)$ different from $h_{\rho,s}(t)$ in the inverse transform (11.5.4)? If so, what common condition needs to be satisfied by $h(t)$ and $\eta(t)$?

11.11. Prove that (11.5.5) is a necessary condition for (11.5.1).

11.12. Show that $H(\omega)$ in (11.5.10) satisfies (11.5.11).

11.13. Let $f(t)$ be an L_2 function. Prove that the set $\{f(t - n), n \in \mathbf{Z}\}$ is linearly independent, except possibly when $f(t)$ is zero almost everywhere. Hint: If the set is not linearly independent, then there exist N and constants $\{a_n, -N \leq n \leq N\}$, not all zero, such that $\sum_{n=-N}^{N} a_n f(t - n) = 0$. Express this equality in the frequency domain and show that it leads to the conclusion $F(\omega) = 0$ almost everywhere.

11.14. Given an L_2 function $f(t)$ such that $\{f(t-n), n \in \mathbf{Z}\}$ are not orthonormal, find a function $\tilde{f}(t)$, related to $f(t)$, such that $\{\tilde{f}(t-n), n \in \mathbf{Z}\}$ are orthonormal. Hint: Use Lemma 11.7.

11.15. Fill in the details in the proof of Lemma 11.12.

11.16. Show that the choice $R(x) = 0$ in (11.6.14) gives $P(e^{j\omega}) \geq 0$ for all ω.

11.17. Consider the condition (11.7.5) for existence and uniqueness of STFT analysis/synthesis pairs in the FIR case. Interpret this condition for the following cases: (a) $M \leq Q = N$; (b) $M = N < Q$; (c) $M > Q > N$. Remark: See Farkash and Raz [1993] for further details.

11.18. Explain how to synthesize the discrete-time signal $\{y_n\}$ from its wavelet coefficients. Write the appropriate equations and draw the corresponding block diagram.

11.19. Write Mathematica procedures `WaveletAnalysis` and `WaveletSynthesis`. The former should implement the wavelet analysis algorithm given in (11.7.5). The latter should implement the synthesis algorithm developed in Prob. 11.18. The inputs to the procedures are (1) the input lists, (2) the filter coefficient lists $\{c_n\}$ and $\{d_n\}$, and (3) the number of stages.

CHAPTER 12

Time-Frequency
Signal Analysis:
Nonlinear Transforms

12.1. INTRODUCTION

In this chapter we continue our discussion of time-frequency signal analysis and deal with nonlinear representations, in particular with *quadratic* ones. Quadratic representations involve products of signal values at different time points. Two simple examples of quadratic representations are the square magnitudes of the short-time Fourier transform and the wavelet transform, respectively. The former is commonly called a *spectrogram*, and the latter has been termed a *scalogram*.

Historically, quadratic time-frequency representations were developed in an attempt to describe the distribution of the signal energy in the time-frequency plane. This is one reason why they are sometimes called *time-frequency distributions* (TFD).[†] It is now known that no TFD can exactly describe the localized time-frequency energy density. Nevertheless, many TFDs have properties that enable good approximations of our concept of localized energy density.

A large number of quadratic TFDs have been developed over the years, starting with the pioneering work of Wigner [1932]. The survey paper by Hlawatsch and Boudreaux-Bartels [1992] lists 21 of them. Most of these belong to a class of distributions that can be derived from a single member, the Wigner-Ville distribution. This class is known as the *Cohen class*. Most of this chapter is devoted to TFDs of the Cohen class. We first treat the Wigner-Ville distribution in a great detail

---

[†] The other reason being the attempts to relate them to probability distributions in quantum mechanics, the science in which they originated.

and give its properties and some of its extensions. Then we introduce the ambiguity function, which is simply the inverse Fourier transform of the Wigner-Ville distribution, but has many interesting properties of its own. After that we present the general Cohen class and some recently developed special cases of it. The two final sections are devoted to a time-frequency representation that is not quadratic and has a completely different focus. This representation, termed the *high-order ambiguity function*, is useful for estimating complex signals with constant amplitude and continuous phase. The discussion of the high-order ambiguity function will lead us back to random signals and their estimation, in the spirit of Chapters 6 through 10 of this book.

12.2. THE WIGNER-VILLE DISTRIBUTION: PART I

In many respects, the Wigner-Ville distribution can be regarded as "the mother of all TFDs." It was introduced by Wigner [1932] for quantum mechanics and later by Ville [1948] for electrical engineering. It can be motivated by quantum mechanics arguments, but this motivation will be skipped.

The *Wigner-Ville distribution* (WVD) of the L_2 function $y(t)$ is given by

$$\text{WVD}\{y(t)\} = W_y(t,\omega) = \int_{-\infty}^{\infty} y(t+0.5\tau)y^*(t-0.5\tau)e^{-j\omega\tau}d\tau. \qquad (12.2.1)$$

A dual formula expresses the WVD in terms of the Fourier transform of the signal (Prob. 12.1):

$$W_y(t,\omega) = \frac{1}{2\pi}\int_{-\infty}^{\infty} Y(\omega+0.5\nu)Y^*(\omega-0.5\nu)e^{j\nu t}d\nu. \qquad (12.2.2)$$

An interesting property of the WVD is that it is a "pure" function of the signal, in the sense that it does not involve an auxiliary (and arbitrary) window function, as do the transforms encountered in Chapter 11. This is just one of many interesting properties that will be presented below.

Example 12.1. Let $y(t)$ be the signal

$$y(t) = \begin{cases} e^{j(\omega_0 t + \mu t^2)}, & -T/2 \le t \le T/2 \\ 0, & \text{otherwise}, \end{cases}$$

where ω_0, μ, and T are fixed. A signal of this form is called *chirp* or *linear FM* (where FM stands for frequency modulation). Its WVD is given as follows.

$$W_y(t,\omega) = \int_{-T+2|t|}^{T-2|t|} e^{j[\omega_0(t+0.5\tau)+\mu(t+0.5\tau)^2]}e^{-j[\omega_0(t-0.5\tau)+\mu(t-0.5\tau)^2]}e^{-j\omega\tau}d\tau$$

$$= \int_{-T+2|t|}^{T-2|t|} e^{j(\omega_0+2\mu t-\omega)\tau}d\tau = 2(T-2|t|)\frac{\sin[(\omega-\omega_0-2\mu t)(T-2|t|)]}{(\omega-\omega_0-2\mu t)(T-2|t|)}.$$

In particular, when T tends to infinity the WVD approaches $2\pi\delta_D(\omega - \omega_0 - 2\mu t)$, so it becomes concentrated on the line $\omega(t) = \omega_0 + 2\mu t$. Note that $\omega_0 + 2\mu t$ is precisely the derivative of the phase of $y(t)$, which is by definition the *instantaneous frequency*. We conclude that, for a chirp signal stretching from $-\infty$ to ∞, the WVD is concentrated along a line in the time-frequency plane, and the equation of the line $\omega(t)$ is the instantaneous frequency of the signal. This interpretation of the WVD is one of the main motivations for its use, but is unfortunately limited to chirp signals. ∎

Claasen and Mecklenbrauker, in their classical papers [1980a, b, c], listed several properties of the WVD that make it suitable for time-frequency signal analysis. Nevertheless, the WVD has at least two major drawbacks, which detract from its usefulness. One will be mentioned after Property W1, while the other will be deferred to the end of this section.

Property W1. The WVD is a real-valued function; that is,

$$W_y^*(t,\omega) = W_y(t,\omega). \tag{12.2.3}$$

Proof.

$$W_y^*(t,\omega) = \int_{-\infty}^{\infty} y^*(t+0.5\tau)y(t-0.5\tau)e^{j\omega\tau}\,d\tau$$

$$= \int_{-\infty}^{\infty} y^*(t-0.5\theta)y(t+0.5\theta)e^{-j\omega\theta}\,d\theta = W_y(t,\omega). \quad ∎$$

This property is necessary for the interpretation of $W_y(t,\omega)$ as an energy distribution. Unfortunately, the WVD is not a nonnegative function. This can be seen even from Example 12.1 (the sinc function assumes both positive and negative values) and can be shown to hold in general (with very few exceptions). The possible negativity of the WVD is disturbing from a theoretical point of view. We would like to deal with energy densities that, like probability densities, are positive, and the WVD is unsatisfactory in this respect.

Property W2. Translation of the signal $y(t)$ in either time or frequency causes a corresponding translation of the WVD; that is,

$$\text{WVD}\{y(t-t_0)\} = W_y(t-t_0,\omega), \quad \text{WVD}\{y(t)e^{j\omega_0 t}\} = W_y(t,\omega-\omega_0). \tag{12.2.4}$$

Proof. The first property follows trivially from the definition. The second follows from the dual formula (11.2.2) upon observing that the Fourier transform of $y(t)e^{j\omega_0 t}$ is $Y(\omega - \omega_0)$. ∎

Property W3. The integral of the WVD over the frequency variable is equal to the instantaneous power at time t. The integral over the time variable is equal to the power density at frequency ω.

$$\frac{1}{2\pi}\int_{-\infty}^{\infty} W_y(t,\omega)d\omega = |y(t)|^2, \quad \int_{-\infty}^{\infty} W_y(t,\omega)dt = |Y(\omega)|^2. \qquad (12.2.5)$$

Proof.

$$\frac{1}{2\pi}\int_{-\infty}^{\infty} W_y(t,\omega)d\omega = \frac{1}{2\pi}\int_{-\infty}^{\infty}\int_{-\infty}^{\infty} y(t+0.5\tau)y^*(t-0.5\tau)e^{-j\omega\tau}d\omega d\tau$$

$$= \int_{-\infty}^{\infty} y(t+0.5\tau)y^*(t-0.5\tau)\delta_D(\tau)d\tau = |y(t)|^2.$$

The second formula is similarly derived from (12.2.2). ∎

Corollary. The integral of the WVD over both the frequency and the time variables equals the total signal energy; that is,

$$\frac{1}{2\pi}\int_{-\infty}^{\infty}\int_{-\infty}^{\infty} W_y(t,\omega)d\omega dt = \int_{-\infty}^{\infty} |y(.)|^2 dt = \frac{1}{2\pi}\int_{-\infty}^{\infty} |Y(\omega)|^2 d\omega. \qquad (12.2.6)$$

The proof follows upon performing the appropriate integrations on (12.2.5). ∎

Property W3 is the "marginal property," since the integrals in (12.2.5) are the marginals of the distribution along the two axes. Cohen [1989] regards the marginal property as the most fundamental one and takes it as a starting point for constructing time-frequency distributions.

Before we continue to list the properties of the WVD, we need to define the instantaneous frequency and group delay of a signal. Let $y(t) = A(t)e^{j\phi(t)}$ be the polar representation of the signal and $Y(\omega) = B(\omega)e^{j\psi(\omega)}$ the polar representation of its Fourier transform. The *instantaneous angular frequency* and *group delay* are defined by

$$\mathrm{IF}(t) = \frac{d\phi(t)}{dt}, \quad \mathrm{GD}(\omega) = -\frac{d\psi(\omega)}{d\omega}. \qquad (12.2.7)$$

According to these definitions, real signals have zero instantaneous frequency, while conjugate symmetrical signals [i.e., such that $y(t) = y^*(-t)$] have zero group delay. While the latter does not pose any problems, the former is somewhat disturbing; we can hardly think of a real sinusoid, say, as having zero instantaneous frequency. This problem can be remedied by replacing real signals by their corresponding *analytic signals*. In fact, this was the approach taken by Ville [1948]. Recall that the analytic signal of a real signal $y(t)$ is obtained by adding to $y(t)$ its Hilbert transform multiplied by j. An equivalent operation is to replace $\{Y(\omega), \omega < 0\}$ by zero, double the values of $\{Y(\omega), \omega \geq 0\}$, and take the inverse Fourier transform of the result. Once $y(t)$ has been replaced by its analytic signal, the definition of $\mathrm{IF}(t)$ makes sense. For example, the IF of a real sinusoid becomes its (usual) frequency.

As we saw in Example 12.1, the WVD of an infinite-duration chirp gives its IF directly. The generalization to an arbitrary signal is given by the following property.

Property W4. The instantaneous frequency and group delay of $y(t)$ are given as the following functions of the WVD.

$$\text{IF}(t) = \frac{\int_{-\infty}^{\infty} \omega W_y(t,\omega)d\omega}{\int_{-\infty}^{\infty} W_y(t,\omega)d\omega}, \quad \text{GD}(\omega) = \frac{\int_{-\infty}^{\infty} t W_y(t,\omega)dt}{\int_{-\infty}^{\infty} W_y(t,\omega)dt}. \tag{12.2.8}$$

Proof.

$$\int_{-\infty}^{\infty} \omega W_y(t,\omega)d\omega =$$

$$\int_{-\infty}^{\infty}\int_{-\infty}^{\infty} \omega A(t+0.5\tau)A(t-0.5\tau)\exp\{j[\phi(t+0.5\tau)-\phi(t-0.5\tau)]\}e^{-j\omega\tau}d\tau\,d\omega =$$

$$j\int_{-\infty}^{\infty}\int_{-\infty}^{\infty} A(t+0.5\tau)A(t-0.5\tau)\exp\{j[\phi(t+0.5\tau)-\phi(t-0.5\tau)]\}$$

$$\frac{\partial e^{-j\omega\tau}}{\partial\tau}d\tau\,d\omega =$$

$$-j\int_{-\infty}^{\infty}\int_{-\infty}^{\infty} \frac{\partial}{\partial\tau}\Big(A(t+0.5\tau)A(t-0.5\tau)\exp\{j[\phi(t+0.5\tau)-\phi(t-0.5\tau)]\}\Big)$$

$$e^{-j\omega\tau}d\tau\,d\omega =$$

$$-j2\pi\frac{\partial}{\partial\tau}\Big(A(t+0.5\tau)A(t-0.5\tau)\exp\{j[\phi(t+0.5\tau)-\phi(t-0.5\tau)]\}\Big)\Big|_{\tau=0}$$

$$= 2\pi A^2(t)\frac{d\phi(t)}{dt}.$$

Also, the denominator of the first formula in (12.2.8) is $2\pi A^2(t)$ by (12.2.5). This proves the first formula in (12.2.8), and the second is obtained in a dual manner. ∎

Property W5. If the signal $y(t)$ is zero outside the interval $[t_1, t_2]$, so is its WVD. If $Y(\omega)$ is zero outside the interval $[\omega_1, \omega_2]$, so is its WVD. In other words, the WVD of a time-limited (band-limited) signal is time limited (band limited) in the same range.

Proof. The limits of the integral over τ in the definition of WVD are from $2\max\{t_1-t, t-t_2\}$ to $2\min\{t_2-t, t-t_1\}$. If $t \notin [t_1, t_2]$, this range is empty. The dual property is proved similarly. ∎

The next formula, due to Moyal [1949], is often quoted in the literature on time-frequency distributions, but it is hard to find evidence for its usefulness. We give it for completeness.

Property W6.

$$\frac{1}{2\pi} \int_{-\infty}^{\infty} \int_{-\infty}^{\infty} W_x(t,\omega) W_y(t,\omega) d\omega\, dt = \left| \int_{-\infty}^{\infty} x(t) y^*(t) dt \right|^2 = |\langle x, y \rangle|^2. \quad (12.2.9)$$

Proof.

$$\frac{1}{2\pi} \int_{-\infty}^{\infty} \int_{-\infty}^{\infty} W_x(t,\omega) W_y(t,\omega) d\omega\, dt =$$

$$\frac{1}{2\pi} \int_{-\infty}^{\infty} \int_{-\infty}^{\infty} \int_{-\infty}^{\infty} \int_{-\infty}^{\infty} x(t+0.5\tau)x^*(t-0.5\tau)y(t+0.5\theta)y^*(t-0.5\theta)e^{-j\omega(\tau+\theta)}$$

$$d\omega\, dt\, d\tau\, d\theta$$

$$= \int_{-\infty}^{\infty} \int_{-\infty}^{\infty} \int_{-\infty}^{\infty} x(t+0.5\tau)x^*(t-0.5\tau)y(t+0.5\theta)y^*(t-0.5\theta)\delta_D(\tau+\theta)dt\, d\tau\, d\theta$$

$$= \int_{-\infty}^{\infty} \int_{-\infty}^{\infty} x(t+0.5\tau)x^*(t-0.5\tau)y(t-0.5\tau)y^*(t+0.5\tau)dt\, d\tau$$

$$= \int_{-\infty}^{\infty} \int_{-\infty}^{\infty} x(\alpha)y^*(\alpha)x^*(\beta)y(\beta)d\alpha\, d\beta = \left| \int_{-\infty}^{\infty} x(t) y^*(t) dt \right|^2. \qquad \blacksquare$$

The next two properties involve the two fundamental linear filtering operations, convolutions in the time and frequency domains.

Property W7. Let $y(t)$ be related to $x(t)$ through the convolution

$$y(t) = \int_{-\infty}^{\infty} h(t-\theta)x(\theta)d\theta.$$

Then

$$W_y(t,\omega) = \int_{-\infty}^{\infty} W_h(t-\theta,\omega) W_x(\theta,\omega)d\theta. \qquad (12.2.10)$$

Proof.

$$W_y(t,\omega) = \int_{-\infty}^{\infty} \int_{-\infty}^{\infty} \int_{-\infty}^{\infty} h(t+0.5\tau-\alpha)x(\alpha)h^*(t-0.5\tau-\beta)x^*(\beta)e^{-j\omega\tau} d\tau\, d\alpha\, d\beta.$$

Make the variable substitutions $\alpha = \theta + 0.5\sigma$ and $\beta = \theta - 0.5\sigma$, $\tau = \gamma + \sigma$. This gives

$$W_y(t,\omega) =$$

$$\int_{-\infty}^{\infty} \int_{-\infty}^{\infty} \int_{-\infty}^{\infty} h(t-\theta+0.5\gamma)h^*(t-\theta-0.5\gamma)x(\theta+0.5\sigma)x^*(\theta-0.5\sigma)e^{-j\omega\gamma}e^{-j\omega\sigma}$$

$$d\theta\, d\gamma\, d\sigma = \int_{-\infty}^{\infty} W_h(t-\theta,\omega) W_x(\theta,\omega)d\theta. \qquad \blacksquare$$

Property W8. Let $y(t)$ be related to $x(t)$ through the product $y(t) = h(t)x(t)$. Then

$$W_y(t,\omega) = \frac{1}{2\pi}\int_{-\infty}^{\infty} W_h(t,\omega-\nu)W_x(t,\nu)d\nu. \qquad (12.2.11)$$

Proof. We have

$$Y(\omega) = \frac{1}{2\pi}\int_{-\infty}^{\infty} H(\omega-\nu)X(\nu)d\nu.$$

Therefore, (12.2.11) follows by same derivation as in the proof of (12.2.10), starting with (12.2.2). ∎

The last property gives a nice relationship between the WVD of a signal and that of its Fourier transform.

Property W9. The WVD of the Fourier transform of $y(t)$ is related to the WVD of $y(t)$ through

$$W_Y(t,\omega) = 2\pi W_y(-\omega,t). \qquad (12.2.12)$$

Proof.

$$W_Y(t,\omega) = \int_{-\infty}^{\infty}\int_{-\infty}^{\infty}\int_{-\infty}^{\infty} y(\alpha)e^{-j\alpha(t+0.5\tau)}y^*(\beta)e^{j\beta(t-0.5\tau)}e^{-j\omega\tau}d\tau\,d\alpha\,d\beta$$

$$= 2\pi\int_{-\infty}^{\infty}\int_{-\infty}^{\infty} y(\alpha)y^*(\beta)e^{-j(\alpha-\beta)t}\delta_D(\omega+0.5\alpha+0.5\beta)d\alpha\,d\beta$$

$$= 4\pi\int_{-\infty}^{\infty} y(\alpha)y^*(-\alpha-2\omega)e^{-j2(\alpha+\omega)t}d\alpha$$

$$= 2\pi\int_{-\infty}^{\infty} y(-\omega+0.5\gamma)y^*(-\omega-0.5\gamma)e^{-j\gamma t}d\gamma = 2\pi W_y(-\omega,t). \qquad∎$$

At this point we should mention some distributions closely related to Wigner-Ville. A distribution proposed by Kirkwood [1933], later by Levin [1967], and finally by Rihaczek [1968] is given by

$$R_y(t,\omega) = \int_{-\infty}^{\infty} y(t)y^*(t-\tau)e^{-j\omega\tau}d\tau = y(t)Y^*(\omega)e^{-j\omega t}. \qquad (12.2.13)$$

This is called the *Rihaczek distribution*.[†] Page [1952] has proposed a distribution based on a causality argument. He defined the instantaneous causal spectrum of a

---

† This is yet another example of the "last entry wins" phenomenon, so common in science.

signal $y(t)$ as $\int_{-\infty}^{t} y(\tau)e^{-j\omega\tau}d\tau$ and used it to define the distribution

$$P_y(t,\omega) = \frac{\partial}{\partial t}\left|\int_{-\infty}^{t} y(\tau)e^{-j\omega\tau}d\tau\right|^2 = 2\mathrm{Re}\left\{y(t)e^{-j\omega t}\int_{-\infty}^{t} y^*(\tau)e^{j\omega\tau}d\tau\right\}.$$
(12.2.14)

Levin [1967] has proposed the anticausal dual of the Page distribution:

$$L_y(t,\omega) = -\frac{\partial}{\partial t}\left|\int_{t}^{\infty} y(\tau)e^{-j\omega\tau}d\tau\right|^2 = 2\mathrm{Re}\left\{y(t)e^{-j\omega t}\int_{t}^{\infty} y^*(\tau)e^{j\omega\tau}d\tau\right\}.$$
(12.2.15)

He also combined the causal and anticausal distributions, thus obtaining the real version of the Rihaczek distribution:

$$PL_y(t,\omega) = 0.5[P_y(t,\omega) + L_y(t,\omega)] = \mathrm{Re}\{y(t)Y^*(\omega)e^{-j\omega t}\} = \mathrm{Re}\{R_y(t,\omega)\}.$$
(12.2.16)

We will return to these distributions in Sec. 12.5, where we will treat them in a more general context.

The second major drawback of the WVD (the first being its possible negative values) is the problem of cross terms. Suppose the signal $y(t)$ is a sum of several signals of different time-frequency characteristics, say $y(t) = \sum_{k=1}^{K} y_k(t)$. In this case it is not difficult to show that

$$W_y(t,\omega) = \sum_{k=1}^{K}\sum_{\ell=1}^{K} W_{k,\ell}(t,\omega),$$

where $W_{k,\ell}(t,\omega)$ is the *cross-Wigner distribution*

$$W_{k,\ell}(t,\omega) = \int_{-\infty}^{\infty} y_k(t+0.5\tau)y_\ell^*(t-0.5\tau)e^{-j\omega\tau}d\tau.$$
(12.2.17)

In many applications, we are interested in the sum $\sum_{k=1}^{K} W_{k,k}(t,\omega)$, which contains only the autoterms. This is because the autoterms depict the time-frequency characteristics of the individual components, while the cross terms may convey misleading information (an example will be given below). Complete removal of the cross terms is impossible in general, but methods have been devised for their attenuation. This subject will be discussed in the next two sections.

Example 12.2. Let $y_1(t) = e^{j\omega_1 t}$ on the interval $[-3T/2, -T/2]$ and zero elsewhere. Let $y_2(t) = e^{j\omega_2 t}$ on the interval $[T/2, 3T/2]$ and zero elsewhere. The four terms of the WVD can be computed in a tedious, but straightforward manner, yielding

$$W_{1,1}(t,\omega) = \begin{cases} \dfrac{2\sin[(3T+2t)(\omega-\omega_1)]}{(\omega-\omega_1)}, & -1.5T \leq t \leq -T \\[4mm] -\dfrac{2\sin[(T+2t)(\omega-\omega_1)]}{(\omega-\omega_1)}, & -T \leq t \leq -0.5T, \end{cases}$$

$$W_{2,2}(t,\omega) = \begin{cases} \dfrac{2\sin[(2t-T)(\omega-\omega_2)]}{(\omega-\omega_2)}, & 0.5T \leq t \leq T \\[3mm] \dfrac{2\sin[(3T-2t)(\omega-\omega_2)]}{(\omega-\omega_2)}, & T \leq t \leq 1.5T, \end{cases}$$

$$W_{1,2}(t,\omega) + W_{2,1}(t,\omega)$$
$$= 4\cos[(\omega_2-\omega_1)t + 2T(\omega-\overline{\omega})]\frac{\sin[(T-2|t|)(\omega-\overline{\omega})]}{(\omega-\overline{\omega})}, \quad |t| \leq 0.5T,$$

where $\overline{\omega} = 0.5(\omega_1 + \omega_2)$. As we see, the autoterms behave as expected. Each occupies the time interval of the corresponding signal and is centered around the corresponding frequency. The two cross terms, however, spoil the picture. They exist on the interval $[-T/2, T/2]$, where the signal is zero, and are centered around the average of the two frequencies. Moreover, the peak value is twice as large as the peak of each autoterm. Clearly, a naive interpretation of the WVD will give false conclusions on the nature of the signal. On the other hand, if there was a way to suppress the cross terms, even a naive interpretation of the result would give the correct information. ∎

12.3. THE WIGNER-VILLE DISTRIBUTION: PART II

In this section we continue our treatment of the WVD and its properties. We first explore the invertibility of the distribution, that is, the recovery of the signal from its WVD. Then we discuss the pseudo-Wigner distribution and finally apply the WVD to discrete-time signals.

To see whether the WVD is invertible, it is natural to undo the Fourier transform first, since the Fourier transform is an invertible operation. Doing so yields

$$y(t+0.5\tau)y^*(t-0.5\tau) = \frac{1}{2\pi}\int_{-\infty}^{\infty} W_y(t,\omega)e^{j\omega\tau}\,d\omega.$$

We can now choose $\tau = 2(t-t_0)$ where t_0 is arbitrary, but fixed. This gives

$$y(2t-t_0)y^*(t_0) = \frac{1}{2\pi}\int_{-\infty}^{\infty} W_y(t,\omega)e^{j2\omega(t-t_0)}\,d\omega. \qquad (12.3.1)$$

As long as $y(t_0)$ is nonzero, we can recover the signal up to a constant complex number for all $2t - t_0$, hence for all t. The magnitude of this number can be recovered by substituting $t = t_0$ in (12.3.1), but the phase cannot be determined. The latter statement is obvious, since modifying $y(t)$ by a constant phase term does not change its WVD. The most obvious choice for t_0 is zero. If it happens that $y(0) = 0$, the right side of (12.3.1) would be identically zero, and then we assign another value to t_0. Unless the signal is identically zero, there must be some t_0 for which this procedure will succeed. In summary, the WVD is invertible, up to a constant phase term.

As we saw in the previous section, the WVD typically exhibits cross terms, which often make the result difficult to interpret. The *pseudo-Wigner distribution* (PWD) [Claasen and Mecklenbräuker, 1980a] is essentially equivalent to the windowing operation in Fourier transform. It serves the triple purpose of reducing the cross terms, attenuating the Fourier transform side lobes (when truncating the signal to a finite duration), and reducing computational complexity in discrete-time processing. The PWD is defined by

$$\mathrm{PW}_y(t,\omega) = \int_{-\infty}^{\infty} y(t+0.5\tau)y^*(t-0.5\tau)h(0.5\tau)h^*(-0.5\tau)e^{-j\omega\tau}d\tau, \quad (12.3.2)$$

where $h(t)$ is some user-chosen window function. As we see, the effect of the window is to localize the integration over τ to the vicinity of $\tau = 0$. Therefore, the distribution at time t is mainly affected by signal values in the vicinity of t.

As in the case of usual Fourier transform, windowing is equivalent to convolution in the frequency domain. More precisely:

Lemma 12.1.

$$\mathrm{PW}_y(t,\omega) = \frac{1}{2\pi} \int_{-\infty}^{\infty} W_h(0,\nu)W_y(t,\omega-\nu)d\nu. \quad (12.3.3)$$

Proof.

$$\int_{-\infty}^{\infty} y(t+0.5\tau)y^*(t-0.5\tau)h(0.5\tau)h^*(-0.5\tau)e^{-j\omega\tau}d\tau$$

$$= \frac{1}{2\pi} \int_{-\infty}^{\infty}\int_{-\infty}^{\infty} y(t+0.5\tau)y^*(t-0.5\tau)W_h(0,\nu)e^{j\nu\tau}e^{-j\omega\tau}d\nu\,d\tau$$

$$= \frac{1}{2\pi} \int_{-\infty}^{\infty} W_h(0,\nu)W_y(t,\omega-\nu)d\nu. \qquad \blacksquare$$

Since the convolution in (12.3.3) is in the frequency domain only, the ability of the PWD to combat cross terms is limited. As is clear from the time domain interpretation of this operation, cross terms can be effectively eliminated only when the individual components are separated in time by intervals larger than the effective duration of the window. An improvement in the attenuation of cross terms can be obtained sometimes by adding smoothing (convolution) in the time domain. The *smoothed pseudo-Wigner distribution* (SPWD) is defined by

$$\mathrm{SPW}_y(t,\omega) = \frac{1}{2\pi} \int_{-\infty}^{\infty}\int_{-\infty}^{\infty} W_h(0,\nu)g(\theta)W_y(t-\theta,\omega-\nu)d\nu\,d\theta, \quad (12.3.4)$$

where $g(t)$ is another window function [which may be equal to $h(t)$ or different]. Yet a more general smoothing operation is

$$\mathrm{SW}_y(t,\omega) = \frac{1}{2\pi} \int_{-\infty}^{\infty}\int_{-\infty}^{\infty} \Phi(\theta,\nu)W_y(t-\theta,\omega-\nu)d\nu\,d\theta, \quad (12.3.5)$$

where $\Phi(t, \omega)$ is some two-dimensional smoothing kernel. This is called a *smoothed Wigner distribution* (SWD), and we will encounter it again in Sec. 12.5 in a more general context. Formula (12.3.4) is clearly a special case of (12.3.5) with a separable kernel.

A side benefit of the smoothing operation (12.3.5) is that a nonnegative distribution can be guaranteed if the smoothing kernel is properly chosen. A sufficient condition is that the smoothing kernel be the WVD of some function. This was discovered by Bertrand et al. [1983] and is proved as follows.

Lemma 12.2. If $\Phi(t, \omega) = W_x(t, \omega)$ for some $x(t)$, then the SWD given by (12.3.5) is nonnegative.

Proof. Under the assumption of the lemma,

$$\mathrm{SW}_y(t, \omega) = \frac{1}{2\pi} \int_{-\infty}^{\infty} \int_{-\infty}^{\infty} \int_{-\infty}^{\infty} \int_{-\infty}^{\infty}$$

$$x(\theta + 0.5\sigma)x^*(\theta - 0.5\sigma)y(t - \theta + 0.5\tau)y^*(t - \theta - 0.5\tau)e^{-j\nu\sigma}e^{-j(\omega-\nu)\tau}d\tau\,d\sigma\,d\nu\,d\theta$$

$$= \int_{-\infty}^{\infty} \int_{-\infty}^{\infty} x(\theta + 0.5\tau)x^*(\theta - 0.5\tau)y(t - \theta + 0.5\tau)y^*(t - \theta - 0.5\tau)e^{-j\omega\tau}d\tau\,d\theta$$

$$= \int_{-\infty}^{\infty} \int_{-\infty}^{\infty} x(\alpha)x^*(\beta)y(t - \beta)y^*(t - \alpha)e^{-j\omega\alpha}e^{j\omega\beta}d\alpha\,d\beta$$

$$= \left| \int_{-\infty}^{\infty} x(\alpha)y^*(t - \alpha)e^{-j\omega\alpha}d\alpha \right|^2 \geq 0. \qquad \blacksquare$$

An interesting corollary of this proof is that the smoothed WVD with the kernel $W_x(t, \omega)$ is simply the spectrogram of $y(t)$ with the window function $x(-t)$ (recall from Sec. 12.1 that the spectrogram is the square magnitude of the STFT).

A common smoothing kernel (which is not, however, a WVD of any function) is the Gaussian $\Phi(t, \omega) = (0.5\pi t_0\omega_0)^{-1/2} \exp\{-(t/t_0)^2 - (\omega/\omega_0)^2\}$. This can be shown to yield a positive SWD when $t_0\omega_0 \geq 1$ (see Prob. 12.4 for a special case). Smoothing kernels that yield positive SWDs have the drawback of not satisfying the marginal property (12.2.5); see Prob. 12.5 for the special case of the spectrogram.

Let us now discuss the Wigner-Ville distribution for discrete-time signals. Assume initially that the signal $y(t)$ is band limited to the frequency range $[-\omega_{\max}, \omega_{\max}]$ and we sample the signal at interval $T \leq \pi/\omega_{\max}$ (later we will see that the sampling rate has to meet a more stringent condition). The following theorem, due to Claasen and Mecklenbräuker [1980b], expresses the WVD of $y(t)$ in terms of its samples.

Theorem 12.1. If $T \leq \pi/\omega_{\max}$, then

$$W_y(t, \omega) = 2T \sum_{k=-\infty}^{\infty} y(t+kT)y^*(t-kT)e^{-j2\omega kT}. \qquad (12.3.6)$$

Proof. By the Nyquist-Shannon interpolation formula for band-limited signals, we have

$$y(t + 0.5\tau) = \sum_{n=-\infty}^{\infty} y(t+nT)\mathrm{sinc}\left(\frac{\pi[0.5\tau - nT]}{T}\right),$$

$$y(t - 0.5\tau) = \sum_{k=-\infty}^{\infty} y(t-kT)\mathrm{sinc}\left(\frac{\pi[0.5\tau - kT]}{T}\right),$$

where $\mathrm{sinc}(x) = (\sin x)/x$. Therefore,

$$W_y(t, \omega) = \sum_{n=-\infty}^{\infty} \sum_{k=-\infty}^{\infty}$$

$$y(t+nT)y^*(t-kT) \int_{-\infty}^{\infty} \mathrm{sinc}\left(\frac{\pi[0.5\tau - nT]}{T}\right)\mathrm{sinc}\left(\frac{\pi[0.5\tau - kT]}{T}\right)e^{-j\omega\tau}d\tau.$$

Also,

$$\mathrm{sinc}\left(\frac{\pi[0.5\tau - nT]}{T}\right) = \frac{1}{2\pi}\int_{-\pi}^{\pi} e^{j(0.5\tau/T - n)\alpha}d\alpha,$$

$$\mathrm{sinc}\left(\frac{\pi[0.5\tau - kT]}{T}\right) = \frac{1}{2\pi}\int_{-\pi}^{\pi} e^{-j(0.5\tau/T - k)\beta}d\beta.$$

So

$$\int_{-\infty}^{\infty} \mathrm{sinc}\left(\frac{\pi[0.5\tau - nT]}{T}\right)\mathrm{sinc}\left(\frac{\pi[0.5\tau - kT]}{T}\right)e^{-j\omega\tau}d\tau$$

$$= \frac{1}{4\pi^2}\int_{-\pi}^{\pi}\int_{-\pi}^{\pi}\int_{-\infty}^{\infty} e^{j[0.5(\alpha-\beta)/T - \omega]\tau}e^{j(k\beta - n\alpha)}d\alpha\, d\beta\, d\tau$$

$$= \frac{2T}{2\pi}e^{-j2\omega kT}\int_{-\pi}^{\pi} e^{j(k-n)\alpha}d\alpha = 2Te^{-j2\omega kT}\delta(k-n), \quad -\frac{\pi}{T} \leq \omega \leq \frac{\pi}{T},$$

and zero elsewhere. Finally,

$$W_y(t, \omega) = 2T \sum_{k=-\infty}^{\infty} y(t+kT)y^*(t-kT)e^{-j2\omega kT}, \quad -\frac{\pi}{T} \leq \omega \leq \frac{\pi}{T}. \qquad \blacksquare$$

As we see, the WVD of a band-limited signal, as a function of ω, has *two periods* on its support interval. By comparison, the usual Fourier transform of the signal has *one* period on this interval. Therefore, $W_y(t, \omega)$ is generally *aliased*; that is, it cannot depict the true frequency contents of the signal. The aliasing is obviously due to the multiplication of the two y-dependent terms in the WVD.

Multiplication in the time domain is equivalent to convolution in the frequency domain, so the product has twice the bandwidth, resulting in violation of the Nyquist condition. Aliasing is avoided in two cases:

1. If the signal is sampled at *twice* the Nyquist rate (i.e., $T \leq 0.5\pi/\omega_{\max}$).
2. If the signal is analytic (i.e., its spectrum is zero for negative frequencies).

Formula (12.3.6) motivated Claasen and Mecklenbrauker to define the WVD of a discrete-time signal $\{y_n\}$ as

$$W_y(n, \theta) = 2T \sum_{k=-\infty}^{\infty} y_{n+k} y_{n-k}^* e^{-j2\theta k}, \qquad (12.3.7)$$

where $\theta = \omega T$. This form of the discrete-time WVD is used either for signals sampled at twice the Nyquist rate or for analytic signals. It follows from this definition that $y_{n+k}y_{n-k}^*$ can be reconstructed from $W_y(n, \theta)$. Taking $k = n - n_0$ for some n_0 enables the reconstruction of $y_{2n-n_0}y_{n_0}^*$. Taking $n = n_0$ enables reconstruction of $y(n_0)$ up to a constant phase term, hence of y_{2n-n_0}. Repeating this once for n_0 even and once for odd yields y_n for all n.

An alternative to the DWVD (12.3.7) has been proposed by Peyrin and Prost [1986]. They considered all three cases of discrete time, discrete frequency, and discrete time and frequency. We will derive only the discrete-time (continuous frequency) version. The other two can be derived in a similar manner.

Let the sampled signal $y(t)$ be represented as $y(t) = \sum_{k=-\infty}^{\infty} y(kT)\delta_D(t-kT)$. Then we can write, using the Fourier representation of the delta function,

$$y(t) = \sum_{k=-\infty}^{\infty} y(kT)\delta_D(t - kT) = \frac{1}{2\pi} \sum_{k=-\infty}^{\infty} y(kT) \int_{-\infty}^{\infty} e^{j\alpha(t-kT)} d\alpha.$$

Substitution in the WVD formula gives

$$W_y(t, \omega) = \frac{1}{4\pi^2} \sum_{k=-\infty}^{\infty} \sum_{\ell=-\infty}^{\infty} y(kT)y^*(\ell T)$$

$$\int_{-\infty}^{\infty} \int_{-\infty}^{\infty} \int_{-\infty}^{\infty} e^{j\alpha(t+0.5\tau-kT)} e^{j\beta(t+0.5\tau-\ell T)} e^{-j\omega\tau} d\tau\, d\alpha\, d\beta$$

$$= \sum_{k=-\infty}^{\infty} \sum_{\ell=-\infty}^{\infty} y(kT)y^*(\ell T)\delta_D[t - 0.5(k+\ell)T]e^{-j2\omega(t-\ell T)}.$$

In particular, for $t = 0.5nT$ (where n is an integer),

$$W_y(0.5nT, \omega) = \sum_{k=-\infty}^{\infty} \sum_{\ell=-\infty}^{\infty} y(kT)y^*(\ell T)\delta_D[0.5nT - 0.5(k+\ell)T]e^{-j2\omega(t-\ell T)}$$

$$= \delta_D(0) \sum_{k=-\infty}^{\infty} y(kT)y^*(nT - kT)e^{-j\omega(2k-n)T}.$$

For $t \neq 0.5nT$ the result is zero, so, in summary,

$$W_y(t, \omega) = \delta_D(t - 0.5nT) \sum_{k=-\infty}^{\infty} y(kT)y^*(nT - kT)e^{-j\omega(2k-n)T}. \qquad (12.3.8)$$

Motivated by (12.3.8), Peyrin and Prost defined the WVD for the discrete-time continuous-frequency case as

$$W_y(0.5n, \theta) = \sum_{k=-\infty}^{\infty} y_k y_{n-k}^* e^{-j\theta(2k-n)}. \qquad (12.3.9)$$

Note that, in contrast with Claasen-Mecklenbrauker's definition, the discrete WVD of Peyrin and Prost is defined both at sampling points and at the midpoints between sampling points. In particular, at the sampling points (even n), the WVD has a period of π/T as a function of ω, so it suffers from the same aliasing problem as (12.3.7). Reconstruction of $\{y_n\}$ from its transform (12.3.9) is discussed in Prob. 12.7.

A pseudo-Wigner distribution for discrete-time signals was developed by Dvir [1990], using the approach of Peyrin and Prost. We give here the result for the continuous-frequency case; the discrete-time-frequency case can be derived in a similar manner. We apply the impulse-train representation to the sampled windowed signal $y(t)h(t - t_0)$, where t_0 is fixed. This yields, as in the previous derivation,

$$\mathrm{PW}_y(t, \omega; t_0) = \delta_D(t - 0.5nT) \sum_{k=-\infty}^{\infty}$$
$$y(kT)h(kT - t_0)y^*(nT - kT)h^*(nT - kT - t_0)e^{-j\omega(2k-n)T}.$$

We now identify t_0 with t, thus centering the window at the time point of interest. This gives

$$\mathrm{PW}_y(t, \omega) = \delta_D(t - 0.5nT) \sum_{k=-\infty}^{\infty}$$
$$y(kT)h(kT - 0.5nT)y^*(nT - kT)h^*(nT - kT - 0.5nT)e^{-j\omega(2k-n)T}.$$

The resulting pseudo-Wigner distribution for discrete-time signals is therefore

$$\mathrm{PW}_y(0.5n, \theta) = \sum_{k=-\infty}^{\infty} y_k y_{n-k}^* h(kT - 0.5nT)h^*(nT - kT - 0.5nT)e^{-j\theta(2k-n)}.$$
$$(12.3.10)$$

Note that, while the signal is sampled at a rate $1/T$, the window is sampled at twice that rate and split into two sequences, at a rate of $1/T$ each. One sequence is used for the even-numbered data points and the other for the odd-numbered ones.

Nuttal [1989] has derived an alias-free version of the discrete WVD, as follows. Suppose the signal $y(t)$ is band limited to $[-\pi/T, \pi/T]$, where T is the sampling interval. Let $\tilde{Y}(e^{j\omega T})$ be the Fourier transform of the sampled signal,

$$\tilde{Y}(e^{j\omega T}) = \sum_{n=-\infty}^{\infty} y(nT)e^{-j\omega Tn}. \tag{12.3.11}$$

Then we know by the Nyquist-Shannon sampling theorem that the Fourier transform of the continuous-time signal is related to that of the sampled signal through

$$Y(\omega) = \begin{cases} T\tilde{Y}(e^{j\omega T}), & -\frac{\pi}{T} \leq \omega \leq \frac{\pi}{T} \\ 0, & \text{otherwise.} \end{cases}$$

Using (12.2.2) and substituting $\lambda = \nu T$, we can write the WVD as

$$W_y(t,\omega) = \frac{T}{2\pi} \int_{-2\pi+2|\omega|T}^{2\pi-2|\omega|T} \tilde{Y}(e^{j(\omega T + 0.5\lambda)})\tilde{Y}^*(e^{j(\omega T - 0.5\lambda)})e^{j\lambda t/T}d\lambda. \tag{12.3.12}$$

Note that (12.3.12) depends only on the sampled signal [via (12.3.11)], but it can be computed for any values of t and ω. Contrary to the aforementioned definitions, here the signal needs to be sampled only at the Nyquist rate, not at twice this rate. Nevertheless, the transform variable t needs to be sampled at intervals of $T/2$ or less to capture the finest details of the WVD. As we have seen, the same holds for the Peyrin-Prost distribution (12.3.9). Problem 12.8 discusses the discrete-time, discrete-frequency version of (12.3.12).

12.4. THE AMBIGUITY FUNCTION

The *ambiguity function* (AF) of a finite energy signal $y(t)$ is defined as

$$A_y(\nu,\tau) = \int_{-\infty}^{\infty} y(t)y^*(t-\tau)e^{j\nu t}dt. \tag{12.4.1}$$

The reason for using new time and frequency variables (ν,τ) rather than (t,ω) will be clear after we state Lemma 12.3 below.

The ambiguity function was introduced by Woodward [1953] in the context of radar applications. Assume that $y(t)$ is a radar pulse transmitted at time zero. Radar pulses are usually narrow-band (modulated) signals centered around some carrier frequency ω_0. The echo from a point target located at distance r from the radar and moving at radial velocity v will be delayed in time and shifted in frequency. The time delay is given by $2r/c$, where c is the speed of light. The frequency shift is due to the Doppler effect and is given (assuming that $|v| \ll c$) by $2v\omega_0/c$. Thus, the echo is given (up to a scale factor) by $y(t - 2r/c)e^{-j(2v\omega_0/c)t}$. As is clear from (12.4.1), the cross correlation between the transmitted and the received signals is just the ambiguity function evaluated at $\tau = 2r/c$ (which is proportional to the range) and $\nu = 2v\omega_0/c$ (which is proportional to the radial velocity).

In the radar context, the ambiguity function depends on the pulse waveform, and its main use is to measure the range-velocity resolution attainable for a given waveform. Suppose that the ambiguity function has an effective width of $(\Delta\tau, \Delta\nu)$ along its two axes. This roughly means that two point targets spread by less than $c\Delta\tau/2$ in range *and* less than $c\Delta\nu/2\omega_0$ in radial velocity cannot be distinguished by the radar, but appear to it as a single target. Two such targets are said to be ambiguous; hence the name ambiguity function.

Having motivated the ambiguity function, let us now modify its definition by symmetrizing it with respect to τ, similarly to the Wigner-Ville distribution. The modified definition is thus

$$\mathrm{AF}\{y(t)\} = A_y(\nu, \tau) = \int_{-\infty}^{\infty} y(t + 0.5\tau)y^*(t - 0.5\tau)e^{j\nu t}dt. \tag{12.4.2}$$

The forms (12.4.1) and (12.4.2) differ only by a factor $e^{j0.5\nu\tau}$. Both are in common use, but we will adhere to the latter. We also have the dual formula (Prob. 12.9)

$$A_y(\nu, \tau) = \frac{1}{2\pi} \int_{-\infty}^{\infty} Y(\omega - 0.5\nu)Y^*(\omega + 0.5\nu)e^{j\omega\tau}d\omega. \tag{12.4.3}$$

Example 12.3. Let $y(t)$ be the chirp signal defined in Example 2.1. Then

$$A_y(\nu, \tau) = \int_{-0.5(T-|\tau|)}^{0.5(T-|\tau|)} e^{j[\omega_0(t+0.5\tau)+\mu(t+0.5\tau)^2]}e^{-j[\omega_0(t-0.5\tau)+\mu(t-0.5\tau)^2]}e^{j\nu t}dt$$

$$= e^{j\omega_0\tau} \int_{-0.5(T-|\tau|)}^{0.5(T-|\tau|)} e^{j(2\mu\tau+\nu)t}dt = e^{j\omega_0\tau}(T-|\tau|)\frac{\sin[0.5(\nu+2\mu\tau)(T-|\tau|)]}{0.5(\nu+2\mu\tau)(T-|\tau|)}.$$

It is common to define $\Delta\tau$ as the point nearest to the origin on the τ-axis for which $A_y(\nu, \tau)$ is zero; $\Delta\nu$ is defined in a dual manner. We thus get

$$\Delta\nu = \frac{2\pi}{T}, \quad \Delta\tau = \begin{cases} T, & \mu T^2 < 4\pi \\ 0.5(T - \sqrt{T^2 - 4\pi/\mu}), & \mu T^2 \geq 4\pi. \end{cases}$$

In particular, if $\mu T^2 \gg 4\pi$, we get the approximation $\Delta\tau \approx \pi/\mu T$ (Prob. 12.11). These results have the following interpretation. The (Doppler) frequency resolution is inversely proportional to the duration. Two Doppler frequencies can be resolved if they differ by at least one cycle during the pulse interval, which is a perfectly intuitive result. The time resolution is determined by the chirp parameter μ. The value $\mu = 0$ corresponds to a nonchirped pulse, in which case the time resolution is equal to the pulse width (another intuitive result). On the other extreme, when μ is very large, the time resolution is inversely proportional to μT, which is approximately equal to the signal bandwidth. Therefore, we can arbitrarily improve the time resolution by increasing the signal bandwidth. This is the famous *pulse compression* property of the chirp waveform, which has found many applications in radar systems. It should be noted, however, that the above analysis does not apply to a mixed time-frequency resolution. For example, it is easily seen from the above expression that the first zero of the ambiguity function along the diagonal line $\nu + 2\mu\tau = 0$ is at $\tau = T$, regardless of μ. Physically, this means that targets whose velocity is proportional to the range by a certain factor (depending on μ) have relatively poor resolution determined only by the signal's duration. ∎

The similarity of the ambiguity function to the WVD is more than superficial. In fact we have:

Lemma 12.3. The ambiguity function is, up to a constant factor, the two-dimensional inverse Fourier transform of the WVD.

Proof. The two-dimensional inverse Fourier transform of the WVD is given by

$$\frac{1}{4\pi^2} \int_{-\infty}^{\infty} \int_{-\infty}^{\infty} \int_{-\infty}^{\infty} y(t+0.5\theta)y^*(t-0.5\theta)e^{-j\omega\theta}e^{j\omega\tau}e^{j\nu t}d\theta\,d\omega\,dt$$

$$= \frac{1}{2\pi} \int_{-\infty}^{\infty} \int_{-\infty}^{\infty} y(t+0.5\theta)y^*(t-0.5\theta)\delta_D(\tau-\theta)e^{j\nu t}d\theta\,dt$$

$$= \frac{1}{2\pi} \int_{-\infty}^{\infty} y(t+0.5\tau)y^*(t-0.5\tau)e^{j\nu t}dt = \frac{1}{2\pi}A_y(\nu,\tau). \qquad (12.4.4)$$

∎

The Fourier relationship between the WVD and the AF implies the following properties, which are the duals of properties W1 through W9. The proofs of most properties are straightforward consequences of (12.4.4), so they are left to the reader (Prob. 12.12).

Property A1. The AF is conjugate symmetric; that is,

$$A_y(\nu,\tau) = A_y^*(-\nu,-\tau). \qquad (12.4.5)$$

Property A2.

$$\text{AF}\{y(t-t_0)\} = A_y(\nu,\tau)e^{-j\nu t_0}, \quad \text{AF}\{y(t)e^{j\omega_0 t}\} = A_y(\nu,\tau)e^{j\omega_0\tau}. \qquad (12.4.6)$$

Property A3.

$$A_y(\nu,0) = \frac{1}{2\pi} \int_{-\infty}^{\infty} Y(\omega-0.5\nu)Y^*(\omega+0.5\nu)d\omega, \qquad (12.4.7a)$$

$$A_y(0,\tau) = \int_{-\infty}^{\infty} y(t+0.5\tau)y^*(t-0.5\tau)dt. \qquad (12.4.7b)$$

Corollary.

$$A_y(0,0) = \int_{-\infty}^{\infty} |y(t)|^2 dt = \frac{1}{2\pi} \int_{-\infty}^{\infty} |Y(\omega)|^2 d\omega. \qquad (12.4.8)$$

Property A4.

$$\text{IF}(t) = -j \frac{\int_{-\infty}^{\infty} \left[\frac{\partial A_y(\nu, \tau)}{\partial \tau} \right]_{\tau=0} e^{j\nu t} d\nu}{\int_{-\infty}^{\infty} A_y(\nu, 0) e^{j\nu t} d\nu}, \quad \text{GD}(t) = j \frac{\int_{-\infty}^{\infty} \left[\frac{\partial A_y(\nu, \tau)}{\partial \nu} \right]_{\nu=0} e^{-j\nu t} d\nu}{\int_{-\infty}^{\infty} A_y(0, \tau) e^{-j\nu t} d\tau}.$$

$$(12.4.9)$$

Property A5. If the signal $y(t)$ is zero outside the interval $[t_1, t_2]$, then $A_y(\nu, \tau)$ is zero for $|\tau| > t_2 - t_1$. If $Y(\omega)$ is zero outside the interval $[\omega_1, \omega_2]$, then $A_y(\nu, \tau)$ is zero for $|\nu| > \omega_2 - \omega_1$.

Property A6.

$$\frac{1}{2\pi} \int_{-\infty}^{\infty} \int_{-\infty}^{\infty} A_x(\nu, \tau) A_y^*(\nu, \tau) d\nu \, d\tau = \left| \int_{-\infty}^{\infty} x(t) y^*(t) dt \right|^2 = |\langle x, y \rangle|^2. \quad (12.4.10)$$

Remark. The special case

$$\frac{1}{2\pi} \int_{-\infty}^{\infty} \int_{-\infty}^{\infty} |A_y(\nu, \tau)|^2 d\nu \, d\tau = \left[\int_{-\infty}^{\infty} |y(t)|^2 dt \right]^2$$

is known as the *constant volume* property of the AF.

Property A7. Let $y(t)$ be related to $x(t)$ through the convolution

$$y(t) = \int_{-\infty}^{\infty} h(t - \theta) x(\theta) d\theta.$$

Then

$$A_y(\nu, \tau) = \int_{-\infty}^{\infty} A_h(\nu, \tau - \sigma) A_x(\nu, \sigma) d\sigma. \quad (12.4.11)$$

Property A8. Let $y(t)$ be related to $x(t)$ through the product $y(t) = h(t) x(t)$. Then

$$A_y(\nu, \tau) = \frac{1}{2\pi} \int_{-\infty}^{\infty} A_h(\nu - \theta, \tau) A_x(\theta, \tau) d\theta. \quad (12.4.12)$$

Property A9. The AF of the Fourier transform of $y(t)$ is related to the AF of $y(t)$ through

$$A_Y(\nu, \tau) = 2\pi A_y(\tau, -\nu). \quad (12.4.13)$$

The role of the ambiguity function in radar theory is discussed in detail in [Rihaczek, 1985]. Other than radar (and active sonar, which is similar in principle to radar), its applications to signal analysis are limited. Gardner [1987, and the references within] has introduced the concept of *cyclic autocorrelation function*

(CAF), which is similar to the ambiguity function. However, while the ambiguity function is defined for finite energy (L_2) signals, the CAF is defined for a class of signals with infinite energy, but finite instantaneous power. The formal definition of the CAF is

$$\text{CAF}\{y(t)\} = R_y(\nu, \tau) = \lim_{T \to \infty} \frac{1}{T} \int_{-T/2}^{T/2} y(t + 0.5\tau) y^*(t - 0.5\tau) e^{-j\nu t} dt, \quad (12.4.14)$$

provided the limit exists and is not identically zero [note that the limit is identically zero if $y(t)$ is an L_2 function]. The CAF is useful for the treatment of *cyclostationary signals*. This topic will not be pursued here; see Gardner [1987] for details.

12.5. THE COHEN CLASS OF DISTRIBUTIONS

Cohen [1966] has proposed a general class of distributions of the Wigner-Ville type. All the distributions mentioned so far in this chapter, as well as many others, are special cases of this class. Other distributions in this class will be mentioned later. An important feature of Cohen distributions is that their properties can be easily inferred from a single entity: the kernel function of the distribution.

The general definition of the Cohen distributions is

$$C_y(t, \omega; \phi) = \frac{1}{2\pi} \int_{-\infty}^{\infty} \int_{-\infty}^{\infty} \int_{-\infty}^{\infty} \phi(\nu, \tau) y(u + 0.5\tau) y^*(u - 0.5\tau) e^{j(\nu u - \nu t - \omega \tau)} du \, d\nu \, d\tau.$$
$$(12.5.1)$$

In general, the kernel $\phi(\nu, \tau)$ is allowed to depend on t and ω, as well as on the signal $y(t)$. However, in many important special cases it depends only on ν and τ.

Comparing (12.5.1) to (12.4.2), we see that the integral over u in (12.5.1) gives the ambiguity function of $y(t)$, so the Cohen distributions can also be written as

$$C_y(t, \omega; \phi) = \frac{1}{2\pi} \int_{-\infty}^{\infty} \int_{-\infty}^{\infty} \phi(\nu, \tau) A_y(\nu, \tau) e^{-j(\nu t + \omega \tau)} d\nu \, d\tau. \quad (12.5.2)$$

Therefore, the distribution $C_y(t, \omega; \phi)$ is obtained by multiplying the ambiguity function of the signal by the kernel and taking the two-dimensional Fourier transform of the result. An equivalent procedure is to compute the WVD of the signal, take its two-dimensional inverse Fourier transform, multiply by the kernel, and take the two-dimensional Fourier transform of the result. This immediately implies that, in the special case where the kernel is independent of t and ω, the Cohen distribution is just the smoothed WVD:

$$C_y(t, \omega; \phi) = \frac{1}{4\pi^2} \int_{-\infty}^{\infty} \int_{-\infty}^{\infty} \Phi(\theta, \nu) W_y(t - \theta, \omega - \nu) d\nu \, d\theta, \quad (12.5.3)$$

where $\Phi(\theta, \nu)$ is the two-dimensional Fourier transform of the kernel.

All the distributions mentioned so far are members of the Cohen class, as follows.

- The WVD corresponds to $\phi(\nu,\tau) = 1$.
- The Rihaczek distribution corresponds to $\phi(\nu,\tau) = e^{j\nu\tau/2}$. This is an immediate consequence of Prob. 12.10.
- The Page and Levin distributions correspond to $\phi(\nu,\tau) = e^{\pm j0.5\nu|\tau|}$, respectively (Prob. 12.13).
- The PWD, SPWD, and SWD are all members of the Cohen class because they are of the form (12.5.3). The kernels $\phi(\nu,\tau)$ are obtained as the two-dimensional inverse Fourier transforms of the corresponding $\Phi(t,\omega)$. In particular, the spectrogram is a member of the Cohen class, whose kernel is the ambiguity function of the window.

Some of the properties 1 through 9 of the WVD apply to other members of the Cohen class. It turns out that the validity of each property can be tested by inspecting the kernel $\phi(\nu,\tau)$, as stated in the following theorem, whose nine parts are in one to one correspondence with Properties W1 through W9 in Sec. 12.2.

Theorem 12.2.

(a) $C_y(t,\omega;\phi)$ is real valued if and only if $\phi(\nu,\tau) = \phi^*(-\nu,-\tau)$.

(b) If $\phi(\nu,\tau)$ is independent of t, $C_y(t,\omega;\phi)$ is translation invariant in time; if $\phi(\nu,\tau)$ is independent of ω, $C_y(t,\omega;\phi)$ is translation invariant in frequency.

(c) (1) $C_y(t,\omega;\phi)$ satisfies the time-marginal property if and only if $\phi(\nu,0) = 1$ for all ν; (2) it satisfies the frequency-marginal property if and only if $\phi(0,\tau) = 1$ for all τ.

(d) (1) $C_y(t,\omega;\phi)$ satisfies the instantaneous frequency property if and only if (c1) holds and $\partial\phi(\nu,\tau)/\partial\tau|_{\tau=0} = 0$ for all ν; (2) it satisfies the group delay property if and only if (c2) holds and $\partial\phi(\nu,\tau)/\partial\nu|_{\nu=0} = 0$ for all τ.

(e) $C_y(t,\omega;\phi)$ satisfies the finite-time-support property if and only if $\int_{-\infty}^{\infty} \phi(\nu,\tau) e^{-j\nu t} d\nu = 0$ for $|t/\tau| > 0.5$; it satisfies the finite-frequency-support property if and only if $\int_{-\infty}^{\infty} \phi(\nu,\tau) e^{-j\omega\tau} d\tau = 0$ for $|\omega/\nu| > 0.5$.

(f) $C_y(t,\omega;\phi)$ satisfies the Moyal formula if and only if $|\phi(\nu,\tau)| = 1$.

(g) $C_y(t,\omega;\phi)$ satisfies the convolution property if and only if $\phi(\nu,\tau_1 + \tau_2) = \phi(\nu,\tau_1)\phi(\nu,\tau_2)$.

(h) $C_y(t,\omega;\phi)$ satisfies the multiplication property if and only if $\phi(\nu_1 + \nu_2,\tau) = \phi(\nu_1,\tau)\phi(\nu_2,\tau)$.

(i) $C_y(t,\omega;\phi)$ satisfies the Fourier transform property if and only if $\phi(\nu,\tau) = \phi(-\tau,\nu)$ for all τ and ν.

Proof.

(a) $C_y(t,\omega;\phi)$ is real if and only if its Fourier transform $(2\pi)^{-1}\phi(\nu,\tau)A_y(\nu,\tau)$ is conjugate symmetric. We have already seen that $A_y(\nu,\tau)$ is conjugate symmetric, so the distribution is real if and only if $\phi(\nu,\tau)$ is conjugate symmetric.

(b) This is an immediate consequence of (12.5.3).

(c)

$$\frac{1}{2\pi}\int_{-\infty}^{\infty} C_y(t,\omega;\phi)d\omega =$$

$$\frac{1}{4\pi^2}\int_{-\infty}^{\infty}\int_{-\infty}^{\infty}\int_{-\infty}^{\infty}\int_{-\infty}^{\infty}\phi(\nu,\tau)y(u+0.5\tau)y^*(u-0.5\tau)e^{j(\nu u-\nu t-\omega\tau)}du d\nu\, d\tau\, d\omega$$

$$=\frac{1}{2\pi}\int_{-\infty}^{\infty}\phi(\nu,0)|y(u)|^2 e^{j\nu(u-t)}du\, d\nu.$$

This is equal to $|y(t)|^2$ if and only if $\phi(\nu,0)=1$ for all ν. The dual property is proved similarly.

(d) This follows from (12.4.9) in a straightforward manner upon replacing $A_y(\nu,\tau)$ by $\phi(\nu,\tau)A_y(\nu,\tau)$ (Prob. 12.14).

(e) The proof relies on Prob. 12.16; see the definitions of $\psi(t,\tau)$, $r_y(t,\tau)$ and $\bar{r}_y(t,\tau)$ there. $r_y(t,\tau)$ is nonzero for $t_1+0.5|\tau|\le t\le t_2-0.5|\tau|$. $C_y(t,\omega;\phi)$ is zero for $t\notin[t_1,t_2]$ if and only if the same holds for $\bar{r}_y(t,\tau)$. But, due to the convolution relationship between $r_y(t,\tau)$ and $\bar{r}_y(t,\tau)$, this holds if and only if $\psi(t,\tau)$ is zero for $t\notin[-0.5|\tau|,0.5|\tau|]$. The dual property is proved similarly.

(f) Since $(2\pi)^{-1}\phi(\nu,\tau)A_y(\nu,\tau)$ is the inverse Fourier transform of $C_y(t,\omega;\phi)$, we have, by Parseval's theorem,

$$(2\pi)^{-1}\langle C_x,C_y\rangle = (2\pi)^{-1}\langle\phi A_x,\phi A_y\rangle.$$

By the Moyal formula (12.4.10) for the AF, we have $(2\pi)^{-1}\langle A_x,A_y\rangle = |\langle x,y\rangle|^2$. Therefore, $(2\pi)^{-1}\langle C_x,C_y\rangle = |\langle x,y\rangle|^2$ for *arbitrary signals* if and only if $|\phi(\nu,\tau)|=1$.

(g) We have by (12.5.2)

$$\int_{-\infty}^{\infty} C_h(t-\theta,\omega;\phi)C_x(\theta,\omega;\phi)d\theta = \frac{1}{4\pi^2}\int_{-\infty}^{\infty}\int_{-\infty}^{\infty}\int_{-\infty}^{\infty}\int_{-\infty}^{\infty}\int_{-\infty}^{\infty}\phi(\nu_1,\tau_1)$$

$$\phi(\nu_2,\tau_2)A_h(\nu_1,\tau_1)A_x(\nu_2,\tau_2)e^{-j(\nu_1 t-\nu_1\theta+\omega\tau_1+\nu_2\theta+\omega\tau_2)}d\tau_1\, d\tau_2\, d\nu_1\, d\nu_2\, d\theta =$$

$$\frac{1}{2\pi}\int_{-\infty}^{\infty}\int_{-\infty}^{\infty}\int_{-\infty}^{\infty}\phi(\nu,\tau_1)\phi(\nu,\tau_2)A_h(\nu,\tau_1)A_x(\nu,\tau_2)e^{-j(\nu t+\omega\tau_1+\omega\tau_2)}d\tau_1\, d\tau_2\, d\nu$$

$$=\frac{1}{2\pi}\int_{-\infty}^{\infty}\int_{-\infty}^{\infty}\int_{-\infty}^{\infty}\phi(\nu,\sigma)\phi(\nu,\tau-\sigma)A_h(\nu,\sigma)A_x(\nu,\tau-\sigma)e^{-j(\nu t+\omega\tau)}$$

$$d\sigma\, d\tau\, d\nu.$$

This is equal to $C_y(\theta,\omega;\phi)$ if and only if

$$\int_{-\infty}^{\infty}\phi(\nu,\sigma)\phi(\nu,\tau-\sigma)A_h(\nu,\sigma)A_x(\nu,\tau-\sigma)d\sigma = \phi(\nu,\tau)A_y(\nu,\tau).$$

However, we have from (12.4.11),

$$\int_{-\infty}^{\infty}A_h(\nu,\sigma)A_x(\nu,\tau-\sigma)d\sigma = A_y(\nu,\tau).$$

Therefore, equality holds if and only if $\phi(\nu, \sigma)\phi(\nu, \tau - \sigma) = \phi(\nu, \tau)$ for all τ, σ, and ν.

(h) The proof is the dual of the proof of (g).

(i) By (12.4.13), we have

$$
\begin{aligned}
C_Y(t, \omega) &= \frac{1}{2\pi} \int_{-\infty}^{\infty} \int_{-\infty}^{\infty} \phi(\nu, \tau) A_Y(\nu, \tau) e^{-j(\nu t + \omega \tau)} d\nu\, d\tau \\
&= \int_{-\infty}^{\infty} \int_{-\infty}^{\infty} \phi(\nu, \tau) A_y(\tau, -\nu) e^{-j(\nu t + \omega \tau)} d\nu\, d\tau \\
&= \int_{-\infty}^{\infty} \int_{-\infty}^{\infty} \phi(-\tau, \nu) A_y(\nu, \tau) e^{-j(-\tau t + \omega \nu)} d\nu\, d\tau.
\end{aligned}
$$

This is equal to $2\pi C_y(-\omega, t)$ if and only if $\phi(-\tau, \nu) = \phi(\nu, \tau)$ for all τ and ν.

∎

Despite the fact that Cohen distributions have been known since 1966, it took over 20 years before they were systematically applied to cross-term reduction in quadratic TFDs. The crucial observation in such an application is that, while the autoterms tend to be concentrated around the origin of the ambiguity function, the cross terms tend to occupy regions in the (ν, τ) plane far from the origin. This observation is due to Flandrin [1984]; see also Martin and Flandrin [1985]. A major contribution was made by Williams and his colleagues, and their work is summarized below.

An early application of interference reduction via the kernel $\phi(\nu, \tau)$ is due to Cohen [1966], who exploited an idea of Born and Jordan [1925] to propose the kernel

$$
\phi(\nu, \tau) = \frac{\sin 0.5\nu\tau}{0.5\nu\tau}. \tag{12.5.4}
$$

The distribution based on this kernel is known as the *Born-Jordan distribution*.

Choi and Williams [1989] proposed the kernel

$$
\phi(\nu, \tau) = e^{-\nu^2 \tau^2 / \sigma}, \tag{12.5.5}
$$

where σ is a parameter controlling the effective width of the kernel in the (ν, τ) plane. This kernel, when multiplying $A_y(\nu, \tau)$, will attenuate its away-from-origin regions, thus attenuating the cross terms of the corresponding Cohen distribution. The parameter σ should ideally be chosen so as to preserve the autoterms of the distribution as much as possible. Comparing (12.5.5) to (12.5.4), we see that the Born-Jordan kernel can be explained by a similar argument (except that it does not have a width-controlling parameter).

Two alternative forms of Choi-Williams distribution are obtained by integrating the definition (12.5.1) with respect to ν or τ:

$$CW(t, \omega) =$$

$$\int_{-\infty}^{\infty} \int_{-\infty}^{\infty} \sqrt{\frac{\sigma}{4\pi\tau^2}} y(u + 0.5\tau) y^*(u - 0.5\tau) \exp\left\{-\frac{\sigma(u-t)^2}{4\tau^2}\right\} e^{-j\omega\tau} du\, d\tau =$$

$$\frac{1}{2\pi} \int_{-\infty}^{\infty} \int_{-\infty}^{\infty} \sqrt{\frac{\sigma}{4\pi\nu^2}} Y(u + 0.5\nu) Y^*(u - 0.5\nu) \exp\left\{-\frac{\sigma(u-\omega)^2}{4\nu^2}\right\} e^{-j\nu t} du\, d\nu.$$

$$(12.5.6)$$

Jeong and Williams [1992] have looked for an alternative to the exponential kernel (12.5.6) with similar (or better) cross-term reduction properties, but such that the support property W5 will be satisfied. They proposed the following procedure for kernel construction. The basic step is to choose a real-valued window function $h(t)$ with the following properties:

R1: $\int_{-\infty}^{\infty} h(t) = 1$.
R2: $h(t) = h(-t)$.
R3: $h(t) = 0$ for $|t| > 0.5$.
R4: $H(\omega)$, the Fourier transform of $h(t)$, is differentiable and has a low-pass characteristic; that is, $|H(\omega)| \ll 1$ for large $|\omega|$.

The kernel is taken as

$$\phi(\nu, \tau) = H(\nu\tau). \qquad (12.5.7)$$

The family of distributions thus obtained has been named the *reduced interference distributions* (RID) by Jeong and Williams. The following theorem summarizes their properties.

Theorem 12.3. The reduced interference distributions satisfy properties W1 through W5.

Proof. The proof relies on Theorem 12.2. W1 follows since $H(\omega)$ is real by R2. W2 is trivial. W3 follows since $H(0) = 1$ by R1. Furthermore, $H(\omega)$ is symmetric and differentiable, so its partial derivatives satisfy part (d) of Theorem 12.2. Therefore, W4 holds. Finally, by R3,

$$\int_{-\infty}^{\infty} \phi(\nu, \tau) e^{-j\nu t} d\nu = \frac{2\pi}{|\tau|} h\left(-\frac{t}{\tau}\right) = 0 \quad \text{for } |t| > 0.5|\tau|,$$

and similarly for $\int_{-\infty}^{\infty} \phi(\nu, \tau) e^{-j\omega\tau} d\tau$. Therefore, W5 holds by part (e) of Theorem 12.2. ∎

The Born-Jordan kernel (12.5.4) is clearly an RID obtained by taking $h(t)$ as the rectangular window function. However, its low-pass characteristic is rather mild. Many window functions used in classical digital signal processing (rather, continuous-time versions thereof) can be used for this purpose. Examples given by Jeong and Williams [1992] include the triangular window, generalized Hamming window, and truncated Gaussian.

We should finally remark that caution is advised when using reduced interference distributions (or the Choi-Williams distribution), since the separation of a signal to multiple components is arbitrary to a large extent. Clearly, there is an infinite number of ways to split a signal into multiple (or even two) components, so the interpretation of the distribution may not be easier than the interpretation of, say, the WVD with its cross terms. It makes sense to employ cross-term reduction only when the multiple components result from some physical mechanism, and they have sufficiently different characteristics (frequencies and/or timings) for the cross-term reduction to be effective.

12.6. HIGH-ORDER AMBIGUITY FUNCTIONS

General nonlinear time-frequency representations are much less common than quadratic distributions, and their theory is much less developed. This section is devoted to a specific nonlinear representation, which we term *high-order ambiguity function* (HAF). The high-order ambiguity function was developed by Peleg and Porat [1991a] for treating constant-amplitude, continuous-phase signals.[†]

To motivate the HAF, let us look again at Example 12.3 and assume that τ is fixed. Then $A(\nu, \tau)$, regarded as a function of ν, has its maximum at $\nu_0 = -2\mu\tau$. This observation can be used to find the value of the chirp parameter μ. Once μ is known, we can form the "reduced" signal $y(t)e^{-j\mu t^2} = e^{j\omega_0 t}$. The Fourier transform of the reduced signal then yields ω_0. In summary, the computation of the ambiguity function for a single τ followed by Fourier transform of the reduced signal yields the parameters of the chirp signal.

Now consider a general *constant-amplitude, polynomial-phase* signal

$$y(t) = b_0 \exp\left\{ j \sum_{m=0}^{M} a_m t^m \right\}, \quad -0.5T \leq t \leq 0.5T. \tag{12.6.1}$$

The HAF, to be introduced shortly, is aimed at treating signals of this type: estimating their parameters, classifying them according to their polynomial degree, separating multiple-component signals of this type, and more. The signal (12.6.1) is a special case of a general constant-amplitude signal $y(t) = b_0 e^{j\phi(t)}$, where $\phi(t)$ is a continuous function. Constant-amplitude signals are common in radar, commu-

---

[†] The original paper of Peleg and Porat referred to this function as the *polynomial transform* (PT). The author has chosen to introduce an alternative terminology for two reasons: (1) the HAF is not necessarily invertible (the original signal may not be recoverable from its HAF), so it does not strictly qualify as a "transform"; (2) the HAF is not a polynomial function of the given signal, so the name may be misleading. Nevertheless, the original notations $\mathcal{P}_M$ and P_M are retained.

nication, and active sonar applications, because they are relatively easy to generate and have a good power efficiency. Continuous phase is used in some cases (e.g., analog FM), while in other cases the phase is discontinuous (e.g., phase-shift keying or direct-sequence spread spectrum).

Let $y(t)$ be a complex-valued signal, and define for any integer q

$$y^{(*q)}(t) = \begin{cases} y(t), & q \text{ even} \\ y^*(t), & q \text{ odd}. \end{cases}$$

For any positive real number τ and any positive integer M, define

$$\mathcal{P}_M[y(t); \tau] = \prod_{q=0}^{M-1} [y^{(*q)}(t - q\tau)]^{\binom{M-1}{q}}. \tag{12.6.2}$$

$\mathcal{P}_M[y(t); \tau]$ is the fundamental operator in the HAF theory. For example, $M = 1, 2, 3$ gives

$$\mathcal{P}_M[y(t); \tau] = \begin{cases} y(t), & M = 1 \\ y(t)y^*(t - \tau), & M = 2 \\ y(t)[y^*(t - \tau)]^2 y(t - 2\tau), & M = 3. \end{cases}$$

The operator $\mathcal{P}_M[y(t); \tau]$ can also be defined inductively, as given by the following lemma.

Lemma 12.4.

$$\mathcal{P}_M[y(t); \tau] = \begin{cases} y(t), & M = 1 \\ \mathcal{P}_{M-1}[y(t); \tau] \{\mathcal{P}_{M-1}[y(t - \tau); \tau]\}^*, & M > 1. \end{cases} \tag{12.6.3}$$

The easy proof is left to the reader (Prob. 12.17). ∎

The Mth-order ambiguity function of $y(t)$ is defined as the Fourier transform of $\mathcal{P}_M[y(t); \tau]$; that is,

$$P_M[y; \omega, \tau] = \int_{-\infty}^{\infty} \mathcal{P}_M[y(t); \tau] e^{-j\omega t} dt. \tag{12.6.4}$$

Comparing this definition to (12.4.1), we see that for $M = 2$ it gives the non-symmetric ambiguity function, with $\omega = -\nu$. This explains the name "high-order ambiguity function" (the sign reversal of the frequency variable is for convenience).

The connection between the HAF and polynomial-phase signals is made in Theorems 12.4 and 12.5 below. Before we can state and prove these theorems, we need some lemmas, as follows.

Lemma 12.5.

$$\sum_{q=0}^{M-1} (-1)^q \binom{M-1}{q} q^k = \begin{cases} 0, & 0 \le k \le M - 2 \\ (-1)^{M-1}(M-1)!, & k = M - 1 \\ 0.5(-1)^{M-1}(M-1)M!, & k = M. \end{cases}$$

For a proof see Feller [1968, p. 58].   ∎

Lemma 12.6.

$$\sum_{q=0}^{M-1}(-1)^q\binom{M-1}{q}(t-q\tau)^m = \begin{cases} 0, & 0\le m\le M-2 \\ (M-1)!\tau^{M-1}, & m=M-1 \\ M!\tau^{M-1}t-0.5(M-1)M!\tau^M, & m=M. \end{cases}$$

The proof follows easily from Lemma 12.5, see Prob. 12.18.   ∎

Lemma 12.7. If $f(t)$ is differentiable up to order M, then

$$f^{(M)}(t) = \lim_{\tau\to 0}\frac{1}{\tau^M}\sum_{q=0}^{M}(-1)^q\binom{M}{q}f(t-q\tau). \qquad (12.6.5)$$

Proof. By induction. For $M=1$, the formula is obvious. Assume that it holds for M, and check for $M+1$.

$$f^{(M+1)}(t) = \lim_{\tau\to 0}\frac{f^{(M)}(t)-f^{(M)}(t-\tau)}{\tau}$$

$$= \lim_{\tau\to 0}\frac{1}{\tau^{M+1}}\left[\sum_{q=0}^{M}(-1)^q\binom{M}{q}f(t-q\tau)-\sum_{q=0}^{M}(-1)^q\binom{M}{q}f(t-(q+1)\tau)\right]$$

$$= \lim_{\tau\to 0}\frac{1}{\tau^{M+1}}\left[\sum_{q=0}^{M}(-1)^q\binom{M}{q}f(t-q\tau)+\sum_{p=1}^{M+1}(-1)^p\binom{M}{p-1}f(t-p\tau)\right]$$

$$= \lim_{\tau\to 0}\frac{1}{\tau^{M+1}}\left[\sum_{q=1}^{M}\left[(-1)^q\left\{\binom{M}{q}+\binom{M}{q-1}\right\}f(t-q\tau)\right]\right.$$

$$\left.+ f(t)+(-1)^{M+1}f(t-(M+1)\tau)\right]=$$

$$\lim_{\tau\to 0}\frac{1}{\tau^{M+1}}\left[\sum_{q=1}^{M}\left[(-1)^q\binom{M+1}{q}f(t-q\tau)\right]+f(t)+(-1)^{M+1}f(t-(M+1)\tau)\right]$$

$$= \lim_{\tau\to 0}\frac{1}{\tau^{M+1}}\sum_{q=0}^{M+1}(-1)^q\binom{M+1}{q}f(t-q\tau). \qquad ∎$$

We are now ready to prove the two main theorems.

Theorem 12.4. If $y(t)$ is the polynomial-phase signal given in (12.6.1), then

$$a_M = \frac{1}{M!\tau^{M-1}}\arg\max_{\omega}|P_M[y;\omega,\tau]|. \qquad (12.6.6)$$

Proof. We get, from Lemma 12.6,

$$\mathcal{P}_M[y(t);\tau] = \prod_{q=0}^{M-1}\left[b_0\exp\left\{j(-1)^q\sum_{m=0}^{M}a_m(t-q\tau)^m\right\}\right]^{\binom{M-1}{q}}$$

$$= b_0^{2^{M-1}}\exp\left\{j\sum_{m=0}^{M}a_m\sum_{q=0}^{M-1}(-1)^q\binom{M-1}{q}(t-q\tau)^m\right\}$$

$$= b_0^{2^{M-1}}\exp\{j[(M-1)!\tau^{M-1}a_{M-1} - 0.5M!(M-1)\tau^M a_M + M!\tau^{M-1}a_M t]\}$$

$$= b_0^{2^{M-1}}\exp\{j\psi\}\exp\{jM!\tau^{M-1}a_M t\}$$

where

$$\psi = (M-1)!\tau^{M-1}a_{M-1} - 0.5M!(M-1)\tau^M a_M.$$

Therefore,

$$P_M[y;\omega,\tau] = b_0^{2^{M-1}}T\exp\{j\psi\}\text{sinc}[0.5(M!\tau^{M-1}a_M - \omega)T];$$

hence (12.6.6) follows.   ∎

Theorem 12.5. Let $y(t)$ be a complex function, differentiable up to order $M-1$ and having constant amplitude. Assume that, for all $\tau > 0$, $\mathcal{P}_M[y(t);\tau]$ is a single tone of frequency $\omega_0(\tau) \neq 0$; that is,

$$\mathcal{P}_M[y(t);\tau] = B\exp\{j[\omega_0(\tau)t + \phi_0(\tau)]\}. \tag{12.6.7}$$

Assume further that

$$\lim_{\tau\to 0}\frac{\omega_0(\tau)}{\tau^{M-1}} \neq 0. \tag{12.6.8}$$

Then the phase of $y(t)$ is a polynomial of degree M; that is, $y(t)$ has the form shown in (12.6.1). The coefficient a_M is given by

$$a_M = \frac{1}{M!}\lim_{\tau\to 0}\frac{\omega_0(\tau)}{\tau^{M-1}}. \tag{12.6.9}$$

Proof. Since $y(t)$ has a constant amplitude, there exists a real function $f(t)$, differentiable to order $M-1$, and a real constant b_0, such that $y(t) = b_0\exp\{jf(t)\}$. Then, as in the proof of Theorem 12.4,

$$\mathcal{P}_M[y(t);\tau] = b_0^{2^{M-1}}\exp\left\{j\sum_{q=0}^{M-1}(-1)^q\binom{M-1}{q}f(t-q\tau)\right\}.$$

By assumption (12.6.7), we get $b_0 = B^{1/2^{M-1}}$ and

$$\sum_{q=0}^{M-1}(-1)^q\binom{M-1}{q}f(t-q\tau) = \omega_0(\tau)t + \phi_0(\tau).$$

Divide both sides by τ^{M-1} and let τ go to zero. By Lemma 12.7, the left side will converge to $f^{(M-1)}(t)$. Hence

$$f^{(M-1)}(t) = \lim_{\tau \to 0} \left\{ \frac{[\omega_0(\tau)t + \phi_0(\tau)]}{\tau^{M-1}} \right\}, \tag{12.6.10}$$

where the limit exists by assumption. Denote

$$f^{(M-1)}(t) = \left[\lim_{\tau \to 0} \frac{\omega_0(\tau)}{\tau^{M-1}} \right] t + \left[\lim_{\tau \to 0} \frac{\phi_0(\tau)}{\tau^{M-1}} \right] = \alpha_1 t + \alpha_0.$$

Since α_1 is assumed to be nonzero [cf. (12.6.8)], the differential equation (12.6.10) has the general solution

$$f(t) = \frac{\alpha_1}{M!} t^M + \frac{\alpha_0}{(M-1)!} t^{M-1} + \sum_{m=0}^{M-2} a_m t^m = \sum_{m=0}^{M} a_m t^m,$$

where

$$a_M = \frac{\alpha_1}{M!} = \frac{1}{M!} \lim_{\tau \to 0} \frac{\omega_0(\tau)}{\tau^{M-1}}. \qquad \blacksquare$$

Corollary. Under the assumptions of Theorem 12.5, a_M is given by

$$a_M = \frac{\omega_0(\tau)}{M! \tau^{M-1}} \tag{12.6.11}$$

for all $\tau > 0$.

Proof. This corollary follows directly from Theorem 12.4. It shows that the highest-order coefficient a_M can be computed from any single value of τ, not just as a limit as τ goes to zero. $\blacksquare$

Theorems 12.4 and 12.5, taken together, show that a constant-amplitude signal has a polynomial phase of degree M if and only if its Mth-order HAF has a spectral line at some nonzero frequency for any positive τ. This makes the HAF a natural tool for dealing with polynomial-phase signals. Furthermore, the constant-amplitude restriction can be relaxed. Suppose $y(t)$ has the form $y(t) = g(t)e^{jf(t)}$, where $f(t)$ is an Mth-order polynomial. Then it follows from (12.6.2) that

$$\mathcal{P}_M[y(t); \tau] = \mathcal{P}_M[g(t); \tau] \mathcal{P}_M[e^{jf(t)}; \tau],$$

so the HAF of $y(t)$ is a frequency domain convolution between the HAF of $g(t)$ and the spectral line of $\mathcal{P}_M[e^{jf(t)}; \tau]$. As long as $\mathcal{P}_M[g; \omega, \tau]$ has its maximum at frequency zero, the maximum of $\mathcal{P}_M[y; \omega, \tau]$ will be at the same frequency as if the amplitude were constant. This means that the HAF satisfies the property given in Theorem 12.4 for signals whose amplitude is not constant, provided that $\mathcal{P}_M[g; \omega, \tau]$ has its maximum at frequency zero.

The discrete-time version of the HAF is defined as

$$\mathcal{DP}_M[y_n;\tau] = \prod_{q=0}^{M-1} [y_{n-q\tau}^{(*q)}]^{\binom{M-1}{q}},\tag{12.6.12}$$

$$DP_M[y;\omega,\tau] = \sum_{n=-\infty}^{\infty} \mathcal{DP}_M[y_n;\tau]e^{-j\omega\Delta n},\tag{12.6.13}$$

where τ is a positive integer. For finite-duration signals, the sum in (12.6.13) is finite and the frequency can be discretized, as the in DFT operation. In this case the computation can be performed by FFT. It can be easily shown that the discrete HAF satisfies Theorem 12.4 (Prob. 12.19). However, Theorem 12.5 is satisfied only if $\tau = 1$ (Prob. 12.20).

12.7. ESTIMATION USING THE HIGH-ORDER AMBIGUITY FUNCTION

We now examine some uses of the HAF to signal estimation and classification. We limit ourselves to signals measured in additive white noise, as was done in Chapter 9 for sinusoidal signals. At the time this book is written, the HAF is at an active development stage, and very little is known about its application to more general signals. Also, since the main use of the HAF appears to be for radar and communication, it is convenient to work exclusively with complex signals. Complex signals are available whenever coherent (in phase and quadrature) demodulation is performed or by converting a real signal to its corresponding analytic signal.

First let us consider signals whose amplitude is constant and whose phase is a polynomial of *known* degree M. The discretized signal thus obeys the model

$$y_n = x_n + u_n = b_0 \exp\left\{ j \sum_{m=0}^{M} a_m \Delta^m n^m \right\} + u_n, \quad 0 \leq n \leq N-1.\tag{12.7.1}$$

Contrary to the model of Chapter 9, $\{u_n\}$ is now *complex* white Gaussian noise with zero mean and variance σ_u^2 (Prob. 2.21).

The following algorithm estimates the coefficients $\{a_m\}$ sequentially, starting at the highest-order coefficient a_M. At each step (except for the first), the effect of the phase term of the higher order is removed.

1. Substitute $m = M$ and $y_n^{(m)} = y_n, 0 \leq n \leq N-1$.
2. Choose a positive integer τ (which may vary for different values of m, more on its choice later). Compute $\hat{a}_m$ by

$$\hat{a}_m = \frac{1}{m!(\Delta\tau)^{m-1}} \arg\max_{\omega} |DP_m[y^{(m)};\omega,\tau]|.\tag{12.7.2}$$

3. Let
$$y_n^{(m-1)} = y_n^{(m)} \exp\{-j\hat{a}_m \Delta^m n^m\}, \quad 0 \le n \le N-1. \qquad (12.7.3)$$

Set $m \leftarrow m - 1$.

4. If $m \ge 1$, go back to step 2; else go to step 5.

5. Estimate a_0 from $y_n^{(0)}$ using the formula

$$\hat{a}_0 = \text{Im}\left\{\log\left[\sum_{n=0}^{N-1} y_n^{(0)}\right]\right\}. \qquad (12.7.4)$$

Remarks.

(a) The discrete transform DP_M can be computed on a discrete frequency grid using FFT. The maximum point can then be found by interpolation, as in frequency estimation from FFT.

(b) The formula (12.7.4) is based on the fact that $y_n^{(0)}$ is ideally given by the constant expression $b_0 \exp\{ja_0\}$ for all $0 \le n \le N-1$ ("ideally" referring to the case when $\{\hat{a}_m, m \ge 1\}$ are equal to their true values).

(c) The choice of τ in step 2 is arbitrary, but in practice it will affect the accuracy of the estimated coefficients. Lower values of τ will increase the number of terms in the HAF sum, but will decrease the resolution, as is clear from (12.7.2). Asymptotic performance analysis of the algorithm as a function of τ has been carried out by Peleg [1992]. He showed that (1) for $M = 2, 3$, the value $\tau = N/M$ gives the lowest asymptotic variance for $\hat{a}_M$; (2) for $4 \le M \le 10$, the value $\tau = N/(M+2)$ gives the lowest asymptotic variance for $\hat{a}_M$. The latter result is based on numerical evaluation of a certain function and was not rigorously proved for an arbitrary M. Also, the optimal choice of τ for $m < M$ has not been determined analytically. Nevertheless, numerous simulations have shown that the choice $\tau = N/m$ for the mth stage is nearly optimal.

(d) Due to the $2\pi/\Delta$ periodicity of (12.6.3), the coefficient a_m can be estimated unambiguously (i.e., is "unaliased") only if

$$|a_m| < \frac{\pi}{m!\Delta(\Delta\tau)^{m-1}}.$$

This requirement imposes a "Nyquist-like" restriction on the sampling interval Δ. It is not directly related to the common Nyquist rate, since the signal bandwidth depends on the coefficients $\{a_m\}$ in a complicated manner.

Next we consider a slightly different case, where the received signal is known to have a polynomial phase, but the degree M is unknown. However, an upper bound on M, say $\overline{M}$, is assumed to be known. As is clear from Theorems 12.4 and 12.5, application of DP_M to a signal whose phase degree is less than M would yield $a_M = 0$. This fact can be used to modify the above algorithm as follows.

1. Start with $m = \overline{M}$ and proceed as in the above algorithm. As long as $\hat{a}_m = 0$, do not perform the phase removal step (12.7.3). A criterion for deciding that $\hat{a}_m = 0$ follows below.

2. When m is reached such that $\hat{a}_m \neq 0$, continue the algorithm exactly as before, including the phase removal step.

This algorithm can be used to classify polynomial phase signals according to their degree, while simultaneously estimating their parameters. The decision that $\hat{a}_m = 0$ can be based on the Cramér-Rao lower bound. This bound is approximately given by [Peleg and Porat, 1991b]

$$\text{CRB}\{\hat{a}_m\} \approx \frac{(2m+1)\binom{2m}{m}^2}{2N^{(2m+1)}\Delta^{2m} \cdot \text{SNR}}. \tag{12.7.5}$$

We can therefore decide that $a_m = 0$ whenever $|\hat{a}_m|$ is below a certain (small) multiple of $[\text{CRB}\{\hat{a}_m\}]^{1/2}$.

Finally, we consider signals whose phase is continuous, but not necessarily polynomial. By Theorem 12.5, $\mathcal{P}_m[y(t); \tau]$ will not be a single tone for any m. However, by the Weierstrass theorem [Rudin, 1976], every continuous function on a closed interval can be uniformly approximated by polynomials. Hence we can expect that, for sufficiently large M_0, the approximation of the phase by $\sum_{m=0}^{M_0} a_m t^m$ will be satisfactory. The above algorithm can therefore be used to obtain polynomial phase approximations for signals having continuous phase on a closed interval. However, a criterion for determining the approximate degree M_0 is required. Such a criterion will follow from the subsequent analysis.

In the above algorithms, the amplitude was assumed constant. As was discussed in Sec. 12.6, the algorithms exhibit some tolerance to time-varying amplitude. Their tolerance to the noise variance will be explored now. This analysis, however, is limited to constant-amplitude signals. Also, we assume that $\tau = N/M$ for the Mth-order function. In this case the discrete HAF is given by

$$DP_M[y; \omega, \tau] = \sum_{n=(M-1)\tau}^{N-1} \left[\prod_{q=0}^{M-1} (y_{n-q\tau}^{(*q)})^{\binom{M-1}{q}} \right] e^{-j\omega n\Delta}. \tag{12.7.6}$$

We wish to derive the first two moments of $DP_M[y; \omega, \tau]$. First, since $E\{w_k^i\} = 0$ for all $i > 0$, we have (Prob. 12.21)

$$E\{y_k^L\} = E\{(x_k + w_k)^L\} = \sum_{i=0}^{L} \binom{L}{i} E\{w_k^i\} x_k^{L-i} = x_k^L, \quad L \geq 0. \tag{12.7.7}$$

Therefore,

$$E\{DP_M[y; \omega, \tau]\} = \sum_{n=(M-1)\tau}^{N-1} \left[\prod_{q=0}^{M-1} (x_{n-q\tau}^{(*q)})^{\binom{M-1}{q}} \right] e^{-j\omega n\Delta} = DP_M[x; \omega, \tau], \tag{12.7.8}$$

so $DP_M[y;\omega,\tau]$ is unbiased.

Next we have

$$|DP_M[y;\omega,\tau]|^2$$

$$= \sum_{n=(M-1)\tau}^{N-1} \sum_{k=(M-1)\tau}^{N-1} \left[\prod_{q=0}^{M-1} (y_{n-q\tau}^{(*q)})^{\binom{M-1}{q}}\right]\left[\prod_{p=0}^{M-1} (y_{k-p\tau}^{(*p)})^{\binom{M-1}{p}}\right]^* e^{-j\omega(n-k)\Delta}.$$

It is easy to see that, when $\tau = N/M$, the data points $y_{n-q\tau}$ and $y_{k-p\tau}$ are distinct, hence uncorrelated, for all $n \neq k$. Using this fact, we get

$$E\{|DP_M[y;\omega,\tau]|^2\}$$

$$= \sum_{n=(M-1)\tau}^{N-1} \sum_{k=(M-1)\tau}^{N-1} \left[\prod_{q=0}^{M-1} (x_{n-q\tau}^{(*q)})^{\binom{M-1}{q}}\right]\left[\prod_{p=0}^{M-1} (x_{k-p\tau}^{(*p)})^{\binom{M-1}{p}}\right]^* e^{-j\omega(n-k)\Delta}$$

$$+ \sum_{n=(M-1)\tau}^{N-1} \left[E\left\{\left[\prod_{q=0}^{M-1} (y_{n-q\tau}^{(*q)})^{\binom{M-1}{q}}\right]\left[\prod_{q=0}^{M-1} (y_{n-q\tau}^{(*q)})^{\binom{M-1}{q}}\right]^*\right\} - |x_n|^{2^M}\right]$$

$$= |DP_M[x;\omega,\tau]|^2 + b_0^{2^M} \sum_{n=(M-1)\tau}^{N-1} \left[E\left\{\left[\prod_{q=0}^{M-1}\left(1 + \frac{u_{n-q\tau}^{(*q)}}{x_{n-q\tau}^{(*q)}}\right)^{\binom{M-1}{q}}\right]\right.\right.$$

$$\left.\left.\cdot \left[\prod_{q=0}^{M-1}\left(1 + \frac{u_{n-q\tau}^{(*q)}}{x_{n-q\tau}^{(*q)}}\right)^{\binom{M-1}{q}}\right]^*\right\} - 1\right]. \tag{12.7.9}$$

Let ξ be a complex Gaussian random variable with zero mean and variance γ. Then (Prob. 12.21)

$$E\{(1+\xi)^L (1+\xi^*)^L\} = E\left\{\sum_{i=0}^{L}\sum_{k=0}^{L}\binom{L}{i}\binom{L}{k}\xi^i(\xi^*)^k\right\} = \sum_{i=0}^{L}\binom{L}{i}^2 i!\gamma^i. \tag{12.7.10}$$

Put $\xi = u_{n-q\tau}^{(*q)}/x_{n-q\tau}^{(*q)}$ and $\gamma = \sigma_u^2/b_0^2 = 1/\text{SNR}$ in (12.7.10), and substitute the result in (12.7.9). This gives

$$E\{|DP_M[y;\omega,\tau]|^2\}$$

$$= |DP_M[x;\omega,\tau]|^2 + \frac{N}{M}b_0^{2^M}\left\{\prod_{q=0}^{M-1}\left[\sum_{i=0}^{\binom{M-1}{q}}\binom{\binom{M-1}{q}}{i}^2 i!\left(\frac{1}{\text{SNR}}\right)^i\right] - 1\right\}$$

$$= |DP_M[x;\omega,\tau]|^2 + \frac{N}{M}b_0^{2^M} K(M,\text{SNR}), \tag{12.7.11}$$

where

$$K(M,\text{SNR}) = \prod_{q=0}^{M-1}\left[\sum_{i=0}^{\binom{M-1}{q}}\binom{\binom{M-1}{q}}{i}^2 i!\left(\frac{1}{\text{SNR}}\right)^i\right] - 1. \tag{12.7.12}$$

The formula (12.7.11) can be interpreted in the following manner. The first term is the energy density of the signal component, and the second is the energy density of the noise component. Thus, the noise energy is uniformly spread over the entire frequency range. The noise will limit the operating range of the algorithms given above. Specifically, the signal energy at the maximum point of the transform has to be sufficiently large with respect to the noise energy to guarantee detection of the true maximum and a reasonably accurate estimate of the frequency of the maximum point. In spectral analysis, it is commonly required that the ratio of the two energies be about 25 or more (that is, about 14 dB). In our case, this requirement can be expressed as

$$|DP_M[x; \omega, \tau]|^2 \geq 25 \frac{N}{M} b_0^{2^M} K(M, \mathrm{SNR}). \tag{12.7.13}$$

The criterion (12.7.13) can be used for determining the operating range of the above algorithms and for choosing the order M_0 in the case of nonpolynomial phase. Assume that both b_0 and σ_u^2 are known. The first algorithm assumes that the polynomial degree M is known. Then, since $|DP_M[x; \omega_0, \tau]|^2 = (N/M)^2 b_0^{2^M}$, we get in this case

$$MK(M, \mathrm{SNR}) \leq \frac{N}{25}. \tag{12.7.14}$$

This condition can be used to test whether the algorithm can be safely applied with given values of N, M, and SNR.

The second algorithm is similar to the first, so the condition (12.7.14) can be used, except that we substitute the upper bound $\overline{M}$ for M.

The case of nonpolynomial phase requires a slightly different treatment. Since the function $\mathcal{DP}_m[x_n; \tau]$ is not a single tone for any m, we do not have a priori knowledge of $\max_\omega |DP_M[x; \omega, \tau]|^2$. Instead, we can set a threshold as in (12.7.13) and compare the measured maximum with this threshold. The description of the algorithm can now be completed as follows. We start at $m = 1$, and for each m we test whether

$$\max_\omega |DP_m[y; \omega, \tau]|^2 \geq 25 \frac{N}{m} b_0^{2^m} K(m, \mathrm{SNR}). \tag{12.7.15}$$

As long as this condition holds, we increase m and repeat. When it fails, we set $M_0 = m - 1$. We then run the basic algorithm downward, starting at M_0, and estimate the phase coefficients. There is no guarantee that the value of M_0 thus found will result in a good approximation of the signal phase. The algorithm may succeed or fail, depending on the nature of the signal, the SNR, and the number of data points.

It is interesting to explore the behavior of the threshold function $K(M, \mathrm{SNR})$. Figure 12.1 shows the values of this function (expressed in decibels) for $0 \leq \mathrm{SNR} \leq 100$ dB and $2 \leq M \leq 8$. As can be seen, when M is 4 or more, there is a sharp increase of $K(M, \mathrm{SNR})$ below a certain SNR. In practice, choosing M such that $K(M, \mathrm{SNR})$ is 0 dB or less will guarantee (12.7.14) for any reasonable value of N. When the SNR is sufficiently large, $K(M, \mathrm{SNR})$ is approximately inversely proportional to the SNR. This approximation can be derived as follows.

$$K(M, \text{SNR}) \approx \prod_{q=0}^{M-1} \left[1 + \binom{M-1}{q}^2 \left(\frac{1}{\text{SNR}} \right) \right] - 1 \approx \frac{1}{\text{SNR}} \sum_{q=0}^{M-1} \binom{M-1}{q}^2$$

$$= \frac{1}{\text{SNR}} \binom{2M-2}{M-1}. \tag{12.7.16}$$

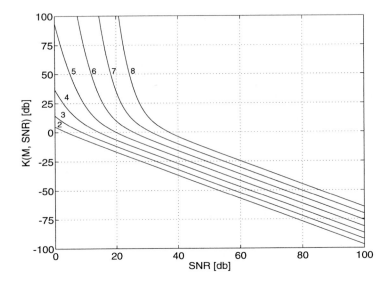

Figure 12.1. The threshold function $K(M, \text{SNR})$ for $2 \leq M \leq 8$.

Example 12.4. This example illustrates the use of the HAF for estimating the kinematic parameters of a moving target by pulse-Doppler radar. We assume the target moves at a constant radial acceleration during the observation interval, so its instantaneous range is given by

$$r(t) = r_0 + \dot{r}_0 t + 0.5 \ddot{r}_0 t^2,$$

where $r_0, \dot{r}_0$, and $\ddot{r}_0$ are, respectively, the range, radial velocity, and radial acceleration at the beginning of the observation interval. Let λ be the radar wavelength and Δ the pulse repetition interval (PRI). Assume once-per-pulse coherent in-phase/quadrature sampling. Then the equivalent discrete-time complex observation is given by

$$y_n = b_0 e^{j(a_0 + a_1 \Delta n + a_2 \Delta^2 n^2)} + u_n,$$

where

$$a_1 = -\frac{4\pi\dot{r}_0}{\lambda}, \quad a_2 = -\frac{2\pi\ddot{r}_0}{\lambda}.$$

For a derivation of this result, see Rihaczek [1985, p. 59].

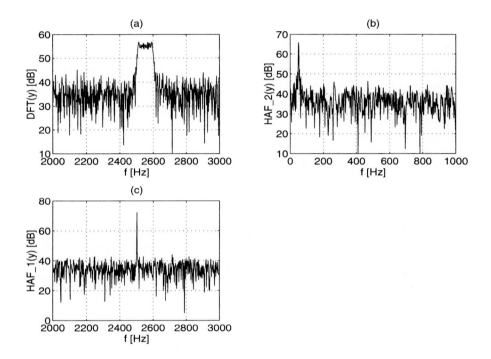

Figure 12.2. (a) The power spectrum of the radar signal in Example 12.4; (b) the second-order HAF of the signal; (c) the first-order HAF of the reduced signal.

As an example, suppose that the radar wavelength is $\lambda = 0.2$ m, and we wish to measure radial velocities in the range ± 400 m/sec. This implies a PRI of $(1/8000)$ sec. Suppose, further, that we wish to achieve a resolution of 0.2 m/sec; that is, the length of the FFT must be 4000 or more. For convenience, we choose $N = 4096$.

Next, assume that the target acceleration can lie in the range ± 40 m/sec^2. The maximum change of velocity during the observation interval is then about 20 m/sec, or about 100 FFT bins. Clearly, the velocity resolution requirement cannot be met with conventional Doppler processing. Figure 12.2(a) illustrates the magnitude of the DFT of the received signal, with $\dot{r}_0 = -250$ m/sec, $\ddot{r}_0 = -20$ m/sec^2, and SNR $= -3$ dB. The part shown in the figure corresponds to

2000 Hz $\leq f_{doppler} \leq$ 3000 Hz, or 200 m/sec $\leq -\dot{r}_0 \leq$ 300 m/sec. The Doppler stretch due to target acceleration is apparent.

Figure 12.2(b) shows the magnitude of the second-order HAF of the signal in this example, computed with $\tau = N/2$. The part shown in the figure corresponds to the frequency range $0 \leq f \leq 1000$ Hz, or $0 \leq -\ddot{r}_0 \leq 390.6$ m/sec^2. The spectral line corresponding to the linear FM caused by the target acceleration is clearly visible. The estimated value of $\ddot{r}_0$ was found to be -19.97 m/sec^2 in this case. Figure 12.2(c) shows the first-order HAF of the reduced signal $y_n^{(1)}$ in the same range as in Figure 12.2(a). Now the spectral line corresponding to the target velocity is free from the stretch seen in Figure 12.2(a). The estimated value of $\dot{r}_0$ was found to be -250.01 m/sec in this case.

The conclusion from this example is that the algorithm based on the HAF estimates the kinematic parameters very well and achieves the resolution requirements at a reasonable computational cost. ∎

12.8. MATHEMATICA PACKAGES

The package `Qtfd.m` contains the single procedure `Wvd[y_]`. This procedure computes the Wigner-Ville distribution of the data sequence `y` using Nuttal's method (12.3.12) (see Prob. 12.8). The number of points on each axis (time and frequency) is twice the number of points in `y`. Any member of the Cohen class can be computed using this procedure by multiplying the inverse two-dimensional Fourier transform of its output with the kernel and taking the two-dimensional Fourier transform of the product.

The package `Haf.m` implements the discrete-time high-order ambiguity function described in Sec. 12.7.

`POperator[data_, tau_, m_]` computes the operator $\mathcal{D}\mathcal{P}_M$ for the sequence data, of order m and delay tau.

`HAF[data_, tau_, delta_, m_, opts___]` computes DP_M for the sequence data, of order m and delay tau. It also estimates the coefficient a_m and removes from the data the highest-order term in preparation for the next HAF. `delta` is the sampling interval of the data. The estimation is done by the procedure `MaxDft[data_]` and consists of two steps. First, a simple interpolation formula is used based on a Hann-windowed DFT (this formula was taken from Parker and Stoneman [1986]). Then, if the option `Refine->True` is used, the estimate is refined by maximizing the DFT (this time without a window) via Newton-Raphson procedure. If we wish to reduce computation time (with a small sacrifice in accuracy), then we can skip the refinement stage by using the default `Refine->False`. The procedure returns {`Y`, `aest`, `datar`}, where `Y` is the HAF of `data`, `aest` is $\hat{a}_m$, and `datar` is the reduced data (after removing the highest-order term).

`PpEst[data_, delta_, m_, opts___]` performs complete estimation of the phase parameters of the polynomial-phase signal `data` by calling `HAF` a total of m times. `delta` is the sampling interval of the data. The delay parameter is chosen internally as $\tau_k = N/k$ for any order k (N being the length of `data`). The option `Refine` is the same as for `HAF`. The procedure returns `a`, the coefficients of the estimated phase polynomial.

`PpData[Nsam_, a_, delta_]` generates `Nsam` data points of a unit-amplitude polynomial-phase signal whose phase polynomial is `a`, and using the sampling interval `delta`.

`AMax[Nsam_, delta_, m_]` gives the maximum absolute values of the coefficients of the polynomial of a polynomial-phase signal such that the estimate obtained by the HAF will be unaliased. `Nsam` is the number of data samples, `delta` the sampling interval, and `m` the degree of the polynomial. The computation assumes that τ_k for the kth-order HAF is chosen as N/k. The procedure returns a list of the maximum permitted values of $\{|a_m|, \ldots, |a_1|\}$.

`Kfunc[m_, snr_]` computes the threshold function $K(M, \text{SNR})$ defined in (12.7.12).

REFERENCES

Bertrand, P., Izrar, B., Nguyen, V. T., and Feix, M. R., "Obtaining Non-negative Quantum Distribution Functions," *Phys. Lett.,* 94, pp. 415–417, 1983.

Born, M., and Jordan, P., "Zur Quantenmechanik," *Z. Phys.,* 34, pp. 858–888, 1925.

Choi, H. I., and Williams, W. J., "Improved Time-frequency Representation of Multicomponent Signals Using Exponential Kernels," *IEEE Trans. Acoustics, Speech, Signal Processing,* 37, pp. 862–871, 1989.

Claasen, T. A. C. M., and Mecklenbrauker, W. F. G., "The Wigner Distribution— A Tool for Time-frequency Signal Analysis; Part One: Continuous-time Signals," *Philips J. Res.,* 35, pp. 217–250, 1980a.

Claasen, T. A. C. M., and Mecklenbrauker, W. F. G., "The Wigner Distribution— A Tool for Time-frequency Signal Analysis; Part Two: Discrete-time Signals," *Philips J. Res.,* 35, pp. 276–300, 1980b.

Claasen, T. A. C. M., and Mecklenbrauker, W. F. G., "The Wigner Distribution—A Tool for Time-frequency Signal Analysis; Part Three: Relations with Other Time-frequency Signal Transformations," *Philips J. Res.,* 35, pp. 372–389, 1980c.

Cohen, L., "Generalized Phase-space Distribution Functions," *J. Math. Phys.,* 7, pp. 781–786, 1966.

Cohen, L., "Time-frequency Distributions—A Review", *Proc. IEEE,* 77, pp. 941–981, 1989.

Dvir, I., "Detection of Transient Signals by the Pseudo-Wigner Distribution," M.Sc. Research thesis, Technion, Haifa, 1990 (in Hebrew).

Feller, W., *An Introduction to Probability Theory and Its Applications,* Wiley, New York, 1968.

Flandrin, P., "Some Features of Time-frequency Representation of Multicomponent Signals," *Proc. Int. Conf. Acoustics, Speech, Signal Processing,* pp. 41.B.1–41.B.4, 1984.

Gardner, W. A., *Statistical Spectral Analysis: A Nonprobabilistic Theory,* Prentice Hall, Englewood Cliffs, NJ, 1987.

Hlawatsch, F., and Boudreaux-Bartels, G. F., "Linear and Quadratic Time-frequency Signal Representations," *IEEE Signal Processing Magazine,* 9, pp. 21–67, 1992.

Jeong, J., and Williams, W. J., "Kernel Design for Reduced Interference Distributions," *IEEE Trans. Signal Processing,* 40, pp. 402–412, 1992.

Kirkwood, J. G., "Quantum Statistics of Almost Classical Ensembles," *Phys. Rev.,* 44, pp. 31–37, 1933.

Levin, M. J., "Instantaneous Spectra and Ambiguity Functions," *IEEE Trans. Information Theory,* IT-13, pp. 95–97, 1967.

Martin, W., and Flandrin, P., "Wigner-Ville Spectral Analysis of Nonstationary Processes," *IEEE Trans. Acoustics, Speech, Signal Processing,* 33, pp. 1461–1470, 1985.

Moyal, J. E., "Quantum Mechanics as a Statistical Theory," *Proc. Cambridge Phil. Soc.,* 45, pp. 99–124, 1949.

Nuttal, A. H., "Alias-free Wigner Distribution Function and Complex Ambiguity Function for Discrete-time Samples," Technical Report TR 8533, *Naval Underwater Systems Center,* New London Laboratory, CT, 1989.

Page, C. H., "Instantaneous Power Spectra," *J. Appl. Phys.,* 23, pp. 103–106, 1952.

Parker, R., and Stoneman, S. A. T., "On the Use of Fast Fourier Transforms when High Frequency Resolution is Required," *J. Sound and Vibrations,* 104, pp. 75–79, 1986.

Peleg, S., Private communication, January 1992.

Peleg, S., and Porat, B., "Estimation and Classification of Signals with Polynomial Phase," *IEEE Trans. Information Theory,* 37, pp. 422–430, 1991a.

Peleg, S., and Porat, B., "The Cramér-Rao Lower Bound for Signals with Constant Amplitude and Polynomial Phase," *IEEE Trans. Signal Processing,* 39, pp. 749–752, 1991b.

Peleg, S., Porat, B., and Friedlander, B., "The Discrete Polynomial Transform, Its Properties and Applications," *Proc. 25th Asilomar Conf. on Circuits, Systems, Computers,* Pacific Grove, CA, 1991.

Peyrin, F., and Prost, R., "A Unified Definition for the Discrete-time, Discrete-frequency and Discrete-time/Frequency Wigner Distributions," *IEEE Trans. Acoustics, Speech, Signal Processing,* 34, pp. 858–867, 1986.

Rihaczek, A. W., *Principles of High-resolution Radar,* Peninsula Publishing, Los Altos, CA, 1985.

Rihaczek, A. W., "Signal Energy Distribution in Time and Frequency," *IEEE Trans. Information Theory,* IT-14, pp. 369–374, 1968.

Rudin, W., *Principles of Mathematical Analysis,* 3rd ed., McGraw-Hill, New York, 1976.

Ville, J., "Théorie et Applications de la Notion de Signal Analytique," *Cables et Transmission,* 2A, pp. 61–74, 1948.

Wigner, E. P., "On the Quantum Correction for Thermodynamic Equilibrium," *Phys. Rev.,* 40, pp. 749–759, 1932.

Woodward, P. M., *Probability and Information Theory with Application to Radar,* Pergamon Press, Oxford, England, 1953.

PROBLEMS

12.1 Use (12.2.1) and the definition of Fourier transform to prove the dual WVD formula (12.2.2).

12.2. Derive the WVD of the signal

$$y(t) = e^{-0.5\gamma t^2} e^{j(\omega_0 t + \mu t^2)}$$

(this is a chirp modulated by a Gaussian envelope). Show that the WVD is positive in this case.

12.3. Verify the expressions in Example 12.2.

12.4. Show that the SWD with $\Phi(t,\omega) = (0.5\pi t_0\omega_0)^{-1/2} \exp\{-(t/t_0)^2 - (\omega/\omega_0)^2\}$ is nonnegative in the special case $t_0\omega_0 = 1$. Hint: Bring (12.3.5) to a form $|\int_{-\infty}^{\infty} \cdots|^2$.

12.5. Consider the spectrogram as a smoothed WVD, and assume that the window function $\gamma(t)$ is normalized to $\int_{-\infty}^{\infty} |g(t)|^2 dt = 1$. Compute the marginals of the distribution according to (12.2.5). Show that the marginal property cannot be satisfied for any window function.

12.6. Show that (12.3.9) coincides with (12.3.7) (up to a scale factor) when n is even (i.e., at the sampling points).

12.7. Explain how the discrete-time signal $\{y_n\}$ can be reconstructed (up to a constant-phase term) from the Peyrin-Prost WVD (12.3.9).

12.8. Suppose we are given a finite number of discrete-time signal samples $\{y_n, 0 \leq n \leq N-1\}$. The signal is assumed to have been sampled at Nyquist rate or higher. The following algorithm is the discrete-time, discrete-frequency version of Nuttal's WVD (12.3.12).

$$\tilde{Y}_k = \sum_{n=0}^{N-1} y_n e^{-j\pi nk/N}, \quad -N \leq k \leq N-1$$

$$W_y(0.5n, k) = \frac{T}{N} \sum_{\ell=-\ell_0}^{\ell_0} \tilde{Y}_{k+\ell} \tilde{Y}^*_{k-\ell} e^{j\pi \ell n/N}, \ 0 \leq n \leq 2N-1, \ -N \leq k \leq N-1,$$

where

$$\ell_0 = \begin{cases} N-k-1, & k \geq 0 \\ N+k, & k < 0. \end{cases}$$

Derive this algorithm from (12.3.12), and justify the choices of the time and frequency sampling points and the range of summation.

12.9 Use (12.4.2) and the definition of Fourier transform to prove the dual ambiguity function formula (12.4.3).

12.10. Show that the nonsymmetric ambiguity function (12.4.1) is, up to a constant factor, the two-dimensional inverse Fourier transform of the Rihaczek distribution (12.2.13)

12.11. Derive the approximation $\Delta\tau \approx \pi/\mu T$ when $\mu T^2 \gg 4\pi$ in Example 12.3.

12.12. Prove properties A1 through A9 of the ambiguity function.

12.13. Show that the Page and Levin distributions are members of the Cohen class with $\phi(\nu, \tau) = e^{\pm j0.5\nu|\tau|}$, respectively.

12.14. Elaborate the proof of part (d) of Theorem 12.2.

12.15. Determine which of the properties W1 through W9 is satisfied by (a) the spectrogram; (b) the pseudo-Wigner distribution; (c) the Choi-Williams distribution.

12.16. Let $r_y(t, \tau) = y(t + 0.5\tau)y^*(t - 0.5\tau)$ [this is called the *instantaneous correlation* of $y(t)$]. Let $\psi(t, \tau) = \int_{-\infty}^{\infty} \phi(\nu, \tau)e^{-j\nu t}d\nu$ and $\bar{r}_y(t, \tau) = (2\pi)^{-1} \int_{-\infty}^{\infty} r_y(u, \tau)\psi(t - u, \tau)du$. Show that the Cohen distribution can be expressed as $C(t, \omega; \phi) = \int_{-\infty}^{\infty} \bar{r}_y(t, \tau)e^{-j\omega\tau}d\tau$.

12.17. Prove Lemma 12.4 (use induction on M).

12.18. Prove Lemma 12.6 using Lemma 12.5.

12.19. State and prove the analog of Theorem 12.4 for the discrete-time HAF.

12.20. Prove that, if the HAF of a constant-amplitude discrete-time signal at $\tau = 1$ is a single tone $B \exp\{j(\omega_0 n + \phi_0)\}$, the signal has polynomial phase of degree M. Find the relationship between ω_0 and the highest-order coefficient of the phase polynomial. Hint: Let the signal phase be f_n. Show that f_n satisfies a difference equation of order $M - 1$ whose characteristic polynomial is $(z - 1)^{M-1}$. Find the right side of the difference equation. Then show that the solution of the difference equation is an Mth-order polynomial and find its highest-order coefficient. See Peleg, Porat, and Friedlander [1991] for more details.

12.21. A standard complex Gaussian random variable Z is defined as $Z = X + jY$, where X and Y are mutually independent Gaussian random variables, each having zero mean and variance 0.5. The probability distribution of Z is, by definition, the joint distribution of X and Y. Thus $f(Z) = \pi^{-1} \exp\{-(X^2 + Y^2)\}$. Prove the following expressions for a standard complex Gaussian random variable Z:

(a) $E\{Z^k\} = 0$ for any positive integer k.

(b) $E\{Z^k (Z^*)^\ell\} = 0$ for any nonnegative distinct integers k and ℓ.

(c) $E\{Z^k (Z^*)^k\} = k!$ for any nonnegative integer k. In particular, the variance of Z, defined as $E\{ZZ^*\}$, is equal to 1.

Hint: Use the material in Chapter 10, for example, the moment generating function or the cumulants-to-moments formula.

APPENDIX A

Notations and Facts

A1. We use the term "vector" to denote a column vector. In rare cases where we need row vectors, we will explicitly refer to them as such. For a complex number a, a^* will denote the complex conjugate of a. For a real or complex matrix (or vector) A, A^T will denote the transpose of A, while A^H will denote the complex conjugate transpose (or *Hermitian transpose*) of A.

A2. The determinant of a square matrix A will be denoted by $|A|$ (not to be confused with the same notation for the absolute value of a scalar).

A3. The Kronecker delta function is denoted by $\delta(n)$. The variable n is an integer and $\delta(0) = 1$, $\delta(n) = 0$ for $n \neq 0$.

A4. We have deviated here from the standard definition of $L_2(\Omega, \mathcal{A}, P)$, which requires only $E(X^2) < \infty$ and does not require the random variables to have $E(X) = 0$. It is not difficult to infer the completeness of our restricted L_2 from that of the "standard" L_2.

A5. The square root of -1 will be denoted by j.

A6. The integrals in this book are Lebesgue integrals, except where indicated. Of course, in many cases they coincide with the usual Riemann integrals.

A7. The z transform of a doubly infinite sequence $\{a_n, n \in \mathbf{Z}\}$ is defined as

$$a(z) = \sum_{n=-\infty}^{\infty} a_n z^{-n}. \tag{a.1}$$

The region of convergence of $a(z)$ is the set of values of z for which $\sum_{n=-\infty}^{\infty} |a_n z^{-n}| < \infty$. It follows from the Cauchy-Hadamard formula that the region of convergence is the open annulus

$$\limsup_{n \to \infty} |a_n|^{1/n} < |z| < \liminf_{n \to \infty} |a_{-n}|^{-1/n}.$$

The inverse z transform is given by

$$a_n = \frac{1}{2\pi j} \oint a(z)z^{n-1}dz, \tag{a.2}$$

where the integration is over a closed contour lying in the interior of the region of convergence of $a(z)$, whose index relative to the point $z = 0$ is 1.

A8. If f is a convex function on $\boldsymbol{R}^1$ and the expectations of both X and $f(X)$ exist, then

$$f(E(X)) \le E(f(X)). \tag{a.3}$$

This is Jensen's inequality. It holds for ordinary integrals (in place of expectations) as well. For a proof see Chung [1974, p. 47].

A9. The Fourier transform of the sequence $\{a_n, n \in \mathbf{Z}\}$ will be defined in agreement with (a.1); that is,

$$a(e^{j\omega}) = \sum_{n=-\infty}^{\infty} a_n e^{-j\omega n}. \tag{a.4}$$

The inverse Fourier transform is, therefore,

$$a_n = \frac{1}{2\pi} \int_{-\pi}^{\pi} a(e^{j\omega})e^{j\omega n}d\omega. \tag{a.5}$$

The Fourier transform exists for any square summable sequence, where the convergence of (a.4) is in the l_2 sense (see App. B).

A10. Let $\{s_n\}$ be a sequence such that $\lim_{n\to\infty} s_n = s$, where $|s| < \infty$. Let m_n be the sequence of partial arithmetic means $m_n = (1/n)\sum_{i=1}^{n} s_n$. Then

$$\lim_{n\to\infty} m_n = s. \tag{a.6}$$

For a proof see Apostol [1974, p. 206].

Let $\{a_n\}$ be a sequence such that $\sum_{n=0}^{\infty} a_n = s$, where $|s| < \infty$. Then

$$\lim_{n\to\infty} \sum_{i=0}^{n-1} \left(1 - \frac{i}{n}\right) a_i = s. \tag{a.7}$$

The left side is known as the *Cesàro sum* of the a_n. Equation (a.7) follows from (a.6) by taking $s_n = \sum_{i=0}^{n-1} a_i$.

If the sequence is not only summable, but satisfies the stronger condition $\sum_{n=0}^{\infty} n|a_n| = b < \infty$, then

$$\left| s - \sum_{i=0}^{n-1} \left(1 - \frac{i}{n}\right) a_i \right| \le \frac{b}{n}. \tag{a.8}$$

This is proved as follows.

$$n\left| s - \sum_{i=0}^{n-1} \left(1 - \frac{i}{n}\right) a_i \right| = \left| \sum_{i=0}^{n-1} i a_i + \sum_{i=n}^{\infty} n a_i \right| \le \sum_{i=0}^{n-1} i|a_i| + \sum_{i=n}^{\infty} n|a_i|$$

$$\leq \sum_{i=0}^{n-1} i|a_i| + \sum_{i=n}^{\infty} i|a_i| = b.$$

A11. The Dirac delta function will be denoted by $\delta_D(t)$.

A12. The identity matrix of dimensions $n \times n$ will be denoted by $\mathbf{1}_n$.

A13. A square root of a symmetric positive definite matrix R is any matrix X satisfying $XX^T = R$. Every symmetric positive definite matrix has a unique square root that (1) is lower triangular and (2) has positive elements on the main diagonal. This square root is called the *Cholesky factor* of R. We will use $R^{1/2}$ to denote any square root of R. We will also use the notations

$$R^{T/2} = (R^{1/2})^T, \quad R^{-1/2} = (R^{1/2})^{-1}, \quad R^{-T/2} = (R^{-1/2})^T.$$

A14. The notation $A \geq B$, where both matrices are symmetric, means that $A - B$ is positive semidefinite. The notation $A > B$ means that $A - B$ is positive definite. The following theorem is used in Chapter 5: Let A and B be positive definite; then $A \geq B$ implies $B^{-1} \geq A^{-1}$. For a proof, see Horn and Johnson [1985, p. 471].

A15. If the equality

$$\frac{\partial}{\partial \theta} \int f(Y, \theta) dY = \int \frac{\partial f(Y, \theta)}{\partial \theta} dY \tag{a.9}$$

holds, we refer to (a.9) as *differentiation under the integral sign*. There are several sufficient conditions for differentiation under the integral sign, depending on the support of $f(Y, \theta)$. A condition adequate for the purpose of this book is as follows. Assume that both sides of (a.9) exist and that for every $\theta_0 \in \Theta$ there exist $\rho(\theta_0) > 0$ and $g_{\theta_0}(Y)$ such that

$$\left| \frac{\partial f(Y, \theta)}{\partial \theta} \right| \leq g_{\theta_0}(Y) \text{ for all } \{\theta : \|\theta - \theta_0\| < \rho(\theta_0)\}$$

and

$$\int g_{\theta_0}(Y) dY < \infty.$$

Then (a.9) holds. This can be proved similarly to the proof of Theorem 10.39 in Apostol [1974, p. 283].

A16. Let $\alpha > 1$ and $\beta > 1$ be related by $\alpha^{-1} + \beta^{-1} = 1$. Hölder's inequality is

$$E|XY| \leq [E|X|^\alpha]^{1/\alpha} [E|Y|^\beta]^{1/\beta}. \tag{a.10}$$

See Billingsley [1986, p. 75] for a proof.

A17. The following simple identity is used throughout Chapter 4.

$$\sum_{t=0}^{N-1} \sum_{s=0}^{N-1} f(t, s) = \sum_{n=-(N-1)}^{-1} \sum_{m=0}^{N+n-1} f(m, m-n) + \sum_{n=0}^{N-1} \sum_{m=0}^{N-n-1} f(m+n, m). \tag{a.11}$$

If $f(t, s)$ is symmetric in t and s, (a.11) simplifies to

$$\sum_{t=0}^{N-1} \sum_{s=0}^{N-1} f(t, s) = \sum_{n=-(N-1)}^{N-1} \sum_{m=0}^{N-|n|-1} f(m + |n|, m). \qquad (a.12)$$

A18. For $0 < \alpha \le 1$, the Lipschitz condition is stronger than continuity, but weaker than differentiability. A differentiable function on a closed interval satisfies the Lipschitz condition of order 1.

A19. The following matrix identity holds for any nonsingular matrix A and vectors $\boldsymbol{u}$, $\boldsymbol{v}$ of the same dimension as A, provided the denominator on the right side is nonzero:

$$(A + \boldsymbol{u}\boldsymbol{v}^T)^{-1} = A^{-1} - \frac{A^{-1}\boldsymbol{u}\boldsymbol{v}^T A^{-1}}{1 + \boldsymbol{v}^T A^{-1}\boldsymbol{u}}. \qquad (a.13)$$

To prove it, multiply the right side by $(A + \boldsymbol{u}\boldsymbol{v}^T)$ and verify that it gives the identity matrix.

A20. The following matrix identities hold whenever all the inverses appearing in them exist:

$$\begin{bmatrix} A & B \\ C & D \end{bmatrix}^{-1} = \begin{bmatrix} A^{-1} & 0 \\ 0 & 0 \end{bmatrix} + \begin{bmatrix} -A^{-1}B \\ 1 \end{bmatrix} (D - CA^{-1}B)^{-1} \begin{bmatrix} -CA^{-1} & 1 \end{bmatrix}$$

$$(a.14)$$

$$\begin{bmatrix} A & B \\ C & D \end{bmatrix}^{-1} = \begin{bmatrix} 0 & 0 \\ 0 & D^{-1} \end{bmatrix} + \begin{bmatrix} 1 \\ -D^{-1}C \end{bmatrix} (A - BD^{-1}C)^{-1} \begin{bmatrix} 1 & -BD^{-1} \end{bmatrix}$$

$$(a.15)$$

The proof is by verification, as in [A19].

A21. An elementary fact from linear algebra (known as *Sylvester's inequality*): the rank of a product of two matrices cannot exceed the rank of either of the two matrices.

REFERENCES

Apostol, T. M., *Mathematical Analysis,* 2nd ed., Addison-Wesley, Reading, MA, 1974.

Billingsley, P., *Probability and Measure,* 2nd ed., Wiley, New York, 1986.

Chung, K. L., *A Course in Probability Theory,* Academic Press, New York, 1974.

Horn, R. A., and Johnson, C. R., *Matrix Analysis,* Cambridge University Press, Cambridge, 1985.

APPENDIX B

Hilbert Spaces

Let $\boldsymbol{R}$ denote the real line. Let $\mathcal{H}$ be a nonempty set. Let $+$ be a binary operation assigning to any two members of $\mathcal{H}$ a member of $\mathcal{H}$, which we call *addition*. Let $\cdot$ be a binary operation assigning to each member of $\boldsymbol{R}$ and to each member of $\mathcal{H}$ a member of $\mathcal{H}$. We call $\cdot$ *scalar multiplication*. A *real vector* (or *linear*) *space* is a triplet $(\mathcal{H}, +, \cdot)$, satisfying the following axioms.

ABELIAN GROUP AXIOMS

(1) $(u + v) + w = u + (v + w)$ (associativity).
(2) $u + v = v + u$ (commutativity).
(3) There exists an element $0 \in \mathcal{H}$ such that $0 + v = v$, $\quad \forall v \in \mathcal{H}$ (existence of additive identity).
(4) For every $v \in \mathcal{H}$ there exists an *inverse* $(-v) \in \mathcal{H}$ such that $v + (-v) = 0$ (existence of inverse).

MULTIPLICATION AXIOMS

(5) $\alpha \cdot (u + v) = \alpha \cdot u + \alpha \cdot v$ (distributivity).
(6) $(\alpha + \beta) \cdot u = \alpha \cdot u + \beta \cdot u$ (distributivity).
(7) $\alpha \cdot (\beta \cdot v) = (\alpha\beta) \cdot v$ (associativity).
(8) $1 \cdot v = v$ (existence of multiplicative identity).

The dot $\cdot$ will usually be omitted, as in the case of multiplication of scalars.

A function $\langle \cdot, \cdot \rangle$ assigning to any two members of $\mathcal{H}$ a member of $\boldsymbol{R}$ is called an *inner product*, if it satisfies the following axioms.

INNER PRODUCT AXIOMS

(9) $\langle u, v \rangle = \langle v, u \rangle$.
(10) $\langle u, u \rangle \geq 0$, with equality if and only if $u = 0$.
(11) $\langle \alpha u + \beta v, w \rangle = \alpha \langle u, w \rangle + \beta \langle v, w \rangle$.

A vector space with an inner product defined on its elements is called an *inner product space*.

The *norm* of a vector in an inner product space is defined as $\|u\| = \sqrt{\langle u, u \rangle}$.

Lemma B.1. For any $u, v \in \mathcal{H}$, we have
(a) $|\langle u, v \rangle| \leq \|u\| \cdot \|v\|$ (Cauchy-Schwarz inequality).
(b) $\|u + v\| \leq \|u\| + \|v\|$ (the triangle inequality).
(c) $\|u + v\|^2 + \|u - v\|^2 = 2\|u\|^2 + 2\|v\|^2$ (the parallelogram equality).

Proof.
(a) If $v = 0$, equality holds trivially. Otherwise,

$$0 \leq \left\langle u - \frac{\langle u, v \rangle}{\|v\|^2} v, u - \frac{\langle u, v \rangle}{\|v\|^2} v \right\rangle = \|u\|^2 - \frac{(\langle u, v \rangle)^2}{\|v\|^2}.$$

Hence,
$$|\langle u, v \rangle| \leq \|u\| \cdot \|v\|.$$

(b) By the Cauchy-Schwarz inequality,

$$\|u + v\|^2 = \|u\|^2 + 2\langle u, v \rangle + \|v\|^2 \leq \|u\|^2 + 2\|u\|\|v\| + \|v\|^2 = (\|u\| + \|v\|)^2.$$

(c) A straightforward computation. ∎

A sequence of vectors $\{v_n, 1 \leq n < \infty\}$ is said to *converge strongly* to a vector v if $\lim_{n \to \infty} \|v_n - v\| = 0$. Strong convergence will be denoted by either of the following:

$$\underset{n \to \infty}{\text{l.i.m.}} \, v_n = v, \quad v_n \to v.$$

The notation l.i.m. stands for *limit in the mean*. Weak convergence is also defined [Akhiezer and Glazman 1963, Sec. 23], but will not be needed here. Therefore, we will usually omit the adjective "strong."

Lemma B.2. The inner product is continuous in the following sense: if $u_n \to u$ and $v_n \to v$, then $\langle u_n, v_n \rangle \to \langle u, v \rangle$. As a special case, the norm is continuous; that is, if $u_n \to u$, then $\|u_n\| \to \|u\|$.

Proof. By the Cauchy-Schwarz inequality,

$$|\langle u, v \rangle - \langle u_n, v_n \rangle| = |\langle u - u_n, v \rangle + \langle u, v - v_n \rangle - \langle u - u_n, v - v_n \rangle|$$
$$\leq \|u - u_n\| \cdot \|v\| + \|u\| \cdot \|v - v_n\| + \|u - u_n\| \cdot \|v - v_n\|.$$

Taking the limit $n \to \infty$ yields zero on the right side, and hence on the left side. ∎

A vector v in an inner product space $\mathcal{H}$ is called a *limit point* of a subset $\mathcal{A}$ of $\mathcal{H}$ if there exists a sequence $\{v_n\}$ in $\mathcal{A}$ that converges to v. The *closure* of a set $\mathcal{A}$, denoted by $\overline{\mathcal{A}}$, is the union of $\mathcal{A}$ and the set of all its limit points. A set $\mathcal{A}$ is said to be *closed* if it contains all its limit points, that is, if it coincides with its closure.

Every subset $\mathcal{A}$ of $\mathcal{H}$ generates a subspace of $\mathcal{H}$ by taking all finite linear combinations of elements of $\mathcal{A}$. This subspace is called the *span* of $\mathcal{A}$ and is denoted by $Sp(\mathcal{A})$. This subspace is the smallest linear space containing all the members of $\mathcal{A}$. The closure of $Sp(\mathcal{A})$, denoted by $\overline{Sp}(\mathcal{A})$, is also a linear space, as can be shown by carrying out the necessary limit operations.

A sequence of vectors $\{v_n, 1 \le n < \infty\}$ is called a *Cauchy sequence* if it satisfies $\lim_{n,m \to \infty} \|v_m - v_n\| = 0$. It is easy to verify, by the triangle inequality, that a sequence that converges to a vector v is a Cauchy sequence, but the converse is not necessarily true. An inner product space is said to be *complete* if every Cauchy sequence in the space converges to a vector in the space.

A *Hilbert space* is a complete inner product space. Older texts used to restrict this definition to the case where the space is infinite dimensional. However, it is more convenient to include finite-dimensional spaces in the definition.

Some examples of Hilbert spaces are:

(1) The n-dimensional Euclidean space $\boldsymbol{R}^n$, for every $n \ge 1$.
(2) The space of all square summable real sequences, that is, sequences $x = \{x_n, 1 \le n < \infty\}$ satisfying $\sum_{n=1}^{\infty} x_n^2 < \infty$. This space is denoted by l_2. The inner product is defined as $\langle x, y \rangle = \sum_{n=1}^{\infty} x_n y_n$.
(3) The space of all real functions on some closed interval $[a, b]$ that are Lebesgue measurable and satisfy $\int_a^b f^2(x)dx < \infty$. This space is denoted by $L_2[a, b]$. The inner product is defined as $\langle f, g \rangle = (b-a)^{-1} \int_a^b f(x)g(x)dx$.
(4) The space of all real functions on the real line $\boldsymbol{R}$ that are Lebesgue measurable and satisfy $\int_{-\infty}^{\infty} f^2(x)dx < \infty$. This space is denoted by L_2. The inner product is defined as $\langle f, g \rangle = \int_{-\infty}^{\infty} f(x)g(x)dx$.

Let $\mathcal{H}$ be a Hilbert space and $\mathcal{A}$ a subset of $\mathcal{H}$. It is easy to check that $\overline{Sp}(\mathcal{A})$ is complete, hence it is a Hilbert subspace of $\mathcal{H}$. Conversely, every Hilbert subspace of $\mathcal{H}$ must be closed.

Two elements u, v of an inner product space are said to be *orthogonal* if $\langle u, v \rangle = 0$. We use the notation $u \perp v$ to indicate the orthogonality of the two vectors. A vector v is orthogonal to a set $\mathcal{A}$ (denoted by $v \perp \mathcal{A}$) if it is orthogonal to every element of the set. Two sets $\mathcal{A}$ and $\mathcal{B}$ are orthogonal (denoted by $\mathcal{A} \perp \mathcal{B}$) if every element of $\mathcal{A}$ is orthogonal to every element of $\mathcal{B}$.

The following theorem gives a fundamental property of Hilbert spaces.

Theorem B.1. Let $\mathcal{H}$ be a Hilbert space and $\mathcal{F}$ a subspace of $\mathcal{H}$. Let x be a fixed element of $\mathcal{H}$. Then there exists a unique element $\hat{x} \in \mathcal{F}$ such that

$$\|x - \hat{x}\| = \min_{f \in \mathcal{F}} \|x - f\|. \tag{b.1}$$

This element satisfies the orthogonality property

$$x - \hat{x} \perp \mathcal{F}. \tag{b.2}$$

Proof. If $x \in \mathcal{F}$, then $\hat{x} = x$, and the theorem is trivial. Therefore, let us assume that $x \notin \mathcal{F}$. Denote

$$\inf_{f \in \mathcal{F}} \|x - f\| = d \geq 0.$$

By the definition of infimum, there exists a sequence $\{f_n\}$ in $\mathcal{F}$ such that

$$\lim_{n \to \infty} \|x - f_n\| = d. \tag{b.3}$$

We want to show that $\{f_n\}$ is a Cauchy sequence. By the parallelogram equality,

$$\begin{aligned}
\|f_m - f_n\|^2 &= \|(x - f_n) - (x - f_m)\|^2 \\
&= 2\|x - f_n\|^2 + 2\|x - f_m\|^2 - 4\left\|x - \frac{1}{2}(f_n + f_m)\right\|^2 \\
&\leq 2\|x - f_n\|^2 + 2\|x - f_m\|^2 - 4d^2.
\end{aligned}$$

By (b.3), the right side tends to zero as $m, n \to \infty$. Hence $\{f_n\}$ is a Cauchy sequence and, since $\mathcal{F}$ is a Hilbert space, it converges to an element of $\mathcal{F}$. Denote this element by $\hat{x}$. We have

$$d \leq \|x - \hat{x}\| \leq \|x - f_n\| + \|\hat{x} - f_n\|.$$

As $n \to \infty$, the first term in the right side tends to d, while the second tends to zero. Hence $\|x - \hat{x}\| = d$; so $\hat{x}$ indeed satisfies (b.1).

To show the orthogonality property (b.2), let $f \in \mathcal{F}$. If $f = 0$, (b.2) holds trivially. Otherwise, define

$$g = \hat{x} + \frac{\langle x - \hat{x}, f \rangle}{\|f\|^2} f \in \mathcal{F}.$$

Then

$$d^2 \leq \|x - g\|^2 = d^2 - \frac{(\langle x - \hat{x}, f \rangle)^2}{\|f\|^2}.$$

Hence, necessarily, $\langle x - \hat{x}, f \rangle = 0$.

It remains to show uniqueness of $\hat{x}$. Suppose that there is another element $\hat{x}' \in \mathcal{F}$ satisfying (b.1). Then both $\hat{x}$ and $\hat{x}'$ satisfy the orthogonality property (b.2). Hence,

$$\|\hat{x} - \hat{x}'\|^2 = \langle \hat{x} - \hat{x}', x - \hat{x}' \rangle - \langle \hat{x} - \hat{x}', x - \hat{x} \rangle = 0.$$

Hence $\hat{x} = \hat{x}'$.   ∎

This theorem is known as the *projection theorem*. The element $\hat{x}$ is called the *(orthogonal) projection* of x on $\mathcal{F}$ and is also denoted by $x|\mathcal{F}$. It is the unique element of $\mathcal{F}$ satisfying the orthogonality condition (b.2). The difference $x - \hat{x}$ is called the *projection error*, or the *residual*. In the context of estimation theory (i.e., when the elements of $\mathcal{H}$ are random variables), $\hat{x}$ is called the *best linear estimator* of x with respect to $\mathcal{F}$. It is easy to check, by the orthogonality of $\hat{x}$ and $x - \hat{x}$, that

$$\|x\|^2 = \|\hat{x}\|^2 + \|x - \hat{x}\|^2. \tag{b.4}$$

The next theorem follows from the projection theorem.

Theorem B.2. Let $\mathcal{H}$ be a Hilbert space and $\mathcal{F}$ a subspace of $\mathcal{H}$. Then there exists a unique Hilbert subspace $\mathcal{G}$ of $\mathcal{H}$ such that $\mathcal{F} \perp \mathcal{G}$ and every $v \in \mathcal{H}$ can be expressed as $v = f + g$, where $f \in \mathcal{F}$ and $g \in \mathcal{G}$.

Proof. Define $\mathcal{G}$ as the set of all vectors in $\mathcal{H}$ that are orthogonal to $\mathcal{F}$. It is easy to check that $\mathcal{G}$ is closed under vector addition and scalar multiplication; hence it is a subspace of $\mathcal{H}$. To establish completeness, let $\{g_n\}$ be a Cauchy sequence in $\mathcal{G}$. Since this sequence is also in $\mathcal{H}$, it converges to some element g in $\mathcal{H}$. By continuity of the inner product (Lemma B.2), we can show that $g \perp \mathcal{F}$; hence $g \in \mathcal{G}$. The property $\mathcal{F} \perp \mathcal{G}$ holds by definition of $\mathcal{G}$. By the projection theorem, every v can be written as $v = \hat{v} + (v - \hat{v})$, where the first term is in $\mathcal{F}$ and the second is in $\mathcal{G}$.

To show uniqueness of $\mathcal{G}$, assume $\mathcal{G}'$ is another Hilbert subspace satisfying the above conditions. By definition of $\mathcal{G}$, necessarily $\mathcal{G}' \subseteq \mathcal{G}$. Suppose there exists $g \in \mathcal{G}$ such that $g \notin \mathcal{G}'$. Let $\hat{g}$ be the projection of g on $\mathcal{G}'$. Then $g - \hat{g} \perp \mathcal{G}'$, but $g - \hat{g} \in \mathcal{G}$; hence $g - \hat{g} \perp \mathcal{F}$. Therefore, $g - \hat{g}$ cannot be represented as a sum of elements of $\mathcal{F}$ and $\mathcal{G}'$, in contradiction to our assumption. ∎

The Hilbert subspace $\mathcal{G}$ is called the *orthogonal complement* of $\mathcal{F}$ in $\mathcal{H}$. Theorem B.2 thus establishes the existence and uniqueness of the orthogonal complement for any Hilbert subspace of a given space. $\mathcal{H}$ is said to be a *direct sum* of $\mathcal{F}$ and $\mathcal{G}$, and we use the notations $\mathcal{H} = \mathcal{F} \oplus \mathcal{G}$, $\mathcal{F} = \mathcal{H} \ominus \mathcal{G}$, and $\mathcal{G} = \mathcal{H} \ominus \mathcal{F}$.

Our next topic will make use of the following lemma.

Lemma B.3. Suppose we are given three Hilbert spaces $\mathcal{F}_s \subseteq \mathcal{F}_l \subseteq \mathcal{H}$ and an element $x \in \mathcal{H}$. Then:
(a) $(x|\mathcal{F}_l)|\mathcal{F}_s = x|\mathcal{F}_s$.
(b) $\|x|\mathcal{F}_s\| \leq \|x|\mathcal{F}_l\|$.

Proof.
(a) On one hand, we have

$$x = x|\mathcal{F}_l + (x - x|\mathcal{F}_l) = (x|\mathcal{F}_l)|\mathcal{F}_s + [x|\mathcal{F}_l - (x|\mathcal{F}_l)|\mathcal{F}_s] + (x - x|\mathcal{F}_l). \tag{b.5}$$

On the other hand,
$$x = x|\mathcal{F}_s + (x - x|\mathcal{F}_s). \tag{b.6}$$

Now
$$[x|\mathcal{F}_l - (x|\mathcal{F}_l)|\mathcal{F}_s] \in \mathcal{F}_l \ominus \mathcal{F}_s \subseteq \mathcal{H} \ominus \mathcal{F}_s$$
$$(x - x|\mathcal{F}_l) \in \mathcal{H} \ominus \mathcal{F}_l \subseteq \mathcal{H} \ominus \mathcal{F}_s.$$

Hence (b.5) and (b.6) are two forms of the decomposition discussed in Theorem B.2. Since that decomposition is unique, (a) is proved.

(b) It clearly follows from (a) and (b.4) that
$$\|x|\mathcal{F}_s\| = \|(x|\mathcal{F}_l)|\mathcal{F}_s\| \le \|x|\mathcal{F}_l\|. \qquad \blacksquare$$

Many problems in Hilbert space theory concern monotone sequences of subspaces of a given space. Let $\mathcal{H}$ be a Hilbert space and $\{\mathcal{F}_n, 1 \le n < \infty\}$ a sequence of subspaces of $\mathcal{H}$. The sequence is said to be *monotone increasing* if $\mathcal{F}_n \subseteq \mathcal{F}_{n+1}$ for all n. It is said to be *monotone decreasing* if $\mathcal{F}_{n+1} \subseteq \mathcal{F}_n$ for all n. For monotone increasing sequences, we define
$$\mathcal{F}_\infty = \overline{\bigcup_{n=1}^{\infty} \mathcal{F}_n}.$$

For monotone decreasing sequences, we define
$$\mathcal{F}_{-\infty} = \bigcap_{n=1}^{\infty} \mathcal{F}_n.$$

Both $\mathcal{F}_\infty$ and $\mathcal{F}_{-\infty}$ are Hilbert spaces.

We first consider monotone decreasing sequences. Let $x \in \mathcal{H}$. Then:

Theorem B.3.
$$\underset{n \to \infty}{\text{l.i.m.}} \; x|\mathcal{F}_n = x|\mathcal{F}_{-\infty}. \tag{b.7}$$

Proof. We first show that $\{x|\mathcal{F}_n, 1 \le n < \infty\}$ is a Cauchy sequence. For $n > m$, we have $\mathcal{F}_n \subseteq \mathcal{F}_m$; hence, by (b.4) and Lemma B.3,
$$\|x|\mathcal{F}_m - x|\mathcal{F}_n\|^2 = \|x|\mathcal{F}_m - (x|\mathcal{F}_m)|\mathcal{F}_n\|^2$$
$$= \|x|\mathcal{F}_m\|^2 - \|(x|\mathcal{F}_m)|\mathcal{F}_n\|^2 = \|x|\mathcal{F}_m\|^2 - \|x|\mathcal{F}_n\|^2.$$

The sequence $\{\|x|\mathcal{F}_n\|\}$ is monotone decreasing and nonnegative, hence it is convergent. Therefore, the right side tends to zero as $m, n \to \infty$, proving that $\{x|\mathcal{F}_n\}$ is indeed a Cauchy sequence. Denote the limit of this sequence by z. It remains to show that $z = x|\mathcal{F}_{-\infty}$. It follows from the proof of Theorem B.1 that it is sufficient to show that $z \in \mathcal{F}_{-\infty}$ and that $x - z \perp \mathcal{F}_{-\infty}$. The first is true because, by its definition, $z \in \mathcal{F}_m$ for all m. To show the second property, let $v \in \mathcal{F}_{-\infty}$. Then

$v \in \mathcal{F}_n$ for all n. Since $x - x|\mathcal{F}_n \perp \mathcal{F}_n$, it follows that $\langle x - x|\mathcal{F}_n, v \rangle = 0$. Hence, by the continuity of the inner product (Lemma B.2),

$$\langle x - z, v \rangle = \lim_{n \to \infty} \langle x - x|\mathcal{F}_n, v \rangle = 0. \qquad\qquad \blacksquare$$

Now let us consider the dual of Theorem B.3 for monotone increasing sequences.

Theorem B.4.

$$\operatorname*{l.i.m.}_{n \to \infty} x|\mathcal{F}_n = x|\mathcal{F}_\infty. \tag{b.8}$$

Proof. The sequence $\{(\mathcal{F}_\infty \ominus \mathcal{F}_n), 1 \leq n < \infty\}$ is monotone decreasing, hence Theorem B.3 applies to it. Before we do that, however, we show that

$$\bigcap_{n=1}^{\infty} (\mathcal{F}_\infty \ominus \mathcal{F}_n) = \{0\}. \tag{b.9}$$

Let v belong to the left side of (b.9). Then $v \in \mathcal{F}_\infty$ and $v \perp \mathcal{F}_n$ for all n. By continuity of the inner product, $v \perp \mathcal{F}_\infty$. Hence, necessarily, $v = 0$, and (b.9) is proved.

Now apply Theorem B.3 to $\{\mathcal{F}_\infty \ominus \mathcal{F}_n\}$ using (b.9).

$$0 = \operatorname*{l.i.m.}_{n \to \infty} x|(\mathcal{F}_\infty \ominus \mathcal{F}_n) = x|\mathcal{F}_\infty - \operatorname*{l.i.m.}_{n \to \infty} x|\mathcal{F}_n. \qquad\qquad \blacksquare$$

REFERENCES

Akhiezer, N. I., and Glazman, M. L., *Theory of Linear Operators in Hilbert Space*, Frederic Ungar, New York, 1963.

APPENDIX C

Asymptotic Theory

THE NOTATIONS o_p and O_p

A sequence of scalar random variables X_n on a common probability space $\{\Omega, \mathcal{A}, P\}$ is said to *converge to zero in probability* if, for all $\delta > 0$,

$$\lim_{n \to \infty} P\{|X_n| \geq \delta\} = 0.$$

An equivalent definition is that, for all $\delta > 0$ and all $\epsilon > 0$, there exists n_0 such that

$$P\{|X_n| \geq \delta\} < \epsilon, \quad \forall n > n_0.$$

The notation $X_n = o_p(1)$ indicates that X_n converges to zero in probability. The sequence X_n is said to converge in probability to a random variable X (on the same probability space) if $X_n - X = o_p(1)$.

The sequence X_n is said to be *bounded in probability* if for all $\epsilon > 0$ there exists $\delta > 0$ such that, for all n,

$$P\{|X_n| \geq \delta\} < \epsilon.$$

The notation $X_n = O_p(1)$ indicates that X_n is bounded in probability.

Let a_n be a sequence of positive real numbers. The notation $X_n = o_p(a_n)$ means that $X_n/a_n = o_p(1)$. The notation $X_n = O_p(a_n)$ means that $X_n/a_n = O_p(1)$.

A sequence of (finite dimensional) vector random variables X_n is said to converge to zero in probability if each of its components is $o_p(1)$ and to be bounded in probability if each of its components is $O_p(1)$. The notations $X_n = o_p(a_n)$ and $X_n = O_p(a_n)$ are similarly defined for vector random variables. It is straightforward to verify that $X_n = o_p(1)$ if and only if $\|X_n\| = o_p(1)$ and $X_n = O_p(1)$ if and only if $\|X_n\| = O_p(1)$.

Lemma C.1. Let X_n and Y_n be two sequences of scalar random variables. Then:
  (a) If $X_n = o_p(1)$ and $Y_n = o_p(1)$, then $X_n Y_n = o_p(1)$.
  (b) If $X_n = o_p(1)$ and $Y_n = O_p(1)$, then $X_n Y_n = o_p(1)$.
  (c) If $X_n = O_p(1)$ and $Y_n = O_p(1)$, then $X_n Y_n = O_p(1)$.

Proof.
  (a) For all $\delta > 0$, $P\{|X_n Y_n| \geq \delta\} \leq P\{|X_n| \geq \sqrt{\delta}\} + P\{|Y_n| \geq \sqrt{\delta}\}$. Taking the limit $n \to \infty$ yields zero on the right side and hence on the left side.
  (b) Choose $\delta > 0$ and $\epsilon > 0$. Then, for all $\delta' > 0$,

$$P\{|X_n Y_n| \geq \delta\} \leq P\{|X_n| \geq \delta'\} + P\{|Y_n| \geq \delta/\delta'\}.$$

Choose δ' such that $P\{|Y_n| \geq \delta/\delta'\} < \epsilon/2$ for all n. Then choose n_0 such that $P\{|X_n| \geq \delta'\} < \epsilon/2$ for all $n > n_0$. These will yield $P\{|X_n Y_n| \geq \delta\} < \epsilon$.
  (c) Choose $\epsilon > 0$ and let δ_1 and δ_2 be such that $P\{|X_n| \geq \delta_1\} < \epsilon/2$ and $P\{|Y_n| \geq \delta_2\} < \epsilon/2$ for all n. Then

$$P\{|X_n Y_n| \geq \delta_1 \delta_2\} \leq P\{|X_n| \geq \delta_1\} + P\{|Y_n| \geq \delta_2\} < \epsilon.$$

Corollaries.
  (a) If $X_n = o_p(a_n)$ and $Y_n = o_p(b_n)$, then $X_n Y_n = o_p(a_n b_n)$.
  (b) If $X_n = O_p(a_n)$ and $Y_n = o_p(b_n)$, then $X_n Y_n = o_p(a_n b_n)$.
  (c) If $X_n = O_p(a_n)$ and $Y_n = O_p(b_n)$, then $X_n Y_n = O_p(a_n b_n)$.
  (d) If $X_n = o_p(a_n)$, then $|X_n|^m = o_p(a_n^m)$.
  (e) If $X_n = O_p(a_n)$, then $|X_n|^m = O_p(a_n^m)$.

Parts (a), (b), and (c) follow from the lemma by straightforward use of the definition of $o_p(a_n)$ and $O_p(a_n)$. Parts (d) and (e) follow by induction on (a) and (c), respectively. ∎

Theorem C.1. Let X_n be a sequence of (scalar or vector) random variables satisfying $X_n = a + o_p(1)$, where a is constant. Let $g(X_n)$ be a (scalar or vector) function of X_n that is continuous at a. Then $g(X_n) = g(a) + o_p(1)$.

Proof. Choose $\epsilon > 0$. By continuity there exists δ such that $\|X_n - a\| < \delta \Rightarrow \|g(X_n) - g(a)\| < \epsilon$. This means that

$$\{\|g(X_n) - g(a)\| \geq \epsilon\} \subseteq \{\|X_n - a\| \geq \delta\};$$

hence

$$P\{\|g(X_n) - g(a)\| \geq \epsilon\} \leq P\{\|X_n - a\| \geq \delta\}.$$

The right side goes to zero as $n \to \infty$; hence so does the left side. ∎

TAYLOR SERIES OF FUNCTIONS OF RANDOM VARIABLES

Let $g(x)$ be a function from $\boldsymbol{R}^K$ to $\boldsymbol{R}$ that is continuous and possesses continuous partial derivatives up to order m. Introduce the notation $g^{(i)}(x;t)$ by the following recursive definition.

$$g^{(0)}(x;t) = g(x)$$

$$g^{(i)}(x;t) = \sum_{k=1}^{K} \frac{\partial g^{(i-1)}(x;t)}{\partial x_k} t_k, \quad i = 1, 2, \ldots, m.$$

Taylor's formula for $g(x)$ is

$$g(x) = \sum_{i=0}^{m-1} \frac{1}{i!} g^{(i)}(x_0; x - x_0) + \frac{1}{m!} g^{(m)}(x_*; x - x_0), \tag{c.1}$$

where x_* is a point on the line segment between x_0 and x. See, for example, Apostol [1974, p. 361] for a proof.

Taylor's formula for random variables is given by the following theorem.

Theorem C.2. Let X_n be a sequence of vector random variables satisfying $X_n = x_0 + O_p(a_n)$, where x_0 is a fixed point in $\boldsymbol{R}^K$ and a_n is a sequence of positive numbers that tends to zero as $n \to \infty$. Let $g(x)$ be a function from $\boldsymbol{R}^K$ to $\boldsymbol{R}$ that is continuous and possesses continuous partial derivatives up to order m. Then

$$g(X_n) = \sum_{i=0}^{m} \frac{1}{i!} g^{(i)}(x_0; X_n - x_0) + o_p(a_n^m). \tag{c.2}$$

Proof. Note that, by the theorem's assumption, $X_n = x_0 + o_p(1)$. Define

$$f(x) = \frac{m!}{\|x - x_0\|^m} \left[g(x) - \sum_{i=0}^{m} \frac{1}{i!} g^{(i)}(x_0; x - x_0) \right]$$

$$= \frac{1}{\|x - x_0\|^m} [g^{(m)}(x_*; x - x_0) - g^{(m)}(x_0; x - x_0)], \quad x \neq x_0,$$

and $f(x_0) = 0$. Then $f(x)$ is continuous at x_0, so by Theorem 1 it satisfies $f(X_n) = f(x_0) + o_p(1) = o_p(1)$. Hence, by the corollary of Lemma C.1, parts (b) and (e),

$$\|X_n - x_0\|^m f(X_n) = O_p(a_n^m) o_p(1) = o_p(a_n^m),$$

which proves (c.2). ∎

CONVERGENCE IN DISTRIBUTION

Let $F_X(\boldsymbol{\alpha})$ denote the cumulative distribution function of the random vector X. Let $X \leq \boldsymbol{\alpha}$ denote the event that each component of X is less or equal to the corresponding component of $\boldsymbol{\alpha}$. A sequence of random vectors X_n is said to converge in distribution to a random vector X if

$$\lim_{n \to \infty} P(X_n \leq \boldsymbol{\alpha}) = P(X \leq \boldsymbol{\alpha}) = F_X(\boldsymbol{\alpha})$$

for all $\boldsymbol{\alpha}$ at which $F_X(\boldsymbol{\alpha})$ is continuous.

The Cramér-Wold theorem enables us to reduce the question of convergence in distribution of vectors to convergence in distribution of scalar random variables. The theorem states that a random sequence of K-dimensional vectors X_n converges in distribution to X if and only if $v^T X_n$ converges in distribution to $v^T X$ for every K-dimensional vector v. For a proof of the Cramér-Wold theorem, see Billingsley [1986, p. 397].

Theorem C.3. Let X_n be a sequence of scalar random variables converging in distribution to X, and let Y_n be another sequence such that $Y_n = X_n + o_p(1)$. Then Y_n converges in distribution to X.

Proof. Let α be a point of continuity of $F_X(\alpha)$. Suppose $\epsilon > 0$ is such that $F_X(\cdot)$ is continuous at both $\alpha + \epsilon$ and $\alpha - \epsilon$. If $X_n \leq \alpha - \epsilon$, then $Y_n \leq \alpha$ or $|X_n - Y_n| \geq \epsilon$. Therefore,

$$P(X_n \leq \alpha - \epsilon) \leq P(Y_n \leq \alpha) + P(|X_n - Y_n| \geq \epsilon).$$

Taking the limit as $n \to \infty$ and using the fact that $X_n - Y_n$ converges to zero in probability yield

$$P(X \leq \alpha - \epsilon) \leq \liminf_{n \to \infty} P(Y_n \leq \alpha).$$

Similarly, if $Y_n \leq \alpha$, then $X_n \leq \alpha + \epsilon$ or $|X_n - Y_n| \geq \epsilon$. Therefore,

$$P(Y_n \leq \alpha) \leq P(X_n \leq \alpha + \epsilon) + P(|X_n - Y_n| \geq \epsilon).$$

Taking the limit as $n \to \infty$ yields

$$\limsup_{n \to \infty} P(Y_n \leq \alpha) \leq P(X \leq \alpha + \epsilon).$$

Combining these two results gives

$$F_X(\alpha - \epsilon) \leq \liminf_{n \to \infty} P(Y_n \leq \alpha) \leq \limsup_{n \to \infty} P(Y_n \leq \alpha) \leq F_X(\alpha + \epsilon). \qquad \text{(c.3)}$$

Since a distribution function is bounded and monotone nondecreasing, it has a countable number of discontinuities at most. Therefore, we can choose a sequence of ϵ's that decreases to zero and such that $F_X(\alpha - \epsilon)$ and $F_X(\alpha + \epsilon)$ are continuous at all points of the sequence. Along this sequence, both sides of (c.3) will converge to $F_X(\alpha)$. This proves that $\lim_{n \to \infty} P(Y_n \leq \alpha)$ exists and is equal to $F_X(\alpha)$. ∎

Corollary. Theorem C.3 holds for random vectors as well.

Proof. For vector sequences X_n and Y_n, $Y_n = X_n + o_p(1)$ clearly implies $v^T Y_n = v^T X_n + o_p(1)$ for all v. Therefore, $v^T Y_n$ converges in distribution to $v^T X$, and the corollary follows from the Cramér-Wold theorem. ∎

Lemma C.2. If a sequence of scalar random variables X_n converges to X in distribution, then $X_n = O_p(1)$.

Proof. Let $\epsilon > 0$. Choose δ_0 sufficiently large such that $P(|X| > \delta_0) < 0.5\epsilon$ and F_X is continuous at δ_0. There exists N such that $|P(|X_n| > \delta_0) - P(|X| > \delta_0)| < 0.5\epsilon$ for all $n > N$. Therefore, $P(|X_n| > \delta_0) < \epsilon$ for all $n > N$. Now for every $1 \leq n \leq N$ there exists δ_n such that $P(|X_n| \geq \delta_n) < \epsilon$. Therefore, if we take $\delta = 1 + \max_{0 \leq n \leq N}\{\delta_n\}$, then $P(|X_n| \geq \delta) < \epsilon$ for all n. Therefore, X_n is bounded in probability, as claimed. ∎

Theorem C.4. Let X_n be a sequence of M-dimensional vectors that converges in distribution to a random vector X. Let A_n be a sequence of square $M \times M$ random matrices converging in probability to a fixed matrix A. Then $A_n X_n$ converges in distribution to AX.

Proof. By the Cramér-Wold theorem, it is sufficient to show that $v^T A_n X_n$ converges in distribution to $v^T AX$ for any M-dimensional vector v. We have

$$v^T A_n X_n = v^T A X_n + v^T (A_n - A) X_n.$$

The first term converges in distribution to $v^T AX$ by the Cramér-Wold theorem. For the second term, we have

$$|v^T (A_n - A) X_n| \leq \|v\| \, \|A_n - A\| \, \|X_n\|.$$

The first factor in this product is constant, the second converges in probability to zero, and the third is bounded in probability by Lemma C.2. Therefore, the left side converges to zero in probability. Hence, by Theorem C.3, the result follows. ∎

Corollary. If A is nonsingular and X is Gaussian with zero mean and covariance Σ, the limiting distribution of $A_n X_n$ has zero mean and covariance $A\Sigma A^T$. ∎

ASYMPTOTIC NORMALITY

If a sequence X_n converges in distribution to a Gaussian random variable, it is said to be *asymptotically normal*. If the limiting distribution has mean μ and variance σ^2, the sequence is said to be asymptotically normal (μ, σ^2). Similarly, a sequence of random vectors is said to be asymptotically normal $(\boldsymbol{\mu}, \Sigma)$ if it converges in distribution to a vector Gaussian random variable with mean $\boldsymbol{\mu}$ and covariance matrix Σ.

The following lemma is very useful in establishing the asymptotic normality of certain sequences.

Lemma C.3. Let X_n be a sequence of random variables and d_n a positive monotone sequence tending to infinity. Denote $\mu_n = EX_n$ and $\sigma_n = (\mathrm{var}\{X_n\})^{1/2}$, and assume that the following hold:

(1) $\lim_{n \to \infty} d_n \sigma_n = \sigma$, where $0 < \sigma < \infty$.

(2) $\lim_{n \to \infty} d_n(\mu_n - \mu) = 0$ for some finite μ.

(3) The sequence $Y_n = (X_n - \mu_n)/\sigma_n$ is asymptotically normal $(0, 1)$.

Then the sequence $Z_n = d_n(X_n - \mu)/\sigma$ is asymptotically normal $(0, 1)$; equivalently, $d_n(X_n - \mu)$ is asymptotically normal $(0, \sigma^2)$.

Proof. We have

$$Y_n - Z_n = Y_n - \frac{d_n}{\sigma}(\sigma_n Y_n + \mu_n - \mu) = \frac{Y_n}{\sigma}(\sigma - d_n \sigma_n) + \frac{d_n(\mu - \mu_n)}{\sigma}.$$

Since Y_n converges in distribution, it is $O_p(1)$ by Lemma C.2. By assumption (1), $\sigma - d_n \sigma_n$ is $o(1)$; hence it is certainly $o_p(1)$. Therefore, by Lemma C.1, part (2), the first term on the right side is $o_p(1)$. The second term is $o(1)$, hence $o_p(1)$, by assumption (2). Therefore, $Y_n - Z_n$ is $o_p(1)$, so by Theorem C.3 Z_n converges in distribution to the limiting distribution of Y_n. ∎

QUADRATIC FORMS IN GAUSSIAN RANDOM VARIABLES

Let $\{u_n, 1 \leq n < \infty\}$ be an i.i.d. sequence of Gaussian random variables with zero mean and unit variance. Let $\{C_n, 1 \leq n < \infty\}$ be a sequence of symmetric matrices (not necessarily positive definite) such that C_n has dimension n. Let

$$x_n = \sum_{i=1}^{n} \sum_{j=1}^{n} u_i C_{n,ij} u_j - \sum_{i=1}^{n} C_{n,ii} = \sum_{i=1}^{n} \sum_{j=1}^{n} u_i C_{n,ij} u_j - \mathrm{tr}\{C_n\}. \qquad (\text{c.4})$$

The x_n are quadratic forms in Gaussian random variables. The following theorems deal with their almost sure convergence and asymptotic normality.

Theorem C.5. Assume that

$$\sum_{n=1}^{\infty} \frac{1}{n^4} [\mathrm{tr}\{C_n^2\}]^2 < \infty. \qquad (\text{c.5})$$

Then $n^{-1} x_n$ converges to zero almost surely.

Proof. Let us first compute Ex_n^4 for a fixed n. Let $C_n = Q_n \Lambda_n Q_n^T$ be the eigenvalue/eigenvector decomposition of C_n, where $\Lambda_n = \mathrm{diag}\{\lambda_{n,1}, \ldots, \lambda_{n,n}\}$ and Q_n is an orthonormal matrix. Then

$$x_n = \sum_{i=1}^{n} w_{n,i}^2 \lambda_{n,i} - \sum_{i=1}^{n} \lambda_{n,i} = \sum_{i=1}^{n} (w_{n,i}^2 - 1)\lambda_{n,i},$$

where $w_{n,i} = \sum_{j=1}^{n} u_j Q_{n,ji}$. For every fixed n, $\{w_{n,i}, 1 \leq i \leq n\}$ are i.i.d. Gaussian with zero mean and unit variance (this follows from the orthonormality of Q_n). Also,

$$Ex_n^4 = \sum_{i=1}^{n} \sum_{j=1}^{n} \sum_{k=1}^{n} \sum_{\ell=1}^{n} E[(w_{n,i}^2 - 1)(w_{n,j}^2 - 1)(w_{n,k}^2 - 1)(w_{n,\ell}^2 - 1)]\lambda_{n,i}\lambda_{n,j}\lambda_{n,k}\lambda_{n,\ell}.$$

Using the well-known formula $Ew^{2m} = (2m)!/(2^m m!)$, it is easy to verify that $E(w_{n,i}^2 - 1)^2 = 2$ and $E(w_{n,i}^2 - 1)^4 = 60$. Now, the expectation on the right side will be zero in all but four cases:

(1) When $i = j$ and $k = \ell$, but $i \neq k$. This will contribute

$$4 \sum_{i=1}^{n} \sum_{k=1}^{n} \lambda_{n,i}^2 \lambda_{n,k}^2 - 4 \sum_{i=1}^{n} \lambda_{n,i}^4 = 4 \left[\sum_{i=1}^{n} \lambda_{n,i}^2 \right]^2 - 4 \sum_{i=1}^{n} \lambda_{n,i}^4.$$

(2) When $i = k$ and $j = \ell$, but $i \neq j$. This will contribute the same as in case (1).

(3) When $i = \ell$ and $j = k$, but $i \neq j$. This will also contribute the same as in case (1).

(4) When all indexes are equal. This will contribute $60 \sum_{i=1}^{n} \lambda_{n,i}^4$.

In summary,

$$Ex_n^4 = 12 \left[\sum_{i=1}^{n} \lambda_{n,i}^2 \right]^2 + 48 \sum_{i=1}^{n} \lambda_{n,i}^4 = 12[\text{tr}\{C_n^2\}]^2 + 48\text{tr}\{C_n^4\}.$$

Also, since $[\sum_{i=1}^{n} \lambda_{n,i}^2]^2 \geq \sum_{i=1}^{n} \lambda_{n,i}^4$,

$$Ex_n^4 \leq 60[\text{tr}\{C_n^2\}]^2.$$

Next, by Chebychev's inequality [Chung, 1974, p. 48],

$$P\{|x_n| \geq \epsilon\} \leq \epsilon^{-4} Ex_n^4,$$

so

$$\sum_{n=1}^{\infty} P\{n^{-1}|x_n| \geq \epsilon\} \leq \epsilon^{-4} \sum_{n=1}^{\infty} \frac{60}{n^4}[\text{tr}\{C_n^2\}]^2 < \infty.$$

Finally, by the Borel-Cantelli lemma [Chung, 1974, p. 74], $n^{-1}x_n$ converges to zero almost surely.

Corollary. Assume that $\{\lambda_{n,i}\}$ are uniformly bounded in magnitude; that is,

$$|\lambda n, i| \leq L, \quad \forall n, i,$$

for some positive L. Then $n^{-1}x_n$ converges to zero almost surely.

Proof. We have in this case

$$\sum_{n=1}^{\infty} \frac{1}{n^4} \left[\sum_{i=1}^{n} \lambda_{n,i}^2 \right]^2 \leq \sum_{n=1}^{\infty} \frac{1}{n^4} [nL^2]^2 = \frac{\pi^2 L^4}{6},$$

so the conclusion of the theorem applies. ∎

Theorem C.6. Assume that the eigenvalues of the matrices C_n satisfy

$$\lim_{n\to\infty} \frac{\sum_{i=1}^{n} |\lambda_{n,i}|^3}{\left[\sum_{i=1}^{n} \lambda_{n,i}^2 \right]^{3/2}} = \lim_{n\to\infty} \frac{\sum_{i=1}^{n} |\lambda_{n,i}|^3}{[\text{tr}\{C_n^2\}]^{3/2}} = 0. \tag{c.6}$$

Then the sequence $\{x_n/[2\text{tr}\{C_n^2\}]^{1/2}\}$ is asymptotically normal $(0,1)$.

Proof. The proof is based on the Liapunov central limit theorem for triangular arrays; see Chung [1974, pp. 196–201]. We first carry out the transformation from the u_i to the $w_{n,i}$, as in the proof of Theorem C.5. Then

$$\frac{x_n}{[2\text{tr}\{C_n^2\}]^{1/2}} = \sum_{i=1}^{n} X_{n,i}, \quad \text{where} \quad X_{n,i} = \frac{\lambda_{n,i}(w_{n,i}^2 - 1)}{\left[2\sum_{k=1}^{n} \lambda_{n,k}^2 \right]^{1/2}}.$$

To show that $\sum_{i=1}^{n} X_{n,i}$ is asymptotically normal $(0,1)$, it is sufficient to show that

$$\lim_{n\to\infty} \sum_{i=1}^{n} E|X_{n,i}|^3 = 0.$$

We have

$$\sum_{i=1}^{n} E|X_{n,i}|^3 \leq \frac{\sum_{i=1}^{n} |\lambda_{n,i}|^3 E[(w_{n,i}^2 + 1)^3]}{\left[2\sum_{i=1}^{n} \lambda_{n,i}^2 \right]^{3/2}} = \frac{28 \sum_{i=1}^{n} |\lambda_{n,i}|^3}{\left[2\sum_{i=1}^{n} \lambda_{n,i}^2 \right]^{3/2}}.$$

By the theorem's assumption the right side goes to zero as $n \to \infty$; hence so does the left side. ∎

REFERENCES

Apostol, T. M., *Mathematical Analysis*, 2nd ed., Addison-Wesley, Reading, MA, 1974.

Billingsley, P., *Probability and Measure*, 2nd ed., Wiley, New York, 1986.

Chernoff, H., "Large Sample Theory: Parametric Case," *Annals of Mathematical Statistics*, 27, pp. 1–22, 1956.

Chung, K. L., *A Course in Probability Theory*, Academic Press, New York, 1974.

Kronecker Products and Liapunov Equations

KRONECKER PRODUCTS

Let the matrices A and B have dimensions $r_a \times c_a$ and $r_b \times c_b$, respectively. The *Kronecker product* of A and B, denoted by $A \otimes B$, is the $r_a r_b \times c_a c_b$ matrix whose element in position $((i-1)r_b + k, (j-1)c_b + \ell)$ is $a_{ij}b_{k\ell}$.

Property K1. The Kronecker product is associative; that is,

$$A \otimes (B \otimes C) = (A \otimes B) \otimes C. \tag{d.1}$$

The proof is by direct verification. The elements of both sides of (d.1) are $a_{ij}b_{k\ell}c_{mn}$, and they appear in the same order. ∎

Remark: The Kronecker product is *not* commutative.

We will use the following notations for repeated Kronecker products:

$$\bigotimes_{i=1}^{n} A_i = A_1 \otimes A_2 \ldots \otimes A_n, \quad A^{\otimes n} = \bigotimes_{i=1}^{n} A.$$

These notations are unambiguous due to the associativity of the Kronecker product.

Property K2.

$$(A \otimes B)^T = A^T \otimes B^T. \tag{d.2}$$

The proof is by direct verification. The element $a_{ij}b_{k\ell}$ appears in position $((j-1)c_b + \ell, (i-1)r_b + k)$ on both sides of (d.2). ∎

Property K3. Let A, B, C, D have dimensions $r_a \times c_a$, $r_b \times c_b$, $r_c \times c_c$, and $r_d \times c_d$, respectively, and assume $c_a = r_c$ and $c_b = r_d$. Then

$$(A \otimes B)(C \otimes D) = (AC) \otimes (BD). \tag{d.3}$$

Proof. The $((i-1)r_b + k, (j-1)c_b + \ell)$th element of $A \otimes B$ is $a_{ij}b_{k\ell}$. The $((j-1)r_d + \ell, (m-1)c_d + n)$th element of $C \otimes D$ is $c_{jm}d_{\ell n}$. Therefore, the $((i-1)r_b + k, (m-1)c_d + n)$th element of the left side of (d.3) is

$$\sum_j \sum_\ell a_{ij}b_{k\ell}c_{jm}d_{\ell n} = \left(\sum_j a_{ij}c_{jm}\right)\left(\sum_\ell b_{k\ell}d_{\ell n}\right),$$

which is the $((i-1)r_b + k, (m-1)c_d + n)$th element of the right side of (d.3). ∎

Corollaries.

$$\left[\bigotimes_{i=1}^{n} A_i\right]\left[\bigotimes_{i=1}^{n} B_i\right] = \bigotimes_{i=1}^{n}(A_i B_i), \tag{d.4}$$

$$\left[\prod_{i=1}^{n} A_i\right] \otimes \left[\prod_{i=1}^{n} B_i\right] = \prod_{i=1}^{n}(A_i \otimes B_i), \tag{d.5}$$

$$(A^m)^{\otimes n} = (A^{\otimes n})^m. \tag{d.6}$$

These are obtained by inductive application of (d.3). ∎

Property K4. If A and B are square and nonsingular, then

$$(A \otimes B)^{-1} = A^{-1} \otimes B^{-1}. \tag{d.7}$$

Proof. By property K3 (omitting the dimensions of the identity matrices),

$$(A \otimes B)(A^{-1} \otimes B^{-1}) = (AA^{-1}) \otimes (BB^{-1}) = \mathbf{1} \otimes \mathbf{1} = \mathbf{1}.$$ ∎

Property K5. If A and B are square and their respective eigenvalues are $\{\lambda_{a,i}, 1 \le i \le r_a\}$ and $\{\lambda_{b,j}, 1 \le j \le r_b\}$, then the eigenvalues of $A \otimes B$ are $\{\lambda_{a,i}\lambda_{b,j}, 1 \le i \le r_a, 1 \le j \le r_b\}$.

Proof. Let $T_a \Lambda_a T_a^{-1}$ and $T_b \Lambda_b T_b^{-1}$ be the Jordan-form decompositions of the two matrices. Then we get, by repeated application of Properties K3 and K4,

$$A \otimes B = (T_a \otimes T_b)(\Lambda_a \otimes \Lambda_b)(T_a \otimes T_b)^{-1}. \tag{d.8}$$

The matrix $\Lambda_a \otimes \Lambda_b$ is upper triangular, with $\{\lambda_{a,i}\lambda_{b,j}, 1 \le i \le r_a, 1 \le j \le r_b\}$ on its main diagonal. Since (d.8) is a similarity transformation, these are the eigenvalues of $A \otimes B$. ∎

Corollary A. The matrix $A \otimes B$ is nonsingular if and only if both A and B are nonsingular.

Corollary B. If both A and B are similar to diagonal matrices, then so is $A \otimes B$. In this case its diagonal form is $\Lambda_a \otimes \Lambda_b$. ∎

Property K6. If $A = U_a \Sigma_b V_a^T$ and $B = U_b \Sigma_b V_b^T$ are the singular value decompositions of A and B, respectively, then

$$A \otimes B = (U_a \otimes U_b)(\Sigma_a \otimes \Sigma_b)(V_a \otimes V_b)^T \qquad (d.9)$$

is the SVD of $A \otimes B$.

Proof. Equation (d.9) is obtained by repeated application of K3 and K2. The matrix $\Sigma_a \otimes \Sigma_b$ is clearly diagonal and its entries are nonnegative. The matrix $U_a \otimes U_b$ is orthogonal, since

$$(U_a \otimes U_b)(U_a \otimes U_b)^T = (U_a U_a^T) \otimes (U_b U_b^T) = \mathbf{1} \otimes \mathbf{1} = \mathbf{1},$$

and similarly for $(V_a \otimes V_b)$. ∎

LIAPUNOV MATRIX EQUATIONS

There are two general types of Liapunov matrix equations, as follows.

$$AX + XB^T = -C \qquad (d.10)$$
$$X - AXB^T = C \qquad (d.11)$$

where A, B, and C are known matrices and X is an unknown matrix. A and B are square, with dimensions $n_a \times n_a$ and $n_b \times n_b$, respectively. C and X have dimensions $n_a \times n_b$. Both equations are clearly linear. We will only be concerned with equations of the second type, sometimes called *discrete Liapunov equations* because of their numerous applications to discrete-time dynamic systems.

Theorem D.1. Let $\boldsymbol{x}$ be the $(n_a n_b)$-dimensional vector obtained by stacking the columns of X in a single column in the natural order. The vector $\boldsymbol{c}$ is constructed from C in a similar manner. Then the Liapunov equation (d.11) is algebraically equivalent to the linear equation

$$(\mathbf{1} - B \otimes A)\boldsymbol{x} = \boldsymbol{c}. \qquad (d.12)$$

Proof. The equation corresponding to the (i, j)th element of (d.11) is

$$x_{ij} - \sum_k \sum_\ell a_{ik} x_{k\ell} b_{j\ell} = c_{ij}.$$

But this is exactly the $((j - 1)n_a + i)$th equation of (d.12). ∎

Theorem D.2. The Liapunov equation (d.11) has a unique solution if and only if the eigenvalues of A and of B satisfy $\lambda_{a,i}\lambda_{b,j} \neq 1$ for all $1 \leq i \leq n_a$ and $1 \leq j \leq n_b$.

Proof. Since (d.11) and (d.12) are equivalent, (d.11) has a unique solution if and only if $\mathbf{1} - B \otimes A$ is nonsingular (i.e., if all its eigenvalues are nonzero). But by Property K5 of the Kronecker product, These eigenvalues are $\{1 - \lambda_{a,i}\lambda_{b,j}\}$; hence the theorem follows. ∎

Theorem D.3. If A and B are stability matrices (i.e., if their eigenvalues are smaller than 1 in magnitude), the unique solution of (d.11) is given by

$$X = \sum_{i=0}^{\infty} A^i C (B^T)^i. \tag{d.13}$$

Proof. If A and B are stability matrices, they satisfy the condition of Theorem D.2; hence the solution is unique. Stability also guarantees the convergence of the sum in (d.13). Finally, substitution of the right side of (d.13) into the left side of (d.11) clearly gives C. ∎

Next consider the case where $A = B$ and C is symmetric and positive semidefinite. Let r be the rank of C. Then C can be expressed as $C = DD^T$, where D is $n \times r$ (this representation is not unique). We call the equation

$$X - AXA^T = DD^T \tag{d.14}$$

a *symmetric Liapunov equation*. If there exists a unique solution of (d.14), it is symmetric. This is because transposition of (d.14) yields $X^T - AX^T A^T = DD^T$, which is the same equation.

Theorem D.4. If A in (d.14) is a stability matrix, the unique solution of (d.14) is positive semidefinite. The solution is positive definite if and only if the pair $\{A, D\}$ is controllable.

Proof. Let $\{d_j, 1 \leq j \leq r\}$ be the columns of D. Substituting $B = A$ and $C = \sum_{j=1}^r d_j d_j^T$ in (d.13) yields, for an arbitrary vector v,

$$v^T X v = \sum_{j=1}^r \sum_{i=0}^{\infty} v^T A^i d_j d_j^T (A^T)^i v = \sum_{j=1}^r \sum_{i=0}^{\infty} (v^T A^i d_j)^2 \geq 0. \tag{d.15}$$

Furthermore, the sum in (d.15) can be zero for $v \neq 0$ if and only if $v^T A^i D = 0$ for all $i \geq 0$. But this can happen if and only if the pair $\{A, D\}$ is uncontrollable. ∎

So far we have considered general Liapunov equations. Let us now consider some specific cases. Denote by Z the $n \times n$ *shift matrix* $[Z]_{i,j} = \delta_{i-j-1}$.

Theorem D.5. The Liapunov equation

$$X - ZXZ^T = DD^T \tag{d.16}$$

has the unique solution

$$X = \sum_{k=1}^{r} L_k L_k^T, \tag{d.17}$$

where L_k is the lower triangular Toeplitz matrix $[L_k]_{i,j} = d_{k,i-j+1}$.

Proof. Since all the eigenvalues of Z are zero, it is a stability matrix. Furthermore, this matrix is nilpotent; that is, $Z^i = 0$ for $i \geq n$. Substitution in (d.13) then yields

$$X = \sum_{k=1}^{r} \sum_{i=0}^{n-1} Z^i d_k d_k^T (Z^T)^i = \sum_{k=1}^{r} L_k L_k^T. \qquad \blacksquare$$

Next consider the case where A and B are top-row companion matrices [cf. (2.12.12)]. Such matrices can be expressed as $A = Z - e_1 a^T$, $B = Z - e_1 b^T$, where e_1 is the first unit vector (with 1 in the first position and zeros elsewhere), and a and b are arbitrary vectors of dimensions n_a and n_b, respectively. Furthermore, assume that the entries of the matrix C are all zero except for the ones in the top row and left column.

Theorem D.6. If A, B are as above, any solution of the Liapunov equation (d.11) must be Toeplitz.

Proof. Suppose X is a solution, so

$$X - (Z - e_1 a^T)X(Z^T - be_1^T) = X - ZXZ^T + e_1 a^T X Z^T + ZX be_1^T - e_1 a^T X be_1^T = C.$$

It is easy to verify that the (i,j)th equation for $i \geq 2$ and $j \geq 2$ is

$$X_{i,j} - X_{i-1,j-1} = 0.$$

Hence X must be constant along its diagonals; that is, it is Toeplitz. $\blacksquare$

Let us further restrict the above special case to $A = B$ and $C = \sigma^2 e_1 e_1^T$, yielding the equation (here we replaced X by R)

$$R - ARA^T = \sigma^2 e_1 e_1^T. \tag{d.18}$$

Theorem D.7. If A is a stability companion matrix, the unique solution R of (d.18) is a symmetric positive-definite Toeplitz matrix. If we denote $[R]_{i,j} = r_{|i-j|}$ and define

$$r_n = -\sum_{i=1}^{n} a_i r_{n-i}, \tag{d.19}$$

then $\{r_0, \ldots, r_n\}$, $\{a_1, \ldots, a_n\}$, and σ^2 satisfy the Yule-Walker equations:

$$
\begin{bmatrix}
r_0 & r_1 & \cdots & r_n \\
r_1 & r_0 & \cdots & r_{n-1} \\
\vdots & & \ddots & \vdots \\
r_n & r_{n-1} & \cdots & r_0
\end{bmatrix}
\begin{bmatrix}
1 \\
a_1 \\
\vdots \\
a_n
\end{bmatrix}
=
\begin{bmatrix}
\sigma^2 \\
0 \\
\vdots \\
0
\end{bmatrix}.
\tag{d.20}
$$

Proof. That R is a symmetric positive-definite Toeplitz matrix follows from Theorems D.4 and D.6. We can write, as in the proof of Theorem D.6,

$$
R - (Z - e_1 a^T) R (Z^T - a e_1^T) = R - Z R Z^T + e_1 a^T R Z^T + Z R a e_1^T - e_1 a^T R a e_1^T
$$
$$
= \sigma^2 e_1 e_1^T.
\tag{d.21}
$$

Equations $(1, j)$, $2 \leq j \leq n$ in (d.21) imply equations 2 through n in (d.20). Equation $n + 1$ in (d.20) is just a restatement of (d.19). Finally, equation $(1, 1)$ in (d.21) implies the first equation in (d.20). ∎

Theorem D.8. If $a(z)$ is a monic stability polynomial of degree n (i.e., if all its roots are inside the unit circle), and the Yule-Walker equation (d.20) is satisfied for $\{r_0, \ldots, r_n\}$ and σ^2, then the Liapunov equation (d.18) holds for the Toeplitz matrix $[R]_{i,j} = r_{|i-j|}$.

Proof. This is the converse of Theorem D.7, and can be proved in a similar manner.

Theorem D.9. Let $a(z)$ be a monic stability polynomial of degree n and R the solution of the Liapunov equation (d.18). Let the lower triangular Toeplitz matrices A_1 and A_2 be defined by $[A_1]_{i,j} = a_{i-j}$, $[A_2]_{i,j} = a_{n-i+j}$. Then R^{-1} is given by

$$
R^{-1} = \sigma^{-2}(A_1 A_1^T - A_2 A_2^T).
\tag{d.22}
$$

This equation is known as the *Gohberg-Semencul formula.*

Proof. Denote $r = [r_1, \ldots, r_n]^T$, $\tilde{r} = [r_n, \ldots, r_1]^T$, $a = [a_1, \ldots, a_n]^T$, and $\tilde{a} = [a_n, \ldots, a_1]^T$. Let S be the matrix

$$
S = \begin{bmatrix} R & \tilde{r} \\ \tilde{r}^T & r_0 \end{bmatrix} = \begin{bmatrix} r_0 & r^T \\ r & R \end{bmatrix}.
$$

By the matrix inversion identities [A19],

$$
S^{-1} = \begin{bmatrix} R^{-1} & 0 \\ 0 & 0 \end{bmatrix} + \Delta^{-1} \begin{bmatrix} -R^{-1}\tilde{r} \\ 1 \end{bmatrix} \begin{bmatrix} -\tilde{r}^T R^{-1} & 1 \end{bmatrix}
$$
$$
= \begin{bmatrix} 0 & 0 \\ 0 & R^{-1} \end{bmatrix} + \Delta^{-1} \begin{bmatrix} 1 \\ -R^{-1}r \end{bmatrix} \begin{bmatrix} 1 & -r^T R^{-1} \end{bmatrix},
\tag{d.23}
$$

where $\Delta = r_0 - r^T R^{-1} r = r_0 - \tilde{r}^T R^{-1} \tilde{r}$. However, the Yule-Walker equation (d.20) implies that $a = -R^{-1}r$, $\tilde{a} = -R^{-1}\tilde{r}$, and $\Delta = \sigma^2$. Therefore, (d.23) can be written as

$$S^{-1} = \begin{bmatrix} R^{-1} & 0 \\ 0 & 0 \end{bmatrix} + \sigma^{-2} \begin{bmatrix} \tilde{a} \\ 1 \end{bmatrix} [\,\tilde{a}^T \quad 1\,]$$

$$= \begin{bmatrix} 0 & 0 \\ 0 & R^{-1} \end{bmatrix} + \sigma^{-2} \begin{bmatrix} 1 \\ a \end{bmatrix} [\,1 \quad a^T\,]. \tag{d.24}$$

Premultiply the two forms of S^{-1} by the matrix Z and postmultiply by Z^T to get

$$R^{-1} + \sigma^{-2}\tilde{a}\tilde{a}^T = ZR^{-1}Z^T + \sigma^{-2}aa^T.$$

Hence

$$R^{-1} - ZR^{-1}Z^T = \sigma^{-2}(aa^T - \tilde{a}\tilde{a}^T).$$

This Liapunov equation is similar to (d.16) and can be solved as in the proof of Theorem D.5 to yield

$$R^{-1} = \sigma^{-2} \sum_{i=1}^{n} Z^i aa^T (Z^T)^i - Z^i \tilde{a}\tilde{a}^T (Z^T)^i = \sigma^{-2}(A_1 A_1^T - A_2 A_2^T). \qquad \blacksquare$$

NUMERICAL SOLUTION OF LIAPUNOV MATRIX EQUATIONS

Consider Eq. (d.11), and assume that B^T is an upper triangular matrix. Let c_k and x_k denote the kth columns of C and X, respectively. It is then easy to verify that the following algorithm solves (d.11) column by column:

$$x_k = (1_{n_a} - B_{k,k}A)^{-1} \left(c_k + \sum_{i=1}^{k-1} B_{k,i} A x_i \right). \tag{d.25}$$

If both A and B have the same dimensions, the number of operations in (d.25) is proportional to n_a^4. Furthermore, if both matrices are triangular, the number of operations can be made proportional to n_a^3 (why?).

If B^T is not upper triangular, it can be expressed as $B^T = Q\overline{B}^T Q^H$, where Q is unitary and $\overline{B}^T$ is upper triangular (both may be complex in general). This is called the *complex Schur form*; see Horn and Johnson [1985, p. 79]. We can then solve the Liapunov equation

$$\overline{X} - A\overline{X}\,\overline{B}^T = CQ$$

and take $X = \overline{X}Q^H$ (show this).

In the special where both A are B are diagonal, the solution of (d.11) via (d.12) is immediate, and there is no need for the above procedure. For example, if $B = A$ and A has distinct eigenvalues, it can be diagonalized first by a similarity transformation and then substituted into (d.12).

AN APPLICATION

Let $\{\{h_t^{(i)}\}, 1 \leq i \leq m\}$ be a collection of m causal and rational impulse response sequences. For each i, let $\{A_i, B_i, C_i\}$ be a minimal state-space realization of $\{h_t^{(i)}\}$; that is, $h_t^{(i)} = C_i A_i^t B_i, t \geq 0$. We are interested in computing the infinite sum

$$H = \sum_{t=t_0}^{\infty} \prod_{i=1}^{m} h_t^{(i)}. \tag{d.26}$$

Sums of this kind appear in Chapter 10; see Secs. 10.4 and 10.5. We now show how to compute (d.26) using Kronecker products and Liapunov equations.

Since the impulse response terms $h_t^{(i)}$ are scalar, their Kronecker product is equal to their usual product; that is, $\prod_{i=1}^{m} h_t^{(i)} = \otimes_{i=1}^{m} h_t^{(i)}$. By applying (d.4), (d.5) and (d.6) to this Kronecker product, we get

$$\bigotimes_{i=1}^{m} h_t^{(i)} = \bigotimes_{i=1}^{m} (C_i A_i^t B_i) = \left[\bigotimes_{i=1}^{m} C_i \right] \left[\bigotimes_{i=1}^{m} A_i \right]^t \left[\bigotimes_{i=1}^{m} B_i \right].$$

Summation over t then yields

$$H = \left[\bigotimes_{i=1}^{m} C_i \right] \left[\bigotimes_{i=1}^{m} A_i \right]^{t_0} \left[\mathbf{1} - \bigotimes_{i=1}^{m} A_i \right]^{-1} \left[\bigotimes_{i=1}^{m} B_i \right]. \tag{d.27}$$

The right side of (d.27) can be computed by solving a Liapunov equation. To this end, let $m = m_1 + m_2$, where m_1 and m_2 are equal if m is even and differ by 1 if it is odd. We can express (d.27) as

$$H = \left[\bigotimes_{i=1}^{m} C_i A_i^{t_0} \right] [\mathbf{1} - \boldsymbol{A}_2 \otimes \boldsymbol{A}_1]^{-1} [\boldsymbol{B}_2 \otimes \boldsymbol{B}_1], \tag{d.28}$$

where $\boldsymbol{A}_2$ is the Kronecker product of the first m_2 matrices $\{A_i\}$, $\boldsymbol{A}_1$ is the Kronecker product of the last m_1 matrices, and $\boldsymbol{B}_2$, $\boldsymbol{B}_1$ are defined in a similar manner. We can now solve the Liapunov equation

$$X - \boldsymbol{A}_1 X \boldsymbol{A}_2^T = \boldsymbol{B}_1 \boldsymbol{B}_2^T$$

and use the solution to compute (d.28), as follows from Theorem D.1.

As was explained above, it is advisable to bring the matrix A_2 to a complex Schur form and then use the algorithm (d.25). Significant simplification takes place when all the matrices $\{A_i\}$ are equal, say to A. In this case it is advisable to transform A to Schur form prior to forming the Kronecker products. Then both $\boldsymbol{A}_1$ and $\boldsymbol{A}_2$ will be lower triangular, and the algorithm (d.25) will be greatly simplified. Finally, if A has distinct eigenvalues, it is preferred to transform it to a diagonal form rather than to a Schur form.

THE MATHEMATICA PACKAGE `Liapunov.m`

The Mathematica package `Liapunov.m` implements some Kronecker product operations and Liapunov equation solutions, as follows.

`Kron[a_, b_]` returns the Kronecker product of the matrices `a` and `b`.

`KronList[list_]` returns the Kronecker product of the matrices in `list`.

`KronPower[a_, n_]` returns the nth Kronecker power of the matrix `a`.

`LiapEq[a_, b_, c_]` solves the Liapunov equation `x - axb = c`, where `a` is lower triangular and `b` is upper triangular, and returns the solution `x`.

`ImAinvB[a_, b_]`, where `a` is a list of square matrices (in Schur form) and `b` is a list of single-column matrices, implements a part of Eq. (d.27) (the product of the last two factors on the right side). `ImAinvB[a_, b_, n_]`, where `a` and `b` are single matrices and `n` is an integer, gives the special case where `a` and `b` are to be raised to the nth Kronecker power. If the `a`'s are diagonal matrices, the computation is much more efficient. For convenience, diagonal matrices are represented by their diagonals only (without the zeros).

`H[{a_, b_, c_}]`, where `a`, `b`, `c`, are lists of matrices, implements Eq. (d.27), with $t_0 = 0$. `H[{a_, b_, c_}, set_]` treats the case discussed in Eq. (10.4.3), where set is the set of τ's (assumed nonnegative). `H[h_, set_]` implements the special case of finite impulse response, as given by (10.4.1). In this case `h` is allowed to be a list of numbers or symbols.

`Arma2SS[{a_,b_}]` returns the state-space matrices corresponding to the ARMA parameter lists `a` and `b`.

`SS2Schur[{a_, b_, c_}]` converts a given state-space realization to a complex Schur form.

`SS2Diag[{a_, b_, c_}]` converts a given state-space realization to a diagonal form. It takes for granted that the matrix `a` is diagonalizable and does not check it. The diagonal form of `a` is returned as a doubly braced list (i.e., a single-row matrix) of the diagonal entries to save space.

REFERENCES

Horn, R. A., and Johnson, C. R., *Matrix Analysis,* Cambridge University Press, Cambridge, 1985.

Author Index

Subject Index

The index uses the following conventions:

- Pages in which mathematical terms are formally defined are indicated in bold type, e.g., **295**.
- A page (or a range of pages) in italic type indicates a primary discussion of the term, e.g., *215–255*.
- Roman type indicates a secondary discussion or an accidental mention of the term. A range of pages in Roman type indicates several mentions of the term (not necessarily related) in the range.